Reuben L. Baumgarten, Ph.D., University of Michigan, is Professor of Organic Chemistry at Lehman College of the City University of New York. He was one of the developers of the Brumlick–Barrett–Baumgarten Framework Molecular Orbital (FMO) Models for student use in the visualization of molecular structures. Professor Baumgarten received the Lehman College 1974–75 Award for Excellence in Teaching.

Organic Chemistry

A BRIEF SURVEY

REUBEN L. BAUMGARTEN

Lehman College of the
City University of New York

THE RONALD PRESS COMPANY · NEW YORK

To Lainie and Steven

Copyright © 1977 by The Ronald Press Company

All Rights Reserved. No part of this book may be reproduced in any form without permission in writing from the publisher.

ISBN 0-8260-0766-X

Library of Congress Catalog Card Number: 74-22533

Printed in the United States of America

Preface

This organic chemistry text has been written primarily for students majoring in biology, home economics, nursing, medical technology and related fields, pharmacy, and agriculture, as well as for liberal arts students minoring in science. It is structured for the one-semester course, and while it does not attempt the comprehensive treatment required for the major courses it more than adequately covers the concepts and principles for an essential grasp of the subject.

Basic principles are stressed in the early part of the text, along with modern concepts of reaction mechanisms and molecular geometry, and their influence on reactions and properties of organic compounds, to provide a theoretical background that will enable students to understand, to deduce, and to reason, rather than just memorize seemingly unrelated facts. A descriptive background has been blended with the theoretical aspects of the subject, to enable students to see how ideas are applied to various problems; this balance should be extremely valuable to all students.

It is believed that when students have progressed through Chapter 6 they will have an understanding of the fundamental ideas of chemical bonding, electronic and steric effects, including resonance, and chemical reactivity and basic classes of organic reaction mechanisms. Nomenclature is introduced early, in Chapter 3, in preparation for the theoretical concepts in Chapters 4, 5, and 6. The importance of nomenclature in organic chemistry can not be overemphasized. A student must be able to name compounds correctly and to draw correct structural formulas if he is to have any success in understanding the subject. Therefore, in addition to its general presentation, nomenclature is briefly discussed whenever a new class of organic compounds is introduced. In Chapter 6 there is a general discussion of the most common types of organic reaction mechanisms, which are applied in subsequent chapters dealing with the different types of organic compounds.

Where the various classes of organic compounds are discussed, starting with the alkanes in Chapter 7 and ending with the amines and diazonium salts in Chapter 16, the generalizations developed in the early chapters are applied with regard to properties and chemical reactivity. Similarities and differences among compounds are noted, and it is shown that although many compounds belong to different classes or families they often react by similar reaction mechanisms.

Stereoisomerism is presented in Chapter 17, so that the following chapter devoted to the chemistry of the carbohydrates can be approached with the principles of optical isomerism fresh in mind. The principles of stereochemistry are applied to some of the reactions given in the preceding chapters, which the student can better appreciate at this stage.

Chapters 18 to 22, from the same theoretical and descriptive perspective, discuss carbohydrates; lipids; amino acids and proteins; nucleic acids, reflecting the importance of and current interest in DNA and RNA; and other natural products. The properties and reactions of these molecules can be easily understood, since their functional groups have already been discussed. The relationship between organic chemistry and biochemistry is stressed, and the material is presented in terms of current concepts.

Synthetic polymers are given a concise presentation in Chapter 23; and Chapter 24 is an overview of spectroscopy, pointing up its importance in elucidating the structure of the simple molecules and macromolecules discussed earlier.

The questions posed at the ends of the chapters are intended to help students in their comprehension of the material; and asterisks call attention to problems and sub-problems that are more challenging. In illustrating certain features of molecular geometry, Framework Molecular Orbital Models, used in conjunction with other forms of representation of the same molecules, facilitate the visualization of spatial relations.

I am much indebted to my parents, whose interest and enthusiasm gave me the courage to pursue this endeavor. The contributions of Mrs. Myrna T. Helfgott, who did such a superb job in the typing of manuscript, and my students, in particular Messrs. John Ioia and Dario Otero, by their interest, enthusiasm, and encouragement, are greatly appreciated. And I would like to here thank my wife, Iris, without whose inspiration, patience, and indulgence none of this would have been possible.

Reuben L. Baumgarten

April, 1977

Contents

1 Electrons and Bonding   3
1 Introduction   2 Structure of the Atom   3 Electronic Configuration of the Atom   4 Chemical Bonding   5 Orbitals, Hybridization, and Shape of Molecules   6 Sigma and Pi Bonds

2 Characteristics of Structure   20
1 Introduction   2 The Carbon Atom   3 Organization of Organic Chemistry   4 Spatial Formulas and Molecular Models   5 Isomerism

3 Nomenclature   28
1 Introduction   2 Systematic Nomenclature   3 Compounds with One Functional Group   4 Some Cyclic Compounds   5 Compounds with More than One Functional Group   6 Other Systems of Nomenclature

4 Acidity, Basicity, and Structure   36
1 Introduction   2 Brønsted–Lowry Theory   3 Lewis Theory   4 Strength of Acids and Bases; K_a and K_b 5 Relative Acid Strength of Binary Acids   6 Relative Acid Strength of Ternary Acids

5 Electronic and Steric Effects   46
1 Introduction   2 Electronic Effects   3 Inductive Effects   4 Resonance   5 Steric Effects

6 Chemical Reactivity   60
Chemical Kinetics and Reaction Mechanisms of Organic Compounds
1 Introduction   2 Factors that Influence Rate of Reaction   3 Reaction Mechanism   4 Theory of Reaction Rates   5 Breaking of Chemical Bonds—Homolytic and Heterolytic Cleavage   6 Types of Organic Reaction   7 A Polar Organic Reaction—Nucleophilic Substitution   8 Free Radical Reactions

7 Saturated Hydrocarbons: Alkanes and Cycloalkanes   74
1 Introduction   2 Structure of Some Alkanes   3 General Formulas of Alkanes—Members of an Homologous Series   4 Nomenclature   5 Definitions   6 Physical Properties   7 Chemical Properties   8 Preparation   9 Conformational Isomerism   10 Cycloalkanes—The Baeyer Strain Theory

8 Unsaturated Hydrocarbons—I: Alkenes   93

1 Introduction 2 Nomenclature 3 Carbon–Carbon Double Bond 4 Geometric Isomerism 5 Physical Properties 6 Chemical Properties 7 Mechanisms for Addition Reactions to Carbon–Carbon Double Bonds 8 Ionic Addition Reactions 9 Free Radical Addition of HBr 10 Cycloaddition Reactions 11 Polymerization 12 Free Radical Halogenation 13 Oxidation Reactions 14 Preparation 15 Synthesis 16 Polyalkenes 17 Conjugate Addition 18 Isoprene

9 Unsaturated Hydrocarbons—II: Alkynes   130

1 Introduction 2 Nomenclature 3 Carbon–Carbon Triple Bond 4 Addition Reactions 5 Acidic Hydrogen 6 Preparation

10 Aromatic Hydrocarbons   138

1 Introduction 2 Structure of Benzene 3 Common Aromatic Compounds—The Hückel $4n + 2$ Rule 4 Nomenclature 5 Electrophilic Aromatic Substitution Reactions 6 Oxidation of the Alkyl Side Chain 7 Orientation in Aromatic Electrophilic Substitution Reactions 8 Thallation 9 Important Derivatives of Benzene

11 Alcohols and Phenols   161

1 Introduction 2 Classification and Nomenclature 3 Preparation of Aliphatic Alcohols 4 Methyl Alcohol 5 Ethyl Alcohol 6 Polyhydric Alcohols 7 Preparation of Phenols 8 Physical Properties 9 The Hydrogen Bond 10 Reactions of the Alcohols 11 Inorganic Esters 12 Oxidation of Alcohols 13 Reactions of Phenols 14 Sulfur Analogs

12 Ethers   178

1 Introduction 2 Preparation 3 Properties 4 Reactions 5 Sulfur Analogs 6 Epoxides

13 Organic Halogen Compounds   189

1 Introduction 2 Organic Monohalogen Compounds 3 Iodides and Fluorides 4 Displacement Reactions of Alkyl Halides 5 Preparation and Use of the Grignard Reagent to Prepare Primary Alcohols

14 Aldehydes and Ketones   197

1 Introduction 2 Nomenclature 3 Preparation of Aldehydes and Ketones 4 Preparation of Formaldehyde, Acetaldehyde, and Acetone 5 Nature of Carbonyl Group 6 Bonding of Carbonyl Group 7 Reactions of Aldehydes and Ketones 8 Nucleophilic Addition Reactions 9 Enol–Keto Tautomerism 10 Halogenation (Substitution of α-Hydrogens) 11 Haloform Reaction—Iodoform Test 12 Halogenation of Ketones in Acid Medium 13 Aldol Condensation 14 Perkin Reaction 15 Cannizzaro Reaction 16 Reactions of Aromatic Aldehydes and Ketones

CONTENTS

15 Carboxylic Acids and Their Derivatives — 228
1 Introduction 2 Nomenclature 3 Acid Strength 4 Physical Properties 5 Preparation 6 Reactions of the Carboxylic Acids and Their Derivatives 7 Salt Formation 8 Other Derivatives of the Carboxylic Acids 9 Reactions of Derivatives 10 Formation of Aldehydes and Ketones from Acyl Halides and Acid Anhydrides 11 Other Reactions of Derivatives 12 Substitution Reactions 13 Preparation of Various Substituted Carboxylic Acids 14 Preparation of Phenolic Acids 15 Maleic and Fumaric Acids

16 Amines, Diazonium Salts, and Dyes — 256
1 Introduction 2 Nomenclature and Classification of Amines 3 Salts of Amines—Nomenclature 4 Physical Properties 5 Basicity 6 Preparation 7 Reactions 8 Electrophilic Substitution Reactions of Aromatic Amines 9 Heterocyclic Amines 10 Preparation and Reactions of Diazonium Salts 11 Synthesis with Diazonium Salts 12 Evidence for the Existence of Benzyne 13 Coupling Reaction of Diazonium Salts 14 Color and Dyes

17 Stereoisomerism—Optical Isomerism — 284
1 Introduction 2 Optical Isomerism 3 Detection of Optical Activity—Plane-Polarized Light 4 Asymmetry and the Tetrahedral Carbon Atom 5 Enantiomerism and the Tetrahedral Carbon Atom 6 Criteria for Optical Activity 7 Fischer Projection Formulas 8 Enantiomers 9 Absolute and Relative Configurations 10 R and S Configurations 11 Molecules with More than One Asymmetric Carbon Atom 12 Resolution of Racemic Mixtures 13 Resolution and Biochemistry 14 Reactions Involving Stereoisomers

18 Carbohydrates — 304
1 Introduction 2 Nomenclature and Classification 3 Fischer–Kiliani Synthesis 4 Monosaccharides 5 Structure Determination of Glucose 6 Evidence for a Cyclic Formula for Glucose 7 Cyclic Structure—Pyranose and Furanose Rings 8 Fructose 9 Reactions of Monosaccharides 10 Disaccharides 11 Lactose 12 Maltose 13 Cellobiose 14 Sucrose 15 Polysaccharides 16 Starch 17 Glycogen 18 Cellulose 19 Photosynthesis in Plants 20 Metabolism of Carbohydrates

19 Lipids—Fats and Oils — 329
1 Introduction 2 Neutral Fats and Oils 3 Glycerides 4 Reactions of Fats and Oils 5 Waxes 6 Drying Oils 7 Soaps and Detergents 8 Saponification Number and Iodine Number 9 Metabolism of Lipids

20 Amino Acids and Proteins — 340
1 Introduction 2 Amino Acids 3 Nomenclature and Classification 4 Synthesis 5 Properties 6 Isoelectric Point 7 Reactions of Amino Acids 8 The Peptide Bond 9 Structure of Polypeptides 10 Synthesis of Polypeptides 11 Classification of Proteins 12 Structure 13 Chemical Behavior 14 Metabolism of Proteins

21 Nucleic Acids   361
1 Introduction 2 Constituents 3 Structure of DNA 4 Primary Structure 5 Secondary and Tertiary Structures 6 Structure of RNA

22 Natural Products   371
1 Introduction 2 Terpenes 3 Monoterpenes 4 Sesquiterpenes 5 Diterpenes and Triterpenes 6 Steroids 7 Cholesterol 8 Important Steroids 9 Biogenesis of Terpenes and Steroids 10 Biogenesis of Terpenes 11 Biogenesis of Steroids 12 Heterocyclic Products 13 Synthesis of Heterocyclic Ring Systems 14 Oxygen Heterocycles 15 Nitrogen Heterocycles 16 Heterocyclic Compounds with More than One Heteroatom

23 Synthetic Polymers   393
1 Introduction 2 Characteristics 3 Classification 4 Addition Polymerization 5 Stereochemical Control of Addition Polymers 6 Condensation Polymerization

24 Spectroscopy and Structure   401
1 Introduction 2 Electromagnetic Spectrum 3 Spectrophotometer 4 UV–VIS Spectroscopy 5 IR Spectroscopy 6 NMR Spectroscopy 7 Mass Spectrometry

Solutions to Selected Problems   426

Index   460

Organic Chemistry

1
Electrons and Bonding

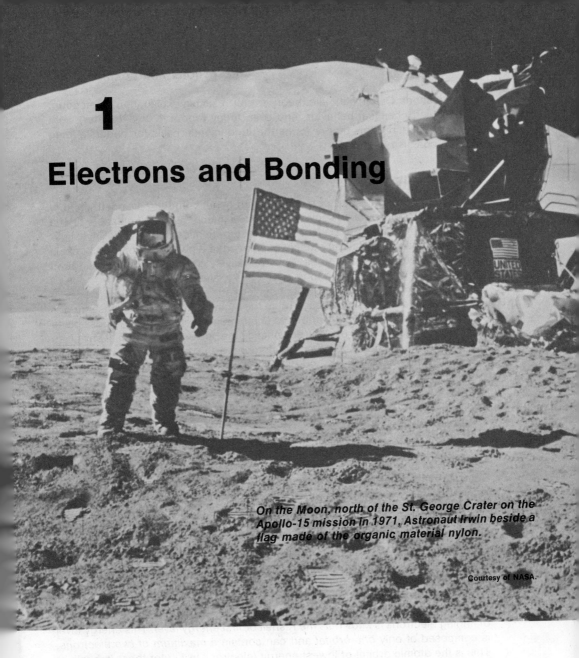

On the Moon, north of the St. George Crater on the Apollo-15 mission in 1971, Astronaut Irwin beside a flag made of the organic material nylon.

Courtesy of NASA.

1.1 Introduction

An understanding of atomic structure and the principles of chemical bonding is essential in the study of organic chemistry. The idea that matter consists of discrete, fundamental particles dates back to about the year 400 B.C., as mentioned in the writings of Democritus, a Greek philosopher, who was introduced to this concept by his teacher, Leucippus. This notion was rejected by Plato and Aristotle, and it wasn't until the time of Sir Isaac Newton (1642–1727) that the concept was suggested once again.

The concept that all matter is composed of small, fundamental particles called **atoms** was proposed in the year 1803 by John Dalton, an English chemist and school teacher. Dalton's atomic theory set forth simple explanations to account for observed chemical behavior, and even today (about 170 years later), with slight modification, this theory forms the fundamental basis for understanding modern chemistry.

1.2 Structure of the Atom

It wasn't until the latter part of the nineteenth century and the twentieth century that certain spectacular discoveries led to a much better elucidation of the structure of the atom. As a result of the numerous chemical and physical investigations, we know today that an atom is composed of a positively charged nucleus containing neutrons and protons, while negatively charged electrons are found outside the nucleus, distributed in space as an **electron cloud.**

Our knowledge of the arrangement of electrons around the nucleus is a result of mathematical calculations based upon the **quantum theory.** As a consequence of the **Heisenberg uncertainty principle** of quantum mechanics, it is impossible to specify the exact location of electrons outside the nucleus of an atom at any given moment. It is possible, however, to determine the **probability** of finding an electron at various distances and directions in space outside the nucleus. The regions where the probability of finding an electron is high correspond to what we call **energy levels** or simply **electronic shells.** These are regions where the electron density is greatest, and within which the electron will be found most of the time.

Every electron can be "characterized" with respect to distance from the nucleus, shape of the electron cloud, orientation in space, and spin. Each energy level can be considered as composed of a series of **sub-levels** or smaller units called **orbitals.** The various orbitals differ from one another with respect to shape or orientation, and an orbital can contain no more than two electrons, according to the **Pauli exclusion principle.**

1.3 Electronic Configuration of the Atom

At this point we should discuss the building up, **aufbau principle,** of the electronic structure of atoms, for a better understanding of the bonding process that occurs when atoms combine to form molecules.

The first region in space outside the nucleus where there is a high probability of finding electrons is known as the K shell or *first energy level*. This energy level is composed of only *one orbital* and can contain *a maximum of two electrons*. This is the atomic orbital of lowest energy (electrons will enter those orbitals of lowest energy first, then proceed to fill up orbitals of increasingly higher energy, until the electronic structure of the atom under consideration is completed); it is spherical in shape, and is called a 1s atomic orbital. The number 1 tells us that the location of the electron is in the *first energy level* outside the nucleus and the letter s tells us that the shape of the orbital is *spherical*. (See Figure 1.1.)

In our discussion of electronic configurations of atoms we shall use numbers (1, 2, 3, . . . , etc.) to refer to different energy levels; those with higher numbers generally possess greater energy content than those with lower numbers. The letters (s, p, d, and f) will be used to denote the shape of a particular orbital within an energy level (i.e., 1s = first-energy-level spherical orbital).

Sec. 1.3 ELECTRONIC CONFIGURATION OF THE ATOM

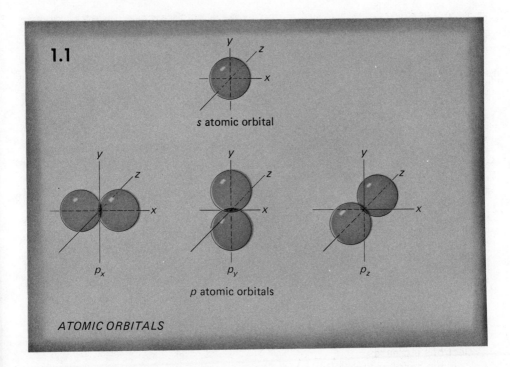

1.1 ATOMIC ORBITALS

The first energy level can hold a maximum of two electrons. Upon completion of this energy level, additional electrons will start to fill up the next higher energy level, the *second energy level* or L shell. The second energy level can contain a maximum of eight electrons, this energy level being composed of four orbitals, each of which can contain two electrons. One of these orbitals is similar to the 1s orbital. It is called a 2s orbital, and like the 1s orbital is spherical in shape, but has a larger radius around the nucleus, since it is part of the second energy level, which extends farther out in space from the nucleus than the first energy level. The remaining three orbitals of the second energy level are of slightly higher energy than the 2s orbital, but all are part of the same overall energy level. These three atomic orbitals are all dumbbell- or pear-shaped, have equivalent energies (orbitals of the same energy are said to be degenerate), and differ only in their orientation in space; they are called p orbitals, and are 2p orbitals since they are in the second energy level.

The shape of the p orbital can be seen in Figure 1.1, where it is represented as two pears on opposite sides of the nucleus. The three p orbitals are oriented along the x-, y-, and z-axes in space, perpendicular to each other (separated by angles of 90°), and are called p_x, p_y and p_z orbitals to differentiate them. In summary, the second energy level can contain eight electrons, and is made up of four orbitals, one spherical in shape (s orbital) and three pear-shaped orbitals perpendicular to each other in space (p orbitals). The third energy level can hold a maximum of eighteen electrons, consisting of nine orbitals, including one s, three p, and five d orbitals. The fourth energy level contains a maximum of

Gilbert Newton Lewis (1875–1946)—professor at the University of California and proponent of the electron pair bond—proposed the *Lewis theory of acids and bases*. (From the Dains Collection, courtesy of the Department of Chemistry, The University of Kansas.)

thirty-two electrons, consisting of 16 orbitals, including one *s*, three *p*, five *d*, and seven *f* orbitals; and the situation becomes more complex as the number of electrons increases.

In the chapters that follow, we need concern ourselves primarily with the first eighteen elements, which are the ones most important in organic chemistry. The electronic structure of any element can be understood by successively filling its atomic orbitals with electrons, proceeding from orbitals of lowest energy to those of higher energies. The atomic number of an element is equal to the number of protons in the nucleus and is also equal to the number of electrons outside the nucleus of the uncharged atom. In Table 1.1 are listed the first eighteen elements of the periodic table with their respective electronic configurations, illustrating the order of build-up of electrons in their atomic orbitals. As mentioned previously,

Table 1.1 Electron Orbital Configurations of Atoms

Element	Atomic number	Electronic configuration	Element	Atomic number	Electronic configuration
H	1	$1s^1$	Ne	10	$1s^2\,2s^2\,2p_x^2\,2p_y^2\,2p_z^2$
He	2	$1s^2$	Na	11	$1s^2\,2s^2\,2p^6\,3s^1$
Li	3	$1s^2\,2s^1$	Mg	12	$1s^2\,2s^2\,2p^6\,3s^2$
Be	4	$1s^2\,2s^2$	Al	13	$1s^2\,2s^2\,2p^6\,3s^2\,3p_x^1$
B	5	$1s^2\,2s^2\,2p_x^1$	Si	14	$1s^2\,2s^2\,2p^6\,3s^2\,3p_x^1\,3p_y^1$
C	6	$1s^2\,2s^2\,2p_x^1\,2p_y^1$	P	15	$1s^2\,2s^2\,2p^6\,3s^2\,3p_x^1\,3p_x^1\,3p_z^1$
N	7	$1s^2\,2s^2\,2p_x^1\,2p_y^1\,2p_z^1$	S	16	$1s^2\,2s^2\,2p^6\,3s^2\,3p_x^2\,3p_y^1\,3p_z^1$
O	8	$1s^2\,2s^2\,2p_x^2\,2p_y^1\,2p_z^1$	Cl	17	$1s^2\,2s^2\,2p^6\,3s^2\,3p_x^2\,3p_y^2\,3p_z^1$
F	9	$1s^2\,2s^2\,2p_x^2\,2p_y^2\,2p_z^1$	Ar	18	$1s^2\,2s^2\,2p^6\,3s^2\,3p_x^2\,3p_y^2\,3p_z^2$

Sec. 1.4 CHEMICAL BONDING

the number represents the energy level the electron is in, and the letter denotes the shape of the particular orbital. The superscript after the letter indicates the number of electrons in the orbital. As mentioned, no more than two electrons can occupy the same orbital (electrons of opposing spins). When only one electron is present in an orbital, it is referred to as an **unpaired electron.** Should an electron of opposite spin become available, the electrons can pair up and form a completely filled orbital.

From Table 1.1 it can be seen that, within a given shell for example, second energy level, the *s* orbital is filled prior to the *p* orbitals. As the *p* orbitals are degenerate orbitals (of equal energies) the electron buildup is such that an electron goes into each of the individual orbitals (p_x, p_y, p_z), a stable electronic arrangement, before these unpaired electrons become paired to completely fill the *p* orbitals. Only the noble gases, for example, He (atomic number 2), Ne (10), and Ar (18) have electronic structures in which the outer or *valence shell of electrons is completely filled*. This is a very stable type of electronic arrangement and will be referred to in the discussion on chemical bonding that follows.

1.4 Chemical Bonding

All atoms desire to attain an electronic structure in which the outer electronic shell is completely filled, similar to the electronic configurations of the noble gases. Atoms react with one another to form molecules. The *attractive forces between atoms* that result in the formation of molecules are referred to as **bonds.** Two basic types of bonds can be formed whenever atoms react, **ionic bonds** and **covalent bonds.**

Ionic bonds, or electrovalent bonds as they are sometimes called, are formed when atoms gain or lose electrons in a chemical reaction. Let us consider the reaction between an atom of sodium (Na) and an atom of chlorine (Cl) in the production of the compound sodium chloride. From the electronic configurations listed in Table 1.1, it can be seen that a sodium atom has one electron in its outer shell and a chlorine atom has seven electrons in its outer shell.

At this point, it is worthwhile to emphasize that usually only the electrons in the outer orbitals of an atom (the valence electrons) are involved in the formation of chemical bonds. This is because when atoms react the first object they come into contact with in space is the *outermost electron cloud* of the attacking atom. For this reason, and for the sake of simplicity, a shorthand notation of writing electronic structures has been developed. This enables us to look only at the outermost electrons in an atom (since these are involved in bond formation) without interference from the other electrons, and thus we can clearly depict and interpret the bonding process that is to occur.

This brings us to the use of so-called **Lewis electron-dot formulas** or **kernel designations** of the atoms. The symbol for the atom is used, and the number of electrons in the outer shell indicated by dots or some other symbol such as a small "ex" (×). For example, the Lewis electron-dot formula representing a sodium atom that has one electron in its outer shell would be Na·. A chlorine atom would be represented as $\overset{\times\times}{\underset{\times\times}{\times}}\text{Cl}^{\times}$. (Note that the electrons are put in pairs of two, since *only two electrons can occupy the same orbital*.)

In order to attain a noble gas type of electronic configuration, it is necessary that the sodium atom lose one electron and the chlorine atom gain one electron to complete its outer shell. An electron is transferred from the sodium atom to the chlorine atom, which results in the formation of charged particles called **ions,** and the bond formed between the charged species is referred to as an **ionic bond.**

$$Na\cdot \; + \; {}^{xx}_{xx}Cl^x \longrightarrow Na^{\oplus} + {}^{xx}_{xx}Cl{:}^{\ominus} \tag{1-1}$$

Since the sodium atom *lost* a negatively charged electron (all atoms are electrically neutral) the resulting sodium ion formed has a **positive charge.** Similarly, the neutral chlorine atom *gained* an electron, and thus the chloride ion formed has a **negative charge.**

Ionic bonds will be formed whenever atoms of greatly differing electronegativities (metal and non-metal) combine. **Covalent bonds,** which involve the *sharing*, rather than the complete transfer, of electrons between atoms, will be formed between non-metals, or atoms of closely similar electronegativities.

The following serve as examples of covalent bonding. In each case, note that by the sharing of electrons, the atoms all attain a noble gas electronic structure. For convenience, dots ($\cdot$) and "ex's" ($\times$) are used for electrons to illustrate that each atom contributes one electron toward the formation of the covalent bond. Each of the single bonds formed contains two electrons, one from each atom.

$$H\cdot \; + \; {}^{\times}H \longrightarrow H{:}H \tag{1-2}$$

$${}^{xx}_{xx}F^{\times} + \cdot\ddot{F}{:} \longrightarrow {}^{xx}_{xx}F{:}\ddot{F}{:} \tag{1-3}$$

$$\underset{\times}{{}^{\times}C^{\times}} + 4H\cdot \longrightarrow H{:}\underset{H}{\overset{H}{C}}{:}H \tag{1-4}$$

Many molecules contain **multiple bonds,** that is so-called *double* and *triple* bonds, formed by the sharing of *four* and *six* electrons, respectively, between atoms. The carbon dioxide molecule, CO_2, contains two double bonds formed by the sharing of four electrons between carbon and each oxygen atom.

$$\underset{\times}{{}^{\times}C^{\times}} + 2{:}\dot{O}{:} \longrightarrow {:}\ddot{O}{::}C{::}\ddot{O}{:} \tag{1-5}$$

The nitrogen molecule, N_2, contains a triple bond. Six electrons, three from each nitrogen atom, form the bond.

$$\underset{\times}{{}^{\times}N^{\times}} + {:}\dot{N}\cdot \longrightarrow {}^{\times}_{\times}N{:::}N{:} \tag{1-6}$$

In all of the previous examples each atom contributed electrons to form the covalent bond. For example, in the fluorine molecule each of the fluorine atoms contributed one electron which was shared between the atoms to produce a single covalent bond. It should be mentioned that there are substances in which one atom of the pair involved in covalent bond formation contributes *both* of the

Sec. 1.4 CHEMICAL BONDING

electrons needed for bond formation. This type of bond is called a **coordinate covalent bond,** or sometimes a **dative** or **semi-polar bond.**

Examples of coordinate covalent bonding can be found in nitric acid

$$H:\overset{\times\times}{\underset{}{O}}:N:\overset{\times\times}{\underset{}{O}}: \\ :\overset{..}{\underset{\times\times}{O}}: \longleftarrow \text{coordinate covalent bond}$$

where the nitrogen contributes both electrons to one of the nitrogen–oxygen bonds, and in the ammonium ion

$$\begin{array}{c} H \\ \underset{\times\times}{} \\ H:N:H \\ \underset{\cdot\times}{} \\ H \end{array} \overset{\oplus}{\longleftarrow} \text{coordinate covalent bond}$$

where the nitrogen shares an available electron pair with a hydrogen ion.

1.4-1 Polar Bonds Covalent molecules containing atoms of similar electronegativities or symmetrical molecules are said to be *non-polar*. For example, let us consider the hydrogen molecule, H:H. The electrons shared between the hydrogen atoms are shared equally. That is to say that, since the hydrogen atoms have the *same electronegativity,* and hence the *same tendency to attract electrons,* the average position of the electrons is found to be midway between the nuclei of the two hydrogen atoms.

However, not all covalent molecules are non-polar. Covalent molecules that are non-symmetrical, and contain atoms of different electronegativities, have **polar bonds.** Hydrogen chloride is a **polar covalent compound,** and in the hydrogen chloride molecule, the shared pair of electrons between the hydrogen and chlorine is *not* equidistant between the hydrogen and chlorine nuclei. Instead, the shared electron pair is actually located much closer to the chlorine end than to the hydrogen end of the molecule. $\left(\underset{\delta\oplus}{H} : \underset{\delta\ominus}{\overset{\times\times}{\underset{\times\times}{Cl}}} :\right)$. This is reasonable if one

remembers that chlorine is a much more electronegative element than hydrogen, and hence has a much greater attraction for electrons than hydrogen has. The result is that the electrons are pulled much closer to the chlorine; and the chlorine end of the molecule, having a richer electron density, is *relatively negative* as compared to the hydrogen end, which is *relatively positive*. This is indicated by the Greek letter delta, δ, the $\delta\oplus$ and $\delta\ominus$ indicating the **polarity** of the bond.

A polar bond is merely a specific type of covalent bonding. A **polar bond** is by no means the same as an **ionic bond;** in the *former* the electron distribution in the molecule is not symmetrical, in the *latter,* there is a complete separation of charge into discrete particles bearing positive or negative charges, called ions. Polar bonds are important in discussing the properties and reactions of many classes of organic compounds, and will be referred to when necessary in subsequent chapters. (A polar molecule possesses a **dipole moment** which is equal to the magnitude of the charge multiplied by the distance between the centers of charge $[\mu = e \times d]$.)

1.5 Atomic Orbitals, Hybridization, and Shape of Molecules

Overlapping of the electron clouds of atomic orbitals leads to the formation of a covalent bond. The orbitals used in bond formation determine the molecular geometry, or the shape of the molecule.

A covalent molecule composed of two atoms, for example, A—B, can only have one geometrical shape. The molecule A—B must be linear with a bond angle of 180°, since the two atoms must lie in a straight line, no matter in which direction you connect them.

The situation becomes more complex the greater the number of atoms in the molecule. In organic chemistry, it is important to be able to know the orientation of the orbitals containing the electrons that form the orbital bonds and the geometry of the molecule under consideration. To achieve these goals it is necessary to introduce the concept of **hybridization.**

Let us consider a molecule containing three atoms, $BeCl_2$. What is the shape of this molecule? Two logical choices immediately come to mind. Is the molecule linear Cl—Be—Cl, or angular $\overset{Be}{Cl \diagdown Cl}$? (The solid line between the Be and a Cl represents a single bond or pair of electrons.)

Inspection of the electronic structures of the Be atom and the Cl atom in Table 1.1 shows their normal or ground state electronic configurations to be as follows: Be, $1s^2\ 2s^2$; and Cl, $1s^2\ 2s^2\ 2p^6\ 3s^2\ 3p^5$. It can be seen that the Cl atom needs one more electron to complete its $3p$ orbital, and attain an electronic configuration identical with that of the noble gas argon.

Since the correct empirical formula of the compound formed between beryllium and chlorine has been shown to be $BeCl_2$, and since each chlorine atom needs one electron to complete its $3p$ orbital, a total of two electrons (one for each chlorine atom) must be contributed by the beryllium atom.

The electronic structure for beryllium is $1s^2\ 2s^2$ and indicates that all the electrons are paired. Bond formation requires the pairing up of unpaired electrons. Two *unpaired* electrons are necessary for the beryllium atom to form bonds with the two chlorine atoms. As previously mentioned, all the orbitals within a given energy level have about the same energy. Thus, it is not surprising that by adding a small amount of energy to a beryllium atom in the ground state it is possible to "promote" an electron from the $2s$ orbital to one of the empty $2p$ orbitals (in the same energy level). This "promotion" process makes the electronic structure of the beryllium atom $1s^2\ 2s^1\ 2p^1$, resulting in the two unpaired electrons necessary for bonding. The beryllium atom is now said to be in the "excited state" and ready for chemical reaction.

Most chemical reactions require that the electrons of an atom be promoted from the ground state to the excited state to get the required number of unpaired electrons necessary for bond formation. Fortunately, the change from ground state to excited state usually requires only a small amount of energy.

The electronic structure of the beryllium atom in the excited state seems to indicate that an s and a p electron are involved in bond formation with the two chlorine atoms. If this were true, then the bonds between the beryllium and chlorine atoms would not be identical; one bond would be formed by the s

Sec. 1.5 ORBITALS, HYBRIDIZATION, AND SHAPE OF MOLECULES

electron of beryllium and the *p* electron of chlorine, and the second bond would be formed by the *p* electron of beryllium and the *p* electron of chlorine.

However, experimental evidence indicates that the two bonds in beryllium chloride are identical. This must be interpreted as indicating that the *same* types of electrons form each bond, which brings up the concept of **hybridization**. The term hybridization as used in chemistry implies that in bond formation electrons often do not behave as "pure" *s*, or "pure" *p*, electrons. Instead, the electrons involved in bond formation in the excited state acquire some of the properties and characteristics of *s* and *p* electrons. In the case of beryllium, we no longer have one pure *s* and one pure *p* electron involved in bonding with the chlorine. We envision two identical electrons in the excited state. These electrons are called *sp* electrons, and the **hybridization** is of the **sp** type (since one *s* and one *p* electron are involved).

Molecules where bonds are formed from *sp*-hybridized electrons are always *linear* in shape with bond angles of 180°. Such is the case in the beryllium chloride molecule, Cl—Be—Cl. This is borne out by experimental evidence.

The concept of hybridization enables us to predict the shape, or geometry, of many molecules. Thus, it is a very useful tool in the hands of the chemist, or anyone who has a fundamental knowledge of electronic structure and the process of bond formation.

The boron trichloride molecule, BCl_3, contains four atoms. From Table 1.1 the electronic structure of boron is seen to be $1s^2\ 2s^2\ 2p_x^1$ in the ground state, and that of chlorine, $1s^2\ 2s^2\ 2p^6\ 3s^2\ 3p^5$. By using the same line of reasoning as was used in the previous discussion on the shape of the beryllium chloride molecule, it can be seen that three unpaired electrons are required of the boron atom to pair up with the three chlorine atoms. (The electronic structure of boron has one unpaired electron in the $2p_x$ orbital. This would perhaps suggest that a compound of the molecular formula BCl should exist. BCl has never been isolated; but the compound containing boron and chlorine that has the formula BCl_3 is well known and has been isolated.)

By promoting a 2s electron to one of the empty 2p orbitals in boron, we obtain an excited state electronic configuration of $1s^2\ 2s^1\ 2p_x^1\ 2p_y^1$ containing three unpaired electrons. Since all the bond angles and bond distances are identical in the boron trichloride molecule, the simplest means of interpreting these facts is by hybridization. The three electrons of the boron atom are said to be **sp^2** hybridized (since one *s* and two *p* electrons are involved), and the shape of the BCl_3 molecule is flat, coplanar, with angles of 120° $\begin{pmatrix} Cl & Cl \\ & B \diagdown_{120°} \\ & Cl \end{pmatrix}$. All molecules having sp^2 hybridization will possess the same molecular geometry as the BCl_3 cited above.

Methane, CH_4, contains five atoms, and is considered the parent compound of organic chemistry. By examining the electronic structures of hydrogen, $1s^1$, and carbon, $1s^2\ 2s^2\ 2p_x^1\ 2p_y^1$, in Table 1.1, it can be seen that carbon has two unpaired electrons in its ground state electronic configuration. The molecular formula of methane, CH_4, indicates that carbon uses four unpaired electrons, one in each bond to a hydrogen atom. The excited state electronic configuration of carbon, $1s^2\ 2s^1\ 2p_x^1\ 2p_y^1\ 2p_z^1$, shows that there are four unpaired electrons available for bonding. The hybridization present in the CH_4 molecule is said to be of the sp^3-type (since one s and three p electrons are involved) and the molecule has the geometric shape of a tetrahedron with bond angles of 109° 28'. (See Figure 1.2.) The carbon atom is in the center of the tetrahedron with the hydrogen atoms at the four corners. Molecules with sp^3 hybridization will have the geometrical shape of a tetrahedron. Whenever a carbon atom has four substituents (four single bonds) attached to it, the hybridization is of the sp^3 type.

1.6 Molecular Orbitals (Sigma and Pi Bonds)

In the compound ethylene, C_2H_4, the carbon atoms are bonded to only three substituents, and a double bond is present $\left(\begin{array}{c}H\\H\end{array}C=C\begin{array}{c}H\\H\end{array}\right)$.

The hybridization of the carbon atoms in ethylene is not the same as in methane. Reference to the excited-state electronic configuration for carbon, $1s^1\ 2s^1\ 2p_x^1\ 2p_y^1\ 2p_z^1$, reveals four unpaired electrons. Since in ethylene there are only *three* substituents attached to each carbon atom, only three of the electrons are hybridized. These are the 2s electron and two of the 2p electrons, forming three hybrid sp^2 electrons. (Thus, ethylene is a flat, coplanar molecule with bond angles of 120°.) The fourth electron is the remaining 2p electron, which is perpendicular to the plane of the sp^2 orbitals.

The four covalent bonds on each carbon atom in ethylene are formed by an

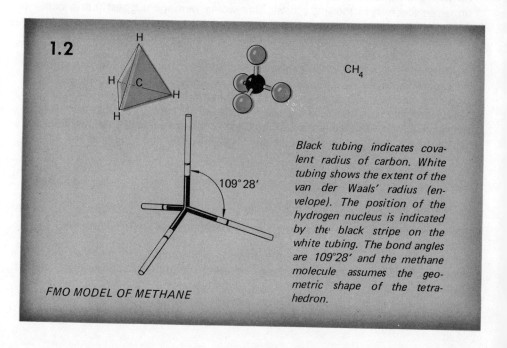

1.2

CH_4

FMO MODEL OF METHANE

109°28'

Black tubing indicates covalent radius of carbon. White tubing shows the extent of the van der Waals' radius (envelope). The position of the hydrogen nucleus is indicated by the black stripe on the white tubing. The bond angles are 109°28' and the methane molecule assumes the geometric shape of the tetrahedron.

Sec. 1.6 MOLECULAR ORBITALS (SIGMA AND PI BONDS)

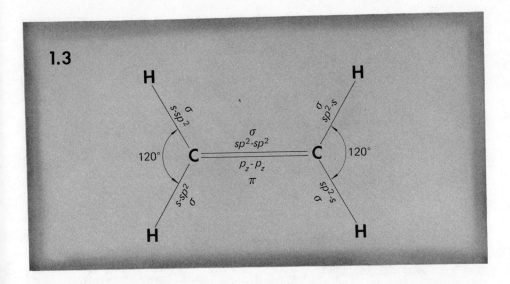

1.3

overlapping of atomic orbitals. The two C—H bonds are formed by the overlapping of two sp^2 electrons from carbon with the $1s$ electron of hydrogen. The C=C is formed by an overlapping of the third sp^2 electron from each of the carbon atoms. The remaining unpaired p electron on each of the carbon atoms overlap to complete the double bond. (See Figure 1.3.)

Once the individual atomic orbitals overlap, the resulting orbitals are said to belong to the molecule as a whole, and are referred to as **molecular orbitals.** Thus, in this example, the overlapping of an sp^2 atomic orbital and an s atomic orbital produces a molecular orbital.

Molecular orbitals can be classified as **sigma (σ) bonds** or **pi (π) bonds.** All bonds that are in the plane of the molecule or involve s electrons (pure or hybridized) are σ bonds. The σ bond is cylindrically symmetrical about a line joining the nuclei (i.e., a carbon–carbon axis). The major electron density in a σ bond lies along the internuclear line between the atoms forming the bond. Molecular orbitals formed by overlapping of p atomic orbitals, whose axes are parallel, are π bonds. The π bond has the shape of two ellipsoids, one above and one below the plane of the molecule, perpendicular to the molecular plane. In a π bond the major electron density is located above and below the plane of the molecule, rather than between the nuclei forming the bond. (See Figure 1.4.)

In ethylene each carbon atom has three σ bonds. The π bond, formed as part of the double bond between the two carbon atoms, is perpendicular to the plane of the molecule. The double bond can be considered as composed of a σ bond and a π bond. (See Figure 1.5.)

The strength of the molecular orbitals is determined by the extent of the overlapping of the electron clouds of the atomic orbitals, which leads to the formation of the chemical bond. Because of less effective overlapping of electron clouds, π bonds are weaker than σ bonds. The π electrons have much greater freedom of movement (above and below the plane of the molecule) than σ-bonded electrons, which are held more tightly between the atomic nuclei. The σ electrons

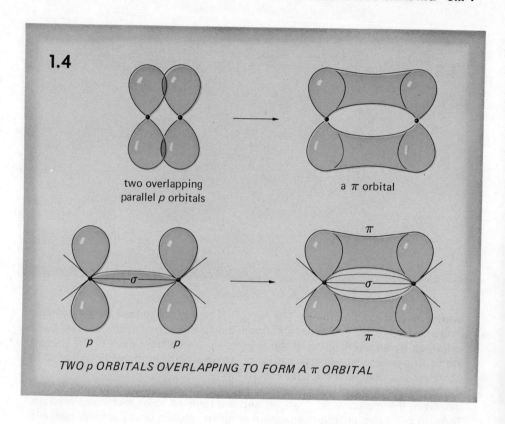

1.4

two overlapping parallel *p* orbitals

a π orbital

TWO *p* ORBITALS OVERLAPPING TO FORM A π ORBITAL

are said to be *localized*. The π electrons are much more "polarizable" than σ electrons, meaning that the π electron cloud can be deformed more easily. This distortion of the electron distribution in one molecule caused by another is called **polarization,** and is important in explaining the reactivity of molecules containing a carbon–carbon double bond.

When molecules contain bonding orbitals that are not restricted to two atoms, but are spread over three or more atoms, the bonding is said to be **delocalized.** This delocalization is particularly important in molecules containing a series of alternating single and double bonds, an arrangement known as a **conjugated system.** In a conjugated system the π clouds of the double bonds present overlap to form a continuous molecular orbital that extends over the entire length of the conjugated system in the molecule. Such an orbital, extending over three or more nuclei, is often referred to as a **polynuclear** orbital. (See Figure 1.6.) The effect of this polynuclear molecular orbital is to spread out the electron density over a large region of the molecule, rather than to concentrate the electron density in one individual π bond, as in ethylene. This delocalization of electron density (or charge) is extremely important when considering the properties of molecules having conjugated systems. We shall refer to this phenomenon when we discuss

Sec. 1.6 MOLECULAR ORBITALS (SIGMA AND PI BONDS)

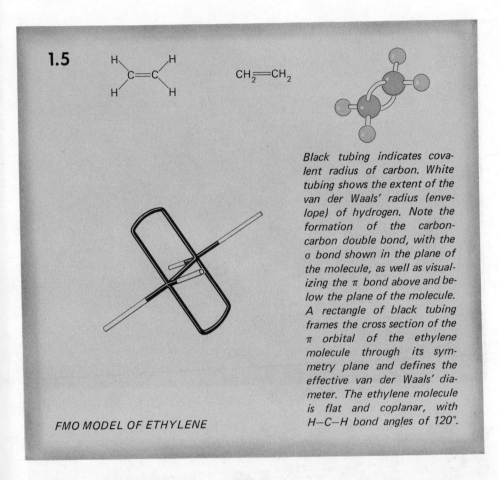

1.5 FMO MODEL OF ETHYLENE

Black tubing indicates covalent radius of carbon. White tubing shows the extent of the van der Waals' radius (envelope) of hydrogen. Note the formation of the carbon-carbon double bond, with the σ bond shown in the plane of the molecule, as well as visualizing the π bond above and below the plane of the molecule. A rectangle of black tubing frames the cross section of the π orbital of the ethylene molecule through its symmetry plane and defines the effective van der Waals' diameter. The ethylene molecule is flat and coplanar, with H—C—H bond angles of 120°.

the chemistry of certain organic compounds, such as 1,3-butadiene (Chapter 8) and benzene (Chapter 10).

Let us now consider the type of bonding present in the acetylene molecule, C_2H_2. Acetylene contains a triple bond, and by similar arguments, as were used in the case of ethylene, the acetylene molecule can be shown to have sp hybridization. (See Figure 1.7.)

The acetylene molecule is linear and has bond angles of 180°. Since only two unpaired electrons from carbon are necessary for bonding to the two substituents of each carbon atom, the remaining two unpaired $2p$ electrons on each carbon overlap to form two π bonds. (See Figure 1.8.) The triple bond consists of *one σ bond* and *two π bonds*. The π bonds are perpendicular to the plane of the molecule. The acetylene molecule appears to be surrounded by a cylindrical cloud of π electrons. All triple bonds can be interpreted as consisting of *one σ bond and two π bonds*.

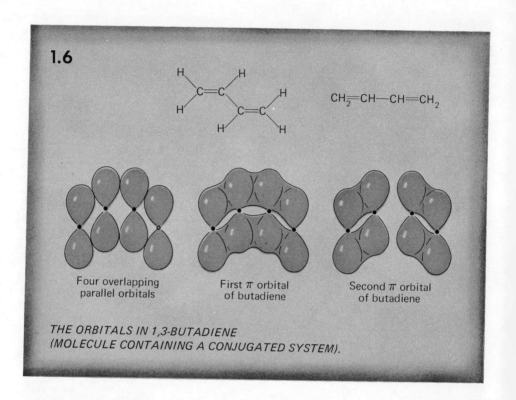

THE ORBITALS IN 1,3-BUTADIENE (MOLECULE CONTAINING A CONJUGATED SYSTEM).

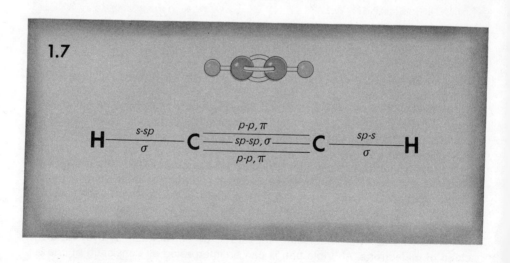

SUMMARY

1.8

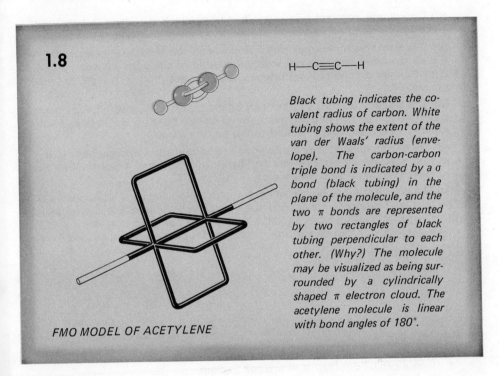

H—C≡C—H

Black tubing indicates the covalent radius of carbon. White tubing shows the extent of the van der Waals' radius (envelope). The carbon-carbon triple bond is indicated by a σ bond (black tubing) in the plane of the molecule, and the two π bonds are represented by two rectangles of black tubing perpendicular to each other. (Why?) The molecule may be visualized as being surrounded by a cylindrically shaped π electron cloud. The acetylene molecule is linear with bond angles of 180°.

FMO MODEL OF ACETYLENE

Summary

- An understanding of atomic structure and chemical bonding is essential in organic chemistry.
- Much of our present knowledge of the electronic structure of the atom is due to the quantum theory.

  (a) The regions outside the nucleus of the atom, where the probability of finding an electron is high, are called energy levels.

  (b) Every energy level is composed of sub-levels, called orbitals, which differ as to the shape of the electron cloud.

  (c) These atomic orbitals are designated by the letters s, p, d, and f.

- The electronic configuration of the atoms is built up according to the aufbau principle, filling those orbitals of lower energy content first and proceeding successively to orbitals of higher energy content.
- Chemical bonding is of two basic types (ionic and covalent); these are most conveniently discussed using Lewis electron-dot formulas.

  (a) Ionic bonds involve a complete transfer of electrons between atoms of greatly differing electronegativities.

  (b) Covalent bonding involves a sharing of electrons between atoms of similar electronegativities, and is of paramount importance in organic chemistry.

  (c) Multiple bonds can be formed by atoms sharing more than one electron pair between them.

(d) Coordinate covalent bonds are bonds in which the electron pair (both electrons) is donated by one of the atoms forming the bond.

(e) Covalent bonds between unlike atoms of somewhat different electronegativities are polar and result in unequal sharing of electrons between atoms.

- Hybridization is important in understanding the geometry, or shape, of covalent molecules.

(a) The valence electrons of the carbon atom can be hybridized in different ways (sp, sp^2, or sp^3) and the geometry of carbon bonds is dependent on the number of substituents attached to the carbon atom.

(b) When four substituents are bonded to a carbon atom as in methane (CH_4), the carbon atom's orbitals are sp^3 hybridized and the molecule has the geometrical shape of a tetrahedron.

- Molecular orbitals can be classified as σ or π bonds. Sigma electrons are said to be localized, whereas π electrons are much more "polarizable." A series of alternating single and double bonds is known as a conjugated system.
- The geometry of carbon–carbon double and triple bonds is discussed.

PROBLEMS

1 Is an s orbital directional in space? Explain.

2 Write the s, p, d, and f electronic designations for each of the following. **(a)** sodium atom; **(b)** sodium ion; **(c)** chlorine atom; **(d)** chloride ion; **(e)** hydride ion.

3 Which of the following substances are ionic, and which are covalent? $LiCl$, F_2, SO_2, CCl_4, Mg_3N_2, PCl_5, C_2H_6, H_2S, NaF, H_2O.

4 Draw Lewis electron-dot formulas for each of the following. **(a)** HF **(b)** NaF **(c)** NH_3 **(d)** MgO **(e)** CH_3Cl **(f)** $SiCl_4$ **(g)** HCN **(h)** $HClO_4$ **(i)** CH_3OH **(j)** $MgCl_2$

5 Arrange the following substances in order of increasing polarity. HBr, HF, HI, HCl.

6 Arrange the following substances in order of increasing polarity, and for each polar covalent molecule, indicate the positive and negative ends of the polar bond. HCl, CH_3Cl, CF_4, ICl, Cl_2, $NaCl$, IF_7, $MgCl_2$.

7 Carbon dioxide is a non-polar molecule. What does this suggest about the shape of the carbon dioxide molecule?

8 *Explain why nitrogen trifluoride, NF_3, is a much less polar molecule than ammonia, NH_3.

9 Using your knowledge of electronic structure and hybridization, predict the geometrical shape of each of the following species. SiF_4, BF_3, $CHCl_3$,

$$\text{H}-\overset{\overset{\text{O}}{\|}}{\text{C}}-\text{H}, \quad HCN, \quad NH_4^{\oplus}, \quad CH_3^{\oplus}, \quad CH_3^{\ominus}.$$

10 For each of the following molecules, indicate the type of electrons (i.e., s, p, sp, sp^2, or sp^3, etc.) that formed each bond, and designate each bond as being either a σ bond or a π bond.

PROBLEMS

(a)
$$H-\underset{\underset{Br}{|}}{C}=CH-C\equiv C-\underset{\underset{}{\overset{\overset{O}{\|}}{}}}{C}-H$$

(b)
$$H_2C=C(CH_3)-CH_2-\underset{\underset{Cl}{|}}{CH}-\underset{\underset{}{\overset{\overset{H}{|}}{C}}}{C}=CHBr$$

(c)
$$H-\overset{\overset{O}{\|}}{C}-H$$

Indicate which of the bonds in these molecules have angles of 120°.

11 Calculate the percentage by weight of all the elements present in the following compounds. (a) C_5H_{12}, (b) $C_3H_6O_2$, (c) $C_6H_6SO_3$.

12 A compound was analyzed and found to contain 24 grams (g) of carbon, 4 g of hydrogen, and 32 g of oxygen. The molecular weight of the compound is 60. Calculate the empirical formula and the true molecular formula of the compound.

13 Upon analysis, a certain compound was found to contain 38.7 % carbon, 9.7 % hydrogen, and 51.6 % oxygen. The molecular weight of the compound is 62. Calculate the empirical formula and the true molecular formula of the compound.

14 * An organic compound has the following percentage composition by weight: C = 24.26 %, H = 4.08 %, Cl = 71.66 %. It was found that 0.1402 liters (l) of the vapor of this compound weighed 0.4416 g at 100° C and a pressure of 740 mm Hg. What is the true molecular formula of this compound?

15 Of the gaseous organic compound propylene 1 l weighs 1.88 g at STP. It has the following composition by weight: C = 85.63 %, H = 14.37 %. Calculate both the empirical and the true molecular formulas for propylene.

16 * When 22.5 g of the sugar *mannose* are dissolved in 250 g of water, the solution has a freezing point of −0.93° C. If the empirical formula of mannose is CH_2O, what is the true molecular formula of mannose?

2
Characteristics of Structure

The Boeing 747, one of the modern giant aircraft made economically viable by the use of the less-refined higher-carbon-content fuel used in its jet engines.

Courtesy of NASA.

2.1 Introduction

Organic chemistry can be defined as the chemistry of substances that contain the element carbon, or perhaps in another way as the chemistry of compounds containing carbon–carbon covalent bonds. The name *organic* was originally used to denote compounds of plant or animal origin. This was particularly true before the year 1800, because of the observation that many carbon-containing compounds were produced by living organisms. Today many organic substances are produced synthetically in the laboratory.

Prior to 1800, the development of organic chemistry was greatly hindered by the belief that organic compounds could be produced only within living organisms. Many organic substances, such as alcohol, were known in ancient times, and it was thought that certain "vital forces" in plants and animals were necessary for their formation.

In 1776 the Swedish chemist Karl Wilhelm Scheele (1742–1786) prepared oxalic acid by the reaction of nitric acid with sucrose. The reaction showed that one organic compound could be made from another, but it was thought that the "vital force" necessary for the reaction was supplied by the sugar.

Sec. 2.3 ORGANIZATION OF ORGANIC CHEMISTRY

Friedrich Wöhler, a German chemist, in 1828 synthesized the organic compound **urea** by heating the inorganic compound ammonium cyanate.

$$NH_4CNO \xrightarrow{\Delta} NH_2-\underset{}{\overset{O}{\underset{\|}{C}}}-NH_2 \qquad (2-1)$$

ammonium cyanate   urea

This synthesis helped disprove the vital-force theory, as did the gradual development of chemistry as a science. The conversion of ammonium cyanate into urea marked the beginning of synthetic organic chemistry as we know it today.

2.2 The Nature of the Carbon Atom

The unique properties and features of the carbon atom, as compared to those of other elements, necessitated a separate branch of chemistry for the study of carbon compounds. Organic chemistry is the division of chemistry dealing exclusively with the compounds of carbon.

The electronic structure of the carbon atom (see Table 1.1) has a kernel designation, $\overset{x}{\underset{x}{\times C \times}}$. The four valence electrons in carbon tend to form covalent bonds in compounds. It is most interesting and important that carbon has the ability to form covalent bonds with other carbon atoms. Carbon atoms have this property to a greater extent than any other element.

The molecules formed consist of short or long chains of carbon, as well as rings of carbon atoms, and result in the large number (more than two million) of different organic compounds known today. Other elements can form covalent bonds between their atoms, but the chains are usually less stable and much more limited in length. The large variety of organic compounds can be envisioned if one further realizes the versatility of the carbon atom to form double and triple bonds in its compounds (Section 1.6).

2.3 Organization of Organic Chemistry

Due to the rather large number and variety of organic compounds known, it is convenient to place the various compounds into a series of groups or **classes.** This enables us to organize the many organic compounds into relatively few categories. Compounds belonging to the same class will have certain similarities in molecular structure, and in their physical and chemical properties, and substances having different characteristics are placed in different classes.

For example, compounds which contain only the elements carbon and hydrogen are referred to as **hydrocarbons.** Hydrocarbons constitute one class of organic compounds; but the hydrocarbons themselves are further subdivided into sub-classes. If only *single bonds* are present in the molecule, the compound is said to be **saturated.** *Saturated hydrocarbons* are called **alkanes** or **paraffins.** Hydrocarbons containing multiple bonds are **unsaturated. Alkenes** or **olefins** are hydrocarbons containing a *double bond;* **alkynes** or **acetylenes** are hydrocarbons containing a *triple bond* in the molecule.

The various classes of organic compounds will be discussed in subsequent chapters throughout the text. In grouping the compounds into various classes, it should be noted that *a group of atoms* within the molecule is responsible for its reactivity. The properties of the molecule, in particular the chemical reactivity, depend primarily on this group of atoms called the **functional group.**

2.3-1 Functional Groups Thus, one can see that the various classes of organic compounds depend on the functional group(s) present in the molecule (e.g., compounds containing a $\mathrm{\backslash C{=}C/}$ are called **alkenes**; those with a $-\overset{\overset{O}{\|}}{C}-OH$ group are **carboxylic acids**; the $-NH_2$ group indicates an **amine**; see Table 2.1).

The actual naming (nomenclature) and writing of structural formulas of organic compounds will be taken up throughout the following chapters. At this point, however, it will be worthwhile to outline a few general principles.

2.4 Representation of Spatial Formulas and Molecular Models

One must remember that all molecular structures are three-dimensional in space. (It is advisable that a student learn the use of a set of molecular models for a true representation of molecular geometry in three dimensions. For the author, the FMO models, used in a number of the figures, have proven quite satisfactory.) Some degree of accuracy will be lost in representation in the two dimensions to which the printed page is limited, but it will be sufficient for our purpose.

Let us consider a compound having the molecular formula C_4H_{10}, and the structural formula shown in Figure 2.1. Ordinary lines (—) represent bonds in the plane of the paper, the dashed lines (- - -) are bonds extending away from the reader, beginning at the plane of the page, and the wedge shaped bonds (▮)

Table 2.1 Some Functional Groups in Relation to Classes of Organic Compounds

Functional group	Class	Functional group	Class
$-\underset{\|}{\overset{\|}{C}}-\underset{\|}{\overset{\|}{C}}-$	Alkane	$-\overset{\overset{O}{\|}}{C}-H$	Aldehyde
$\mathrm{\backslash C{=}C/}$	Alkene	$-\overset{\overset{O}{\|}}{C}-$	Ketone
$-C{\equiv}C-$	Alkyne	$-\underset{\|}{\overset{\|}{C}}-X$ (X = halogen)	Halide
(benzene ring)	Arene (aromatic)	$-\overset{\overset{O}{\|}}{C}-OH$	Carboxylic acid
$-\underset{\|}{\overset{\|}{C}}-OH$	Alcohol	$-\overset{\overset{O}{\|}}{C}-NH_2$	Amide
$-\underset{\|}{\overset{\|}{C}}-O-\underset{\|}{\overset{\|}{C}}-$	Ether	$-NH_2$	Amine

Sec. 2.4 REPRESENTATION OF SPATIAL FORMULAS AND MODELS

2.1

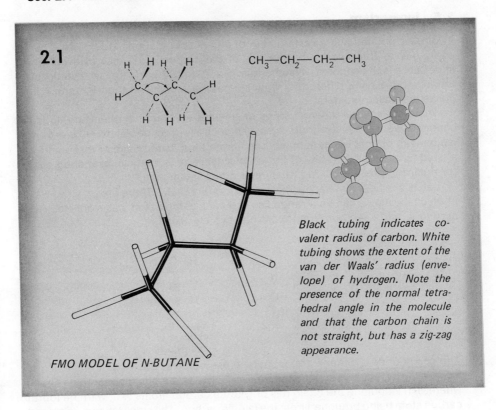

FMO MODEL OF N-BUTANE

Black tubing indicates covalent radius of carbon. White tubing shows the extent of the van der Waals' radius (envelope) of hydrogen. Note the presence of the normal tetrahedral angle in the molecule and that the carbon chain is not straight, but has a zig-zag appearance.

are bonds extending towards the reader from the plane of the page. This type of representation will be useful in discussing certain aspects of organic chemistry. However, the same C_4H_{10} formula will more often be drawn as

$$\begin{array}{c} H\ H\ H\ H \\ | \ | \ | \ | \\ H-C-C-C-C-H \\ | \ | \ | \ | \\ H\ H\ H\ H \end{array}$$

where the bonds (—) are single bonds (or a pair of electrons), or as

$$-\overset{|}{\underset{|}{C}}-\overset{|}{\underset{|}{C}}-\overset{|}{\underset{|}{C}}-\overset{|}{\underset{|}{C}}-$$

where the hydrogen atoms are omitted for simplicity. The C_4H_{10} can also be represented as $CH_3-CH_2-CH_2-CH_3$, or, in its most condensed form, as $CH_3CH_2CH_2CH_3$. It will be seen in the following chapters that organic compounds can contain continuous chains of carbon atoms, branched chains, or even rings. The representation of the structural formulas will always follow the fundamental concepts presented above.

2.5 Isomerism

As already pointed out, a vast number of organic compounds are known. This is partly due to the electronic structure of the carbon atom, but of even more importance is the concept of **isomerism**. Compounds having the same molecular formula, but having different arrangements of the atoms in their structural formulas are called **isomers**.

There are several different types of isomerism in organic chemistry. In later chapters, we will concern ourselves with such types as **geometrical (cis-trans) isomerism, optical isomerism,** and **conformational isomerism;** at this point, we should take a look at a few of the simpler types of isomerism: **branched chain** (or **nuclear**), **positional,** and **functional group isomerism**.

Let us consider a molecule having the formula C_4H_{10}. There are only two ways in which the atoms can be arranged so that each carbon atom has eight electrons around it, and each hydrogen atom has two electrons.

n-butane   isobutane or
                  2-methylpropane

The compounds n-butane and isobutane are isomers. Each has the molecular formula C_4H_{10}, but in n-butane the four carbon atoms are in a continuous chain, whereas in isobutane the chain has only three carbon atoms in it, with the fourth carbon atom being branched from the middle carbon. The small letter n- is placed before the name of a compound to indicate that all the carbon atoms are in a continuous chain; an i- (iso) is used to indicate chain branching (methyl group) on the carbon atom adjacent to the carbon at the end of the chain. These two compounds (n-butane and isobutane) are examples of **branched chain isomerism**.

Although compounds may have the same molecular formula, they may have several Lewis electron-dot formulas. For example, two can be drawn for C_3H_7Cl.

n-propyl chloride   i-propyl chloride
(1-chloropropane)   (2-chloropropane)

It can be seen that the position of the chlorine atom is different in these molecules, being attached to the carbon at the end of the carbon chain in n-propyl chloride, and to the middle carbon atom in i-propyl chloride. Compounds such as these, differing only in the position of a substituent in the molecule are called **positional isomers**.

Two substances, ethyl alcohol and dimethyl ether, have the molecular formula C_2H_6O (see Figure 2.2). From the structural formulas it can be seen that the arrangement of the atoms is different in the two structures, and hence that these

Sec. 2.5 ISOMERISM

2.2

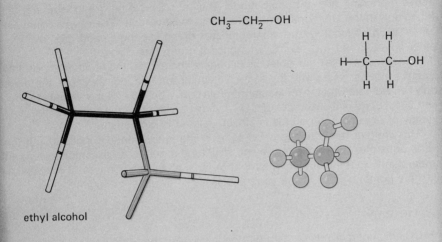

ethyl alcohol

dimethyl ether

These two compounds, despite having the same molecular formula, are completely different substances, and are examples of positional isomers. The black tubing indicates the covalent radius of carbon, and the white tubing shows the extent of the van der Waals' envelope. The red tubing (blue in the figure) represents the covalent radius of oxygen in the C—O and O—H bonds. The two unshared electron pairs on the oxygen atom in the alcohol and ether are indicated by two pieces of solid red tubing attached to the oxygen atom in the molecule.

compounds have different physical and chemical properties. This can be demonstrated rather easily in the laboratory: ethyl alcohol is a liquid that boils at 78° C and reacts with metallic sodium to produce hydrogen gas; dimethyl ether is a gas and does *not* react with sodium. Clearly then, the two structural formulas shown above represent different compounds. They are **functional group isomers,** differing only in the functional group present in the molecule. The ethyl alcohol has an —OH functional group characteristic of all alcohols, and the dimethyl ether contains the —C—O—C— functional group (see Table 2.1).

The more complex the molecule becomes, the greater the number of possible isomers that can exist with the same molecular formula. For example, there are only two isomers with the molecular formula C_4H_{10}, but 75 with the formula $C_{10}H_{22}$, and 366,319 possible isomers of $C_{20}H_{42}$. One can see that the great number and variety of organic compounds is closely related to isomerism.

This chapter has touched upon some of the fundamental concepts, and the basic principles of nomenclature and structural formulas so essential toward an understanding of the subject matter of organic chemistry. Chapter 3 will take us into the nomenclature of the various classes of organic compounds.

Summary

- Organic compounds contain carbon in covalent bonding and such elements as hydrogen, oxygen, and nitrogen.

    (a) Many organic compounds are found in natural products that either were or are living organisms.

    (b) Some organic substances are produced synthetically in the laboratory.

- The nature of the carbon atom, its electronic structure, and its ability to form covalent bonds with other carbon atoms result in a large variety of organic compounds.

- Organic chemistry can be organized and divided into various classes of compounds depending on the functional group present in the molecule.

    The functional group will give certain characteristics and properties to any molecule containing it. It is responsible for the chemical reactivity of a given class of compounds.

- Molecular structures are three-dimensional. It is important to be able to visualize structures in three dimensions.

- Isomers are compounds having the same molecular formula, but different structural arrangement or spatial orientation of the atoms within the molecule.

    The different types of isomerism largely account for the great number and variety of organic compounds known.

PROBLEMS

1   Match the compounds in column A with the appropriate functional group listed in column B.

PROBLEMS

A	B	A	B
$CH_3-CH_2-\overset{\overset{O}{\|\|}}{C}-H$	Alkane	$CH_3-CH=CH-CH_3$	Halide
CH_3-CH_2-Br	Alkene	$CH_3-CH_2-NH_2$	Aldehyde
CH_3-CH_2-OH	Alkyne	$CH_3-\overset{\overset{O}{\|\|}}{C}-NH_2$	Ketone
$CH_3-C\equiv C-CH_3$	Arene (aromatic)	$CH_3-CH_2-O-CH_2-CH_3$	Carboxylic acid
$CH_3-CH_2-CH_3$	Alcohol	$CH_3-\overset{\overset{O}{\|\|}}{C}-CH_3$	Amide
(naphthalene)	Ether	$CH_3-\overset{\overset{O}{\|\|}}{C}-OH$	Amine

2 Draw the structural formulas for the five isomers having the formula C_6H_{14}.

3 *Draw the structural formulas for all four isomers having the formula C_3H_5Cl.

4 Indicate which type of isomerism is illustrated by the following pairs of compounds.

(a) $CH_3-CH_2-CH_2-CH_2-CH_3$ $CH_3-\underset{\underset{CH_3}{\|}}{\overset{\overset{CH_3}{\|}}{C}}-CH_3$

(b) $CH_3-CH_2-CH_2-CH_2OH$ $CH_3-CH_2-\underset{\underset{OH}{\|}}{CH}-CH_3$

(c) $CH_3-C\equiv C-CH_3$ $CH_3-CH_2-C\equiv C-H$

(d) $CH_3-\overset{\overset{O}{\|\|}}{C}-CH_3$ $CH_2=CH-CH_2OH$

(e) $CH_3-CH=CHCl$ $Cl-CH_2-CH=CH_2$

5 Can a compound contain more than one functional group in a molecule? If the answer is yes, give some specific examples from the compounds listed in Problem 4.

3
Nomenclature

A new organic chemicals plant near Baton Rouge, with a production capacity of 500 million pounds a year of substances subject to the rules of nomenclature of organic compounds.

Courtesy of Gulf Oil Corporation.

3.1 Introduction

In order to distinguish individuality among human beings each of us is given a different name. The increasingly large number of organic compounds, ranging from simple to exceedingly complex molecules, requires a comprehensive system of nomenclature so that each structural formula has an individual name that refers to just one molecular structure. A system of nomenclature was essential to the development of chemistry as a science. Chemists have been meeting periodically since 1892 when a conference was held at Geneva, in an attempt to devise a system of nomenclature to be used universally, so that everyone would be able to read and understand the chemical literature. The rules adopted and set

Sec. 3.2 THE SYSTEMATIC NOMENCLATURE

forth by these conferences on the nomenclature of organic compounds are referred to as the IUPAC system (International Union of Pure and Applied Chemistry).

3.2 The Systematic Nomenclature

Basically this system of naming organic compounds considers the structure as derived from a simple carbon chain skeleton with modifications. The most fundamental skeleton containing a continuous chain of carbon atoms is found in the **unbranched alkanes,** and this class of compounds is the foundation for the IUPAC system of nomenclature.

Table 3.1 contains the names of some unbranched alkanes, which serve as roots for the names of other classes of organic compounds. The rules for naming compounds according to the IUPAC system are as follows:

1 The saturated hydrocarbons are named **alkanes.**

2 The naming of branched chain compounds is based on the name of the alkane corresponding to the longest continuous chain of carbon atoms in the compound; for example, in the compound $\underset{1}{CH_3}-\underset{2}{\underset{|}{\overset{Cl}{CH}}}-\underset{3}{CH_2}-\underset{4}{CH_2}-\underset{5}{\underset{|}{\overset{CH_3}{CH}}}-\underset{6}{CH_2}-\underset{7}{CH_3}$ *the longest continuous chain of carbon atoms is seven;* the last part of the name of this compound is therefore **heptane.**

3 The carbon atoms in the longest continuous chain are numbered in such a manner that substituents other than hydrogen are given numbers corresponding to their position in the carbon chain. Since the carbon chain can be numbered starting from either end of the chain, the direction of numbering used is the one whereby the substituents have the lowest numbers possible; for example, in the compound cited in rule 2 above, numbering from left to right gives the substituents numbers of (Cl) 2 and (CH_3) 5, whereas if the chain had been numbered from right to left, the numbers of the substituent carbon atoms would have been 3 and 6.

In connection with the naming of substituents, it should be mentioned that groups which are derived from *alkanes* are called *alkyl groups*. The name is taken from the parent hydrocarbon with the same number of carbon atoms by changing the *-ane* ending to *-yl*. Thus the CH_3 group is derived from the alkane of one

Table 3.1 Nomenclature for Straight-chain Alkanes

Molecular formula	Structure	Name	Molecular formula	Structure	Name
CH_4	CH_4	Methane	C_8H_{18}	$CH_3-(CH_2)_6-CH_3$	*n*-Octane
C_2H_6	CH_3-CH_3	Ethane	C_9H_{20}	$CH_3-(CH_2)_7-CH_3$	*n*-Nonane
C_3H_8	$CH_3-CH_2-CH_3$	Propane	$C_{10}H_{22}$	$CH_3-(CH_2)_8-CH_3$	*n*-Decane
C_4H_{10}	$CH_3-(CH_2)_2-CH_3$	*n*-Butane	$C_{11}H_{24}$	$CH_3-(CH_2)_9-CH_3$	*n*-Undecane
C_5H_{12}	$CH_3-(CH_2)_3-CH_3$	*n*-Pentane	$C_{12}H_{26}$	$CH_3-(CH_2)_{10}-CH_3$	*n*-Dodecane
C_6H_{14}	$CH_3-(CH_2)_4-CH_3$	*n*-Hexane	$C_{13}H_{28}$	$CH_3-(CH_2)_{11}-CH_3$	*n*-Tridecane
C_7H_{16}	$CH_3-(CH_2)_5-CH_3$	*n*-Heptane	$C_{20}H_{42}$	$CH_3-(CH_2)_{18}-CH_3$	*n*-Eicosane

carbon atom, CH_4, by simply removing one hydrogen atom. The name for CH_4 is *methane* (Table 3.1) and the CH_3 group is called a *methyl group*. In the same manner the alkane of two carbon atoms, C_2H_6 is named *ethane,* and the C_2H_5 group derived from it is named an *ethyl group*. This concept will be discussed in more detail in the chapter on the alkanes. The naming of some common substituent groups is shown in Table 3.2. The compound

$$CH_3-\underset{1}{C}H-\underset{2}{C}H-\underset{3}{C}H_2-\underset{4}{C}H_2-\underset{5}{C}H-\underset{6}{C}H_2-\underset{7}{C}H_3$$
$$\qquad\;\;|\qquad\qquad\qquad\qquad|$$
$$\qquad\;\;Cl\qquad\qquad\qquad\;\;CH_3$$

would be correctly named 2–chloro–5–methylheptane according to the rules outlined above. The longest continuous chain of carbon atoms is seven, hence the name *heptane*. The Cl is on the second carbon, and the CH_3 group on carbon number 5. Their positions are indicated by the numbers 2 and 5 in the name. Dashes are used to separate the different substituents present in the molecule. The compound

$$CH_3-\underset{2}{\overset{CH_3}{\underset{CH_3}{C}}}-\underset{3}{\overset{CH_3}{\underset{CH_3}{C}}}-CH_3$$

is correctly named 2,2,3,3-tetramethylbutane. When two similar side chains are present (e.g., two methyl groups), the prefix *di-* is added (dimethyl) and the number of the carbon atom indicating the location of the substituents is separated by a comma. For three similar groups, *tri-* is used, and for four, *tetra-*, etc. Thus, in the 2,2,3,3-tetramethylbutane, there are four CH_3 groups present, two on carbon number 2 and two on carbon number 3.

Note that the longest continuous chain of carbon atoms need not always be written on a horizontal line.

Table 3.2 Nomenclature of Some Functional Groups

Substituent group	Name	Substituent group	Name
F—	Fluoro	$CH_3-CH_2-CH_2-$	n-Propyl (normal propyl)
Cl—	Chloro	$CH_3-CH-CH_3$	i-Propyl (isopropyl)
Br—	Bromo		
I—	Iodo		
—OH	Hydroxy	$CH_3-\underset{CH_3}{\overset{CH_3}{C}}-$	t-Butyl (tertiary butyl)
—NO_2	Nitro		
—NH_2	Amino		
—CN	Cyano		
CH_3-	Methyl	$CH_3-\underset{CH_3}{\overset{CH_3}{C}}-CH_2-$	Neopentyl
CH_3-CH_2-	Ethyl		

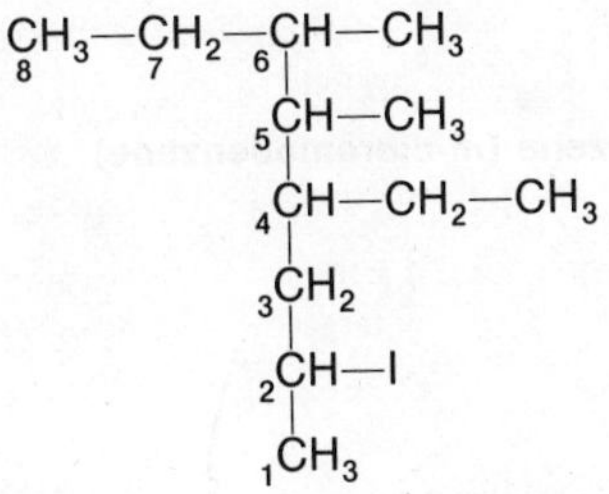

Sec. 3.4 NOMENCLATURE OF SOME CYCLIC COMPOUNDS

$$\text{CH}_3{-}\underset{7}{\text{CH}_2}{-}\underset{6}{\text{CH}}{-}\text{CH}_3$$
with branches giving 2-iodo-4-ethyl-5,6-dimethyloctane

(structure numbered 1–8 from bottom CH₃ upward, with substituents: 2-I, 4-ethyl (CH₂CH₃), 5-CH₃, 6-CH₃)

2-iodo-4-ethyl-5,6-dimethyloctane

The nomenclature of the different classes of organic compounds will be discussed in more detail in the appropriate sections. The basic rules of the IUPAC system are used in the naming of all compounds; that is to say, one must look for the longest continuous chain of carbon atoms and use the lowest numbers possible to indicate the position of any substituents. The only difference in the naming of the different classes of organic compounds is in the ending of the name.

3.3 Nomenclature of Organic Compounds Containing One Functional Group

The names of all saturated hydrocarbons (alkanes) end in *-ane*. The name of a compound containing a double bond ends in *-ene* rather than *-ane* (e.g., $CH_3{-}CH{=}CH_2$ is named prop*ene*).

A compound containing an OH group ends in *-ol*; a $-\overset{\overset{\displaystyle O}{\|}}{C}-H$ group in *-al*, etc. Table 3.3 illustrates some typical examples.

3.4 Nomenclature of Some Cyclic Compounds

Many organic compounds are in the form of rings. Since a *ring of carbon atoms* has no end, the assignment of numbers in a **cyclic compound** is done so that the functional group is on position 1. When more than one substituent is present, the numbering around the ring is in the direction to give the substituents the lowest numbers.

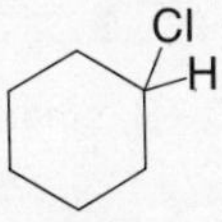

chlorocyclohexane (*cyclo* indicates a ring compound)

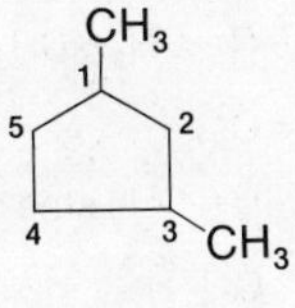

1,3-dimethylcyclopentane

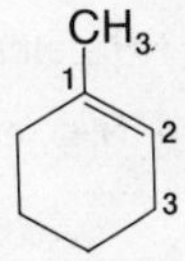

1-methylcyclohexene

1,3-dibromobenzene (*m*-dibromobenzene)

(benzene ring with H's or benzene ring symbol is called benzene)

Table 3.3 Nomenclature of Some Classes of Organic Compounds

Functional group	Class of compounds	Name	Example
—C—C—	Saturated hydrocarbon	Alk-*ane*	CH_3—CH_2—CH_3, prop-*ane*
C=C	Olefin	Alk-*ene*	CH_3—$C(CH_3)$=CH_2, 2-methylprop-*ene*
—C≡C—	Acetyl-*ene*	Alk-*yne*	H—C≡C—CH_3, prop-*yne*
—OH	Alcohol	Alkan-*ol*	CH_3OH, methan-*ol*
—OR	Ether	Alkyl-*ether*	CH_3—O—CH_2—CH_3, methoxyethane (methylethyl-*ether*)
—C(=O)—H	Aldehyde	Alkan-*al*	CH_3—C(=O)—H, ethan-*al*
—C(=O)—	Ketone	Alkan-*one*	CH_3—C(=O)—CH_2—CH_3, butan-*one*
—COOH	Acid	Alkan-*oic acid*	CH_3—C(=O)—OH, ethan-*oic acid*
—$CONH_2$	Amide	Alkan-*amide*	CH_3—C(=O)—NH_2, ethan-*amide*
—NH_2	Amine	Alkyl-*amine*	CH_3—CH_2—NH_2, ethyl-*amine*

3.5 Nomenclature of Compounds Containing More than One Functional Group

If more than one functional group is present in a molecule, the principal functional group receiving the lowest number is given by the following order of preference: acids, then aldehydes or ketones, alcohols, amines, ethers, acetylenes, olefins, halogens; for example

$$Cl-\underset{3}{CH_2}-\underset{2}{CH}=\underset{1}{CH_2}$$ 3-chloropropene (*not* 1-chloro-2-propene)

$$CH_3-\underset{|}{\overset{OH}{CH}}-CH_2-\overset{O}{\underset{\|}{C}}-H$$ butanal-3-ol

A more detailed discussion of nomenclature will be presented as the situation requires.

3.6 Other Systems of Nomenclature

The outstanding advantage of the IUPAC naming system is that, under the system, no two compounds can be given the same name. Although it is generally accepted that the IUPAC system is best for naming organic compounds, there are other systems of nomenclature that are also used.

Occasionally compounds are named as a derivative of the simplest compound of a series, for example

$$CH_3-\underset{|}{\overset{CH_3}{\underset{H}{C}}}-CH_3$$ IUPAC: 2-methylpropane; or, as a derivative of methane: trimethylmethane

Sometimes compounds are named by the group present in the molecule, for example

$$CH_3-CH_2-CH_2-Cl$$ IUPAC name: 1-chloropropane; or: *n-propyl* chloride

Common names that have nothing to do with the substituents present are used for certain compounds, for example

$$CH_3-CH=CH_2$$ IUPAC name: propene
Common name: propylene

$$CH_3-\overset{O}{\underset{\|}{C}}-CH_3$$ IUPAC name: propanone
Common name: acetone

$$CH_3-\overset{O}{\underset{\|}{C}}-OH$$ IUPAC name: ethanoic acid
Common name: acetic acid

In order to be able to draw the correct molecular structures, to name compounds, and most important, to read and comprehend organic chemistry, it is essential that the student familiarize himself with the fundamental principles of all the different systems of organic nomenclature.

Nomenclature is perhaps the most important concept for the student of organic chemistry; it will be virtually impossible to write and discuss various reactions unless the student first learns to draw the correct structural formulas, and write the correct names, for the compounds of interest.

Summary

• The IUPAC (International Union of Pure and Applied Chemistry) system of nomenclature is the best and most generally used systematic way of naming organic compounds.

• The IUPAC system is based on naming compounds as being derived from simple unbranched alkanes.

• Alkyl functional groups are derived from alkanes by the removal of one hydrogen atom; the -ane ending is changed to -yl. Other functional groups are shown in Table 3.2.

• The rules for the systematic nomenclature of various classes of organic compounds are illustrated throughout the chapter.

• Other systems of nomenclature are discussed briefly. These include naming compounds as derivatives of the simplest compound of a series and use of so-called common names for some compounds.

PROBLEMS

1 Draw the correct structural formula for each of the following compounds.
   (a) 2-bromobutane
   (b) ethanal
   (c) ethyl n-propylamine
   (d) 2,3-dibromopentane
   (e) 3-hexanol
   (f) 4-methyl-2-hexyne
   (g) 3-chlorobutanoic acid
   (h) propionamide
   (i) 2,2,4-trimethyl-3-i-propyl-5-iododecane
   (j) 1,3-dinitrobenzene
   (k) t-butyl alcohol
   (l) 2-bromopentanal
   (m) ethoxyethane
   (n) 3-methyl-2-heptanone
   (o) 3-chlorocyclopentene
   (p) i-propyl iodide

2 Give the correct IUPAC name for each of the following compounds.

   (a) $CH_3-CH(CH_3)-\underset{\underset{\underset{CH_3}{\overset{|}{CH_2}}}{\underset{|}{\overset{|}{CH_2}}}}{\overset{\overset{Cl}{|}}{C}}-C(CH_3)_2-CH_2-CH_2-CH_3$

   (b) $CH(CH_2CH_3)=C(CH_3)_2$

   (c) [cyclohexene with CH$_3$ and Cl substituents]

PROBLEMS

(d) 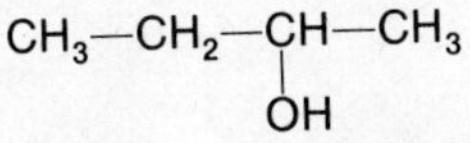 $CH_3-CH_2-\underset{\underset{OH}{|}}{CH}-CH_3$

(e) $CH_3-O-CH_2-CH_2-CH_3$

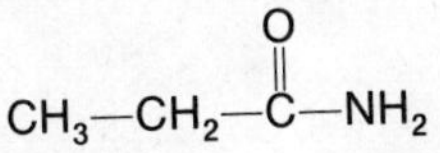

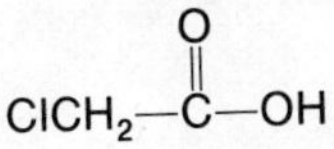

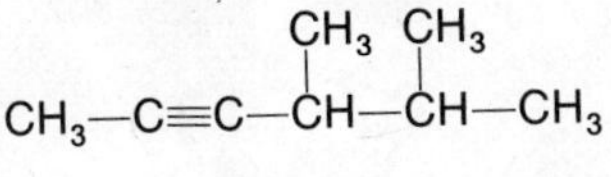

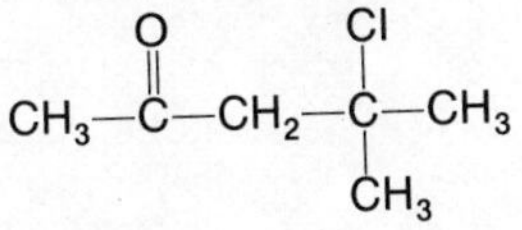

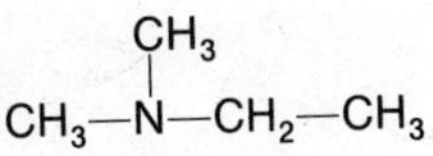

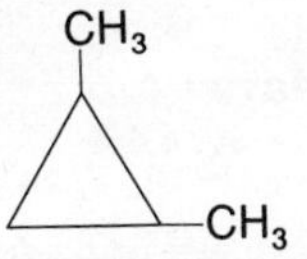

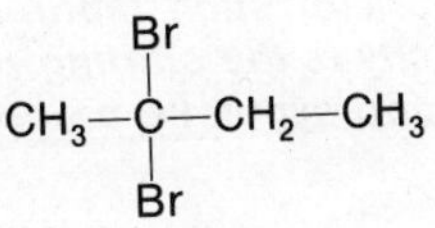

(f)  4-chloro-nitrobenzene (NO$_2$ and Cl para on benzene ring)

(g) 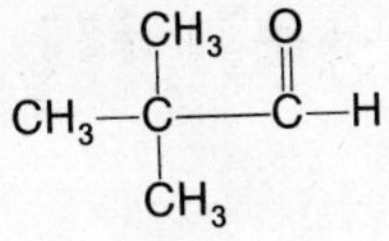 $CH_3-\underset{\underset{CH_3}{|}}{\overset{\overset{CH_3}{|}}{C}}-\overset{\overset{O}{\|}}{C}-H$

(h) $CH_3-CH_2-\overset{\overset{O}{\|}}{C}-NH_2$

(i) $ClCH_2-\overset{\overset{O}{\|}}{C}-OH$

(j) $CH_3-C\equiv C-\underset{}{\overset{\overset{CH_3}{|}}{CH}}-\overset{\overset{CH_3}{|}}{CH}-CH_3$

(k) $CH_3-\overset{\overset{O}{\|}}{C}-CH_2-\underset{\underset{CH_3}{|}}{\overset{\overset{Cl}{|}}{C}}-CH_3$

(l) $CH_3-\overset{\overset{CH_3}{|}}{N}-CH_2-CH_3$

(m) cyclopropane with two CH$_3$ groups

(n) $CH_3-\underset{\underset{Br}{|}}{\overset{\overset{Br}{|}}{C}}-CH_2-CH_3$

(o) $CH_3-\underset{\underset{CH_3}{|}}{CH}-\underset{\underset{Br}{|}}{\overset{\overset{CH_3}{|}}{C}}-CH_2-\overset{\overset{O}{\|}}{C}-OH$

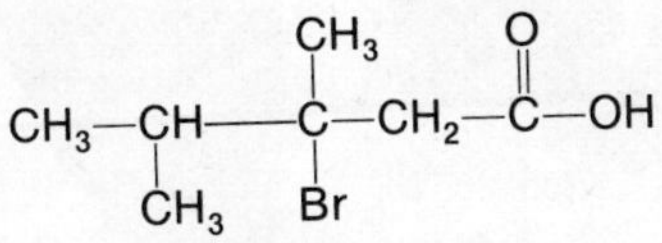

3 Explain what is wrong with the IUPAC name of each of the following compounds, and give a correct name in each case.

(a) 4-methylhexane
(b) 1-methylbutane
(c) 2,3,3-trichlorobutane
(d) 3-ethylbutane
(e) 1,3-diethylcyclopropane
(f) 3,3-diethylbutane
(g) 3-methyl-3-butene
(h) 1-bromo-2-cyclohexene

4 Draw the correct structural formulas for all the isomeric heptanes C$_7$H$_{16}$. Name each isomer by the IUPAC naming system, and as a derivative of methane.

5 Draw the correct structural formula of the following compounds.
(a) diethyl-*t*-butyl methane
(b) ethyl-di-*n*-propyl-*i*-propylmethane.

4
Acidity, Basicity, and Structure

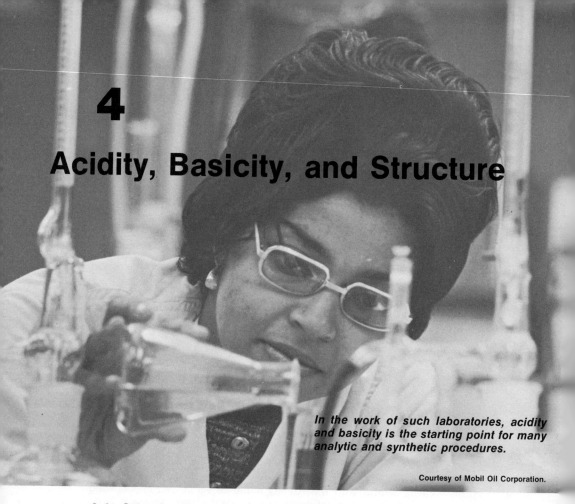

In the work of such laboratories, acidity and basicity is the starting point for many analytic and synthetic procedures.

Courtesy of Mobil Oil Corporation.

4.1 Introduction

A discussion of acid–base theory is pertinent at this point. The strength of acids and bases depends on the nature of the atoms and types of bonds present in the molecule in question. The concepts used in this chapter will be extended to explain the reactivity of many organic compounds in subsequent chapters.

Numerous theories of acids and bases have been developed. Most significant here are (1) the Brønsted–Lowry, and (2) the Lewis theory of acids and bases.

4.2 The Brønsted–Lowry Theory

According to the Brønsted–Lowry theory, an acid is any substance that can donate a proton, and a base is any substance that can accept a proton. Let us consider the general acid–base reaction.

$$\underset{\text{acid}}{HY} + \underset{\text{base}}{B} \rightleftharpoons \underset{\substack{\text{conjugate}\\\text{acid}}}{HB^{\oplus}} + \underset{\substack{\text{conjugate}\\\text{base}}}{Y^{\ominus}} \quad (4\text{–}1)$$

According to the theory, HY is an acid since it is the proton donor; B is the base because it is the proton acceptor. The loss of a proton by HY produces the

Sec. 4.3 THE LEWIS THEORY

species $Y^{\ominus}$, and the subsequent gain of a proton by B produces $HB^{\oplus}$. The products of the reaction, $HB^{\oplus}$ and $Y^{\ominus}$, are referred to as the conjugate acid and the conjugate base. The $Y^{\ominus}$ is the conjugate base of acid HY, and $HB^{\oplus}$ is the conjugate acid of base B. The reader can easily see that a conjugate acid and base (or conjugate base and acid) differ from each other only by the loss or gain of a proton. Some typical examples of acid-base reactions are:

(1) $\underset{\text{acid}}{HCl} + \underset{\text{base}}{H_2O} \rightleftharpoons \underset{\substack{\text{conj.}\\\text{acid}}}{H_3O^{\oplus}} + \underset{\substack{\text{conj.}\\\text{base}}}{Cl^{\ominus}}$ (4-2)

(2) $\underset{\text{base}}{NH_3} + \underset{\text{acid}}{H_2O} \rightleftharpoons \underset{\substack{\text{conj.}\\\text{acid}}}{NH_4^{\oplus}} + \underset{\substack{\text{conj.}\\\text{base}}}{OH^{\ominus}}$ (4-3)

(3) $\underset{\text{base}}{HCO_3^{\ominus}} + \underset{\text{acid}}{H_2O} \rightleftharpoons \underset{\substack{\text{conj.}\\\text{acid}}}{H_2CO_3} + \underset{\substack{\text{conj.}\\\text{base}}}{OH^{\ominus}}$ (4-4)

(4) $\underset{\text{acid}}{H_2O} + \underset{\text{base}}{H_2O} \rightleftharpoons \underset{\substack{\text{conj.}\\\text{acid}}}{H_3O^{\oplus}} + \underset{\substack{\text{conj.}\\\text{base}}}{OH^{\ominus}}$ (4-5)

(5) $\underset{\text{base}}{HNO_3} + \underset{\text{acid}}{H_2SO_4} \rightleftharpoons \underset{\substack{\text{conj.}\\\text{acid}}}{H_2NO_3^{\oplus}} + \underset{\substack{\text{conj.}\\\text{base}}}{HSO_4^{\oplus}}$ (4-6)

If an acid is a *strong acid* (ionizes to a great extent to produce $H_3O^{\oplus}$ ions), the *conjugate base related to it is weak,* and vice versa. Thus, it is an experimental fact that hydrochloric acid, HCl, is essentially 100 % ionized in aqueous solution, and is regarded as a strong acid. The conjugate base related to HCl is $Cl^{\ominus}$, chloride ion, which is weak. Conversely, HCN is only slightly ionized in water, and hence a weak acid; $CN^{\ominus}$, the conjugate base, is a strong base.

Reaction 5 is of interest because *nitric acid,* HNO_3, behaves as a *base* relative to sulfuric acid. The reader should be aware that the strengths of acids and bases are relative, and that when one compares nitric acid and sulfuric acid, the latter is the stronger acid. Therefore, in reaction 5, HNO_3 behaves as a base relative to H_2SO_4. This reaction is a key step in the nitration of aromatic compounds, a reaction to be discussed in a later chapter.

4.3 The Lewis Theory

Whereas the Brønsted-Lowry theory emphasizes the role of the proton in acid-base reactions, the Lewis theory deals with the role of electron pairs in acid-base reactions. An acid is defined under the Lewis theory as any substance that can accept an electron pair, and a base is any substance that can donate a pair of electrons. The reaction of boron trichloride with ammonia to form a boron trichloride-ammonia addition compound illustrates these points.

$$\underset{}{\overset{:\ddot{C}l:}{\underset{:\ddot{C}l:}{:\ddot{C}l:B}}} + {\overset{\times}{\times}}NH_3 \longrightarrow Cl_3B\overset{\delta^{\ominus}\;\;\delta^{\oplus}}{\underset{\times}{\times}}NH_3 \qquad (4-7)$$

In this reaction the BCl_3 has only six electrons around the B atom, and therefore *can accept a pair of electrons* to form an octet; thus the BCl_3 molecule behaves as a *Lewis acid*. The NH_3 has a *pair of unshared electrons on the nitrogen atom available for bonding*, which it *donates* to the empty orbital in BCl_3; thus the NH_3 behaves as a *Lewis base*. The addition compound formed is termed a *Lewis salt* and has a coordinate covalent bond between the B and N atoms.

Since the Lewis theory is based on the role of electron pairs in acid–base reactions, it can be seen that any substance having an unshared electron pair can, possibly, behave as a Lewis base. Thus, it will be seen that many organic compounds containing nitrogen or oxygen atoms that have unshared electron pairs available for bonding are potential Lewis bases.

Substances classified as bases under the Lewis theory are also bases under the Brønsted–Lowry theory. This can be understood if we recall that a Lewis base, which is an electron-pair donor, is capable of accepting a proton due to the attraction of the positive proton for the negatively charged electrons.

The number of substances that can be classified as acids under the Lewis theory is much more numerous and varied in molecular structure than those considered acids under the Brønsted–Lowry theory. The reason for this is that under the Brønsted–Lowry theory acids must be limited to include only substances that contain hydrogen and are capable of donating a proton. The Lewis theory is much more general, in that it emphasizes electron pairs, and can classify as acids many species that do not contain hydrogen. Such compounds as $AlCl_3$, $FeBr_3$, $ZnCl_2$, $SnCl_2$, SO_3, and BF_3, which would not be thought of as acids in the Brønsted–Lowry sense, are typical Lewis acids. These compounds all have a central atom that can accept a pair of electrons. The Lewis theory allows for much more flexibility in the definition of acidic substances than does the Brønsted–Lowry theory, and is the most useful of the acid–base theories as applied to organic reactions.

4.4 Strength of Acids and Bases; K_a and K_b

The quantitative experimental determination of the relative strengths of acids and bases in aqueous solution can be made by examining the following equilibria.

$$HX + H_2O \rightleftharpoons H_3O^\oplus + X^\ominus \quad \left(K_a = \frac{[H_3O^\oplus][X^\ominus]}{[HX]}\right) \quad (4-8)$$
acid

$$B + H_2O \rightleftharpoons BH^\oplus + OH^\ominus \quad \left(K_b = \frac{[BH^\oplus][OH^\ominus]}{[B]}\right) \quad (4-9)$$
base

The numerical values of the equilibrium constants K_a and K_b measure the strength of the acid and base respectively. K_a signifies the ability of an acid to donate a proton to a base, K_b refers to the ability of a base to accept a proton from an acidic substance.

For strong acids and bases the numerical values of K_a and K_b are high since the equilibrium lies far to the right, indicating that the species are highly ionized (the numerator of the equilibrium constant is much larger than the denominator). For weak acids and bases, K_a and K_b will be numerically small, since the degree of ionization is much less and the equilibrium lies towards the left-hand side of the

Sec. 4.5 RELATIVE ACID STRENGTH OF BINARY ACIDS

equation. The relative strengths of acids and bases can be compared by simply examining the numerical values of their equilibrium (ionization) constants K_a or K_b in the same solvent at the same temperature.

4.5 Relative Acid Strength of Binary Acids

Binary acids are molecules composed of two different elements, one of which is hydrogen (e.g., HCl, HF, H_2S, and H_2Te), and can be represented by the general formula, HX. The acidity of an acid, HX, will depend on how strongly ionized the acid is in solution, or the extent of the equilibrium reaction $HX + H_2O \rightleftharpoons H_3O^\oplus + X^\ominus$ as measured by K_a. Thus, the more easily a proton can be removed from an acid, the stronger the acid.

The question of predicting the relative acidity of a series of compounds requires a theoretical explanation as to which species will lose protons more easily, and why. With respect to binary acids, let us first consider the series of acids formed by the elements of the second period (horizontal row) in the periodic table from group IVA to group VIIA, namely CH_4, NH_3, H_2O, and HF.

Since the acid strength depends on removal of a proton from the HX molecule, one simple explanation to account for relative acidities is based on the electronegativity of the elements in the series. The electronegativity of the elements is in the order $F > O > N > C$. Since F is the most electronegative element in the series, it tends to hold on to electrons most strongly. This weakens the polar covalent bond between hydrogen and fluorine, $H^{\delta\oplus} {:}F{:}^{\delta\ominus}$ (the fluorine atom wants to have a negative charge due to its high electronegativity) and facilitates proton removal. In other words, F seeks to complete its octet of electrons, which would convert the F atom to a stable fluoride ion, F^-, and at the same time facilitate removal of a proton by a base. Although HF is a weak acid as compared with HCl or H_2SO_4, it is the strongest acid of the series cited above.

The other elements, namely O, N, and C, are less electronegative than F. Hence they do not attract electrons as strongly, and therefore the hydrogen atom still has a fair share of the electrons in the bond, H:X. The result is that the H is removed as a proton with greater difficulty. The order of acid strength in the series is predicted to be in the order of electronegativity, $HF > H_2O > NH_3 > CH_4$. This is substantiated by the experimental values of the equilibrium constants (K_a HF = 6.8×10^{-4}, K_a $H_2O = 1 \times 10^{-14}$). The equilibrium constants for NH_3 and CH_4 behaving as acids are so small that one can assume NH_3 and CH_4 are either extremely weak as acids or not acidic at all.

Let us now examine the acids of the elements in group VIIA of the periodic table; HF, HCl, HBr, and HI. If we were to predict the order of acidity of these acids according to the principle of electronegativity, the sequence would be $HF > HCl > HBr > HI$. However, it can be shown by experiment that the actual order is just the reverse, $HI > HBr > HCl > HF$.

This illustrates that the prediction of acid strength that is in accord with experimental data is more complex than a simple explanation based on relative electronegativities. Indeed, there are numerous explanations that must be used to explain acid strength; and which criteria are to be used as a basis depends on the compounds in question.

The situation concerning the binary acids of group VIIA (hydrogen halides) differs from the one discussed previously (HF, H_2O, NH_3, CH_4). In considering group VIIA, the elements are all in the same vertical column of the periodic table, rather than in a horizontal row (period). Experimentally it can be shown that the order of acid strength increases as we go down the group from F to I, that is HI > HBr > HCl > HF. For a plausible explanation of this sequence, one must examine the relative sizes of the elements (ions) in the series. As one proceeds down a group in the periodic table, the number of electronic energy levels increases, and the elements of higher atomic number have a larger atomic (or ionic) radius. Thus, I is larger than Br, which is larger than Cl or F (I > Br > Cl > F).

That HI is the strongest acid of the series can be interpreted to mean that, because of the size of the large I atom, the H atom can not get close enough to the I atom to form a strong bond.

The rather weak bond between the H and I can be broken easily, and the H removed as a proton. Since F is the smallest anion, the H and F can get closer to each other and the bond between H and F is relatively strong (since the H and F electron clouds can overlap more effectively than in HI) (H)(F). The shorter HF bond distance (0.92 Å) as compared to that of HI (1.62 Å) makes it more difficult to remove a proton, and hence HF, which has the smallest internuclear distance in the group, is the weakest acid.

Suffice it to say that when we compare binary acids in a given horizontal period of the periodic table, electronegativity should be used as a basis for interpretation, and the stronger acids will be those containing the more electronegative elements towards the right-hand side of the periodic table. In a given group (vertical column) of the periodic table the trend should be based on size (not electronegativity), and the stronger acid will be the one containing the element of largest size (furthest down the group). It is difficult to predict the relative strength of acids that are not in the same group or period; it is customary to consult tables of ionization constants or acid–base charts for data of interest.

4.6 Relative Acid Strength of Ternary Acids

Ternary acids are molecules composed of three different elements, hydrogen, oxygen, and a third element, and can be represented by the general formula, H—O—X, (e.g., $HClO_4$, H_2SO_4, HNO_3, and H_3PO_4).

In a binary acid, HX, the proton must be removed from the X atom to which it is bonded directly. In all ternary acids, the hydrogen atom is bonded to the oxygen atom, which in turn is bonded to atom X. Since in all ternary acids, protons are removed from the oxygen atom(s), variations in acidity are influenced mainly by the nature and properties of the X atom. $HOX + H_2O \rightleftharpoons H_3O^\oplus + OX^\ominus$.

Sec. 4.6 RELATIVE ACID STRENGTH OF TERNARY ACIDS

The X atom must exert some influence that is felt in the O—H bond in the X—O—H molecule in order to effect proton removal and, hence, acid strength. If the species, X, is an electron-withdrawing group it will tend to pull electrons towards itself from the oxygen atom of the O—H portion of the molecule. Since the oxygen atom itself is highly electronegative, it will pull more strongly on the electrons it shares with hydrogen in the O—H bond. This will weaken the O—H bond, hydrogen acquiring a partial positive charge, and make proton removal more facile; the compound will be more acidic. In other words, the presence of *electron-withdrawing substituents in a molecule will increase acidity.*

$$\overset{\delta\oplus}{H} \longrightarrow O \longrightarrow \overset{\delta\ominus}{X} \qquad \text{(X is electron-withdrawing substituent)}$$

On the other hand, if X is an electron-releasing substituent, it will donate its electrons, increasing the electron density around the oxygen atom, which in turn will form a stronger bond to the hydrogen atom. This will make proton removal more difficult and result in a weaker acid. *Electron-donating substituents will decrease the acidity* of a molecule.

$$\overset{\delta\ominus}{H} \longleftarrow O \longleftarrow \overset{\delta\oplus}{X} \qquad \text{(X is electron-donating substituent)}$$

From this discussion it can be seen that an electrical effect was transmitted from atom to atom throughout the chain of the molecule by deformation of the electron clouds of the various bonds. This type of electrical effect is known as an **inductive effect**.

When discussing the role of inductive effects in determining acidities, reactivities of compounds, etc., the element hydrogen is used as a reference standard substance. Therefore, a substituent will be electron-withdrawing or electron-donating as compared to hydrogen.

4.6-1 The Inductive Effect and Acid Strength. Numerous explanations can be used to explain the relative acidities of ternary acids. Perhaps the simplest and most fundamental interpretation is based on the electronegativity of the central atom in the molecule. This criterion gives the correct answer most of the time.

For example, one might ask which is the stronger acid, nitric acid, HNO_3, or perchloric acid, $HClO_4$? The structural formulas of the two acids are as follows.

nitric acid   perchloric acid

The central atoms in the two acids are N and Cl, respectively. Since Cl is more electronegative than N, the electron density will be pulled towards the Cl to a greater extent than towards the N. The net result of the inductive effect towards the central atom is that the H—O bond is weakened more in perchloric acid than in nitric acid, and thus perchloric acid is a stronger acid (see previous discussions). It may also be argued that since perchloric acid has four oxygen atoms to three for nitric acid, the electronegativity of the additional oxygen atom tends to help weaken the H—O bond to a greater extent. So, although both perchloric and nitric acids are strong acids being essentially 100 % ionized in water solution, perchloric acid is the stronger of the two.

Let us now consider the following group of ternary acids: $HClO_3$ (chloric), $HBrO_3$ (bromic), and HIO_3 (iodic). The structural formulas of these acids show that the halogen is the central atom in each case.

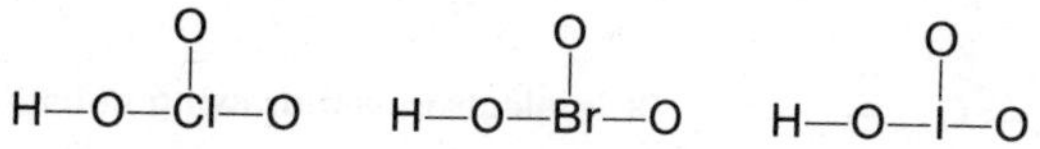

$$\begin{array}{ccc} \text{O} & \text{O} & \text{O} \\ \| & \| & \| \\ \text{H—O—Cl—O} & \text{H—O—Br—O} & \text{H—O—I—O} \\ \text{chloric acid} & \text{bromic acid} & \text{iodic acid} \end{array}$$

Since Cl is the most electronegative of the central atoms and I the least, one would expect the order of acid strength, according to our previous line of reasoning, to be $HClO_3 > HBrO_3 > HIO_3$. This order is indeed borne out by experimental evidence. Note that the order of acid strength shown above for ternary acids in the same group of the periodic table is in reverse of the order obtained for binary acids ($HI > HBr > HCl > HF$), where atomic size rather than electronegativity is used as a standard.

We can compare the relative strengths of many different ternary acids; in each case it can be shown that it is the electronegativity of the central atom, by virtue of the transmission of its inductive effect through the chain of atoms in the molecule, which determines the acid strength.

We shall now apply some of the concepts set forth in this chapter to organic compounds, and we shall make use of them throughout the book.

Let us first consider the relative acid strength of the organic compound methyl alcohol (methanol), CH_3—OH, as compared to that of water, H—OH. In light of our previous discussion, it should be apparent that methyl alcohol is a weaker acid than water. This can be attributed to the inductive effect of the methyl (CH_3) group present in the methanol molecule. The methyl group is more electron-donating than the hydrogen present in the water molecule. In the methanol molecule, since the methyl group tends to donate electrons, the electron flow is towards the hydroxyl (OH) end of the molecule. This strengthens the O—H bond, making proton removal more difficult than in water, and decreases the acidity of the methanol.

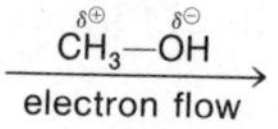

$$\overset{\delta^{\oplus}}{CH_3}—\overset{\delta^{\ominus}}{OH} \xrightarrow{\text{electron flow}}$$

Organic compounds that contain a carboxyl group ($-\overset{\overset{\text{O}}{\|}}{C}-OH$) are acidic and are called carboxylic acids (see Table 2.1). The acid–base reaction of a carboxylic acid with water can be represented by the following equation.

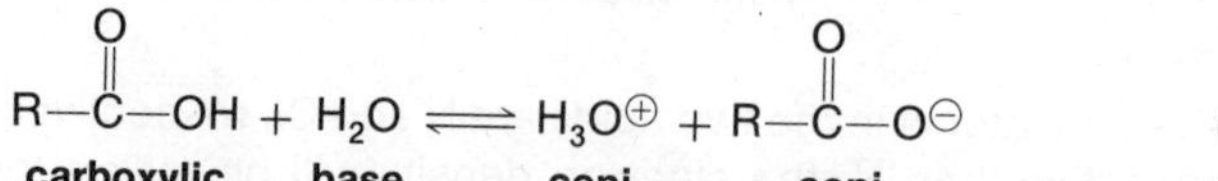

$$\underset{\substack{\text{carboxylic} \\ \text{acid}}}{R-\overset{\overset{\text{O}}{\|}}{C}-OH} + \underset{\text{base}}{H_2O} \rightleftharpoons \underset{\substack{\text{conj.} \\ \text{acid}}}{H_3O^{\oplus}} + \underset{\substack{\text{conj.} \\ \text{base}}}{R-\overset{\overset{\text{O}}{\|}}{C}-O^{\ominus}} \qquad (4\text{--}10)$$

(R = any alkyl group)

Sec. 4.6 RELATIVE ACID STRENGTH OF TERNARY ACIDS

The equilibrium constant for this reaction can be written as

$$K_a = \frac{[H_3O^{\oplus}]\left[R-\overset{\overset{\displaystyle O}{\|}}{C}-O^{\ominus}\right]}{\left[R-\overset{\overset{\displaystyle O}{\|}}{C}-OH\right]}$$

and is a measure of the acid strength. The carboxylic acids are relatively weak compared to the inorganic acids HCl, H_2SO_4, and HNO_3.

However, let us consider what the order of acid strength would be for the following carboxylic acids:

$$CH_3-\overset{\overset{\displaystyle O}{\|}}{C}-OH \quad \text{acetic acid}$$

$$Cl-CH_2-\overset{\overset{\displaystyle O}{\|}}{C}-OH \quad \text{monochloroacetic acid}$$

$$CH_3-CH_2-\overset{\overset{\displaystyle O}{\|}}{C}-OH \quad \text{propionic acid}$$

Structurally, the only difference among the three acids is that an H atom on the CH_3 group of acetic acid has been replaced by a Cl atom in monochloroacetic acid, and by another CH_3 group in propionic acid. Thus, the relative acidities can be explained by the *inductive* effects of the Cl and CH_3 groups as compared to H.

Since Cl is strongly electronegative, one would suspect that it is an electron-withdrawing group. The Cl atom withdraws electrons from the carbon atom to which it is attached, which in turn pulls electrons from the carboxyl (COOH) end of the molecule. This weakens the O—H bond, facilitates proton removal, and increases acidity.

$$\overset{\delta\oplus}{Cl}-\overset{\overset{\displaystyle H}{|}}{\underset{\underset{\displaystyle H}{|}}{C}}-\overset{\overset{\displaystyle O}{\|}}{C}-\overset{\delta\ominus}{OH}$$
$$\xleftarrow{\text{electron flow}}$$

One would then predict that monochloroacetic acid is a stronger acid than acetic acid on the basis that Cl has an electron-withdrawing inductive effect compared to H.

The inductive effect explains why propionic acid is a slightly weaker acid than acetic acid. The CH_3 group is a weak electron-donating group (as are all alkyl groups), which pushes electrons towards the carboxyl end of the molecule.

$$\overset{\delta\oplus}{CH_3}-\overset{\overset{\displaystyle H}{|}}{\underset{\underset{\displaystyle H}{|}}{C}}-\overset{\overset{\displaystyle O}{\|}}{C}-\overset{\delta\ominus}{OH}$$
$$\xrightarrow{\text{electron flow}}$$

The increased electron density on the oxygen atom strengthens the O—H bonds, makes proton removal more difficult, and decreases acidity.

The experimental K_a values for the three acids are

$$CH_3-\overset{\overset{O}{\|}}{C}-OH \qquad (K_a = 1.8 \times 10^{-5})$$

$$Cl-CH_2-\overset{\overset{O}{\|}}{C}-OH \qquad (K_a = 1.5 \times 10^{-3})$$

$$CH_3-CH_2-\overset{\overset{O}{\|}}{C}-OH \qquad (K_a = 1.4 \times 10^{-5})$$

The numerical values of K_a indicate that a 1-molar solution of monochloroacetic acid should be an acid approximately 100 times stronger than 1-molar acetic acid, whereas propionic acid should be slightly weaker than acetic acid. These conclusions were predicted theoretically by our knowledge of the inductive effect. The inductive effect is a powerful tool, which will be used to help explain and interpret many facets of organic chemistry.

Summary

- Numerous theories of acids and bases have been developed. The Brønsted-Lowry theory and the Lewis theory are two representative ways of defining acids and bases.
- The Lewis theory is perhaps the most general theory and the most important with respect to the understanding of various concepts in organic chemistry.

    The Lewis theory is based on the role of electron pairs. A Lewis Acid is defined as an electron-pair acceptor, and a Lewis Base is an electron-pair donor.
- The strength of acids and bases is determined by the degree of ionization in some solvent such as water. The position of the equilibrium of the ionization is measured by the numerical value of the equilibrium constant, K_a or K_b.

    Strong acids and bases have high numerical values for K_a and K_b, whereas weak acids and bases have small numerical values for K_a and K_b.
- The strength of binary acids increases as one goes across the periodic table horizontally from left to right in any given period. This can be explained on the basis of the electronegativity of the elements in the series.

    The strength of binary acids in any group (vertical column) of the periodic table increases as one goes down the group. This is mainly attributed to the size of the elements in the group.
- The relative strengths of ternary acids depend primarily on the electronegativity (and positive charge) of the central atom attached to the oxygen atom(s).

    Inductive effects are important in understanding this phenomenon.
- Electron-withdrawing substituents in a molecule increase acidity, whereas electron-donating substituents will decrease acidity of a molecule.

PROBLEMS

1. What is the formula of the conjugate base of each of the following? (a) H_2S, (b) NH_3, (c) $OH^{\ominus}$, (d) CH_3OH, (e) CH_3NO_2.
2. What is the formula of the conjugate acid of each of the following? (a) $HSO_4^{\ominus}$, (b) $NH_2^{\ominus}$, (c) $I^{\ominus}$, (d) $C_6H_5NH_2$, (e) $C_6H_5O^{\ominus}$.
3. Substances that can behave as either an acid or a base are said to be amphoteric or amphiprotic. Write an equation in terms of the Brønsted–Lowry theory for each of the following species to illustrate that the substance can be either an acid or a base. (a) $HPO_4^{\ominus\ominus}$, (b) $HCO_3^{\ominus}$, (c) CH_3CH_2OH. (HINT: Consider the self-ionization of water, $H_2O + H_2O \rightleftharpoons H_3O^{\oplus} + OH^{\ominus}$.)
4. Perchloric acid, $HClO_4$, is a stronger acid than nitric acid. Write an acid–base equation according to the Brønsted–Lowry theory for the reaction between these two substances. Label the acid, base, conjugate acid, and conjugate base in the reaction.
5. Malonic acid, $HOOC-CH_2-COOH$, contains two acidic hydrogens on the carboxyl groups, which ionize in two individual steps. Write equations for each ionization step. The ionization constant, k_1a for ionization step A is 1.4×10^{-3}, while the ionization constant, k_2a, for ionization step B is 2×10^{-6}. A comparison of the numerical values of k_1a and k_2a indicates that the extent of reaction of step A is about 700 times that of step B. Explain.
6. Does the fact that an acid is polyprotic (i.e., H_3PO_4, H_2SO_4, etc.) necessarily mean that the acid is stronger than a monoprotic acid (i.e., HNO_3, HCl, HCN, etc.)? Explain.
7. According to the Brønsted–Lowry theory of acids and bases, an acid is defined as a proton donor. Arrange the following sets of compounds in order of increasing acidity. (a) H_3AsO_4, HNO_3, H_3PO_4, H_3SbO_4 (b) HNO_3, H_2CO_3, $HClO_4$, H_2SO_4 (c) AsH_3, PH_3, NH_3, SbH_3
8. A Lewis base is defined as an electron-pair donor. Arrange the following species in order of increasing basicity according to the Lewis theory: $NH_2^{\ominus}$, $CH_3^{\ominus}$, $F^{\ominus}$, $OH^{\ominus}$.
9. Arrange the following compounds in increasing order of acidity (ability to donate a proton): ethanoic acid, 3-chloropropanoic acid, 2-hydroxypropanoic acid, 2,2,2-trifluoroethanoic acid, butanoic acid, 2-cyanoethanoic acid.
10. Write an equation for the Lewis acid–base reaction between: (a) BF_3 and $CH_3CH_2-\ddot{O}-CH_2CH_3$, and (b) SO_3 and pyridine,

5
Electronic and Steric Effects

Electron micrograph (enlargement 25,000 diameters) of an rDNA molecule—the two strands are bonded together, appearing (mid-right) as a heavier single strand, through the electronic and steric effects of hydrogen bonding.

From *Scientific American*, August, 1973; courtesy of Dr. Pieter C. Wensink.

5.1 Introduction

The role of molecular structure as an influence on the properties and reactivity of functional groups is seen throughout the various classes of organic compounds. The electronic and steric effects within the molecules are largely responsible for similarities and differences between compounds. In this chapter, for example, we shall see that by changing the nature of the group R in carboxylic acids,

$$R-\overset{\overset{O}{\|}}{C}-OH,$$

one can change the acid strength considerably. We will first discuss **electronic effects** in molecules in terms of *inductive* and *resonance* effects; then we will turn to **steric effects,** which involve the bulk of the atoms within the molecule.

5.2 Electronic Effects

It is advantageous to consider the influence and role of various substituents in determining the properties and reactivity of a molecule. This can be done by an examination of the electronic structure and electronic effects within the molecule.

The electronic effects of the substituents (functional groups) can be divided into two general categories; **inductive effects** and **resonance effects.** These effects manifest themselves by the ability of a substituent to either donate electrons to another region in the molecule, or to attract electrons to itself from the

Sec. 5.3 INDUCTIVE EFFECTS

other portions of the molecule. The tendency of a substituent to donate or withdraw electrons is a characteristic inherent in the nature of each particular substituent.

The *inductive effect* is an electrostatic effect that is transmitted through chains of atoms along single bonds. The cause of this effect may arise from ionic charges or from the action of dipoles due to the polarization of a polar covalent bond within the molecule.

The *resonance effect* on the other hand acts along a π-electron system, involving multiple bonds, unshared electrons, and incompletely filled outer electronic orbitals.

5.3 Inductive Effects

The nature of the inductive effect has already been discussed in Chapter 4. This effect is mainly due to the presence of a permanent dipole (polar covalent bond) in a molecule, and subsequent distortion of the bonding electrons throughout the chain of atoms in the molecule. Some groups can be classified as having an electron-withdrawing inductive effect as compared to hydrogen. These groups are acid-strengthening, and a partial list includes the following.

$$
\begin{array}{lll}
-NO_2 & -F & -OH \\
-CN & -Cl & -OR \\
-COOH & -Br & -NH_2 \\
-COOR & -I & -NR_3^{\oplus} \\
>C=O & & -SH \\
-SO_3H & -C_6H_5 & \text{(weak)}
\end{array}
$$

Groups that have an electron-releasing inductive effect are acid weakening, and some representative types are $-\overset{\overset{\displaystyle O}{\|}}{C}-O^{\ominus}$, $-CH_3$, and (represented by an R) other alkyl groups (weak).

The influence of inductive effects on the relative acidities of monochloroacetic acid, acetic acid, and propionic acid has been discussed in Section 4.6–1. While in this series the acid strength as measured by K_a varied by a factor of about 100, the influence of the inductive effect can be very profound. Consider for example the following group of acids where the acid strength varies by a factor of about 10,000.

$Cl_3-C-COOH$
trichloroacetic acid
($K_a = 2.0 \times 10^{-1}$)

$Cl_2-CH-COOH$
dichloroacetic acid
($K_a = 5.0 \times 10^{-2}$)

$ClCH_2-COOH$
monochloroacetic acid
($K_a = 1.5 \times 10^{-3}$)

CH_3-COOH
acetic acid
($K_a = 1.8 \times 10^{-5}$)

The presence of the Cl atom(s) on the α-carbon atom (adjacent to the carboxyl group) increases the numerical value of K_a, and hence the acidity. This is because of the electron-withdrawing inductive effect of the Cl group. The more chlorine atoms present, the stronger the acid. Thus, trichloroacetic acid is the strongest of the four acids shown above.

Butyric acid, $CH_3-CH_2-CH_2-\overset{\overset{O}{\|}}{C}-OH$ ($K_a = 1.6 \times 10^{-5}$) is only slightly weaker than acetic acid, $CH_3-\overset{\overset{O}{\|}}{C}-OH$ ($K_a = 1.8 \times 10^{-5}$). This is due to substitution of more alkyl groups in place of the α-hydrogen atoms in acetic acid. Alkyl groups are only weakly electron releasing and have only a slight effect on the acidity. The inductive effect decreases with distance, and therefore only operates effectively over one or two bonds' distance from any functional group.

The diminishing of the inductive effect with distance can be seen by a comparison of the following acids.

$CH_3-CH_2-CH_2-\overset{\overset{O}{\|}}{C}-OH$
butanoic (butyric) acid
($K_a = 1.6 \times 10^{-5}$)

$CH_3-CH_2-\underset{\underset{Cl}{|}}{CH}-\overset{\overset{O}{\|}}{C}-OH$
2-chlorobutanoic acid
($K_a = 1.4 \times 10^{-3}$)

$CH_3-\underset{\underset{}{}}{\overset{\overset{Cl}{|}}{CH}}-CH_2-\overset{\overset{O}{\|}}{C}-OH$
3-chlorobutanoic acid
($K_a = 8.9 \times 10^{-5}$)

$Cl-CH_2-CH_2-CH_2-\overset{\overset{O}{\|}}{C}-OH$
4-chlorobutanoic acid
($K_a = 3.0 \times 10^{-5}$)

The greater the distance between the Cl and the carboxyl group, the weaker is the inductive effect of the Cl group. The chloro acids are all stronger than butyric acid itself, but as the distance between the Cl and the COOH groups increases, the inductive effect becomes smaller, until it is almost negligible (compare K_a's for butyric acid and 4-chlorobutanoic acid).

The most common functional group that characterizes basic substances in organic chemistry is the $-NH_2$ group or amino group (present in amines). The unshared electron pair on the nitrogen atom makes amines typical Lewis bases.

The organic compound methylamine, CH_3-NH_2 ($K_b = 4.4 \times 10^{-4}$) is a stronger base than ammonia, NH_3 ($K_b = 1.8 \times 10^{-5}$). The reason for this is that the methyl group has an electron-releasing inductive effect relative to hydrogen. The electron density is thus increased around the nitrogen atom making the

$$\underset{\text{electron flow}}{\overset{\delta\oplus \quad \cdot\cdot \ \delta\ominus}{CH_3-NH_2} \longrightarrow}$$

unshared electron pair more available for bonding than in ammonia, and thus increasing the basic strength. Electron-releasing groups are acid weakening or base strengthening.

5.4 Resonance

The purpose of drawing a structural or electronic formula of a molecule is to have a picture on paper that accurately represents all characteristics and properties of the substance. However, this can not be done for all molecules. Occasionally, one structural representation is not enough to depict all features of a substance. This usually arises when pi bonds are present in a molecule, and when a so-called "conjugated system" of alternating single and double bonds is present. In order to be able to represent the properties of these molecules, one needs to discuss the phenomenon of resonance.

Resonance is not limited only to organic compounds; it occurs in inorganic species as well (e.g., HNO_3, NO_3^-, O_2, O_3, SO_2, SO_3, $CO_3^=$, NO). The resonance effect is most frequently discussed for the organic compound benzene, C_6H_6.

In 1865, the German chemist, Friedrich August Kekulé (1829–1896) suggested on the basis of experimental evidence that benzene has two structures that rapidly interconvert to each other, with a sort of rapid equilibrium between:

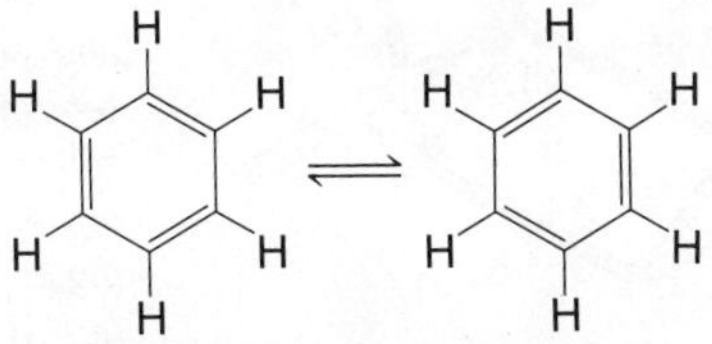

Kekulé formulas for benzene

The cyclic hexagonal structure, and alternating single and double bonds (conjugated system) were deduced by Kekulé on the basis of experimental evidence. However, the Kekulé formula for benzene did not explain all the experimental facts; in particular, the failure of benzene to react with permanganate solution, a reaction characteristic of double bonds. Kekulé explained this by suggesting that the bonds alternated positions very rapidly so that each carbon–carbon bond was not actually a single or double bond but something in between.

Modern day technology has developed such techniques as X-ray diffraction, which enable chemists to determine molecular geometry (position of atoms, internuclear distances, bond angles, etc., in molecules) very accurately. Experimental measurement has determined the carbon–carbon single-bond distance to be 1.54 Å and the carbon–carbon double-bond distance to be 1.34 Å. In benzene, *all* of the carbon–carbon bonds are 1.40 Å, a value intermediate between the single- and double-bond distances cited above. This experimental fact can be interpreted to mean that all the bonds in the benzene ring are identical, and are not characterized as true single or double bonds. In current terminology, the two Kekulé formulas for benzene are actually considered as resonance forms, differing only in the electronic arrangement in the molecule.

According to molecular orbital theory (Chapter 1) the bonding type of the carbon atoms in benzene is *sp²* hybridization with bond angles of 120°. Each carbon atom also has a *p* electron perpendicular to the plane of the ring. The six *p* electrons (one from each carbon atom) overlap, forming three π bonds which

5.1 C₆H₆

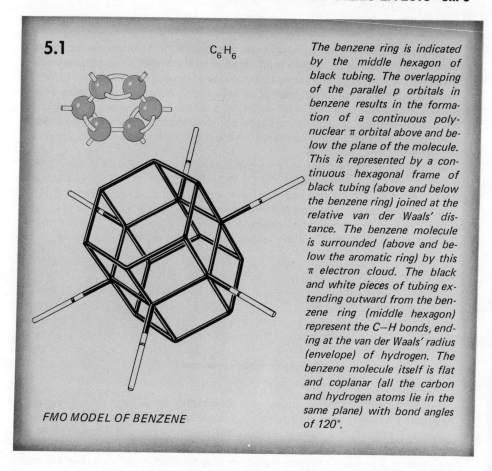

FMO MODEL OF BENZENE

The benzene ring is indicated by the middle hexagon of black tubing. The overlapping of the parallel p orbitals in benzene results in the formation of a continuous polynuclear π orbital above and below the plane of the molecule. This is represented by a continuous hexagonal frame of black tubing (above and below the benzene ring) joined at the relative van der Waals' distance. The benzene molecule is surrounded (above and below the aromatic ring) by this π electron cloud. The black and white pieces of tubing extending outward from the benzene ring (middle hexagon) represent the C—H bonds, ending at the van der Waals' radius (envelope) of hydrogen. The benzene molecule itself is flat and coplanar (all the carbon and hydrogen atoms lie in the same plane) with bond angles of 120°.

actually form a continuous electron cloud above and below the plane of the benzene ring. (See Figure 5.1.)

Examination of the two Kekulé formulas for benzene reveals that they differ only in the arrangement of the π electrons. Neither of these structures is the actual formula for the benzene molecule, since neither can account for all the observed properties of benzene. Each structure contributes to the overall characteristics of benzene, and so is called a *resonance form*. The actual formula of benzene is really a combination of all the contributing structures. It is intermediate between all the resonance forms that can be drawn for it, and is called a *resonance hybrid*. It should be noted that the *contributing structures do not exist*. Since our knowledge in representing plausible structures is so limited that we are unable to draw *one* single formula to account for *all* the properties of benzene, by using the formulas of all the contributing resonance forms we are in a better position to picture and comprehend the properties of the benzene molecule. The *resonance hybrid* is the only form of benzene which *actually exists*.

The *resonance forms* of such substances as benzene are *indicated by a double-headed arrow* between them.

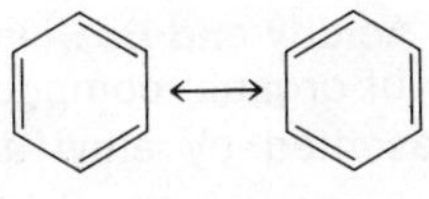

Sec. 5.4 RESONANCE

The *resonance hybrid* may be represented as [benzene with dashed circle], where the dashed lines represent the bonds in which the electron density is distributed.

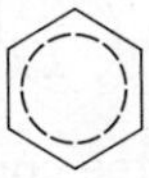

According to resonance theory, the electron density is distributed over as large a region in a molecule as possible. The spreading of the electron density over a large area, rather than concentrating the electrons in a small space, makes the resulting molecule more stable. Since π electrons are delocalized in a conjugated system, they can be spread out over the molecule, decreasing the electron density at any one spot and producing a more stable species than in other molecules (resonance stabilization).

5.4-1 Drawing Resonance Structures

The drawing of resonance structures is fundamental to the understanding of organic chemistry. A curved arrow is used to indicate the movement of an electron pair from one atom to another atom in the molecule. It should be understood that the *movement of an electron pair to an atom necessitates at the same time the movement of another electron pair from that atom,* for the atom to maintain its proper valence (e.g., a carbon atom should have four bonds attached to it, not five, which would result by simply adding an electron pair). All resonance structures of a compound should have the same number of paired and unpaired electrons. The following examples illustrate these points.

Benzene

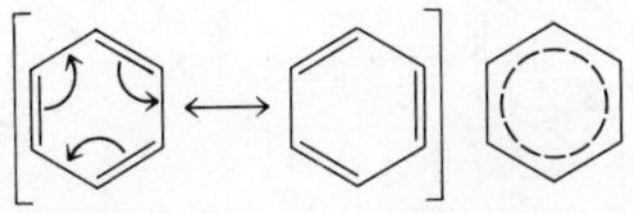

Carboxylate ion

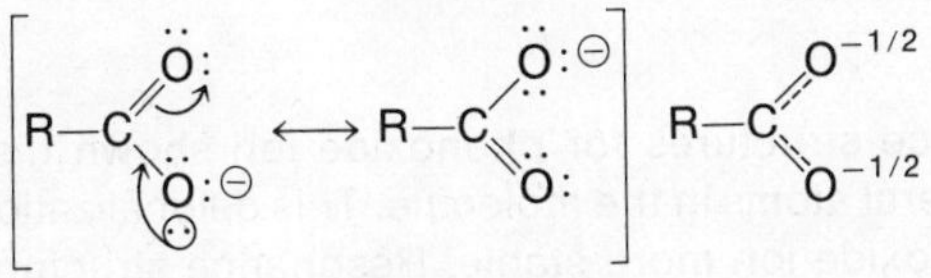

Chlorobenzene

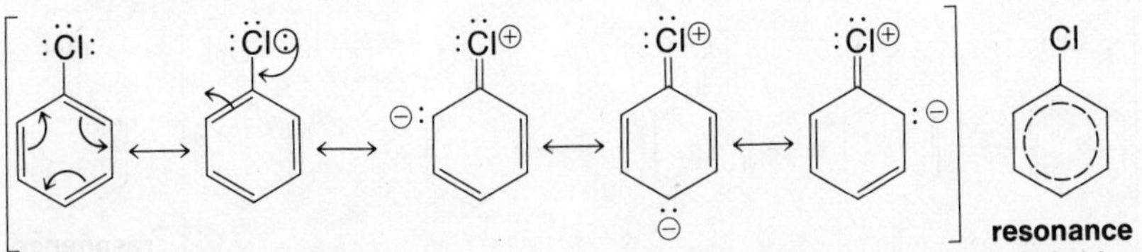

resonance hybrid

5.4-2 Resonance Effects on Acidity and Basicity

Let us consider the relative acid strength of three classes of organic compounds, the alcohols, carboxylic acids, and phenols, as represented by ethyl alcohol, C_2H_5OH, acetic acid, CH_3—COOH, and phenol [C$_6$H$_5$OH], respectively. The K_a values for the acid-base reaction of these three compounds with water are as follows.

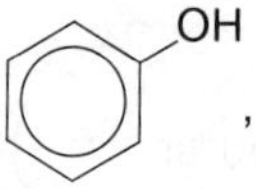

$$C_2H_5OH + H_2O \rightleftharpoons H_3O^\oplus + C_2H_5O^\ominus \quad (K_a = 1 \times 10^{-16}) \quad (5\text{-}1)$$

$$CH_3COOH + H_2O \rightleftharpoons H_3O^\oplus + CH_3COO^\ominus \quad (K_a = 1.8 \times 10^{-5}) \quad (5\text{-}2)$$

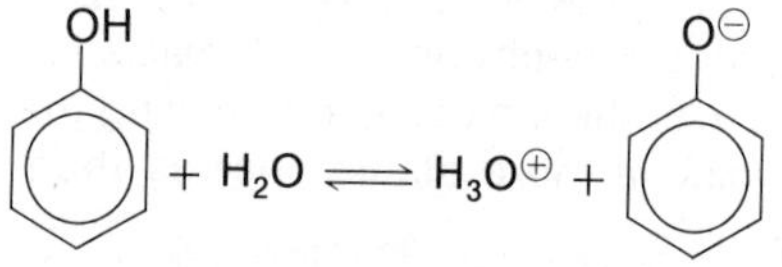

$$[PhOH] + H_2O \rightleftharpoons H_3O^\oplus + [PhO^\ominus] \quad (K_a = 1 \times 10^{-10}) \quad (5\text{-}3)$$

The question that arises is why should the OH of the carboxyl group donate protons more easily than the OH group in alcohols or phenols? From the K_a values one can see that acetic acid is about 100,000 times as strong an acid as phenol, which is itself a stronger acid than ethyl alcohol by a factor of one million.

The simplest explanation for the great difference in acidity among these compounds is based on resonance theory.

Let us first compare ethyl alcohol and phenol. The phenoxide ion formed by loss of a proton from phenol is resonance stabilized, as shown in the resonance structures below.

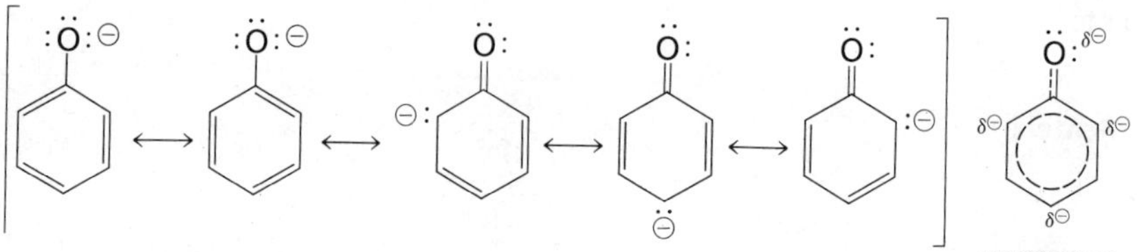

[Resonance structures of phenoxide ion, ending with resonance hybrid]

The resonance structures for phenoxide ion shown distribute the negative charge over several atoms in the molecule. This delocalization of electron density makes the phenoxide ion more stable. Resonance structures can be drawn for phenol itself; however, the last three structures indicate charge separation from a

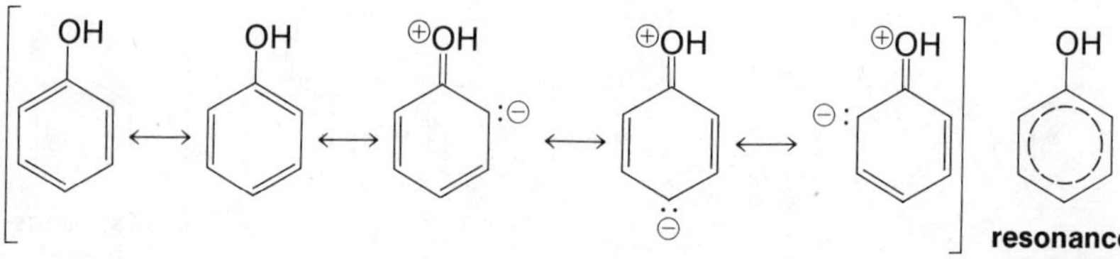

[Resonance structures of phenol, ending with resonance hybrid]

Sec. 5.4 RESONANCE

neutral molecule rather than charge redistribution, and require higher energy, thus being only minor contributors to the resonance hybrid. No resonance structures can be drawn for the stabilization of the ethoxide ion formed when ethyl alcohol loses a proton. The reaction of phenol with water proceeds to a greater extent than the corresponding reaction of ethyl alcohol with water, since a more stable species is being formed. Therefore, phenol is more acidic than ethyl alcohol, and phenols as a class of compounds are more acidic than the alcohols.

The acidity of the carboxylic acids, such as acetic acid, is due to the resonance stabilization of the carboxylate ion. The contributing structures for acetate ion are:

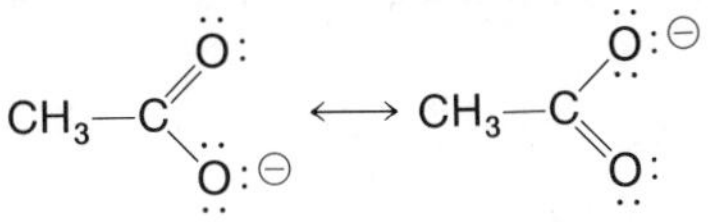

$$CH_3-C\begin{matrix}O:\\O:^{\ominus}\end{matrix} \longleftrightarrow CH_3-C\begin{matrix}O:^{\ominus}\\O:\end{matrix}$$

The resonance forms of acetic acid require charge separation as in the case of phenol.

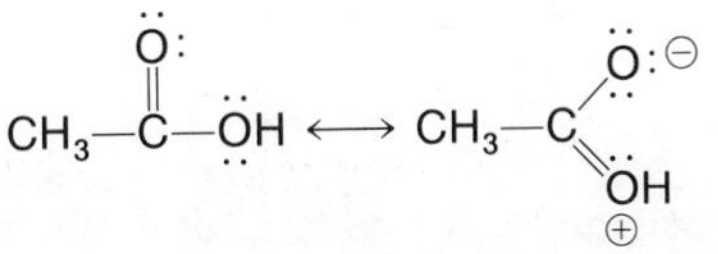

$$CH_3-\overset{\overset{\overset{..}{O}:}{\|}}{C}-\overset{..}{O}H \longleftrightarrow CH_3-C\begin{matrix}O:^{\ominus}\\OH^{\oplus}\end{matrix}$$

The resonance stability of the acetate ion which has two equivalent contributing structures is much greater than the stability of acetic acid where the resonance forms require charge separation. For this reason the acetic acid reacts with water to produce the more stable acetate ion.

In summary: because of resonance effects carboxylic acids are stronger acids than phenols, which in turn are stronger than alcohols. In the carboxylic acids another factor to consider is the electronegative character of the oxygen of the

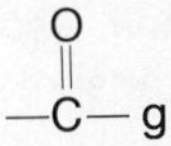

$-\overset{\overset{O}{\|}}{C}-$ group, which attracts electrons and also helps proton removal. This effect, in addition to the resonance effect, accounts for the greater acidity of carboxylic acids as compared to phenols.

The base strength of aniline and methylamine as shown by a comparison of K_b for the following reactions

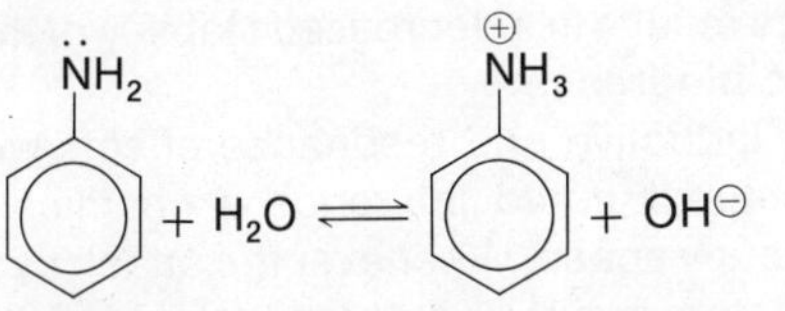

$$\underset{\text{aniline}}{C_6H_5\ddot{N}H_2} + H_2O \rightleftharpoons C_6H_5\overset{\oplus}{N}H_3 + OH^{\ominus} \quad (K_b = 4 \times 10^{-10}) \quad (5\text{-}4)$$

$$\underset{\text{methylamine}}{CH_3\ddot{N}H_2} + H_2O \rightleftharpoons CH_3\overset{\oplus}{N}H_3 + OH^{\ominus} \quad (K_b = 4 \times 10^{-4}) \quad (5\text{-}5)$$

indicates that methylamine is one million times as strong a base as aniline. The basicity is determined by the ability to donate an unshared electron pair on the nitrogen atom to another species. The difference in the basicity of the two compounds by a factor of one million cannot be entirely attributed to the difference in inductive effects between the methyl and phenyl groups.

A resonance effect is responsible for the bulk of the difference in basicity.

No resonance stabilization is found in methylamine, CH_3NH_2, or the methylammonium ion, $CH_3NH_3^{\oplus}$. Resonance forms can be drawn for aniline and the anilinium ion.

It can be seen that the two contributing forms for the anilinium ion are equivalent structures, whereas the last three structures for aniline involve charge separation and are of higher energy content.

Although the last three resonance forms of aniline are less important than the first two, they decrease the electron density on the nitrogen atom. The decrease in the electron density on the nitrogen makes the unshared electron pair less available for bonding in aniline than in methylamine where resonance is absent. Thus, aniline is a weaker base than methylamine.

5.5 Steric Effects

The atoms in all molecules occupy a volume characteristic of their size. Crowding of these atoms into too small a space results in a decreased stability of the system, and is called **steric strain** or **steric hindrance.**

Even long before the nature of inductive and resonance effects was understood, it was known that a substituent could influence a chemical reaction primarily because of its ability to occupy space. Whenever the attack by a reagent on the reactive functional group or atom in a molecule is blocked by other atoms in the molecule, the molecule is said to be sterically hindered. Usually steric hindrance in a molecule will result in a slowing up in the rate of the reaction. Occasionally, if the steric hindrance is severe due to the presence of very bulky

Sec. 5.5 STERIC EFFECTS

groups in the molecule, the normal course of the reaction may be affected, resulting in the formation of a product entirely different from the one expected.

Steric effects depend solely on the size of the groups or atoms and the amount of space they occupy, and are completely independent of electronic effects. The reader should have some understanding of the relative bulk and size of various atoms or groups within a molecule. This requires a knowledge of bond lengths, covalent radii, and atomic volumes of the individual atoms. Much of this information can be predicted to some degree from the periodic table. It might be advisable for the reader to consult a general chemistry text to refresh his knowledge of the subject. This will enable him to be able to predict and visualize the relative sizes and bulk of various groups within a molecule. For example, a t-butyl group is larger in size and occupies more space than an isopropyl group, which in turn is larger than an ethyl or methyl group; (e.g., $CH_3-\underset{\underset{CH_3}{|}}{\overset{\overset{CH_3}{|}}{C}}- > CH_3-CH-CH_3 > CH_3-CH_2- > CH_3-$).

An example of a steric effect can be seen in the Lewis acid–base reaction between trimethylboron and a series of amines. The equilibrium constants for the reaction, in Table 5.1, will indicate the effect due to the bulkiness of the groups.

The reaction of the amines with water ($R-NH_2 + H_2O \rightleftharpoons RNH_3^{\oplus} + OH^{\ominus}$) is measured by the K_b values in the table. The significance of the relatively constant numerical value of K_b as the R group changes from methyl (CH_3) to t-butyl [$(CH_3)_3-C-$] is due to the fact that the reaction involves the amine and the very small proton from water. No steric factors are evident here. However, changing R from methyl to t-butyl in the reaction with trimethylboron, decreases the value of K, the equilibrium constant for the Lewis acid–base salt, by a factor of about 280. This decrease in K can be attributed to steric factors due to the increasing bulkiness of the R groups. As the groups become larger there is **steric hindrance** in the Lewis salt formed between R and the methyl groups on the boron. This is

Table 5.1 Steric Effects in the Reaction of Trimethylboron with Amines

$$R-\overset{..}{N}H_2 + \underset{\underset{CH_3}{|}}{\overset{\overset{CH_3}{|}}{B}}-CH_3 \rightleftharpoons R-\underset{\underset{H}{|}}{\overset{\overset{H}{|}}{N^{\oplus}}}:{}^{\ominus}\underset{\underset{CH_3}{|}}{\overset{\overset{CH_3}{|}}{B}}-CH_3$$

R	K	$K_b \times 10^4$
CH_3-	28	4.2
CH_3-CH_2-	13	4.3
$(CH_3)_2-CH-$	2.7	4.3
$(CH_3)_3-C-$	0.1	2.8

$$K = \frac{[RNH_2^{\oplus}B(CH_3)_3^{\ominus}]}{[RNH_2][B(CH_3)_3]} \qquad K_b = \frac{[RNH_3^{\oplus}][OH^{\ominus}]}{[RNH_2]}$$

especially true where R is *t*–butyl. These steric problems are not present in the amines themselves, and result only when the adduct is formed; the addition of the amine and trimethylboron by formation of a B—N bond creates steric interference. The decreased stability of the adduct because of increased steric hindrance is reflected in the lowering of the value of the equilibrium constant, K. As the groups become bulkier there is a decreasing tendency to form the addition compound, typical of steric effects. This is just one example of the importance of steric effects in organic reactions. Very often, steric factors are just as important as resonance and inductive effects, and will be discussed later with respect to appropriate reactions.

Summary

• Electronic effects can be divided into inductive and resonance effects.

• The inductive effect is an electrostatic effect transmitted through chains of atoms along single bonds.

    (a) Some groups have an electron-withdrawing inductive effect compared to hydrogen. Others have an electron-releasing (donating) inductive effect.

    (b) The inductive effect decreases with distance and only operates effectively over one or two bonds' distance from any functional group.

    (c) Examples of the influence of the inductive effect on acidity and basicity of molecules are discussed.

• The resonance effect acts along a pi electron system, involving multiple bonds, unshared electrons, and incompletely filled outer electronic orbitals.

• Resonance is found in compounds whose electrons can be moved and arranged in more than one low-energy state. The dispersing of the electron density makes the resulting molecule more stable.

• Benzene, which has a continuous cyclic conjugated system, is especially stable because of resonance. The so-called resonance hybrid (lowest energy state) is the only form of benzene that actually exists.

• Resonance effects in molecules can influence the acidity and basicity of substances, such as of phenols, carboxylic acids, and amines.

• Steric effects due to the size and bulk of substituents in molecules are important.

    (a) Crowding of atoms into too small a space decreases the stability of a system by what is called steric hindrance.

    (b) Steric hindrance in a molecule can slow down the rate of a reaction or even affect the nature and amount of the product formed in some cases.

PROBLEMS

1 Draw suitable resonance contributing structures for each of the following.
    (a) HNO_3, **(b)** SO_3, **(c)** $NO_3^{\ominus}$, **(d)** NO_2 **(e)**   Br

PROBLEMS

2 For each of the following pairs, which species is more stable? Explain each choice.

(a) C$_6$H$_5$—O$^\ominus$ CH$_3$—CH$_2$—O$^\ominus$

(b) CH$_3$—C(=O)—O$^\ominus$ CH$_3$—CH$_2$—O$^\ominus$

(c) C$_6$H$_5$—CH=CH—CH=CH$_2$ C$_6$H$_5$—CH$_2$—CH$_2$—CH=CH$_2$

(d) CH$_2$=CH—CH=CH$_2$ CH$_3$—CH$_2$—CH$_2$—CH$_3$

3 How can you explain the fact that the two carbon–oxygen bonds in sodium acetate, Na$^\oplus$CH$_3$—CO$_2{}^\ominus$, have the same length?

4 How do you account for the fact that the α hydrogens in an aldehyde (such as propanal, CH$_3$—CH$_2$—C(=O)—H, with β and α labeled) are weakly acidic?

5 Which one out of the following pairs of compounds is the stronger acid or base?

(a) 4-NO$_2$-C$_6$H$_4$-COOH   4-CH$_3$-C$_6$H$_4$-COOH

(b) 4-F-C$_6$H$_4$-OH   2,4,6-(NO$_2$)$_3$-C$_6$H$_2$-OH

(c) 4-NO$_2$-C$_6$H$_4$-NH$_2$ 4-CH$_3$-C$_6$H$_4$-NH$_2$

(d) CH$_3$NH$_2$

(e) C$_6$H$_5$-NH$_2$ 4-CH$_3$-C$_6$H$_4$-N(CH$_3$)$_2$

6 Arrange the following sets of compounds in order of increasing acidity.

(a)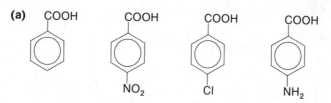

(b) butanoic acid, 3-bromobutanoic acid, 2-bromobutanoic acid, 4,4-dichlorobutanoic acid, 4-bromobutanoic acid.

(c) propanoic acid, ethanoic acid, 2-cyanoethanoic acid, 2,2,2-trichloroethanoic acid, 6-chloroheptanoic acid, 2,2-dichloropropanoic acid.

7 Arrange the following sets of compounds in order of increasing basicity.

(a) [structures: aniline, p-nitroaniline, cyclohexylamine, p-methoxyaniline, diphenylamine]

(b) CH_3NH_2, $CH_3-\underset{H}{N}-CH_3$, [aniline], $(CH_3)_3N$, [N-methylaniline]

8 *The ortho-hydroxybenzoic acid, [salicylic acid structure] ($k_a = 1.05 \times 10^{-3}$) is about 100 times stronger an acid than benzoic acid, [benzoic acid structure] ($k_a = 6.3 \times 10^{-5}$), whereas p-hydroxybenzoic acid, [p-hydroxybenzoic acid structure] ($k_a = 2.6 \times 10^{-5}$) is a weaker acid than benzoic acid. Explain. (HINT: What are the possibilities for hydrogen bonding in these molecules?)

PROBLEMS

9 * The compound quinuclidine, [structure of quinuclidine with CH₂–CH₂ bridge and N], reacts over 700 times more rapidly than triethylamine, $(CH_3CH_2)_3N$, with isopropyl iodide $CH_3-\overset{H}{\underset{|}{C}}-CH_3$.

Explain. Would you expect quinuclidine or triethylamine to react to a greater extent with $B(CH_3)_3$ in a Lewis acid–base reaction?

10 * Explain why cyclopentadiene, [structure of cyclopentadiene], is a stronger acid than benzene.

6
Chemical Reactivity:
Chemical Kinetics and Reaction Mechanisms of Organic Compounds

The Apollo-Saturn V lifting off at Kennedy Space Center in 1971, the chemical reactivity of its rocket fuel generating 7½ million pounds of thrust as it sends the 363-foot-high space vehicle on the first phase of its flight to the moon.

Courtesy of NASA.

6.1 Introduction

Chemical kinetics is the branch of chemistry that concerns itself with the study of the velocity of chemical reactions and with the investigation of the mechanisms by which they proceed. *How rapidly and by what type of mechanism does a reaction take place?* The branch of chemistry known as thermodynamics considers only the energy relationship between the reactants and products of a chemical reaction, but does not attempt to describe the various stages through which the reactants have to pass to be transformed into the final products. Chemical kinetics supplies information that can be used to find the mechanism for the conversion of reactants to products.

Sec. 6.3 THE REACTION MECHANISM

Not all reactions lend themselves to kinetic study. Some reactions (ionic reactions, such as ionization and precipitation) are so rapid that they seem instantaneous. Other reactions are quite slow and may require months or even years before any perceptible change occurs. Between these two extremes are many reactions whose velocities (rates) are measurable, and whose kinetic data will be valuable in elucidating the reaction mechanism.

6.2 Factors that Influence Rate of Reaction

Various factors affect the rate of chemical reaction. The rate of reaction depends not only on the nature of the reactants, but also on the temperature and the concentration of the reactants. An increase in temperature leads to an increase in reaction velocity; in fact, a 10° C rise in temperature increases the rate of many reactions by a factor of 2 or more (to a very rough approximation). An increase in the initial concentration of reactants also results in a faster rate in most reactions. In addition, many reactions are influenced by the presence of **catalysts,** substances having the ability to accelerate or decelerate the rates of certain chemical reactions. Also, certain **photochemical reactions** caused by the impingement of light, are greatly stimulated when light of an appropriate frequency is allowed to pass through the reaction mixture.

6.3 The Reaction Mechanism

What is meant by the phrase "reaction mechanism"? When we write such a chemical equation as $A + B \longrightarrow C + D$, we are merely writing the formulas of the reactants we are starting with, and the products into which the reactants are transformed. We are saying nothing about the path or the steps involved in the conversion of A and B into C and D. Most chemical reactions proceed by a series of steps, rather than one single step. For example, $A + B$ might first react to produce substance E, which might change into F and G, after which F and G are finally transformed into C and D.

(1) $\qquad A + B \xrightarrow{\text{slow}} E \qquad$ (6–1)

(2) $\qquad E \xrightarrow{\text{fast}} F + G \qquad$ (6–2)

(3) $\qquad F + G \xrightarrow{\text{fast}} C + D \qquad$ (6–3)

$\qquad\qquad\overline{A + B \longrightarrow C + D} \qquad$ (6–4)

The three steps shown above describe what we call the reaction mechanism for the conversion of $A + B$ to $C + D$. The overall result is simply the conversion of the reactants ($A + B$) into the products ($C + D$). Data obtained from an investigation of the rate of this reaction provide evidence that the conversion of $A + B$ to $C + D$ does indeed proceed by the reaction mechanism indicated by the three steps shown above.

In reactions that proceed by a mechanism of more than one step, the slow step (in this case, step 1) is the so-called rate-determining step of the reaction. The slow step in any reaction serves as a bottleneck; for once it is completed the subsequent steps occur at a faster rate than the slow step, and the reaction

Jacobus Hendricus van't Hoff (1852–1911)—Dutch physical chemist and recipient of the first Nobel Prize in chemistry in 1901—laid the basis for stereochemistry by his studies of the asymmetric carbon atom. (Courtesy of the Rijks Museum in Leiden.)

proceeds toward completion. Thus, *the slow step in any reaction sequence determines the overall rate of the reaction.*

The substances, *E, F,* and *G,* which are produced during the course of the reaction, are called **reaction intermediates** if they are formed in detectable amounts or if they can actually be isolated from the reaction mixture during the course of the reaction. If their presence cannot be detected by experimental techniques (because they are so reactive that they exist only for a very, very small amount of time) but evidence assumes their existence in order to formulate a logical reaction mechanism, then *E, F,* and *G* are called **transition states** for the reaction. A system in the transition state is called an **activated complex.** The transition state represents an energy maximum between two stages of a reaction. Some of these substances may be true reaction intermediates, while others may only be transition states, to account for the conversion of reactants to products in a plausible manner.

6.4 Theory of Reaction Rates

In reactions involving two or more molecules it is logical to assume that before reaction can occur the molecules must come in contact with each other, that is to say, they must collide. The rate at which the reaction proceeds depends *not* on the frequency of the molecular collisions but the frequency of so-called **effective collisions** leading to reaction. Only those collisions between molecules possessing a certain amount of energy (required by the particular reaction in question) and a proper orientation at the time of collision will lead to reaction. This quantity of energy will vary for every chemical reaction, and is known as the **activation energy** characteristic of that reaction. Thus, only molecules possessing an energy equal to, or greater than, the activation energy required for a particular reaction will lead to the formation of products when the molecules collide.

Let us consider the hypothetical reaction of *A + B* to form *C + D.* This

Sec. 6.4 THEORY OF REACTION RATES

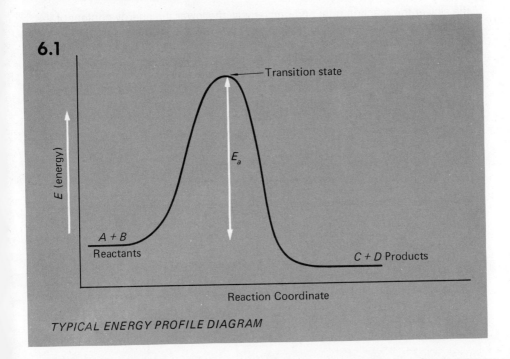

TYPICAL ENERGY PROFILE DIAGRAM

reaction can be represented pictorially by an **energy profile diagram** (see Figure 6.1) that describes the transformation of the reactants into the products.

The energy of the system is plotted on the y-axis, and the reaction coordinate that represents the progress of the reactants toward the formation of products, on the x-axis. The energy of the system is shown at three stages: initially before reaction has begun; at the transition state; and after the reaction has been completed. The transition state is the point of highest energy in the diagram.

The difference in height between the energy of the reactants and the transition state represents the energy of activation, E_a, required by the molecules before reaction can occur. In other words, the molecules must have sufficient energy to reach the transition state before they can be converted into products.

A boy rolling a ball up a hill provides an interesting analogy. Energy is required to push the ball to the top of the hill. Once the ball is at the top it will roll down the other side into the valley on its own. If the ball is pushed only part way up the hill it will fall back to the point it initially started from. With reference to the energy profile diagram, the transition state corresponds to the point on the top of the hill. If the molecules ($A + B$) have sufficient energy (the energy of activation) to reach this point, the molecules will simply continue on their path down to the valley to form the products $C + D$. If, however, $A + B$ have only a small fraction of the energy required, they will rise only a small distance up the hill before falling back to the starting point, and remain as the original starting materials, $A + B$.

In other words, every reaction has an energy barrier through which it must pass in order for the reactants to form products. The higher the energy of the transition state is above the starting materials, the greater the energy barrier and the slower the reaction rate; and conversely, the lower the energy content of the transition state, the faster the reaction will go, since a greater proportion of the molecular collisions will have the required energy to lead to reaction. Catalysts affect the rate of chemical reactions by changing the energy of the transition state. For example, a catalyst can increase the rate of a reaction by lowering the energy of activation (energy barrier) of the transition state. The exact mechanism of catalyst function in a reaction is not completely understood at present. It can only be said that the catalyst changes the time the reaction would take if no catalyst were present.

6.5 The Breaking of Chemical Bonds— Homolytic and Heterolytic Cleavage

In any chemical reaction the conversion of reactants into products must require the breaking of bonds between atoms, and the formation of new bonds in the products. The bonds are broken in one of two ways: by **homolytic** or by **heterolytic cleavage.**

Consider a covalent molecule $A:B$ where two electrons are shared between atoms A and B. If the chemical bond between A and B is broken symmetrically, that is to say each atom (A and B) gets one electron $A:B \longrightarrow A\cdot + B\cdot$, this is known as **homolytic cleavage.** The substances $A\cdot$ and $B\cdot$ each have an odd, unpaired electron, and are called free radicals. Any substance that has an odd, unpaired electron is known as a **free radical.** A free radical located on a carbon atom contains seven electrons about that carbon core $\left(-\overset{|}{\underset{|}{C}}\cdot\right)$ and is electrically neutral. Most free radicals are extremely reactive (due to their odd, unpaired electron) and want to complete their outer electronic orbitals.

The bond between A and B can also be broken unsymmetrically, so that either A or B gets both of the shared electrons: $A:B \longrightarrow A^{\oplus} + B:^{\ominus}$; or $A:^{\ominus} + B^{\oplus}$. This is known as **heterolytic cleavage.** The substances produced have positive and negative charges and are **ions.** If A or B is a carbon atom bearing a positive charge it is called a **carbonium ion.** Carbonium ions are extremely reactive since they contain only six electrons around the carbon core $\left(-\overset{|}{\underset{|}{C}}{}^{\oplus}\right)$ and want to complete their outer shell of electrons. A carbon atom with a negative charge is called a **carbanion** $\left(-\overset{|}{\underset{|}{C}}:^{\ominus}\right)$. Carbanions have a complete outer shell of electrons, but must be stabilized by sharing their unshared electron pair with some electron-deficient moiety.

Thus, when bonds are broken, either free radicals or ions must be produced. These are the reactive species that are formed in the intermediate steps of reactions leading to the conversion of reactants into products. In general then, all reactions must proceed through the formation of either free radicals or ions, and so we can classify all reaction mechanisms as belonging to one of two large classes:

Ionic reactions proceed through an ionic mechanism and usually occur at

Sec. 6.7 NUCLEOPHILIC SUBSTITUTION

room temperature, in the dark (without the need for light), and in solution. Polar solvents, such as water and alcohol, favor the formation of carbonium ions and carbanions.

Free-radical reactions, on the other hand, usually occur at elevated temperatures, in the presence of sunlight, and in the vapor state. If a reaction does not proceed readily in the dark, this is often taken as evidence that the reaction probably involves free radicals.

6.6 Types of Organic Reaction

All reaction mechanisms can be further subdivided as belonging to certain categories: **substitution reactions, elimination reactions, addition reactions,** etc. In a substitution reaction, a substituent in a molecule is replaced by another one; in an elimination reaction, something is removed from a molecule; in an addition reaction, something is added to a molecule.

Reaction mechanisms can also be classified as to the type of reagent that is attacking the substrate in question. The reagents are classified according to their electron densities, and are called **nucleophiles** or **electrophiles**. These terms will be defined and illustrated with some examples in the following section.

6.6-1 Nucleophilic and Electrophilic Reagents

The term **nucleophilic reagent** or **nucleophile** means a "nucleus-seeking" group. The translation is not immediately clear, since the group is not actually seeking a nucleus, but is actually seeking a region of low electron density to which it is able to supply electrons. A nucleophile can be more accurately defined as an electron-pair donor. Typical nucleophilic reagents are $OH^{\ominus}$, $Br^{\ominus}$, $I^{\ominus}$, $\ddot{N}H_3$, $H_2\ddot{O}$:, and carbanions, all of which have unshared pairs of electrons available for bonding.

An **electrophilic reagent** or **electrophile** can be defined as an "electron-seeking" group, and the term applies to substances deficient in electrons, which are seeking sites of high electron density. Some typical electrophilic reagents are $H^{\oplus}$, $NO_2^{\oplus}$, $Br^{\oplus}$, $AlCl_3$, $ZnCl_2$, $FeBr_3$, BF_3, and carbonium ions, all of which have an empty electronic orbital capable of accepting a pair of electrons.

Most organic reactions proceed by an ionic (polar) mechanism. In these reactions an electron pair is transferred from one species to another during the course of the reaction; some substances behave as **electron-pair donors,** others are **electron-pair acceptors.**

According to our previous discussion, it should be apparent that electron-pair donors (or Lewis bases) can also be called nucleophilic reagents, and electron-pair acceptors (or Lewis acids) are actually electrophilic reagents. Thus, reaction mechanisms can be classified as being nucleophilic or electrophilic, depending on which type of reagent is involved in the reaction.

6.7 A Polar Organic Reaction— Nucleophilic Substitution

In a **nucleophilic substitution reaction,** a nucleophilic reagent, N:, attacks the carbon atom of the substrate, displacing some substituent. The group that is displaced (called the leaving group) takes its bonding electrons. The new bond is formed between N and the carbon using the electrons supplied by the nucleophilic reagent.

$$N:^- + \overset{|}{\underset{|}{C}}\!\!-\!\!X \longrightarrow N-\overset{|}{\underset{|}{C}}- + X: \quad (6-5)$$

A typical example of this type of reaction is the reaction of methyl bromide with hydroxide ion to form methyl alcohol and bromide ion.

$$H-\ddot{O}:^{\ominus} + \overset{\delta\oplus}{C}H_3-\overset{\delta\ominus}{\ddot{B}r}: \longrightarrow \left[H-\overset{\delta\ominus}{\ddot{O}}-----\overset{H\ \overset{\delta\oplus}{\ }\ H}{\underset{H}{C}}-----\overset{\delta\ominus}{\ddot{B}r} \right] \longrightarrow H\ddot{O}-CH_3 + :\ddot{B}r:^{\ominus} \quad (6-6)$$

transition state

The hydroxide ion, being the nucleophilic reagent in this reaction, donates a pair of electrons to the carbon atom, displacing the bromide ion.

The bond between the oxygen of the $OH^{\ominus}$ and the carbon atom is starting to form at the same time as the C—Br bond is being broken. This reaction is a simple one-step reaction, and can be represented as in the energy profile diagram of Figure 6.2. As the transition state is approached there are *two* species ($OH^{\ominus}$ and CH_3Br) reacting, and the reaction is said to be bimolecular. In the transition state one may consider that both OH and Br are somewhat, and about equally, attached to the methyl (CH_3) group. The reaction is called a **bimolecular nucleophilic substitution reaction**, S_N2, which means *nucleophilic substitution–bimolecular*. Kinetic evidence shows that the rate of this reaction is proportional to both the

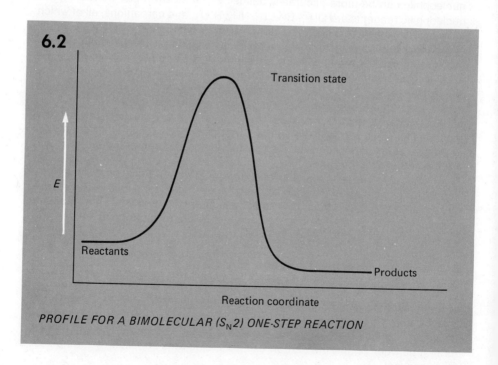

PROFILE FOR A BIMOLECULAR (S_N2) ONE-STEP REACTION

Sec. 6.7 NUCLEOPHILIC SUBSTITUTION

concentration of $OH^\ominus$ (nucleophile) and CH_3Br (substrate); and the reaction is said to follow **second-order kinetics.** The so-called rate expression for the reaction can be expressed as: Rate = $k[CH_3Br][OH^\ominus]$. The kinetic data support the concept that both species ($OH^\ominus$ and CH_3Br) are involved in the formation of the transition state.

In S_N2 reactions the nucleophilic reagent attacks the carbon atom (substrate) from the side of the molecule opposite the leaving group. As the new bond is being formed and the bond to the leaving group breaks, the other three bonds attached to the carbon atom **invert** (like an umbrella turned inside out in a rainstorm) so that in the product the new bond is formed on the opposite side of the molecule to where the leaving group was originally present. The S_N2 reactions are then said to proceed with **inversion of configuration.** (The arrangement of atoms that characterizes a particular stereoisomer is called its **configuration.** This will be discussed in Chapter 17.)

6.7-1 Steric Effects in S_N2 Reactions

Let us see how steric factors can affect the rate of an S_N2 reaction. Consider the reaction, $R-Br + I^\ominus \longrightarrow R-I + Br^\ominus$, run under S_N2 reaction conditions. The rate of this reaction will depend on various factors among which are the nature of the nucleophilic reagent ($I^\ominus$), the leaving group ($Br^\ominus$), the role of the solvent, and the R group in the alkyl halide. The stronger the nucleophilic reagent, and the better the leaving group (the more easily displaced), the faster will be the rate of the reaction. The nature of the R portion of the alkyl halide also influences the rate of the reaction. For example, in considering the reaction $R-Br + I^\ominus \longrightarrow R-I + Br^\ominus$, the rate decreases rather dramatically from $R = CH_3$ to $R = (CH_3)_3C$. Data obtained from kinetic studies on this reaction indicate that the relative rates of various alkyl groups towards iodide ion under S_N2 reaction conditions are as shown in Table 6.1.

Thus, when R is a *t*-butyl group, the rate of the reaction is approximately 150,000 times slower than when R is a methyl group. The differences in rate between these S_N2 reactions seem to be due primarily to steric factors in the R group. The bulkier groups (i.e., *t*-butyl) cause crowding, raising the energy of activation required to form the transition state, and thus slow down the rate of the reaction.

Table 6.1 Relative Reaction Rate for ($RBr + I^\ominus \longrightarrow R-I + Br^\ominus$) under S_N2 Conditions

Alkyl group (R)	Relative rate
CH_3-	150
CH_3-CH_2-	1
$CH_3-CH-CH_3$	0.01
CH_3-C-CH_3 $\quad\;\; \vert$ $\quad\;\; CH_3$	0.001

$$\left[\text{I} \text{-----} \underset{R}{\overset{R}{\underset{|}{C}}} \text{-----} \text{Br} \right] \quad R = H, CH_3, \text{(etc.)}$$

transition state

The crowding in the transition state can easily be seen. It will be noted that when the nucleophile (I$^\ominus$) and leaving group (Br$^\ominus$) are both associated with carbon, *five* bonds are directly attached to the carbon atom in the transition state. This is in contrast to the four bonds present on the carbon atom in the original alkyl halide. Since atoms occupy space, their steric requirements will greatly influence the stability of the transition state. It follows that, in going from four substituents in the alkyl halide to five substituents in the transition state, the transition state will be more crowded and sterically hindered than the reactants. Therefore, it should not be too surprising that the more bulky R groups should tend to form a less stable transition state, which is reflected in a slower rate of the reaction. (In the case of CH_3Br used as an alkyl halide, the transition state contains three small hydrogen atoms in contrast to $(CH_3)_3$—C—Br, where the transition state contains three bulkier methyl groups attached to the carbon atom). This is a case where steric factors have a pronounced effect on the rate of the reaction. All S_N2 reactions are subject to steric factors and the steric hindrance due to the size of the alkyl (R) groups in the substrate. The generalization can be made that all S_{N2} reactions are greatly affected by steric factors; the more substitution in the R group, the slower the reaction rate.

6.7–2 The S_N1 Mechanism Some nucleophilic substitution reactions occur by an alternate mechanism—a two-step mechanism, which has two transition states. The transition state with the higher energy determines the rate of the reaction (see Figure 6.3). The feature of the two-step mechanism which distinguishes it from the S_N2 (one-step) reaction is that there is only one species involved in the formation of the transition state which determines the rate of the reaction. These reactions are called **unimolecular nucleophilic substitution,** S_N1 reactions.

Usually, substances able to form relatively stable carbonium ions will react by the S_N1 mechanism. Consider the reaction of *t*–butyl bromide with hydroxide ion to produce *t*–butyl alcohol.

Step 1 $CH_3-\underset{CH_3}{\overset{CH_3}{\underset{|}{\overset{|}{C}}}}-Br \xrightarrow{\text{slow}} CH_3-\underset{CH_3}{\overset{CH_3}{\underset{|}{\overset{|}{C^\oplus}}}} + Br^\ominus$ (6–7)

Step 2 $CH_3-\underset{CH_3}{\overset{CH_3}{\underset{|}{\overset{|}{C^\oplus}}}} + OH^\ominus \xrightarrow{\text{fast}} CH_3-\underset{CH_3}{\overset{CH_3}{\underset{|}{\overset{|}{C}}}}-OH$ (6–8)

In the slow or rate-determining step ionization occurs to produce a carbonium ion which then reacts with the nucleophilic reagent to produce the product. Since the nucleophilic reagent enters into the reaction after the rate-determining step, it

Sec. 6.7 NUCLEOPHILIC SUBSTITUTION

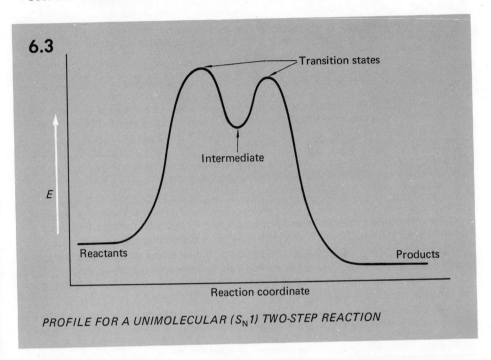

6.3

PROFILE FOR A UNIMOLECULAR (S_N1) TWO-STEP REACTION

does not affect the overall rate of the reaction. Kinetic data show that the rate of the reaction will depend *only* on the concentration of *t*-butyl bromide in this case, and the reaction follows *first-order* kinetics. Thus, S_N1 reactions are called *nucleophilic substitution-unimolecular,* and the rate of these reactions is independent of the concentration of nucleophilic reagent used. The rate expression for the reaction can be expressed as: Rate = $k[(CH_3)_3$—C—Br].

Tertiary alkyl halides, which can usually form more stable carbonium ions than primary halides, normally react by the S_N1 mechanism, whereas primary halides tend to react by the S_N2 mechanism. (The relative stability of carbonium ions is in the order: tertiary > secondary > primary, or $R_3C^{\oplus} > R_2CH^{\oplus} > RCH_2^{\oplus}$ (R = any alkyl group), and can be attributed to the electron-donating inductive effect of the alkyl groups attached to the carbon atom bearing the positive charge. The stability of carbonium ions will be discussed in more detail in Section 8.8-2.)

6.7-3 Electronic Effects and other Factors in the S_N1 Reaction

In contrast to the S_N2 reaction, the rate of the S_N1 reaction does not depend on steric factors, but is dependent on the stability of the carbonium ion formed during the course of the reaction. Thus, compounds that can form more stable carbonium ions will react at a faster rate under S_N1 reaction conditions than substances that form less stable carbonium ions. Since the stability of carbonium ions (see Section 8.8-2) is related to electronic effects, *the rate of S_N1 reactions is affected largely by electronic effects*, that is by the tendency of substituents to donate or withdraw electrons, whereas *the rate of S_N2 reactions is dependent on steric effects.*

Table 6.2 Relative Reaction Rate for (RBr + $H_2O \longrightarrow$ R—OH + HBr) under S_N1 Conditions

Alkyl group (R)	Relative rate
CH$_3$—C(CH$_3$)—CH$_3$ (with CH$_3$ above)	10^6
CH$_3$—CH—CH$_3$	45
CH$_3$—CH$_2$—	1.7
CH$_3$—	1.0

The rate of S_N1 reactions depends on various factors, such as the nature (strength) of the nucleophilic reagent, and the nature of the leaving group. Of prime importance is the polarity and ionizing power of the solvent used, since more polar solvents tend to promote ionization and favor S_N1 reactions.

The first step in the S_N1 mechanism is the rate-determining step, which leads to the formation of a carbonium ion.

$$RX \longrightarrow \left[\overset{\delta\oplus}{R} \text{------} \overset{\delta\ominus}{X} \right] \longrightarrow R^\oplus + X^\ominus \qquad (6\text{-}9)$$

transition state   carbonium ion

From the above equation it can be seen that in the transition state the R (alkyl) portion of the alkyl halide has started to acquire a positive charge, which finally forms a carbonium ion with a full positive charge. Since charge separation starts with the formation of the transition state and ultimately leads to the formation of a fully charged species, a carbonium ion, it is not surprising that S_N1 reactions proceed more rapidly in more polar or ionizing solvents.

The nature of the R group in the alkyl halide also influences the rate of the S_N1 reaction, since it determines the type of carbonium ion formed. The influence of the R group (through its electronic effects in stabilizing the carbonium ion formed) on the rate of an S_N1 reaction is illustrated by the data, in Table 6.2, obtained for the reaction RBr + $H_2O \longrightarrow$ R—OH + HBr, run in a polar solvent (formic acid) under S_N1 reaction conditions. Note that a change in the structure of the R group can affect the reaction rate, and that the trend is in the opposite direction to what is observed when studying the influence of R groups on the reaction rate in S_N2 reactions (e.g., a *t*-butyl group reacts one million times more rapidly than a methyl group in this S_N1 reaction).

Other examples of ionic reactions include nucleophilic addition, electrophilic addition, elimination, electrophilic aromatic substitution, and oxidation–reduction. These will be discussed in the appropriate sections.

6.8 Free Radical Reactions

Although most organic reactions are polar or ionic, a large number of organic reactions are **free radical reactions.** Free radicals are species containing one (or

Sec. 6.8 FREE RADICAL REACTIONS

more) unpaired electrons. The stability of free radicals is in the same order as that of carbonium ions, namely, $R_3C\cdot > R_2CH\cdot > RCH_2\cdot$. They are exceedingly reactive substances since they want to complete their outer shell of electrons. Free radical reactions usually occur in the presence of sunlight (or light of some wavelength) and at high temperatures. They are chain reactions, and once started continue until all the reactants are consumed, or until something occurs to *break the chain* (e.g., free-radical inhibitor entering the reaction).

All reactions that proceed by free radical mechanisms take place by a chain reaction, which involves three basic steps: (1) initiation, (2) propagation, and (3) termination.

The free-radical halogenation of an alkane in the presence of sunlight will illustrate these points. The net reaction is given by the sum of the individual steps of the chain. The reaction of methane with chlorine to produce methyl chloride in the presence of sunlight is a typical reaction.

Initiation $\quad Cl_2 \xrightarrow{h\nu} 2Cl\cdot \quad$ (reactive free radical generated) $\quad$ (6–10)

Propagation
$$Cl\cdot + CH_4 \longrightarrow HCl + CH_3\cdot \quad (6\text{–}11)$$
$$CH_3\cdot + Cl_2 \longrightarrow CH_3Cl + Cl\cdot \quad (6\text{–}12)$$

Sum of propagation steps $\quad CH_4 + Cl_2 \longrightarrow CH_3Cl + HCl \quad (6\text{–}13)$

Termination
$$Cl\cdot + Cl\cdot \longrightarrow Cl_2 \quad (6\text{–}14)$$
$$CH_3\cdot + CH_3\cdot \longrightarrow C_2H_6 \quad (6\text{–}15)$$
$$CH_3\cdot + Cl\cdot \longrightarrow CH_3Cl \quad (6\text{–}16)$$

The initiation step starts the reaction by producing reactive $Cl\cdot$ free radicals. The $Cl\cdot$ free radicals are generated by the chlorine molecule absorbing energy sufficient to break the Cl—Cl bond and generate two reactive $Cl\cdot$ free radical fragments. The chlorine free radical, wanting an electron to complete its electronic structure, reacts with a methane molecule, abstracting a hydrogen atom (with its electron) from the methane, to produce HCl and a reactive methyl free radical ($CH_3\cdot$). The methyl free radical in turn reacts with more chlorine molecules, pulling off a chlorine with one of its electrons to produce CH_3Cl and a $Cl\cdot$ free radical. The $Cl\cdot$ free radical keeps the chain going by reacting with more methane, as shown in the first propagation step. The reaction sequence will continue until all the reactants are consumed to form the products.

The termination steps shown above (6.14, 6.15, and 6.16) will stop the reaction since they will consume the free radicals before they can react to produce the desired products.

The initiation of free radicals can occur by heat, light, or by certain chemicals that serve as free radical initiators. The amazing thing about free radical reactions is that a *single* free radical (once formed) can produce very large numbers of molecules of product by the chain reaction sequence.

In this chapter we have illustrated in a somewhat simplified fashion some of the fundamental types of reaction mechanisms by which organic compounds react. From a practical standpoint, we have not included all the possible classes of reaction mechanisms. Reactions proceeding by mechanisms involving carbanions, cycloaddition, and rearrangements, for example, will be discussed in subsequent chapters.

Despite distinct differences in the molecular structure of organic compounds, their numerous reactions proceed only by several fundamental types of reaction mechanisms.

Summary

• Chemical kinetics is the branch of chemistry that involves the study of reaction rates and the elucidation of reaction mechanisms.

• Reaction rates are affected by such factors as the nature of the reactants, temperature, concentration of reactants, and catalysts.

• The reaction mechanism describes the path (all the individual steps) in the conversion of the reactants into products.

   (a) The slowest step in the reaction sequence is the rate-determining step.

   (b) The transition state is a point of maximum potential energy.

   (c) A true reaction intermediate is a substance formed in detectable amounts, which can be isolated out of the reaction mixture.

• According to reaction rate theory, the rate at which a reaction proceeds depends on the number of effective collisions and the (energy barrier) activation energy required by the reaction.

• Organic reactions involve the breaking and forming of covalent bonds.

   (a) Bonds are broken by either homolytic or heterolytic cleavage.

   (b) Homolytic bond cleavage produces free radicals; heterolytic bond cleavage produces charged particles (in the case of carbon atoms, either carbonium ions or carbanions).

• It is very useful to examine organic reactions as to type of reaction mechanism.

• Reaction mechanisms can be classified according to the reactive species formed (free radicals, carbonium ions or carbanions) during the course of the reaction.

   (a) Free-radical reactions.

   (b) Polar or ionic reactions.

   (c) Reaction mechanisms can also be classified by such as substitution, addition, elimination reactions.

• Reagents can be classified as nucleophilic and electrophilic depending on their electron densities.

• Nucleophilic substitution reactions proceed by either a one-step mechanism (S_N2) or by a two-step mechanism (S_N1).

   (a) The S_N2 reaction is influenced by steric factors.

   (b) The S_N1 reaction is affected mainly by electronic factors.

• Free-radical reactions are like chain reactions, involving three basic steps: (1) initiation, (2) propagation, and (3) termination.

PROBLEMS

1 Classify each of the following reactions as belonging to a particular category (i.e., nucleophilic addition, electrophilic aromatic substitution, elimination, or free-radical addition, etc.).

PROBLEMS

(a) 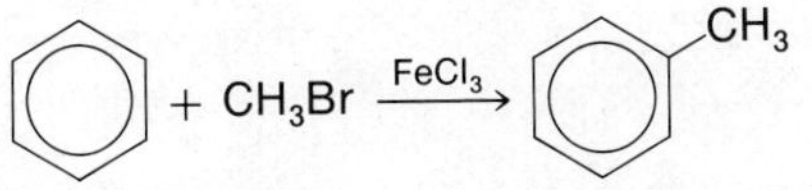 C6H5–H + CH$_3$Br $\xrightarrow{FeCl_3}$ C6H5–CH$_3$

(b) CH$_2$=CH$_2$ $\xrightarrow{H_2}{Pt}$ CH$_3$–CH$_3$

(c) CH$_3$–CH$_2$–O$^\ominus$Na$^\oplus$ + CH$_3$I $\longrightarrow$ CH$_3$–CH$_2$–OCH$_3$ + NaI

(d) CH$_3$–CH(Br)–CH$_3$ + alc. KOH $\longrightarrow$ CH$_3$–CH=CH$_2$ + KBr + H$_2$O

(e) CH$_3$–CH$_2$OH + conc. H$_2$SO$_4$ $\xrightarrow{\Delta}$ CH$_2$=CH$_2$

(f) CH$_3$–CO–CH$_3$ + HCN $\longrightarrow$ CH$_3$–C(OH)(CN)–CH$_3$

(g) CH$_2$=CH$_2$ + HI $\longrightarrow$ CH$_3$–CH$_2$I

(h) CH$_3$–CH$_3$ + Cl$_2$ $\xrightarrow{h\nu}$ CH$_3$–CH$_2$Cl + HCl

(i) CH$_3$CH$_2$Br + SH$^\ominus$ $\longrightarrow$ CH$_3$–CH$_2$–SH + Br$^\ominus$

(j) CH$_3$–CO–Cl + NH$_3$ $\longrightarrow$ CH$_3$–CO–NH$_2$ + HCl

(k) H–C≡C–H + Br$_2$ $\longrightarrow$ HBrC=CBrH

(l) (CH$_3$)$_3$C–Br + alc. KOH $\longrightarrow$ (CH$_3$)$_2$C=CH$_2$

2 Classify each of the following reactions and write *all* the steps for the proposed reaction mechanism, showing how the reactants were converted into the products in each case.

(a) CH$_3$–CH$_2$Br + OH$^\ominus$ $\longrightarrow$ CH$_3$CH$_2$OH + Br$^\ominus$

(b) CH$_3$Br + Br$_2$ $\xrightarrow{h\nu}$ CH$_2$Br$_2$ + HBr

(c) (CH$_3$)$_3$C–I + SH$^\ominus$ $\longrightarrow$ (CH$_3$)$_3$C–SH + I$^\ominus$

7

Saturated Hydrocarbons: Alkanes and Cycloalkanes

An oil tanker, part of the trans-oceanic lifeline of the contemporary industrialized world, carrying oil rich in alkanes and cycloalkanes.

Courtesy of the Mobil Oil Corporation.

7.1 Introduction

The most fundamental class of organic compounds is that of the **saturated hydrocarbons**. As the name implies, saturated hydrocarbons are compounds containing only the elements carbon and hydrogen, and only single bonds are present in the molecule. The saturated hydrocarbons are considered the parent compounds from which all other organic compounds can be derived. Because of their lack of reactivity with many chemical reagents the saturated hydrocarbons are called paraffins (from the Greek, meaning little affinity). Indeed, many of the higher-molecular-weight members of the series are constituents of lubricating oils, tar, vasoline and paraffin wax. Natural gas is composed mainly of methane, CH_4, and petroleum contains numerous hydrocarbons.

7.2 Structure of Some Alkanes

The simplest saturated hydrocarbon is methane, CH_4, which is considered the parent compound of organic chemistry (see also the FMO model of methane in Figure 1.2).

$$H-\underset{\underset{H}{|}}{\overset{\overset{H}{|}}{C}}-H$$

Sec. 7.4 NOMENCLATURE OF ALKYL GROUPS

The next member of the series is ethane, C_2H_6 (or CH_3-CH_3).

$$\begin{array}{c} H \quad H \\ | \quad\; | \\ H-C-C-H \\ | \quad\; | \\ H \quad H \end{array}$$

Then comes propane, C_3H_8 (or $CH_3-CH_2-CH_3$), butane, C_4H_{10}, and so on, as shown in Table 3.1. Although we write the structural formulas with the chain of carbon atoms in a straight line for simplicity, note that actually the carbon atoms are in a sort of zig-zag arrangement; for example, butane, C_4H_{10} (or $CH_3-CH_2-CH_2-CH_3$) is somewhat more precisely depicted as

$$CH_3\diagdown_{CH_2}\diagup^{CH_2}\diagdown_{CH_3}$$

(see also the FMO model of butane shown in Figure 2.1). Each member of the saturated hydrocarbons is characterized by the *-ane* ending in its name. The naming of the series has been discussed in Section 3.2 on nomenclature. The class of saturated hydrocarbons containing a continuous chain or branched chains of carbon atoms is collectively referred to as the **alkanes.** Alkanes having only a single chain of carbon atoms, and no branching, are called normal hydrocarbons. The small letter *n* stands for *normal* in the naming (e.g., $CH_3-CH_2-CH_2-CH_2-CH_3$ is called *n*-pentane).

7.3 General Formulas of Alkanes— Members of an Homologous Series

Every alkane can be represented by the general formula C_nH_{2n+2}, where *n* represents the number of carbon atoms in the molecule. When $n = 1$, the formula of the corresponding alkane is CH_4, $n = 2$, C_2H_6, $n = 3$, C_3H_8, etc. It can be seen that each alkane differs from the next member of the series by a CH_2 unit, called a **methylene group.** A series of compounds having molecular structures that only differ by the same unit (e.g., CH_2) throughout is referred to as an **homologous series.** Since the members of an homologous series have very similar structures, one would expect the compounds to have somewhat similar physical and chemical properties. This is true of the alkanes as it is of every class of organic compounds.

7.4 Nomenclature of Alkyl Groups

The student should refer to Section 3.2 to review the rules for the nomenclature of organic compounds, with particular emphasis on the alkanes. The branched-chain compounds often contain groups derived from the alkanes known as **alkyl groups.** The alkyl groups are formed from the corresponding alkane by removing a hydrogen atom. The **methyl group,** $H-\overset{\overset{\displaystyle H}{|}}{\underset{\underset{\displaystyle H}{|}}{C}}-$ (or CH_3-), is formed from methane, CH_4; the **ethyl group,** C_2H_5-, is formed from ethane, C_2H_6.

$$\begin{array}{c} H\;\;H \\ |\;\;\;| \\ H-C-C- \\ |\;\;\;| \\ H\;\;H \end{array} \quad (\text{or } CH_3-CH_2-)$$

ethyl group

Two groups can be formed from propane, $CH_3-CH_2-CH_3$; removal of a hydrogen atom from one of the carbon atoms at the end of the chain forms C_3H_7,

$$\text{the } \begin{array}{c} H\;\;H\;\;H \\ |\;\;\;|\;\;\;| \\ H-C-C-C- \\ |\;\;\;|\;\;\;| \\ H\;\;H\;\;H \end{array} \quad (\text{or } CH_3-CH_2-CH_2-) \text{ group called the } \textbf{normal propyl group}$$

(*n*-propyl group); removal of a hydrogen atom from the middle carbon atom gives

$$\begin{array}{c} H\;\;H\;\;H \\ |\;\;\;|\;\;\;| \\ H-C-C-C-H \\ |\;\;\;\;\;\;\;| \\ H\;\;\;\;\;\;H \end{array} \quad (\text{or } CH_3-\underset{|}{CH}-CH_3) \text{ called the } \textbf{isopropyl group} \text{ (}i\text{-propyl group).}$$

The next higher member of the alkanes has the formula C_4H_{10}. It will be recalled from Section 2.5 that there are two isomers of the molecular formula C_4H_{10}.

$$\underset{1}{CH_3}-\underset{2}{CH_2}-\underset{3}{CH_2}-\underset{4}{CH_3} \qquad \overset{1}{CH_3}-\overset{2}{\underset{|}{CH}}-\overset{3}{CH_3} \\ \underset{4}{CH_3}$$

n-butane 2-methylpropane
(or isobutane)

The removal of a hydrogen atom from one of the end carbon atoms (carbon 1 or 4) in *n*-butane produces the ***n*-butyl group,** C_4H_9- ($CH_3-CH_2-CH_2-CH_2-$). A second group, also of the formula C_4H_9- can be formed by removal of a hydrogen atom from carbon 2 or 3, the **secondary butyl group,** $CH_3-\underset{|}{CH}-CH_2-CH_3$ (*sec*-butyl group).

The isomer 2-methylpropane, commonly called isobutane (*iso*- usually means chain branching when used in a name of a compound) can produce two other alkyl groups of formula C_4H_9-. For example, removal of a hydrogen atom from any of the three carbon atoms designated by numbers 1, 3, or 4 forms the **isobutyl group,** $CH_3-\underset{|}{CH}-CH_3$ (*i*-butyl group). Removal of a hydrogen from carbon 2
$\phantom{\text{group,} CH_3-C}CH_2-$

gives the **tertiary butyl group,** $CH_3-\underset{|}{\overset{|}{C}}-CH_3$ (*t*-butyl group). Thus, a total of four
$\phantom{\text{gives the tertiary butyl group,} CH_3-}CH_3$

different groups can be derived from the alkanes of molecular formula C_4H_{10}. Although each of these groups has the formula C_4H_9 and is a "butyl" group, to indicate the differences in their structural formulas the symbols *n*-, *sec*-, *t*-, and *i*- are used for normal, secondary, tertiary, and iso, respectively.

7.5 Some Definitions

The reader will notice the use of normal, secondary, tertiary and iso- in the naming of the various alkyl groups. *Normal,* meaning straight continuous chain, and *iso-*, meaning branched chain, have already been defined. Frequently, when referring to groups or carbon atoms in a molecule, the terms *primary, secondary,* and *tertiary* are used. These terms describe in more detail the nature of the group or more specifically the type of carbon atom in the molecule from which the hydrogen atom has been removed to form the group. Every carbon atom in a molecule can be designated as being primary, secondary, or tertiary.

A primary carbon atom is one that is attached to only one other carbon atom; a secondary carbon atom is attached to two other carbon atoms, and a tertiary carbon atom is attached to three other carbon atoms in a molecule.

With reference to the alkyl groups we just discussed, the ethyl group, $CH_3—CH_2—$, can be considered a primary (1°) group since the carbon atom that has lost the hydrogen is attached to only one other carbon atom in the molecule; and the *n*–propyl group, $CH_3—CH_2—CH_2—$, is also a primary group for the same reason. However, the *i*–propyl group, $CH_3—CH—CH_3$, is a secondary (2°) group,

since the central carbon atom (from which the hydrogen atom has been removed) is attached to two other carbon atoms.

The *n*–butyl group, $CH_3—CH_2—CH_2—CH_2—$, and the *i*–butyl group,

$$CH_3—\overset{H}{\underset{CH_3}{C}}—CH_2—,$$

are both primary (1°) groups; the secondary butyl group, $CH_3—CH—CH_2—CH_3$, as the name indicates, is a secondary (2°) group; while the

tertiary butyl group, $CH_3—\overset{CH_3}{\underset{|}{C}}—CH_3$, is tertiary (3°) since the central carbon atom is

attached to three other carbon atoms in the molecule. The symbols 1°, 2°, and 3° designate primary, secondary, and tertiary carbon atoms, respectively.

The methyl group, $CH_3—$, cannot be defined under this classification. However, if one desires to, he can designate every carbon atom in a molecule as being primary, secondary, or tertiary (or 4°, quaternary to designate a carbon atom attached to four other carbon atoms).

The importance of primary, secondary, and tertiary carbon atoms and the roles they play in reaction mechanisms will be discussed in the appropriate sections in the text.

In naming the *alkyl groups* the name is derived from the parent category *alkane* simply by changing the *-ane* ending to *-yl*. (e.g., the CH_3—CH_2— group derived from eth*ane* is called eth*yl*). Only the more common alkyl groups have been discussed here. Generally speaking the letter R is used to designate any alkyl group (R = CH_3, C_2H_5, etc.); the alkanes are represented by the general formula R—H, the alkyl halides by R—X (X = halogen).

7.6 Physical Properties of the Alkanes

Since all the alkanes are non-polar molecules it is plausible to expect that the members of the series are insoluble in water and have relatively low boiling points. Under conditions of normal temperature and pressure, the first four alkanes, from methane through *n*-butane, are all gases, those of higher molecular weight are liquids up to nineteen carbon atoms, and alkanes of more than twenty carbon atoms are waxy solids (paraffins). In the case of isomers, the branched-chain alkane usually has a lower boiling point than the straight-chain isomer.

CH_3—CH_2—CH_2—CH_2—CH_3

n-pentane
(b.p. 36° C)

CH_3—CH—CH_2—CH_3
          |
         CH_3

2-methylbutane
(b.p. 28° C)

         CH_3
          |
CH_3—C—CH_3
          |
         CH_3

2,2-dimethylpropane
(neopentane),
(b.p. 10° C)

7.7 Chemical Properties of the Alkanes

The alkanes as a class of organic compounds are extremely unreactive. They do not react with most chemical reagents, such as acids and bases. They are not oxidized by strong oxidizing agents, such as permanganate ion. Alkanes do react with the halogens, and with oxygen. Since the alkanes are all members of a homologous series and have similar properties, once the reaction of an individual alkane is known, we can assume that the reaction is characteristic of all alkanes. Occasionally, however, one comes across a compound, in a series of compounds, that is an exception to the rule because of some complex feature of its molecular structure.

The reactions of alkanes with halogens and oxygen are classified as **free radical substitution reactions** (Chapter 6). The major characteristic of the reactions of alkanes is that in every case a hydrogen atom of the alkane is replaced or substituted by another atom or group of atoms from the attacking reagent. The reactions occur in the presence of sunlight, or at elevated temperatures, and proceed by a free radical mechanism.

7.7–1 Halogenation of Alkanes The **free radical halogenation reaction** has already been discussed in Section 6.8, using the chlorination of methane as an illustrative example to show the initiation, propagation, and termination steps of a free-radical mechanism. It would be well to refer to Section 6.8 and review this reaction.

Sec. 7.7 CHEMICAL PROPERTIES OF THE ALKANES

The reaction of alkanes with chlorine or bromine does not occur in the dark, and proceeds very slowly at room temperature, but in the presence of sunlight or at higher temperatures the reaction goes well. This is because the halogen molecules absorb energy under the latter conditions and split into the very reactive fragments known as free radicals.

Once again let us write the steps in the mechanism for the chlorination of methane as a typical example of the halogenation of an alkane.

(1)
$$Cl:Cl \xrightarrow{\Delta \text{ (heat), or } h\nu \text{ (sunlight)}} 2\, Cl\cdot \qquad (7\text{-}1)$$

chlorine molecule → chlorine atoms (free radicals)

(2)
$$CH_4 + Cl\cdot \longrightarrow CH_3\cdot + HCl \qquad (7\text{-}2)$$

methane → methyl free radical

(3)
$$CH_3\cdot + Cl:Cl \longrightarrow CH_3Cl + Cl\cdot \quad (\text{etc.}) \qquad (7\text{-}3)$$

methyl chloride

The chain reaction continues until all the reactants are consumed. The overall reaction can be summarized as

$$CH_4 + Cl_2 \xrightarrow[h\nu]{\Delta \text{ or}} CH_3Cl + HCl \qquad (7\text{-}4)$$

and the net result is that a chlorine atom has been substituted for a hydrogen atom in the methane molecule to form methyl chloride.

The reaction usually will not stop at this stage, but will continue to give a mixture of side products, replacing, one by one, all the hydrogen atoms on the methane to form methylene chloride, chloroform, and finally carbon tetrachloride.

$$CH_3Cl + Cl_2 \longrightarrow CH_2Cl_2 + HCl \qquad (7\text{-}5)$$
dichloromethane or methylene chloride

$$CH_2Cl_2 + Cl_2 \longrightarrow CHCl_3 + HCl \qquad (7\text{-}6)$$
trichloromethane or chloroform

$$CHCl_3 + Cl_2 \longrightarrow CCl_4 + HCl \qquad (7\text{-}7)$$
tetrachloromethane or carbon tetrachloride

The mixture of products limits the synthetic utility of the free-radical halogenation reaction. However, the reaction can be controlled so as to form methyl chloride as the major product of the reaction, with very little in the way of side-reactions, by starting the reaction sequence with a large excess of methane.

The limitation of the halogenation of alkanes as a useful reaction in the laboratory is that mixtures of products are obtained from other members of the series. For example, the reader should consider the monochlorination of propane and conclude that two products, 1-chloropropane and 2-chloropropane are formed.

In our discussion of the halogenation reaction we mentioned only chlorine and bromine; the other halogens, fluorine and iodine, were omitted. Direct halogenation with fluorine is an exothermic reaction that proceeds with explosive violence and cannot be used in the laboratory. Reaction of alkanes with iodine is of no practical value since only a negligible amount of the alkyl iodide is formed. The dissociation of iodine molecules into iodine atoms (free radicals) occurs very readily even at room temperature. However, the iodine atoms (free radicals) are comparatively unreactive and do not readily react with alkanes. The formation of alkyl fluorides and iodides will be discussed in Section 13.3.

7.7-2 Oxidation of Alkanes The *combustion of alkanes* is perhaps the most important reaction of this class of compounds, since it is the basis for the use of hydrocarbons as fuels. The reaction of alkanes with excess oxygen at high temperature produces carbon dioxide and water as products along with a large quantity of heat.

$$CH_4 + 2O_2 \longrightarrow CO_2 + 2H_2O + \text{(heat)} \qquad (7-8)$$

Natural gas, which is mainly methane, and gasoline, which is a mixture of alkanes (pentanes through dodecanes), undergo this same reaction with oxygen.

Incomplete combustion occurs when a limited supply of oxygen is used. In this case carbon monoxide or even carbon is formed rather than carbon dioxide.

$$2CH_4 + 3O_2 \longrightarrow 2CO + 4H_2O \qquad (7-9)$$

or

$$CH_4 + O_2 \longrightarrow C + 2H_2O \qquad (7-10)$$

Carbon monoxide is a deadly poisonous gas; the carbon black produced is used as an ink pigment, and as an additive to rubber in the production of automobile tires.

7.7-3 Pyrolysis or Cracking Although the alkanes are extremely stable compounds (the least reactive of any class of organic compounds), when heated to extremely high temperatures in the absence of air the molecules break apart or "crack" into smaller fragments.

$$CH_4 \xrightarrow{\Delta} C + 2H_2$$
$$(4\ H)$$

$$CH_3-CH_2-CH_3 \xrightarrow{\Delta} CH_3-CH_3 + CH_4 + CH_3-CH=CH_2 + CH_2$$
$$= CH_2 + H_2 + C + \text{(etc.)}$$

Sec. 7.8 PREPARATION OF THE ALKANES

This so-called **cracking** or **pyrolysis reaction** produces a mixture of lower molecular weight compounds. The reaction is extremely important in the petroleum industry, and is used to convert high-molecular-weight hydrocarbons into lower-molecular-weight hydrocarbons (from five to twelve carbon atoms), which can be used for gasoline. Both saturated and unsaturated hydrocarbons are produced in the cracking process.

7.8 Preparation of the Alkanes

Commercially, a mixture of alkanes can be obtained by fractional distillation of petroleum. In the laboratory, however, the general methods used for preparing or **synthesizing** alkanes include: (1) the reduction of alkyl halides and (2) the coupling of alkyl halides with organometallic compounds.

The synthesis of organic compounds in the laboratory is of primary concern to the organic chemist, the objective being to prepare as pure a product as possible, free from contaminants and side products. Thus, only reactions having few or no side reactions should be used in organic synthesis.

Reactions should be examined and learned for an entire class of compounds, rather than for just one or two compounds in the series. The easiest way to do this is to write *general equations* whenever possible, noting the class of compounds involved in the reaction. Index cards may be helpful in placing reactions in the same category on one card (i.e., Preparation of Alkanes, Reactions of Alkanes, Reactions of Alkenes, etc.).

7.8-1 Reduction of Alkyl Halides The **reduction of alkyl halides** converts alkyl halides into alkanes. Reducing agents, such as hydrogen (or zinc and hydrochloric acid, or zinc and acetic acid, which react to produce hydrogen in the reaction mixture) are used

$$CH_3-CH_2Br \xrightarrow[CH_3COOH]{Zn^{\oplus}} HBr + CH_3-CH_3 \qquad (7-11)$$

ethyl bromide   **ethane**

The alkane contains the same number of carbon atoms as the alkyl halide, the hydrogen replacing the halogen atom. The general equation is

$$R-X + [H_2] \longrightarrow R-H + HX \qquad (7-12)$$

alkyl halide   **alkane**

The reaction of a Grignard reagent (see Section 13.5) with water to produce an alkane is in effect an example of a reduction reaction of an alkyl halide.

$$RMgX + HOH \longrightarrow R-H + Mg(OH)X \qquad (7-13)$$

Grignard reagent   **alkane**

Generally, pure alkanes are produced by this reaction.

7.8-2 Coupling of Alkyl Halides with Dialkyllithium Compounds

The most versatile method of synthesizing alkanes by the coupling of two alkyl groups was developed several years ago by Professor E. J. Corey at Harvard University and Professor Herbert House at Massachusetts Institute of Technology.

An alkyl halide is reacted with lithium metal to produce an alkyl lithium compound, which is then converted by treatment with a cuprous halide to a lithium dialkylcopper compound.

$$RX + Li \longrightarrow R\text{—}Li \xrightarrow{CuX} R_2CuLi \qquad (7\text{-}14)$$

alkyl halide   alkyl lithium   lithium dialkylcopper

The coupling reaction to produce an alkane involves the reaction of another alkyl halide and the dialkylcopper compound. Both symmetrical and non-symmetrical alkanes can be synthesized by this reaction sequence.

$$R'X + R_2CuLi \longrightarrow R\text{—}R' + R\text{—}Cu + LiX \qquad (7\text{-}15)$$

alkyl halide (works best when R' is a primary group)   lithium dialkylcopper   alkane

The preparation of 2-methylhexane from isopropyl bromide and n-butyl bromide will illustrate the complete synthetic sequence.

$$CH_3\text{—}\underset{Br}{\overset{H}{C}}\text{—}CH_3 \xrightarrow{Li} CH_3\text{—}\underset{Li}{\overset{H}{C}}\text{—}CH_3 \xrightarrow{CuI}$$

i-propyl bromide   i-propyllithium

$$\left(CH_3\text{—}\overset{H}{\underset{|}{C}}\text{—}CH_3\right)_2 CuLi \xrightarrow{CH_3\text{—}CH_2\text{—}CH_2\text{—}CH_2Br}_{n\text{-butyl bromide}}$$

lithium di-i-propylcopper

$$CH_3\text{—}CH_2\text{—}CH_2\text{—}CH_2\text{—}\underset{CH_3}{\overset{H}{C}}\text{—}CH_3$$

2-methylhexane

In 1855, the French chemist Charles Adolphe Wurtz discovered a coupling reaction of alkyl halides with organometallic compounds. The **Wurtz Reaction** is between an alkyl halide and sodium metal; it usually results in the formation of a mixture of alkanes, and is limited to the synthesis of symmetrical alkanes, (R—R);

$$2\,RX + 2\,Na \longrightarrow 2\,NaX + R\text{—}R.$$

Sec. 7.9 CONFORMATIONAL ISOMERISM

7.9 Conformational Isomerism

It is an established experimental fact that under ordinary conditions *almost totally unrestricted free rotation of carbon atoms occurs in carbon–carbon single bonds.* Thus, it would seem that the ethane molecule for example, would have an infinite number of rotational orientations depending on the angle of rotation of one carbon atom with respect to the other. The various rotational arrangements of the atoms in the molecule are called **conformations** or **conformational isomers.**

The ethane molecule does have an infinite number of conformations. However, under ordinary conditions such as normal atmospheric pressure and room temperature, essentially free rotation about the carbon–carbon single bond is possible, so that one conformation can easily convert into another conformation. For this reason only one type of ethane molecule is known to us. Under conditions of extremely low temperature, rotational barriers are such that free rotation is somewhat restricted (the kinetic or thermal energy of the molecule decreases), and evidence for the existence of various conformations of ethane has been obtained.

Under ordinary conditions only two conformations of ethane are important. One conformation corresponds to the minimum amount of energy associated with the rotation about the carbon–carbon single bond. This is the most stable of the conformational isomers and is called the **staggered form.** The other important conformation is associated with the energy maximum, is the least stable conformation, and is called the **eclipsed form.**

7.9-1 Newman Projection Formulas Conformational isomers are best represented by using Newman projection formulas. Newman projection formulas can be drawn from structural formulas as shown in Figures 7.1 and 7.2. The molecule

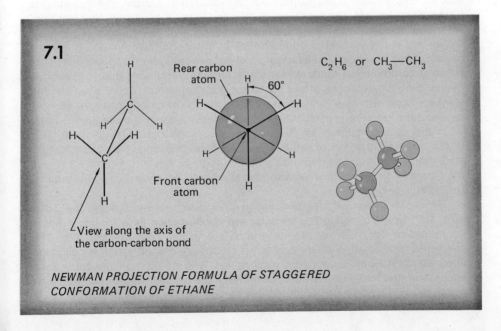

7.1

NEWMAN PROJECTION FORMULA OF STAGGERED CONFORMATION OF ETHANE

7.2 C_2H_6 or $CH_3\text{—}CH_3$

NEWMAN PROJECTION FORMULA OF ECLIPSED CONFORMATION OF ETHANE

should be viewed along the carbon–carbon bond axis looking from the front to the back of the molecule. In the Newman formula the front carbon atom is represented by the point of intersection of the three other bonds. The rear carbon, which is directly behind the front carbon, is indicated by a circle. The three bonds from the circle are for the substituents attached to the rear carbon. In the case of ethane all the substituents on both carbons are hydrogen atoms. Figure 7.1 shows that in the staggered conformation of ethane the two bonds on the two carbon atoms are separated by an angle of 60°, while Figure 7.2 shows that in the eclipsed conformation there is steric interference (repulsion) due to the close proximity on the hydrogen atoms on adjacent carbon atoms. For this reason, the staggered conformation is more stable than the eclipsed conformation. At room temperature the ethane molecule exists predominantly in the staggered form.

In the case of ethane all the staggered conformations are identical. This is not always true; for example, the *n*-butane molecule has two different staggered conformations depending on the angle of rotation. Figure 7.3 shows the Newman projection formulas for the staggered conformations of *n*-butane, $CH_3\text{—}CH_2\text{—}CH_2\text{—}CH_3$.

The conformation with the two methyl groups on opposite sides of the molecule, 180° apart, is called the **trans** or **anti conformation.** It is the most stable conformation because the methyl groups are as far apart as possible. The staggered conformation where the methyl groups are only 60° apart is called the **skew conformation.** Skew conformations are less stable than the *trans* form because of greater steric interference between the two methyl groups, which are separated by an angle of 60° compared to 180° in the *trans* conformation.

Sec. 7.10 CYCLOALKANES—THE BAEYER STRAIN THEORY

7.3

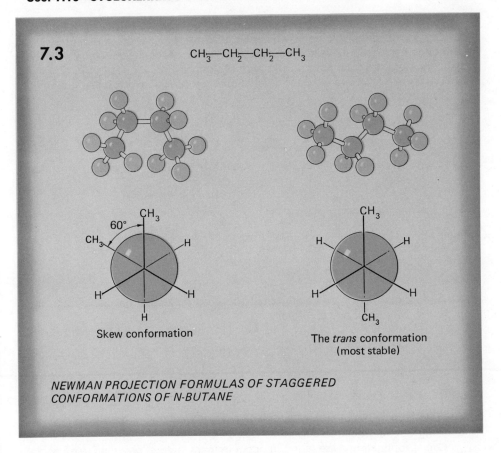

NEWMAN PROJECTION FORMULAS OF STAGGERED
CONFORMATIONS OF N-BUTANE

7.10 Cycloalkanes—The Baeyer Strain Theory

The cycloalkanes are compounds that are saturated hydrocarbons in the form of rings. The general formula for the cycloalkanes is C_nH_{2n}, or two hydrogens less than the corresponding open-chain alkane. All the cycloalkanes undergo the characteristic reactions of the alkanes except for cyclopropane and cyclobutane.

The first member of the cycloalkanes is cyclopropane, C_3H_6, a three-membered ring.

$$\underset{H_2C\underline{\qquad}CH_2}{\overset{CH_2}{\triangle}} \quad \text{or} \quad \triangle \qquad \text{cyclopropane}$$

The cyclopropane molecule is planar with C—C—C bond angles of 60°. The 60° bond angle is considerably less than the normal tetrahedral angle of 109° 28′, which singly bonded carbon atoms tend to assume. This creates a strain in the ring (Baeyer strain theory) and there is a tendency to relieve this strain by the ring opening to form the normal tetrahedral angle. Thus, in chemical reactions, cyclopropane rings will open whenever possible to alleviate the ring strain. Although cyclopropane will undergo a substitution reaction with chlorine in the presence of ultraviolet light, to produce cyclopropyl chloride with the ring still intact, bromine will open the ring by an addition reaction to form a straight-chain dibromo compound,

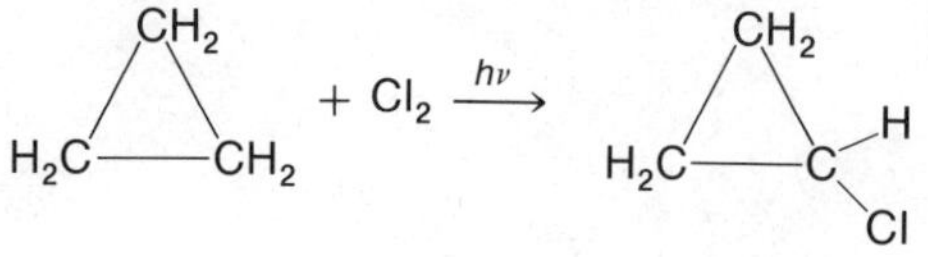

$$\triangle + Cl_2 \xrightarrow{h\nu} \text{cyclopropyl chloride} \quad (7\text{-}16)$$

$$\triangle + Br_2 \xrightarrow{h\nu} CH_2\text{-}CH_2\text{-}CH_2 \quad (7\text{-}17)$$
$$\text{(ring splits open)} \qquad \text{1,3-dibromopropane}$$
with Br on carbons 1 and 3.

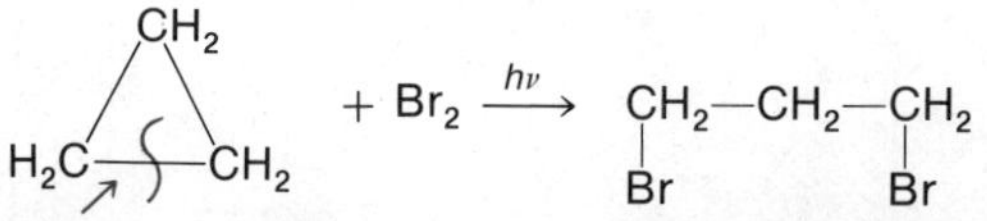

Cyclobutane, C_4H_8, is a somewhat puckered molecule, deviating from coplanarity, with bond angles of 90°.

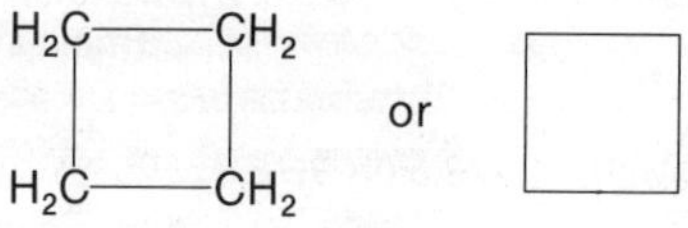

cyclobutane

The ring strain present in the cyclobutane ring is less than in the cyclopropane, and ring opening occurs, but not as readily.

Cyclopropane rings are found to occur in many natural products (e.g., some compounds obtained from chrysanthemums). Cyclobutane rings are found in pine oil, and a four-membered ring containing a nitrogen atom in place of one carbon atom occurs in penicillin.

Cyclopentane, C_5H_{10}, is a nearly planar molecule with bond angles of 108°

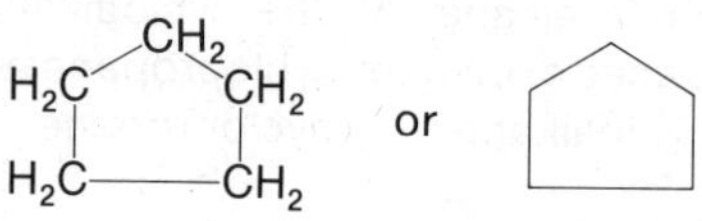

cyclopentane

and cyclohexane, C_6H_{12}, is slightly non-planar and has bond angles of 120°.

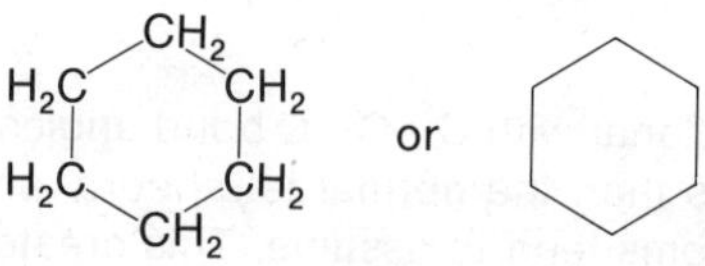

cyclohexane

Since the bond angles in cyclopentane and cyclohexane are relatively close to the

Sec. 7.10 CYCLOALKANES—THE BAEYER STRAIN THEORY

normal tetrahedral angle of 109° 28' one would expect five- and six-membered rings to be extremely stable and to occur in many compounds. This is found to be the case, five- and six-membered rings being by far the most abundant in organic compounds. There is relatively little ring strain present in cyclopentane or cyclohexane and no tendency for these rings to open. The reactions of these compounds are similar to the substitution reactions of the open-chain alkanes; for example

$$\text{cyclopentane} + Cl_2 \xrightarrow{h\nu} \text{chlorocyclopentane} + HCl \qquad (7\text{-}18)$$

7.10-1 Conformational Isomerism in Cyclohexane Completely unrestricted rotation about carbon–carbon single bonds is not possible in ring compounds. Nevertheless, conformational isomers of cyclohexane exist and are most important in the molecular structure of many natural products and biologically active substances.

In order to relieve the slight ring strain (120° bond angle vs. 109° 28') present in the cyclohexane ring, the molecule assumes a puckered shape, and forms two non-planar conformations, called the **chair** and **boat forms** because of their shape (see Figure 7.4). The **chair form** is the **staggered conformation** while the **boat form** is the **eclipsed conformation.** Figure 7.5 shows FMO models of the chair and boat conformations of cyclohexane. The interference of the two hydrogen atoms on carbons 1 and 4 in the boat form can be seen. The staggered conformation of the hydrogens in the chair form of the cyclohexane ring makes it

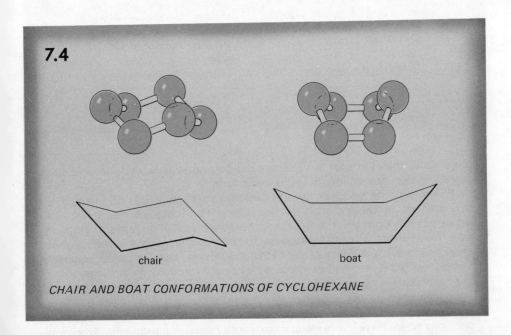

7.4

chair   boat

CHAIR AND BOAT CONFORMATIONS OF CYCLOHEXANE

7.5 C_6H_{12}

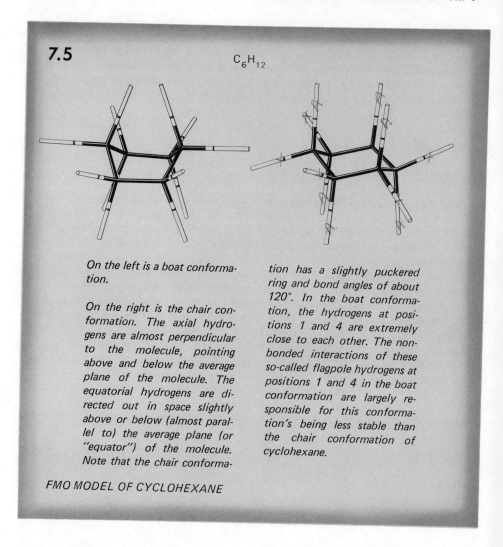

On the left is a boat conformation.

On the right is the chair conformation. The axial hydrogens are almost perpendicular to the molecule, pointing above and below the average plane of the molecule. The equatorial hydrogens are directed out in space slightly above or below (almost parallel to) the average plane (or "equator") of the molecule. Note that the chair conformation has a slightly puckered ring and bond angles of about 120°. In the boat conformation, the hydrogens at positions 1 and 4 are extremely close to each other. The nonbonded interactions of these so-called flagpole hydrogens at positions 1 and 4 in the boat conformation are largely responsible for this conformation's being less stable than the chair conformation of cyclohexane.

FMO MODEL OF CYCLOHEXANE

more stable than the boat form. Under ordinary conditions, cyclohexane molecules will exist predominantly in the chair form.

Examination of the chair form of cyclohexane (Figure 7.5) reveals that the hydrogen atoms can be divided into two categories. Six of the bonds to hydrogen atoms point straight up or down almost perpendicular to the plane of the molecule. These are called **axial hydrogens.** The other six hydrogens lie slightly above or slightly below the plane of the cyclohexane ring, and are called **equatorial hydrogens.** Molecular models indicate that axial hydrogens on the same side of the cyclohexane ring are very close to each other. The steric interference becomes important when bulkier substituents are attached in the axial positions. These substituents, because of steric requirements, prefer to be in the equatorial positions where there is less steric interference between substituents. Sometimes

SUMMARY

this is possible, since there are actually two chair conformations of cyclohexane. One form can interconvert by flipping inside out, into the other form, the result being that all the substituents that were axial in the first conformation are now equatorial in the second chair conformation, and *vice versa*. When more than one substituent is present on a cyclohexane ring, the molecule will adopt a conformation where as many bulky substituents as possible are equatorial rather than axial. The importance of the axial and equatorial positions, and of conformational isomers, can best be seen in the chemistry of natural products. Cyclohexane rings are constituent parts in many natural-product molecules that have a biological origin and some physiological function.

Cyclic compounds are known with as many as thirty or more carbon atoms in the ring. The larger ring compounds, especially the cyclic ketones, are used in the manufacture of perfumes. Although five- and six-membered rings are by far the most common, larger ring compounds occur in some natural products. Some rings containing more than six atoms are nearly strainless due to permissive puckering of the ring (Mohr–Sasche principle).

Summary

- The most fundamental class of organic compounds is the saturated hydrocarbons, called paraffins or alkanes.
- Alkanes contain only carbon and hydrogen and single covalent bonds.
    (a) The parent compound of the series is methane.
    (b) The general formula of an alkane is C_nH_{2n+2}.
    (c) Alkanes are all members of an homologous series, which differ from each other only by $-CH_2-$ (methylene) groups.
- The IUPAC system of nomenclature discussed in Chapter 3 applies to the alkanes.

    Alkyl groups are derived from alkanes by removal of one hydrogen atom. Alkyl groups can be classified as being primary, secondary, or tertiary.
- The alkanes are non-polar molecules.
    (a) They are insoluble in water and have relatively low boiling points.
    (b) Branched-chain isomers have lower boiling points than the straight-chain compounds.
- Chemical properties of the alkanes:
    (a) The alkanes are extremely unreactive.
    (b) They undergo free-radical substitution reactions, where a hydrogen atom on the alkane is replaced by another substituent: halogenation and oxidation.
- Pyrolysis or cracking breaks the molecule apart into smaller fragments.
- Alkanes can be prepared by:
    (a) Reduction of alkyl halides.
    (b) Coupling of alkyl halides with dialkyllithium compounds.
- Essentially free and unrestricted rotation of atoms with respect to each other about the carbon–carbon single bond leads to conformational isomers.
    (a) Under ordinary conditions rotation from one conformation to another occurs without difficulty, so that no conformational isomers can be separated.

(b) Low-temperature studies have produced experimental evidence for the existence of the staggered and eclipsed conformations in molecules. The staggered conformation is the most stable, while the eclipsed form is the least stable.

(c) Newman Projection formulas are valuable in discussing conformational isomers.

- The cycloalkanes are saturated hydrocarbons in the form of rings. The general formula for this class of compounds is C_nH_{2n}.

  (a) The Baeyer strain theory explains the chemistry of small three- and four-membered ring compounds.

  (b) Five- and six-membered rings are relatively free from strain and are much more stable.

  (c) Cyclohexane rings exist mainly in the boat and chair conformations, the chair being the more stable conformation. The substituents on the cyclohexane ring are in the axial or equatorial positions, the bulky substituents preferring the equatorial position.

  (d) Larger ring compounds are nearly strainless due to puckering of the ring.

PROBLEMS

1 Draw correct structural formulas for each of the following compounds.
  (a) 3-methyl-4-bromoheptane
  (b) 2,2,4-trimethyl-3-iodopentane
  (c) 1,1-dichloro-2-methylcyclopentane
  (d) methylisobutyl sec-butyl-methane
  (e) cyclohexyl bromide

2 Give the correct IUPAC name for each of the following compounds, and designate each carbon atom in the compound as being primary, secondary, or tertiary.

  (a) $CH_3-CH_2-\underset{\underset{CH_3}{|}}{\overset{\overset{H}{|}}{C}}-\underset{\underset{CH_3}{|}}{\overset{\overset{Cl}{|}}{C}}-\underset{\underset{H}{|}}{\overset{\overset{CH_3}{|}}{C}}-CH_3$

  (b) $CH_3-\underset{\underset{CH(CH_3)_2}{|}}{CH}-CH_2-CH_2-CH_3$

  (c) $CH_3-CHBr-CH_2-C(CH_3)_3$

  (d) cyclobutane with Cl substituents on adjacent corners

  (e) $CH_3-CHCl-C(Cl)_2-CH(C_2H_5)-CH_3$

3 Draw all the isomers having the formula $C_4H_8Br_2$, and name each isomer by the IUPAC naming system.

4 Arrange the following compounds in order of increasing boiling point: n-pentane, 2-methylpentane, n-hexane, 2,2-dimethylpentane, ethane.

5 Complete each of the following equations, by drawing the structural formula(s) of the product(s). If no reaction occurs, write N.R.

PROBLEMS

(a) $CH_3-CH_3 + Br_2 \xrightarrow{h\nu}$

(b) (ethane) + Na $\longrightarrow$

(c) $C_3H_8 + O_2 \xrightarrow{\Delta}$

(d) ⬡ + $Cl_2 \xrightarrow{\Delta}$

(e) $CH_3-CH_2-CH_2-CH_3 + KMnO_4 \xrightarrow{H_2SO_4}$

(f) $CH_3-\underset{H}{\overset{CH_3}{\underset{|}{\overset{|}{C}}}}-CH_2-CH_2-CH_3 + I_2 \xrightarrow{h\nu}$

(g) lithium di-i-propylcopper + ethyl bromide $\longrightarrow$

(h) $CH_3-CH_2I + Zn + HCl \longrightarrow$

(i) $CH_3-CH_2-CH_2MgBr + H_2O \longrightarrow$

(j) (2-methylpentane) + HCl $\longrightarrow$

6 Starting with isopropylbromide, write equations showing how you would prepare (a) propane, and (b) 2,3-dimethylbutane.

7 A certain compound, A, has a molecular formula of $C_6H_{13}Br$. When it is reacted with magnesium metal in anhydrous ether, a Grignard reagent is formed which upon hydrolysis with water and acid produces the compound n-hexane as a product. When compound A is reacted with sodium metal in a Wurtz reaction, the product obtained is 4,5-diethyloctane. From this information draw the correct structural formula of compound A. What is the correct name of A by the IUPAC naming system?

8 Consider the free radical bromination of ethane to form ethyl bromide (bromoethane). Write out *all* the steps for the mechanism of this reaction.

9 * Draw the correct structural formula of each of the following alkanes.

Alkane	Molecular Weight	Substitution Products
A	16	Forms 1 monochloro
B	56	Forms 1 monochloro
C	58	Forms 2 monochloro
D	72	Forms 1 monochloro
E	72	Forms 3 monochloro
F	72	Forms 4 monochloro
G	86	Forms 2 monochloro

10 * Below are listed some bond dissociation energies which are required to break the bonds into their respective atoms.

C—H	102 kcal/mole	H—Cl	102.7	Cl—Cl	57.8
C—Cl	81	H—I	71.4	I—I	36.2
C—I	53				

Write a series of equations for all the steps in the mechanism for the reaction of methane with chlorine to form methyl chloride. From the bond dissociation energy data, calculate how much energy is absorbed or liberated by this process.

Using the bond dissociation energy data, can you find a plausible explanation as to why methane does *not* react readily with iodine?

11 Draw Newman projection formulas for the important conformations of each of the following compounds. Indicate which is the most stable and which is the least stable conformation. **(a)** *n*–propyl chloride **(b)** 1,2–dichloroethane * **(c)** cyclohexane

12 Which of the following conformational isomers is the more stable? Explain.

13 Write equations for the reaction of **(a)** cyclopropane, and **(b)** cyclopentane, with bromine in the presence of sunlight. How do you explain the different types of products formed in these two reactions?

14 * Draw the structural formula of the chair form of **(a)** 1,2-dimethyl cyclohexane, and **(b)** 1,3-di-*t*-butylcyclohexane. Which is the *most stable* conformational isomer in each case? Explain why these are the most stable conformations.

8

Unsaturated Hydrocarbons—I: Alkenes

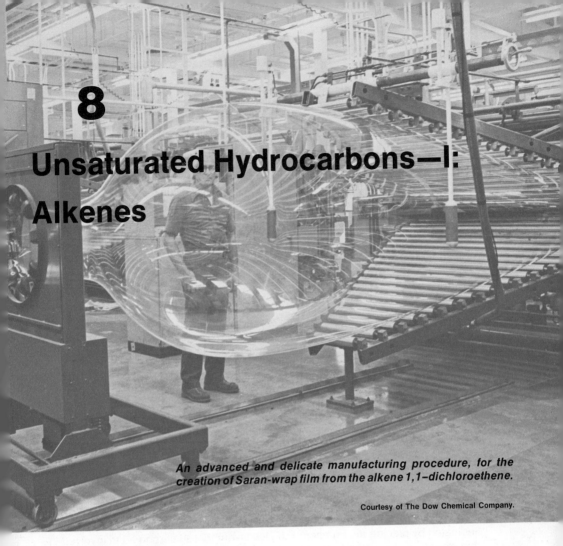

An advanced and delicate manufacturing procedure, for the creation of Saran-wrap film from the alkene 1,1-dichloroethene.

Courtesy of The Dow Chemical Company.

8.1 Introduction

The unsaturated hydrocarbons contain only the elements carbon and hydrogen and are characterized by the presence of carbon–carbon multiple bonds in the molecule. The **alkenes** or **olefins** contain a carbon–carbon double bond, while the **alkynes** or **acetylenes** have a carbon–carbon triple bond as a functional group (see Chapter 9). The unsaturated hydrocarbons, in contrast to the alkanes, are extremely reactive compounds. The alkenes and alkynes undergo reactions with halogens, hydrogen, acids, potassium permanganate, ozone, diborane, etc., and undergo polymerization reactions. Polyethylene and polypropylene plastics as well as synthetic rubber are representative products of the unsaturated hydrocarbons.

8.2 Nomenclature of Alkenes

Hydrocarbons that are alkenes have the general formula C_nH_{2n} (like the cycloalkanes) and one or more carbon–carbon double bond(s) present in the molecule.

Table 8.1 Nomenclature of Alkenes

Compound	IUPAC name	Common name
$CH_2{=}CH_2$	ethene	ethylene
$CH_3{-}CH{=}CH_2$	propene	propylene
$\overset{4}{C}H_3{-}\overset{3}{C}H_2{-}\overset{2}{C}H{=}\overset{1}{C}H_2$	1-butene (since double bond is between carbons 1 and 2)	
$\overset{1}{C}H_3{-}\overset{2}{C}H{=}\overset{3}{C}H{-}\overset{4}{C}H_3$	2-butene (since double bond is between carbons 2 and 3)	
$\overset{3}{C}H_3{-}\underset{\underset{CH_3}{\|}}{\overset{2}{C}}{=}\overset{1}{C}H_2$	methylpropene	isobutylene
$\overset{1}{C}H_3{-}\overset{2}{C}H{=}\overset{3}{C}H{-}\underset{\underset{CH_3}{\|}}{\overset{4}{C}H}{-}\overset{5}{C}H_2{-}\overset{6}{C}H_3$	4-methyl-2-hexene	
(3-chlorocyclohexene structure)	3-chlorocyclohexene	
$\overset{1}{C}H_2{=}\overset{2}{C}H{-}\overset{3}{C}H{=}\overset{4}{C}H_2$	1,3-butadiene	

The rules for the nomenclature of the alkenes are essentially those mentioned in our discussion of the IUPAC system (Chapter 3). The names are derived from the corresponding alkane, but the ending is changed from -*ane* to -*ene,* indicating the presence of a double bond. The longest continuous chain of carbon atoms containing the double bond is so numbered as to give the carbon atoms of the double bond the lowest numbers possible, and the position of the double bond is indicated by placing the number of the lower-numbered carbon atom involved in the double bond either before or after the name of the compound. Common names rather than the IUPAC names are frequently used to name some alkenes. This is particularly true of the first few members of the series. The examples in Table 8.1 illustrate these points.

8.3 The Carbon–Carbon Double Bond

The nature of the carbon–carbon double bond has been discussed in Chapter 1, and can be considered as being composed of a σ and a π bond. The delocalization of the π electrons is reponsible for the reactivity of the alkenes. The rather loosely held π electrons undergo reactions with electron-seeking (electrophilic) reagents either by ionic or free-radical addition across the double bond. These reactions will be discussed in the sections on the reactions of alkenes.

8.4 Geometric Isomerism

The carbon–carbon double bond, in contrast to the single bond, has restricted rotation. The reason for this can be easily understood if one recalls that the π

Sec. 8.4 GEOMETRIC ISOMERISM

bond (of the double bond) is formed by the overlapping of the two *p* atomic orbitals. A rotation of one carbon atom of the double bond with respect to the other carbon atom through an angle of 90° prevents the *p* orbitals from overlapping sufficiently. In other words, the rotation would break the π bond, a process requiring a large amount of energy. This does not occur under ordinary conditions; the double bond is not broken, as the energy involved with the rotational barrier is high, and rotation is therefore restricted.

As a consequence of the restricted rotation about a carbon–carbon double bond the carbon atoms and their substituents must all lie in the same plane. The molecule is flat because the atoms are coplanar. (See the FMO model of ethylene in Figure 1.3; notice that the two carbon atoms and four hydrogen atoms all lie in the same plane.)

When two different substituents are attached to each of the carbon atoms of the double bond there can be two different spatial arrangements (isomers) of the molecule. These structural isomers (stereoisomers—see Chapter 17) are a direct result of the restricted rotation and planarity of the double bond. They are called **geometric isomers** or **cis** and **trans isomers.** When similar substituents are on the same side of the double bond the isomer is called the *cis* isomer; when similar substituents are on opposite sides or corners of the double bond, it is called the *trans* isomer.

$$\underset{\text{cis}}{\overset{A}{\underset{B}{>}}C=C\overset{A}{\underset{B}{<}}} \qquad \underset{\text{trans}}{\overset{A}{\underset{B}{>}}C=C\overset{B}{\underset{A}{<}}}$$

In our earlier discussion of the nomenclature of alkenes we erroneously named a compound as 2-butene. In light of our present discussion we should be aware that there are actually two geometric isomers of 2-butene, depending on the relative positions of the H and CH_3 groups (see Figure 8.1). The correct naming of these isomers is

$$\underset{cis\text{-2-butene}}{\overset{H}{\underset{H_3C}{>}}C=C\overset{H}{\underset{CH_3}{<}}} \qquad \underset{trans\text{-2-butene}}{\overset{H}{\underset{H_3C}{>}}C=C\overset{CH_3}{\underset{H}{<}}}$$

It should be pointed out that the presence of a double bond in a molecule does not necessarily mean that geometric isomers are present. Ethylene and propylene do not have any geometrical isomers since two of the substituents (in this case hydrogen atoms) attached to one of the doubly bonded carbons are identical. Cis–trans isomerism is possible only when two different groups are attached to each carbon atom (e.g., in 2-butene).

One would expect that *trans* isomers would be more stable than the corresponding *cis* isomer of an alkene. This is generally true since the bulkier substituents are further apart in the *trans* isomer, thereby decreasing the steric hindrance in the molecule. *Cis* and *trans* isomers have different physical properties and may differ in some chemical properties.

8.1

$CH_3—CH=CH—CH_3$

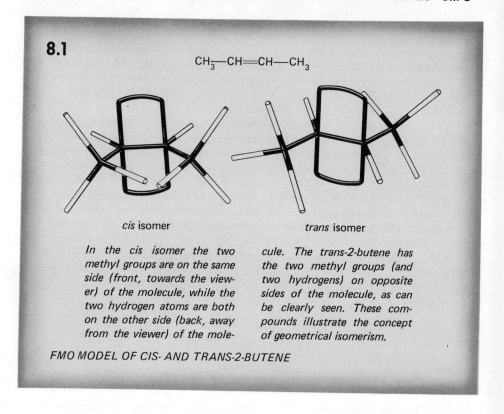

cis isomer   trans isomer

In the cis isomer the two methyl groups are on the same side (front, towards the viewer) of the molecule, while the two hydrogen atoms are both on the other side (back, away from the viewer) of the molecule. The trans-2-butene has the two methyl groups (and two hydrogens) on opposite sides of the molecule, as can be clearly seen. These compounds illustrate the concept of geometrical isomerism.

FMO MODEL OF CIS- AND TRANS-2-BUTENE

Restricted rotation about carbon–carbon single bonds occurs in many ring compounds. For example, rotation about the carbon–carbon single bond in cyclopropane, cyclobutane, or cyclohexane rings would tend to break the ring open. Thus, geometrical isomerism occurs in ring compounds where restricted rotation is present, since substituents can be either on the same or on opposite sides of the ring (as indicated by the use of heavy lines (—), and dashed lines (---)

cis-1,2-dichlorocyclobutane   trans-1,2-dichlorocyclobutane

for bonding substituents to the ring carbon atoms). Some ring compounds exhibit optical isomerism in addition to cis–trans isomerism. (Optical isomerism will be discussed in Chapter 17).

8.5 Physical Properties of the Alkenes

The physical properties of the alkenes are very similar to those of the alkanes. One interesting point is that the cis isomer of a compound has a larger dipole

Sec. 8.6 CHEMICAL PROPERTIES OF THE ALKENES

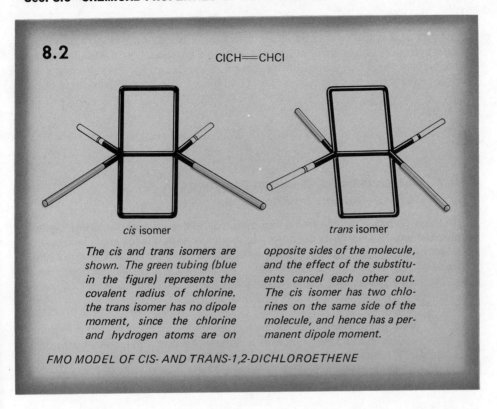

8.2

ClCH=CHCl

cis isomer — The cis and trans isomers are shown. The green tubing (blue in the figure) represents the covalent radius of chlorine. the trans isomer has no dipole moment, since the chlorine and hydrogen atoms are on

trans isomer — opposite sides of the molecule, and the effect of the substituents cancel each other out. The cis isomer has two chlorines on the same side of the molecule, and hence has a permanent dipole moment.

FMO MODEL OF CIS- AND TRANS-1,2-DICHLOROETHENE

moment than the corresponding *trans* isomer. If the *trans* isomer has the same groups on each of the carbon atoms it has no dipole moment, since the groups cancel each other (e.g., *cis*- and *trans*-1,2-dichloroethene; see Figure 8.2). *Cis* and *trans* isomers are sometimes distinguished from each other by determining the dipole moments of the compounds, the *cis* isomer having the larger dipole moment.

8.6 Chemical Properties of the Alkenes

The characteristic reaction of the alkenes is addition across the double bond, in contrast to the substitution reaction of the alkanes. This can be illustrated by the reactions of ethane and ethene with chlorine.

SUBSTITUTION $\quad CH_3-CH_3 + Cl_2 \xrightarrow{h\nu} CH_3-CH_2Cl + HCl \quad$ (8-1)

ethyl chloride

ADDITION $\quad CH_2=CH_2 + Cl_2 \longrightarrow \underset{\underset{Cl}{|}}{\overset{\overset{Cl}{|}}{CH_2-CH_2}} \quad$ (8-2)

1,2-dichloroethane

Since π bonds are energetically weaker than most σ bonds, addition of a reagent across the double bond will break the π bond of the double bond and form two new σ bonds, an energetically favorable process. Also, the equilibrium of the addition reaction favors the formation of products. For these reasons the predominant reaction of alkenes is an addition reaction across the double bond.

$$\mathrm{}\!}\mathrm{C{=}C} + XY \longrightarrow -\underset{X}{\overset{Y}{\mathrm{C{-}C}}}- \qquad (8\text{-}3)$$

8.7 Mechanisms for Addition Reactions to Carbon–Carbon Double Bonds

Addition across a double bond can occur by either an **ionic** or a **free-radical mechanism**. The reactivity of carbon–carbon double bonds towards electrophilic (electron-seeking) reagents is due principally to the polarizability of the π electrons of the double bond. Since the π electrons are not held as tightly as σ electrons by the atomic nuclei, they are more susceptible to attack by many electrophilic reagents (ionic and free radical).

8.7-1 Ionic Mechanism The ionic mechanism for addition across a carbon–carbon double bond takes place in a series of steps. First, the electrophilic reagent, $E^{\oplus}$, attacks the π cloud of the double bond from a direction perpendicular to the plane of the molecule, forming a carbonium ion. The carbonium ion

$$\overset{E^{\oplus}}{\underset{}{\searrow}}\mathrm{C}\underset{\sigma}{\overset{\pi}{=}}\mathrm{C} \xrightarrow{\text{slow}} -\overset{E}{\underset{\oplus}{\mathrm{C{-}C}}}- \qquad (8\text{-}4)$$

has only six electrons around the carbon atom, and therefore combines rapidly with some nucleophilic reagent, such as a negatively charged ion ($N{:}^{\ominus}$), which can supply an electron pair to form a stable octet. Usually, the nucleophile enters the molecule from the side of the molecule opposite to where the electrophilic reagent attacked. This mode of **trans addition** results because of steric factors.

$$-\overset{E}{\underset{\oplus}{\mathrm{C{-}C}}}- + N{:}^{\ominus} \xrightarrow{\text{fast}} -\overset{E}{\underset{N}{\mathrm{C{-}C}}}- \qquad (8\text{-}5)$$

The net result is addition of elements EN across the double bond, and can be summarized as

$$\mathrm{C{=}C} + EN \xrightarrow{\text{slow}} -\overset{E}{\underset{\oplus}{\mathrm{C{-}C}}}- \xrightarrow{N{:}^{\ominus},\ \text{fast}} -\overset{E}{\underset{N}{\mathrm{C{-}C}}}- \qquad (8\text{-}6)$$

Sec. 8.7 ADDITION REACTIONS TO DOUBLE BONDS

There is evidence strongly suggesting that the free carbonium ion pictured above is not the actual reaction intermediate in all cases of ionic addition to alkenes. In brominations (reaction of an alkene with bromine), for example, it is fairly certain that if the free carbonium is formed at all, it very rapidly cyclizes to form a cyclic bromonium ion.

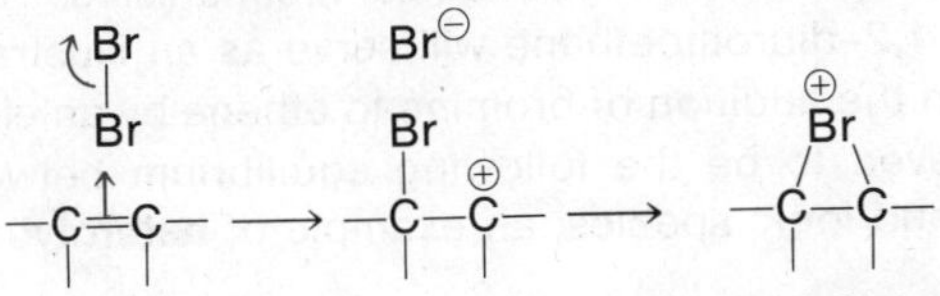

$$-\overset{|}{C}=\overset{|}{C}- \xrightarrow{Br_2} -\overset{|}{\underset{|}{C}}-\overset{|}{\underset{|}{C}}^{+}- \xrightarrow{} -\overset{|}{\underset{|}{C}}\overset{Br^+}{\underset{}{-}}\overset{|}{\underset{|}{C}}- \quad (8\text{-}7)$$

cyclic bromonium ion

Recent research investigations have provided stereochemical evidence (Chapter 17) for the formation of a cyclic carbonium ion in other electrophilic addition reactions of the alkenes.

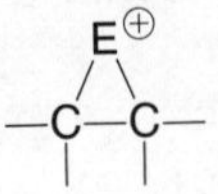

$$-\overset{|}{C}\overset{E^+}{-}\overset{|}{C}- \quad \text{cyclic carbonium ion}$$

The nucleophile can then approach the cyclic carbonium ion from the side opposite to E, resulting in *trans* addition.

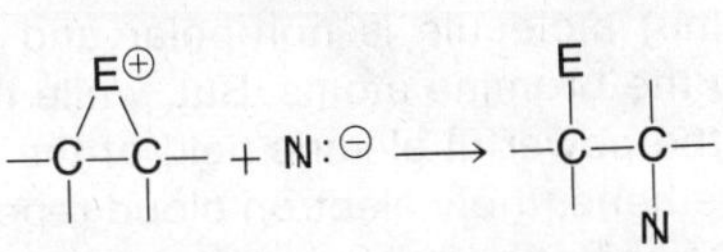

$$-\overset{|}{C}\overset{E^+}{-}\overset{|}{C}- + N\!:^{\ominus} \longrightarrow -\overset{E}{\underset{|}{C}}-\overset{|}{\underset{N}{C}}- \quad (8\text{-}8)$$

8.7-2 Free-Radical Mechanism The free-radical addition across the carbon–carbon double bond proceeds in a manner analogous to ionic addition, except that the free radical takes only *one* of the electrons from the alkene. The steps of the free-radical mechanism are basically the same as those shown in Section 6.8, namely a chain reaction involving an initiation-propagation-termination sequence.

INITIATION $\quad\quad\quad\quad\quad ROOR \xrightarrow{\Delta} 2RO\cdot \quad\quad\quad\quad (8\text{-}9)$

PROPAGATION $\quad\quad\quad\quad RO\cdot + EN \longrightarrow RON + E\cdot \quad\quad (8\text{-}10)$

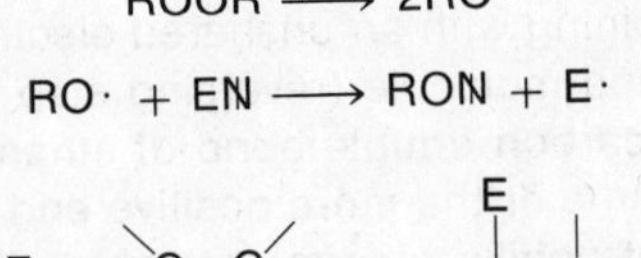

$$E\cdot + \;\;\diagup\!\!\!C\!=\!C\!\diagdown\;\; \longrightarrow -\overset{E}{\underset{|}{C}}-\overset{|}{\underset{\cdot}{C}}- \quad (8\text{-}11)$$

$$-\overset{E}{\underset{|}{C}}-\overset{|}{\underset{\cdot}{C}}- + EN \longrightarrow -\overset{E}{\underset{|}{C}}-\overset{|}{\underset{N}{C}}- + E\cdot \quad (8\text{-}12)$$

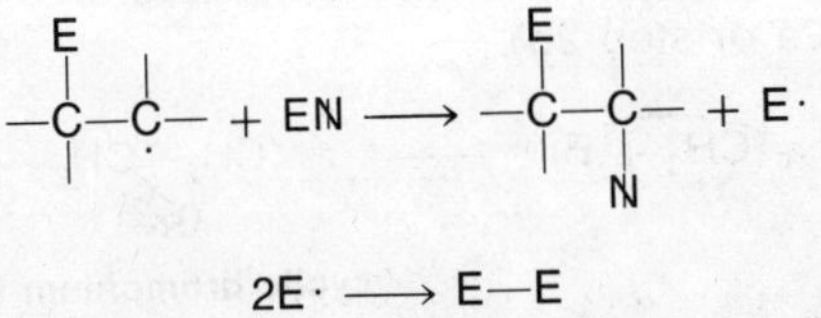

TERMINATION $\quad\quad\quad\quad 2E\cdot \longrightarrow E\text{—}E \quad\quad\quad\quad (8\text{-}13)$

The net result of the free-radical addition reaction is addition of EN across the double bond as in the ionic mechanism.

8.7–3 Electrophilic Addition of Bromine to Ethene

Let us now consider the addition reaction of a halogen across the carbon–carbon double bond with respect to both the ionic and the free-radical mechanisms. The bromination of ethene to produce 1,2-dibromoethane will serve as an illustrative example.

The initial step in the addition of bromine to ethene by an electrophilic (ionic) mechanism is believed to be the following equilibrium between the covalent bromine molecule and ionic species; an example of **heterolytic bond cleavage.**

(1) $$Br-Br \rightleftharpoons :\ddot{B}r^{\oplus} + :\ddot{B}r:^{\ominus} \qquad (8-14)$$

However, since the bromine molecule has only a slight tendency to ionize, the above equilibrium lies towards the left-hand side of the equation. It is more reasonable that the non-polar halogen molecule may become partially polarized as the halogen, X_2, approaches the double bond, a region of high electron density.

$$\begin{array}{c} \diagdown C \diagup \\ \| \quad \overset{\delta\oplus}{\longrightarrow} \overset{\delta\ominus}{Br-Br} \\ \diagup C \diagdown \end{array}$$

The halogen (bromine) molecule is non-polar and the electron density is shared equally between the bromine atoms. But, while the bromine molecule is under the influence of the powerful electric field of the π cloud of the carbon–carbon double bond in ethene, the π electron cloud repels the electron cloud of the bromine molecule, which has a similar negative charge. This repulsion distorts the electron cloud in the bromine molecule, and in effect makes the bromine atom that is nearer the double bond relatively positive and the bromine atom, further away, relatively negative. The alkene (ethene) has distorted the electron cloud or **polarized** the halogen (bromine) molecule.

The partially positively charged bromine ($Br^{\delta\oplus}$) of the polarized bromine molecule is a good electrophile. The bromonium ion $:\ddot{B}r^{\oplus}$, is a very reactive species also, since it has only six electrons in its outer shell. It is a strong electrophilic reagent desiring to complete its outer electronic shell with eight electrons, either by combining with an unshared electron pair of a bromide ion, $Br^{\ominus}$, to reform the bromine molecule (reversing step 1), or by attacking the π electrons in the carbon–carbon double bond of ethene.

The bromonium ion, $Br^{\oplus}$, or the more positive end of the polarized bromine molecule acts as an electrophilic reagent, attacking the electron-rich carbon–carbon double bond in the alkene. This results in the formation of a cyclic bromonium ion by the electrophile attaching itself to both of the doubly bonded carbon atoms (step 2a or step 2b).

(2a) $$CH_2\!=\!\!CH_2 + Br^{\oplus} \longrightarrow \begin{array}{c} CH_2-CH_2 \\ \diagdown \ \diagup \\ Br^{\oplus} \end{array} \qquad (8-15)$$

cyclic bromonium ion

Sec. 8.7 ADDITION REACTIONS TO DOUBLE BONDS

or

(2b) $\quad CH_2{=}CH_2 + \overset{\delta\oplus}{Br}-\overset{\delta\ominus}{Br} \longrightarrow \underset{\underset{Br\oplus}{|}}{CH_2-CH_2} + Br^{\ominus}$ (8-16)

Current experimental evidence excludes the idea of a free carbonium ion's being formed during the course of the reaction, and instead, suggests the participation of a cyclic bromonium ion that can be attacked by any nucleophile present to form the product.

The cyclic bromonium ion produced in step 2a or step 2b, desiring an electron pair, can react with either a bromide ion or a bromine molecule to produce the product.

(3a) $\quad \underset{\underset{\oplus}{Br}}{CH_2-CH_2} + Br^{\ominus} \longrightarrow \underset{Br}{\overset{Br}{\underset{|}{CH_2-CH_2}}}$ (8-17)

or

(3b) $\quad \underset{\underset{\oplus}{Br}}{CH_2-CH_2} + Br_2 \longrightarrow \underset{Br}{\overset{Br}{\underset{|}{CH_2-CH_2}}} + Br^{\oplus}$ (8-18)

Because of steric factors, the nucleophile (i.e., $Br^{\ominus}$, Br_2, etc.) attacks the cyclic bromonium ion from the side opposite the first bromine atom (step 3a or step 3b), and attaches itself to the carbon atom having the lowest electron density. The reaction usually results in *trans* addition across the carbon–carbon double bond. The overall reaction can be written as

$$CH_2{=}CH_2 + Br_2 \longrightarrow \underset{Br}{\overset{Br}{\underset{|}{CH_2-CH_2}}} \quad (8\text{-}19)$$

**1,2-dibromoethane
(ethylene dibromide)**

There is a great deal of experimental evidence that in aqueous solution the halogenation of alkenes does indeed occur by an ionic mechanism, as shown in the bromination of ethene. One of the classical experiments that gave evidence in support of this mechanism was the bromination of ethylene in an aqueous solution containing sodium chloride. The products obtained from the reaction mixture included 1,2-dibromoethane, 1-bromo-2-chloroethane, and ethylene bromohydrin. But, no 1,2-dichloroethane could be found among the products.

This can be explained most simply and logically by an ionic mechanism leading to the formation of a cyclic bromonium ion, which can react with a nucleophilic reagent (i.e., $Br^{\ominus}$, $Cl^{\ominus}$, H_2O, etc.) to complete its outer electronic shell. Reaction of the intermediate bromonium ion with bromide ion, chloride ion, and water produces the observed products.

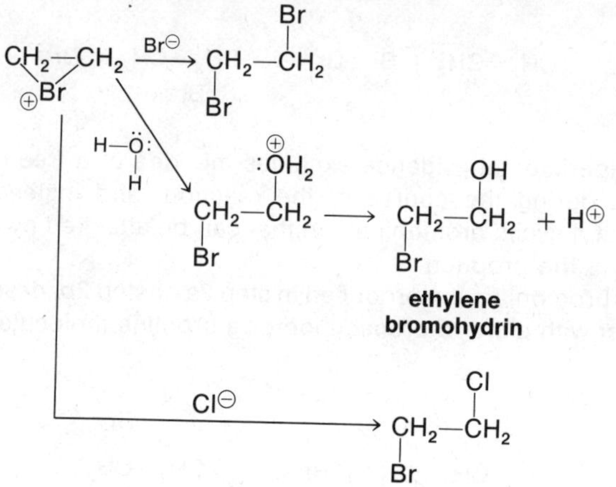

$$\underset{\oplus Br}{CH_2-CH_2} \xrightarrow{Br^\ominus} \underset{Br}{CH_2}-\overset{Br}{CH_2} \qquad (8-20)$$

$$\underset{H}{H-\overset{..}{O}:} \searrow \underset{Br}{CH_2}-\overset{\oplus\!\!:OH_2}{CH_2} \longrightarrow \underset{Br}{CH_2}-\overset{OH}{CH_2} + H^\oplus \qquad (8-21)$$

ethylene bromohydrin

$$Cl^\ominus \longrightarrow \underset{Br}{CH_2}-\overset{Cl}{CH_2} \qquad (8-22)$$

The possibility that the 1-bromo-2-chloroethane can be formed by reaction of the 1,2-dibromoethane with chloride ion can be eliminated since no displacement of Br by Cl occurs when this reaction is run.

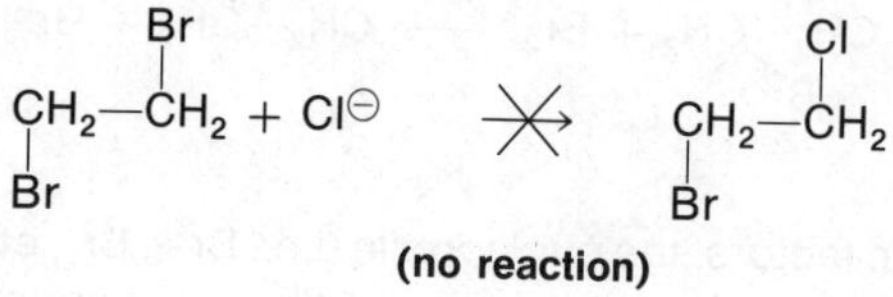

$$\underset{Br}{CH_2}-\overset{Br}{CH_2} + Cl^\ominus \quad \cancel{\longrightarrow} \quad \underset{Br}{CH_2}-\overset{Cl}{CH_2} \qquad (8-23)$$

(no reaction)

The fact that no 1,2-dichloroethane is formed as a product is most significant. If the proposed mechanism is correct a chloronium ion, $Cl^\oplus$, would have to be formed in the course of the reaction to lead to the production of this product. This is impossible since the sodium chloride present furnishes only chloride, $Cl^\ominus$, ions, and no $Cl^\oplus$ ions.

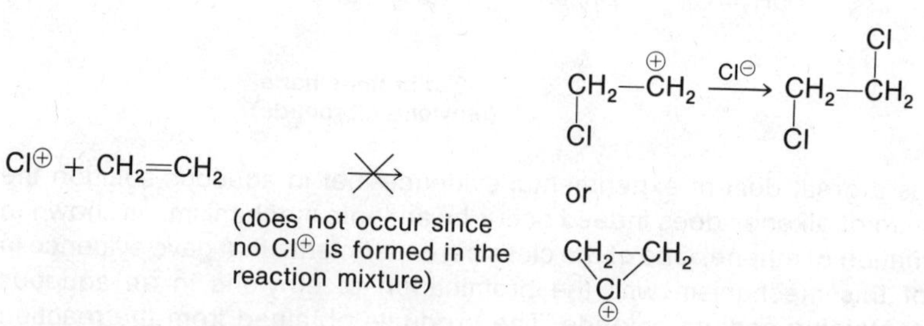

$$Cl^\oplus + CH_2{=}CH_2 \quad \cancel{\longrightarrow}$$
(does not occur since no $Cl^\oplus$ is formed in the reaction mixture)

$$\underset{Cl}{CH_2}-\overset{\oplus}{CH_2} \xrightarrow{Cl^\ominus} \underset{Cl}{CH_2}-\overset{Cl}{CH_2}$$

or $\qquad (8-24)$

$$CH_2-CH_2 \atop \underset{\oplus}{\diagdown Cl \diagup}$$

It has also been shown that no exchange of Cl for Br takes place in the reaction of 1,2-dibromoethane with $Cl^\ominus$ ions, and so our picture of the steps of the ionic mechanism is essentially correct.

It might be mentioned that the sodium ion, $Na^\oplus$, will not attack the π cloud of the carbon-carbon double bond. Although the sodium ion is positively charged, it has a complete valence shell of electrons ($Na^\oplus$; electronic structure is 2,8) and

Sec. 8.8 IONIC ADDITION REACTIONS

therefore is a weak electrophilic reagent. In contrast, the hydrogen ion, $H^\oplus$, which desires an electron pair to complete its valence shell, will attack the π cloud of a carbon–carbon double bond. This will be seen in our discussion of addition of halogen acids to alkenes.

It should be noted that although some electrophilic addition reactions to alkenes proceed through the formation of a cyclic bridged ion (similar to the cyclic bromonium ion), ionic addition reactions do occur that involve free, classical carbonium ions' being formed during the course of the reaction.

The free-radical bromination of ethene proceeds by the steps outlined in free-radical chain reactions. The first step is an example of **homolytic bond cleavage.**

INITIATION $\qquad Br\!-\!Br \xrightleftharpoons{h\nu \text{ or } \Delta} 2:\!\ddot{Br}\cdot$ (8-25)

PROPAGATION $\qquad CH_2\!=\!CH_2 + :\!\ddot{Br}\cdot \longrightarrow CH_2\!-\!CH_2\cdot$ (8-26)
$\qquad\qquad\qquad\qquad\qquad\qquad\qquad\qquad\;\;\;|$
$\qquad\qquad\qquad\qquad\qquad\qquad\qquad\qquad\;\;Br$

$\qquad CH_2\!-\!CH_2\cdot + Br_2 \longrightarrow CH_2\!-\!CH_2 + :\!\ddot{Br}\cdot$ (8-27)
$\;\;|\qquad\qquad\qquad\qquad\qquad\;\;|\;\;\;\;\;|$
$\;Br\qquad\qquad\qquad\qquad\qquad Br\;\;Br$

TERMINATION $\qquad CH_2\!-\!CH_2\cdot + :\!\ddot{Br}\cdot \longrightarrow CH_2\!-\!CH_2$ (8-28)
$\qquad\qquad\qquad\qquad\;|\qquad\qquad\qquad\qquad\;|\;\;\;\;\;|$
$\qquad\qquad\qquad\qquad Br\qquad\qquad\qquad\;\;Br\;\;Br$

8.8 Ionic Addition Reactions

Many substances will undergo electrophilic addition across the double bond of an alkene by the ionic mechanism. Included in this category are such reagents as the halogens, hydrogen halides, hypohalous acids, and water.

8.8-1 Halogenation The bromination of ethylene has already been discussed as an example of the halogenation reaction. Chlorine adds across the carbon–carbon double bond in the same manner as bromine.

$$CH_3\!-\!CH\!=\!CH_2 + Cl_2 \longrightarrow CH_3\!-\!\underset{\underset{Cl}{|}}{CH}\!-\!CH_2Cl \qquad (8\text{-}29)$$

1,2-dichloropropane

The reaction of alkenes with fluorine is too vigorous to be of practical use, and iodine does not react readily with alkenes. The product produced by the addition of a halogen across a carbon–carbon double bond results in two halogen atoms becoming attached to two adjacent carbon atoms. Since the halogens are in the same "vicinity," the product is referred to as a **vicinal dihalide.**

$$\diagup\!\!\!\mathrm{C}\!\!=\!\!\mathrm{C}\diagdown + X_2 \longrightarrow -\underset{X}{\overset{X}{\mathrm{C}}}-\underset{|}{\mathrm{C}}- \qquad (8\text{-}30)$$

alkene + halogen   vicinal dihalide
(Br$_2$ or Cl$_2$)

The reaction of alkenes with bromine is useful in testing a compound for unsaturation. Bromine, a red-brown liquid, will react with alkenes, which are colorless, to produce a colorless dibromide addition product. Bromine in carbon tetrachloride (as a solvent) is an excellent reagent for detecting unsaturation in organic compounds. The solution will change from red-brown to colorless when added to an alkene, indicating the presence of a carbon–carbon double bond.

8.8-2 Hydrohalogenation (Ionic Mechanism)—Markovnikov's Rule The addition of halogen acids (hydrohalogenation) to an alkene occurs quite readily. The proton (H$^{\oplus}$) is a good electrophilic reagent that attacks the π electrons of the double bond. The steps in the mechanism for the hydrohalogenation reaction are

$$\diagup\!\!\!\mathrm{C}\!\!=\!\!\mathrm{C}\diagdown + \overset{\oplus}{\mathrm{H}}X^{\ominus} \longrightarrow -\underset{H}{\mathrm{C}}-\overset{\oplus}{\mathrm{C}}- \xrightarrow{X^{\ominus}} -\underset{H}{\mathrm{C}}-\overset{X}{\mathrm{C}}- \qquad (8\text{-}31)$$

alkene   hydrogen   alkyl halide
halide

Of the halogen acids, hydroiodic acid adds most readily to double bonds, and hydrochloric acid is the least reactive. This is in line with the relative acid strength discussed in Section 4.6 (HI > HBr > HCl). HF does not add readily to alkenes. Why?

The addition of a hydrogen halide to a symmetrical olefin, such as ethylene, is relatively simple, and forms a single product.

$$CH_2\!\!=\!\!CH_2 + HBr \longrightarrow CH_3\!-\!CH_2\!-\!Br \qquad (8\text{-}32)$$
ethyl bromide

However, addition of an unsymmetrical reagent to an unsymmetrical olefin introduces the possibility of two alternatives. Let us consider the reaction of propylene with hydrogen bromide.

$$CH_3\!-\!CH\!\!=\!\!CH_2 + HBr \longrightarrow CH_3\!-\!CH_2\!-\!CH_2Br \quad (\text{or} \quad CH_3\!-\!\underset{Br}{CH}\!-\!CH_3)$$

n-propyl bromide   i-propyl bromide

$$(8\text{-}33)$$

It is found experimentally that *i*-propyl bromide is the major product of the reaction.

Sec. 8.8 IONIC ADDITION REACTIONS

The Russian chemist Vladimir Markovnikov studied many addition reactions and proposed the following rule to account for the orientation in addition reactions of alkenes: *When an unsymmetrical reagent adds to an unsymmetrical alkene, the negative portion of the reagent becomes attached to the double bonded carbon having the least number of hydrogens.* The addition of HBr to propylene, forming *i*-propyl bromide, is an illustration of **Markovnikov's rule.**

The explanation of Markovnikov's rule in terms of modern organic chemistry is based on the theory that, in electrophilic addition reactions to unsymmetrical alkenes, the reaction proceeds so that the most stable carbonium ion is formed during the course of the reaction.

Let us once again examine the reaction of propylene with HBr in terms of the accepted ionic mechanism. The ionization of HBr produces protons (HBr $\longrightarrow$ H$^{\oplus}$ + Br$^{\ominus}$). The proton (electrophilic reagent) can react with the propylene in one of two ways, producing two possible carbonium ions.

$$CH_3-CH=CH_2 + H^{\oplus} \longrightarrow$$

$$CH_3-CH_2-CH_2^{\oplus} \quad \text{or} \quad CH_3-\overset{\oplus}{C}H-CH_3 \qquad (8\text{-}34)$$

n-propyl carbonium ion *i*-propyl carbonium ion

The nucleophile, Br$^{\ominus}$, attacks the carbon atom bearing the positive charge to form the product.

$$CH_3-\overset{\oplus}{C}H-CH_3 + Br^{\ominus} \longrightarrow CH_3-\underset{Br}{CH}-CH_3 \qquad (8\text{-}35)$$

i-propyl carbonium ion

Since *i*-propyl bromide is the predominant product, rather than *n*-propyl bromide, the interpretation is that the *i*-propyl carbonium ion is a more stable species than the *n*-propyl carbonium ion.

Investigation of numerous organic reactions which proceed by ionic mechanisms has led chemists to believe that a more substituted carbonium ion has greater stability than a less substituted carbonium ion. The stability of carbonium ions decreases as one goes from tertiary to secondary to primary.

STABILITY $\qquad R_3C^{\oplus} > R_2CH^{\oplus} > RCH_2^{\oplus}$
$\qquad\qquad\qquad\qquad\quad 3° \qquad\quad 2° \qquad\quad 1°$

If in the course of a reaction more than one carbonium ion can be formed, the most stable species will be formed (3° > 2° > 1°) leading to the formation of products (i.e., $CH_3-\overset{\oplus}{C}H-CH_3$ rather than $CH_3-CH_2-CH_2^{\oplus}$).

The great utility of Markovnikov's rule is that it allows one to predict the products of addition reactions. One need only look for the most stable carbonium ion that can be formed in the course of reaction. The reason for the greater stability of the more highly substituted carbonium ion is that the alkyl groups present have an electron-releasing inductive effect and tend to disperse the positive charge of the ion, producing a more stable species. The presence of electron-withdrawing substituents would produce a less stable carbonium ion, and addition would appear to go anti-Markovnikov. This is illustrated in the reaction of HBr with 3,3,3-trifluoropropene (F is an electron-withdrawing group).

$$F_3C-CH=CH_2 + H^\oplus \longrightarrow F_3\overset{\delta\ominus\delta\oplus}{C}-CH_2-\overset{\oplus}{C}H_2 \text{ (more stable than } F_3\overset{\delta\ominus\delta\oplus}{C}-\overset{\oplus}{C}H-CH_3)$$
3,3,3-trifluoropropene
$$\downarrow Br^\ominus$$
$$F_3C-CH_2-CH_2Br \tag{8-36}$$

8.8-3 Halohydrin Formation The addition of hypohalous acids such as HOCl, HOBr, and HOI to an alkene produces a halohydrin. The addition proceeds by an ionic mechanism, the positive halogen attacking the π cloud of the double bond, followed by the carbonium ion reacting with the solvent, water. Loss of a proton produces the halohydrin.

$$\text{>C=C<} + X^\oplus \longrightarrow -\underset{X}{\overset{|}{C}}-\overset{\oplus}{\underset{|}{C}}- \xrightarrow{H_2\ddot{O}:} -\underset{X}{\overset{|}{C}}-\underset{|}{\overset{:\overset{\oplus}{O}H_2}{C}}- \xrightarrow{-H^\oplus} -\underset{X}{\overset{|}{C}}-\underset{|}{\overset{OH}{C}}- \tag{8-37}$$

The reaction can be represented by the general equation

$$\underset{\text{alkene}}{\text{>C=C<}} + \underset{\substack{\text{hypohalous} \\ \text{acid}}}{H\overset{\ominus}{O}X^\oplus} \longrightarrow \underset{\text{halohydrin}}{-\underset{X}{\overset{|}{C}}-\underset{|}{\overset{OH}{C}}-} \tag{8-38}$$

A specific example is

$$CH_3-CH=CH_2 + HOCl \longrightarrow \underset{\substack{\text{propylene} \\ \text{chlorohydrin}}}{CH_3-\underset{OH}{\overset{Cl}{\underset{|}{CH}}}-\overset{|}{CH_2}} \tag{8-39}$$

Note that the addition proceeds in accordance with Markovnikov's rule.

8.8-4 Hydration The addition of water across a double bond is called hydration, the resulting product being an alcohol. Since water is ionized very slightly there are not enough protons available to attack the π cloud of the double bond. For this reason an acid catalyst, such as sulfuric acid, must be used to furnish protons

Sec. 8.9 FREE-RADICAL ADDITION OF HBr

to form the intermediate carbonium ion. (The $HSO_4^{\ominus}$ ion of the H_2SO_4 is a weak nucleophile, and will not compete with the water for the carbonium ion.)

$$\text{\Large$\rangle$}C=C\text{\Large$\langle$} + H^{\oplus} \longrightarrow -\underset{H}{\overset{|}{C}}-\overset{\oplus}{\underset{|}{C}}- \xrightarrow[-H^{\oplus}]{H_2O} -\underset{H}{\overset{|}{C}}-\overset{OH}{\underset{|}{C}}- \quad (8\text{--}40)$$

alkene   alcohol

Actually, an alkyl hydrogen sulfate formed by the sulfuric acid adding across the double bond can also be isolated as a side product. This reaction may be understood more easily if we write the formula of sulfuric acid as $H^{\oplus}$—$\overset{\ominus}{O}SO_3H$, rather than H_2SO_4, and water as $H^{\oplus}$—$\overset{\ominus}{O}H$. The addition will then be seen to proceed in accordance with Markovnikov's rule.

$$CH_3-CH=CH_2 + H\overset{\oplus}{O}SO_3H^{\ominus} \longrightarrow CH_3-\underset{OSO_3H}{\overset{|}{CH}}-CH_3 \quad (8\text{--}41)$$

isopropyl hydrogen sulfate

$$CH_3-CH=CH_2 + \overset{\oplus}{H}-\overset{\ominus}{O}H \xrightarrow{H_2SO_4} CH_3-\underset{OH}{\overset{|}{CH}}-CH_3 \quad (8\text{--}42)$$

2-propanol
(i-propyl alcohol)

8.9 Free-Radical Addition of HBr

The ionic reaction of HBr with propylene will result in the formation of *i*-propyl bromide as the major product.

$$CH_3-CH=CH_2 + HBr \longrightarrow CH_3-\underset{Br}{\overset{|}{CH}}-CH_3 \quad (8\text{--}43)$$

i-propyl bromide

However, the addition of HBr in the presence of a peroxide catalyst proceeds by a free-radical mechanism, the result being the abnormal anti-Markovnikov's addition product.

$$CH_3-CH=CH_2 + HBr \xrightarrow{H_2O_2} CH_3-CH_2-CH_2Br \quad (8\text{--}44)$$

n-propyl bromide

The position of the H and Br in the product are contrary to Markovnikov's rule. The hydrogen peroxide serves as a free radical initiator in the reaction.

INITIATION
$$HO-OH \xrightarrow{h\nu} 2HO\cdot \quad (8\text{--}45)$$

$$HO\cdot + HBr \longrightarrow H_2O + Br\cdot \quad (8\text{--}46)$$

The bromine free radical then attacks the π cloud of the double bond in propylene. Two new free radicals can be formed; however, the more substituted free radical (as in the case of carbonium ions) is the more stable (3° > 2° > 1°) and is formed more readily.

PROPAGATION
$$CH_3-CH=CH_2 + Br\cdot \longrightarrow CH_3-\dot{C}H-CH_2Br \quad (8\text{-}47)$$
$$\begin{pmatrix} \text{more stable than} \\ CH_3-CH-CH_2\cdot \\ | \\ Br \end{pmatrix}$$

$$CH_3-\dot{C}H-CH_2Br + HBr \longrightarrow CH_3-CH_2-CH_2Br + Br\cdot \quad (8\text{-}48)$$

The n-propyl bromide is the final product formed. The chain reaction will continue as long as the Br· free radical is regenerated in the propagation steps, and stop when all the reactants are consumed, or by some other termination step such as combining of free radicals.

TERMINATION
$$Br\cdot + Br\cdot \longrightarrow Br_2 \quad (8\text{-}49)$$

$$2CH_3-\dot{C}H-CH_2Br \longrightarrow \begin{matrix} CH_3-CH-CH_2Br \\ | \\ CH_3-CH-CH_2Br \end{matrix} \quad (8\text{-}50)$$

This "abnormal" addition reaction works only with HBr in the presence of a free radical initiator, and not with HCl or HI. The synthetic utility of this reaction, in contrast to the ionic addition, is that one can place a substituent in different positions in a molecule by varying the reaction conditions. For example

$$R-CH=CH_2 \xrightarrow{HBr} R-\underset{Br}{CH}-CH_3 \quad (8\text{-}51)$$
$$\downarrow HBr, H_2O_2$$
$$R-CH_2-CH_2Br \quad (8\text{-}52)$$

8.10 Cycloaddition Reactions

A third type of mechanism of addition to carbon–carbon double bonds is the so-called **cycloaddition mechanism**. The cycloaddition mechanism occurs less frequently than ionic or free-radical addition, and, as can be seen from its name, implies the intermediate formation of a cyclic transition state.

$$\text{C=C} + XY \longrightarrow \begin{bmatrix} -C\!=\!C- \\ || \\ X\text{---}Y \end{bmatrix} \longrightarrow \begin{matrix} -C-C- \\ | | \\ X Y \end{matrix} \quad (8\text{-}53)$$

(--- represents the breaking and forming of bonds)

As in the other addition mechanisms, the π bond of the double bond is broken and two new σ bonds are formed in the product. The mechanism presented here appears to be a simple one-step four-centered addition reaction; actually the

Sec. 8.10 CYCLOADDITION REACTIONS

reaction is usually of a more complex nature, but this oversimplification will serve our purposes. A most important point to consider is that reactions occurring by a cycloaddition mechanism will result in *cis* **addition** of the reagent XY in the product, since, in the cyclic transition state formed, all the substituents must enter from the same side of the molecule. This is in contrast to the *trans* **addition** observed in most ionic addition reactions.

Catalytic hydrogenation, hydroboration, addition of carbenes, and the Diels-Alder reaction are typical *cis* cycloaddition reactions of the alkenes.

8.10-1 Catalytic Hydrogenation The addition of hydrogen to an alkene will occur only in the presence of a catalyst, under pressure. Finely divided metals, such as platinum, palladium, or nickel, adsorbed on charcoal, are the most effective catalysts. The function of the catalyst in the reaction is not completely understood. It is known that the presence of a catalyst lowers the activation energy required by the reaction, but the exact mechanism of the hydrogenation reaction is not understood.

Most chemists believe that the hydrogen is adsorbed on the metal catalyst surface in the form of hydrogen atoms.

$$H_2 + \underset{\text{metal catalyst}}{\sim\!\sim\!\sim\!\sim} \longrightarrow \overset{H\ H}{\underset{\sim\!\sim\!\sim\!\sim}{|\ \ |}} \tag{8-54}$$

The alkene is also adsorbed on the surface, then addition of the hydrogen atoms to the double bond occurs (*cis* addition), followed by desorption of the product from the surface of the catalyst.

$$\overset{}{\underset{}{>}}C=C\overset{}{\underset{}{<}} + \overset{H\ H}{\underset{\sim\!\sim\!\sim}{|\ \ |}} \longrightarrow \overset{-C=C-}{\underset{\sim\!\sim\!\sim}{\overset{H\ H}{|\ \ |}}} \longrightarrow \underset{\text{metal catalyst}}{\sim\!\sim\!\sim} + \underset{\text{alkane}}{-\overset{H\ H}{\underset{|\ \ |}{C-C}}-} \tag{8-55}$$

The product of the hydrogenation of an alkene is an alkane.

$$\underset{\text{ethene}}{CH_2\!=\!CH_2} + H_2 \xrightarrow{Pt} \underset{\text{ethane}}{CH_3\!-\!CH_3} \tag{8-56}$$

8.10-2 Hydroboration The addition of diborane (B_2H_6) in the presence of a suitable solvent (tetrahydrofuran or diglyme) to a carbon–carbon double bond is called **hydroboration**. The diborane adds as if it were the monomer, BH_3, resulting in a hydrogen atom and a boron atom adding across the double bond. The hydroboration reaction was first developed by Herbert C. Brown, now at Purdue University.

The B in the BH_3 behaves as an electrophilic reagent and the hydrogen portion of the molecule (in the form of a hydride) as a nucleophilic reagent. The addition is believed to proceed by the cycloaddition mechanism and follows Markovnikov's rule. The net result is addition of a hydride ion to the carbon atom of the double bond having the least number of hydrogens, and a BH_2 fragment to the other carbon atom of the double bond.

$$\diagup C=C\diagdown + BH_3 \xrightarrow[\text{or THF}]{\text{diglyme}} \left[\begin{array}{c} -\overset{|}{\underset{|}{C}}=\overset{|}{\underset{|}{C}}- \\ H\cdots BH_2 \end{array} \right] \longrightarrow -\overset{|}{\underset{H}{C}}-\overset{|}{\underset{BH_2}{C}}- \qquad (8\text{-}57)$$

The reaction proceeds until all the B-H bonds have reacted, resulting in the formation of a **trialkylborane**.

$$-\overset{|}{\underset{H}{C}}-\overset{|}{\underset{BH_2}{C}}- + 2 \diagup C=C\diagdown \longrightarrow \left(-\overset{|}{\underset{H}{C}}-\overset{|}{\underset{|}{C}}- \right)_3 B \qquad (8\text{-}58)$$
trialkylborane

The vast importance of the hydroboration reaction is due to the fact that the trialkylboranes can be decomposed by treatment with hydrogen peroxide in the presence of base to form alcohols. Boric acid is the other oxidation product.

$$\left(-\overset{|}{\underset{H}{C}}-\overset{|}{\underset{|}{C}}- \right)_3 B + H_2O_2 \xrightarrow{OH^\ominus} -\overset{|}{\underset{H}{C}}-\overset{|}{\underset{OH}{C}}- + H_3BO_3 \qquad (8\text{-}59)$$
alcohol

The hydroboration reaction of alkenes leads to the formation of an alcohol, but more important, the *cis* addition of diborane, followed by peroxide oxidation, produces the alcohol which would be formed by the anti-Markovnikov addition of water to an alkene. Contrast the reaction of propylene with water in the presence of sulfuric acid as a catalyst, and the hydroboration reaction.

$$CH_3-CH=CH_2 + H_2O \xrightarrow{H^\oplus} CH_3-\underset{\underset{OH}{|}}{CH}-CH_3 \qquad (8\text{-}60)$$
i-propyl alcohol (by normal Markovnikov addition)

$$CH_3-CH=CH_2 + B_2H_6 \xrightarrow{THF} \left(CH_3-\overset{|}{\underset{H}{C}}-\overset{|}{\underset{|}{C}}- \right)_3 B$$
$$\downarrow H_2O_2, NaOH$$
$$CH_3-CH_2-CH_2OH \qquad (8\text{-}61)$$
n-propyl alcohol

Sec. 8.10 CYCLOADDITION REACTIONS

The overall reaction (hydroboration followed by oxidation) can be considered as anti-Markovnikov addition of water to a carbon–carbon double bond. The addition of diborane does actually follow Markovnikov's rule; it is the oxidation step that produces the abnormal result.

8.10-3 Carbene Addition Carbenes are extremely reactive species formed as reaction intermediates. The simplest carbene is methylene, $:CH_2$, formed by the decomposition of diazomethane or ketene on exposure to ultraviolet light.

$$CH_2N_2 \xrightarrow{h\nu} N_2 + :CH_2 \qquad (8\text{-}62)$$
diazomethane   methylene

$$CH_2{=}C{=}O \xrightarrow{h\nu} CO + :CH_2 \qquad (8\text{-}63)$$
ketene   methylene

Methylene contains a carbon atom with only six valence electrons, as do all carbenes. Since carbon atoms want to complete their valence shells with eight electrons, carbenes are very reactive towards nucleophilic species.

Another example of a carbene is dichlorocarbene, $:CCl_2$, formed by reacting chloroform with a strong base such as potassium t-butoxide.

$$CHCl_3 + (CH_3)_3{-}C{-}O^\ominus \rightleftharpoons (CH_3)_3{-}COH + :CCl_3^\ominus \qquad (8\text{-}64)$$
chloroform t-butoxide

$$:CCl_3^\ominus \longrightarrow :CCl_2 + Cl^\ominus \qquad (8\text{-}65)$$
dichlorocarbene

Carbenes add to the double bond in alkenes by cycloaddition to produce cyclopropane derivatives. The insertion of the carbene to form a three-membered ring is of great synthetic value, since it is difficult to introduce a cyclopropane ring in molecules by other synthetic methods.

Some typical examples of carbene addition reactions are

$$CH_3{-}CH{=}CH_2 + CH_2N_2 \xrightarrow{h\nu} CH_3{-}\underset{\underset{CH_2}{|}}{CH}{-}CH_2 \qquad (8\text{-}66)$$
    (or $CH_2{=}C{=}O$)
propene   methylcyclopropane
(addition of $:CH_2$)

$$\bigcirc\!\!\!| + CHCl_3 \xrightarrow{(CH_3)_3CO^\ominus K^\oplus} \text{[bicyclic dichlorocyclopropane]} \qquad (8\text{-}67)$$
(addition of $:CCl_2$)

8.10-4 The Diels-Alder Reaction The Diels-Alder reaction, named after its discoverers, was the first example of a cycloaddition reaction to be reported. The reaction involves the addition of a cyclic diene like 1,3–butadiene, to an alkene such as maleic anhydride.

112 UNSATURATED HYDROCARBONS—I: ALKENES Ch. 8

$$\text{1,3-butadiene} + \text{maleic anhydride} \longrightarrow \text{a Diels-Alder adduct} \qquad (8\text{-}68)$$

The product has a new six-membered ring formed by the *cis* addition, and the reaction is energetically very favorable due to the relative stability of six-membered rings. In fact, the reaction conditions for Diels–Alder reactions are often so mild that many of them can be run at room temperature. The mechanism is believed to be more complex than the simple cycloaddition picture we present here.

$$ \qquad (8\text{-}69)$$

The Diels–Alder reaction is particularly useful in synthesizing six-membered rings with a *cis* ring fusion. This becomes extremely important when considering the synthesis of many biologically active natural products such as steroids.

8.11 Polymerization of Alkenes

The addition of an alkyl group to an alkene is known as alkylation. Many alkenes combine with themselves (self-alkylation) and form a giant molecule, consisting of many repeating units of the small alkene molecule. This process is known by the more familiar term **polymerization** (*poly-* = many, *-mer* = repeating units).

It is sometimes possible to select reaction conditions and a proper catalyst so as to be able to control the number of units in the product. One can then form a *di*mer (two units), *tri*mer (3 units) etc., or *poly*mer. The polymerization process is an addition reaction that may proceed by an ionic or free-radical mechanism, depending on the reaction conditions.

For example, 2-methylpropene (isobutylene) undergoes self-polymerization in the presence of cold sulfuric acid (as a catalyst) by an ionic mechanism. Two molecules of 2-methylpropene combine to form the dimer, 2,4,4-trimethyl-2-pentene and some 2,4,4-trimethyl-1-pentene as a minor product.

$$2\ CH_3-\underset{CH_3}{\overset{CH_3}{C}}=CH_2 \xrightarrow[10^\circ C]{H_2SO_4} CH_3-\underset{CH_3}{\overset{CH_3}{\underset{|}{C}}}-CH_2-\overset{CH_3}{\underset{}{C}}=CH_2 + CH_3-\underset{CH_3}{\overset{CH_3}{\underset{|}{C}}}-CH=\overset{CH_3}{\underset{}{C}}-CH_3 \qquad (8\text{-}70)$$

2,4,4-trimethyl-1-pentene   2,4,4-trimethyl-2-pentene

The mechanism for the addition reaction is ionic in nature and follows Markovnikov's rule.

Sec. 8.11 POLYMERIZATION OF ALKENES

$$CH_3-\underset{\underset{CH_3}{|}}{C}=CH_2 + H^\oplus \longrightarrow CH_3-\underset{\underset{CH_3}{|}}{\overset{CH_3}{|}}{\overset{\oplus}{C}}-CH_3 \quad (8\text{-}71)$$

$$CH_3-\underset{\underset{\oplus}{|}}{\overset{CH_3}{|}}{C}-CH_3 + CH_3-\underset{\underset{CH_3}{|}}{C}=CH_2 \longrightarrow CH_3-\underset{\underset{\oplus}{|}}{\overset{CH_3}{|}}{C}-CH_2-\underset{\underset{CH_3}{|}}{\overset{CH_3}{|}}{C}-CH_3 \quad (8\text{-}72)$$

$$CH_3-\underset{\underset{\oplus}{|}}{\overset{CH_3}{|}}{C}-CH_2-\underset{\underset{CH_3}{|}}{\overset{CH_3}{|}}{C}-CH_3 \xrightarrow{-H^\oplus} CH_3-\underset{\underset{CH_3}{|}}{C}=CH-\underset{\underset{CH_3}{|}}{\overset{CH_3}{|}}{C}-CH_3 + CH_2=\underset{\underset{CH_3}{|}}{\overset{CH_3}{|}}{C}-CH_2-\underset{\underset{CH_3}{|}}{\overset{CH_3}{|}}{C}-CH_3$$

(major product)   **(minor product)**

(8-73)

(The more substituted alkene is more stable and is the major reaction product.)

The 2,4,4-trimethyl-2-pentene can be hydrogenated to 2,2,4-trimethylpentane (isooctane) which has excellent anti-knock properties in gasoline (high octane rating).

$$CH_3-\underset{\underset{CH_3}{|}}{C}=CH-\underset{\underset{CH_3}{|}}{\overset{CH_3}{|}}{C}-CH_3 + H_2 \xrightarrow{Pt} CH_3-\underset{\underset{H}{|}}{\overset{CH_3}{|}}{C}-CH_2-\underset{\underset{CH_3}{|}}{\overset{CH_3}{|}}{C}-CH_3 \quad (8\text{-}74)$$

isooctane

The formation of macromolecules or polymers is extremely important in the manufacture of plastics, synthetic textiles such as nylon, and in naturally occurring substances such as starch, cellulose, and proteins.

The polymerization of the alkenes produces many substances of commercial importance. For example, poly(tetrafluoroethylene), ($F_2C=CF_2$), better written as $-(CF_2-CF_2)_n$, is sold under the trade name of Teflon in cookware.

The polymer **polyethylene,** a plastic, is formed by heating ethylene under pressure in the presence of certain catalysts. This polymerization process is a free-radical chain reaction.

$$R^\times + CH_2::CH_2 \longrightarrow R^\times CH_2:CH_2\cdot \quad (8\text{-}75)$$

$$R^\times CH_2:CH_2\cdot + CH_2::CH_2 \longrightarrow R^\times CH_2:CH_2:CH_2:CH_2\cdot \quad \text{(etc.)} \quad (8\text{-}76)$$

In the course of the reaction, chain-branching can occur, the net result being a giant macromolecule, with some chain-branching.

Recently, the chemist Karl Ziegler developed a series of catalysts which permit the ethylene to be polymerized by an ionic mechanism under conditions of low pressure and low temperature. The polyethylene formed is nearly linear and has a higher melting point and greater strength than the polymer produced by the free-radical polymerization.

$-(CH_2-CH_2)_n-$ (n = large number)
polyethylene

These so-called Ziegler catalysts (e.g., a mixture of R_3Al and $TiCl_4$) are also effective in polymerizing other simple alkenes under milder reaction conditions (e.g., polypropylene plastic, $-(CH_2-\underset{|}{\overset{CH_3}{CH}})_n-$ is made commercially by this process).

8.12 Free Radical Halogenation of Alkenes—Substitution of Halogen

The *halogenation* of alkenes *at high temperatures* in the gaseous state proceeds by a free radical mechanism resulting in a **substitution reaction** instead of the ionic addition reaction which occurs at room temperature.

For example, ethylene can be chlorinated at a temperature of 400° C to 1-chloroethene (vinyl chloride). A chlorine atom has been substituted for a hydrogen atom.

$$CH_2{=}CH_2 + Cl_2 \xrightarrow{400°\ C} \underset{\substack{\text{1-chloroethene}\\\text{(vinyl chloride)}}}{CH_2{=}CHCl} + HCl \qquad (8\text{-}77)$$

The vinyl group, $CH_2{=}CH-$, occurs in polyvinyl plastics produced by polymerization reactions of compounds containing this functional group.

More complex alkenes than ethylene will also undergo the free radical halogenation reaction. However, substitution will occur on the carbon atom adjacent to the carbon–carbon double bond.

$$CH_3-CH{=}CH_2 + Cl_2 \xrightarrow{600°\ C} \underset{\substack{\text{3-chloro-1-propene}\\\text{(allyl chloride)}}}{Cl-CH_2-CH{=}CH_2} + HCl \qquad (8\text{-}78)$$

The substitution of a halogen for a hydrogen atom on the carbon atom adjacent to the carbon–carbon double bond is called **allylic halogenation.** (The $-CH_2-CH{=}CH_2$ group is named an allyl group.) The allylic position is attacked preferentially to the other positions because the allyl free radical is the most stable of the free radicals that can be formed in the course of the reaction. Allylic halogenation can be accomplished under much milder conditions by using certain *N*-haloamides as reagents instead of the halogens. The compound *N-bromosuccinimide* (NBS),

$$\begin{array}{c} CH_2-C{\overset{O}{\underset{}{\diagup\!\!\!\diagdown}}} \\ |\qquad\quad N-Br \\ CH_2-C{\underset{O}{\diagdown\!\!\!\diagup}} \end{array}$$

Sec. 8.13 OXIDATION REACTIONS OF ALKENES

is the most popular reagent *used for allylic bromination of alkenes* at room temperature. The overall reaction sequence can be represented as

$$CH_3-CH=CH_2 + NBS \longrightarrow BrCH_2-CH=CH_2 + (succinimide) \quad (8-79)$$

8.13 Oxidation Reactions of Alkenes

Alkenes can be oxidized by a variety of chemical reagents such as ozone and potassium permanganate.

8.13-1 Ozonolysis—Structure Determination of Alkenes The reaction of alkenes with ozone is quite rapid and quantitative. The ozone is passed through a solution containing an alkene and reacts with the double bond, forming an unstable product called an **ozonide**. Ozonides are often explosive and are not isolated, but instead are decomposed by reduction with zinc and acetic acid to form stable products; aldehydes, $R-\overset{\overset{\displaystyle O}{\|}}{C}-H$, or ketones, $R-\overset{\overset{\displaystyle O}{\|}}{C}-R'$.

$$\underset{}{\overset{}{>}}C=C\underset{}{\overset{}{<}} + O_3 \longrightarrow (ozonide) \xrightarrow[CH_3COOH]{Zn} \underset{}{\overset{}{>}}C=O + O=C\underset{}{\overset{}{<}} + H_2O \quad (8-80)$$

aldehydes or ketones

The ozonolysis reaction involves cleavage of the carbon–carbon double bond in the alkene to form aldehydes or ketones (*-lysis* = cleavage reaction).

The proposed mechanism for the ozone addition is indicated by the following steps:

$$\overset{}{>}C=C\overset{}{<} + :\overset{\oplus}{O}:\overset{}{\underset{O}{\overset{}{\cdot\cdot}}}\overset{\ominus}{O}: \longrightarrow \left[\overset{}{>}\underset{O-O}{\overset{}{C-C}}\overset{}{<}\underset{O}{}\right] \overset{rearr.}{\rightsquigarrow} \left[\overset{}{>}\underset{O-O}{\overset{O}{C}\overset{}{C}}\overset{}{<}\right] \xrightarrow[H_2O]{Zn, H^\oplus} \overset{}{>}C=O + O=C\overset{}{<}$$

ozonide

(8-81)

The net result of the ozonolysis reaction is the cleavage of the alkene into two fragments, each of which has an oxygen atom attached to the original double bonded carbon atoms.

$$CH_3-\underset{CH_3}{\overset{CH_3}{C}}=\underset{CH_3}{\overset{H}{C}}-CH_3 \xrightarrow[2.\ Zn\ +\ CH_3COOH,\ H_2O]{1.\ O_3} CH_3-\overset{\overset{\displaystyle O}{\|}}{C}-CH_3 + CH_3-\overset{\overset{\displaystyle O}{\|}}{C}-H \quad (8-82)$$

acetone   **acetaldehyde**

The utility of the ozonolysis reaction is that the aldehydes and ketones formed can be easily transformed into derivatives that can be identified, and related back to the structure of the original alkene. For example, suppose a certain alkene whose formula was established as C_5H_{10} undergoes ozonolysis to form two products identified as

$$CH_3-\underset{H}{\overset{CH_3}{C}}-\overset{O}{\overset{\|}{C}}-H \text{ and } \underset{H}{\overset{H}{C}}=O.$$

The structure of the original alkene must have been $CH_3-\underset{H}{\overset{CH_3}{C}}-C=C-H$, since the underlined carbon atoms in the products were the double bonded carbon atoms in the alkene. If the alkene, C_5H_{10}, formed $CH_3-CH_2-\overset{O}{\overset{\|}{C}}-H$ and $CH_3-\overset{O}{\overset{\|}{C}}-H$ as products, the structure of the alkene would be $CH_3-CH_2-\overset{H}{\underset{|}{C}}=\overset{H}{\underset{|}{C}}-CH_3$. Thus, the reaction of an alkene with ozone is extremely important in elucidating the structural formulas of isomeric alkenes.

8.13-2 Oxidation with Permanganate—Glycol Formation Potassium permanganate is another oxidizing agent that can cleave double bonds. In the cold, alkenes are oxidized by dilute alkaline solutions of potassium permanganate to **glycols** without bond cleavage.

$$CH_2=CH_2 + KMnO_4 \xrightarrow[OH^\ominus]{cold} \underset{OH}{CH_2}-\underset{OH}{CH_2} + MnO_2\downarrow \quad (8\text{-}83)$$
$$\text{(brown)}$$
ethylene glycol

The oxidation of the alkene to the glycol is accompanied at the same time by the reduction of the permanganate, which is *purple,* to a *brown* precipitate of MnO_2. This color change can be used as a characteristic test for alkenes (Baeyer test) to distinguish them from the alkanes, which are not oxidized by permanganate.

At higher temperatures in neutral solution the $KMnO_4$ will cleave the double bond to form aldehydes and ketones (as in the ozonolysis reaction). The aldehydes, $R-\overset{O}{\overset{\|}{C}}-H$, are usually oxidized further by the powerful permanganate oxidizing agent and end up as carboxylic acids, $R-\overset{O}{\overset{\|}{C}}-OH$. Under mild reaction conditions the permanganate will oxidize alkenes to glycols, whereas more vigorous reaction conditions produce carboxylic acids and ketones.

$$CH_3-\overset{CH_3}{\underset{}{C}}=\overset{H}{\underset{}{C}}-CH_3 + KMnO_4 \xrightarrow[OH^\ominus]{cold} CH_3-\underset{OH}{\overset{CH_3}{C}}-\underset{OH}{\overset{H}{C}}-CH_3 \quad (8\text{-}84)$$

2-methylbutane-2,3-diol
(a glycol)

Sec. 8.14 PREPARATION OF ALKENES

$$CH_3-\underset{acetone}{\overset{O}{\underset{\|}{C}}-CH_3} + CH_3-\underset{acetic\ acid}{\overset{O}{\underset{\|}{C}}-OH} \quad \left(from\ CH_3-\overset{O}{\underset{\|}{C}}-H\right) \quad \xleftarrow{\underset{hot}{KMnO_4}} \quad (8\text{-}85)$$

8.14 Preparation of Alkenes

The principal type of reaction used to form a double bond in a molecule is classified as an **elimination reaction.** Two substituents are removed from adjacent carbon atoms, one of which leaves an electron pair behind, resulting in the formation of an additional bond between the carbon atoms.

$$-\underset{Y}{\overset{|}{C}}-\underset{Z}{\overset{|}{C}}- \xrightarrow{-YZ} -\overset{|}{C}=\overset{|}{C}- \qquad (8\text{-}86)$$

The species YZ is eliminated and usually is a stable molecule, such as water or a hydrogen halide.

8.14-1 Mechanism of Elimination Reactions There are two fundamental mechanisms for elimination reactions, called E_1 and E_2 mechanisms. The E_1 mechanism (elimination-unimolecular) as the symbol indicates, is a two-step mechanism involving ionization to produce a carbonium ion (the rate-determining step), followed by a fast step (usually loss of a small molecule or a proton) to form the alkene.

E1 MECHANISM

$$-\underset{H\ \ X}{\overset{|\ \ \ |}{C-C}}- \xrightarrow{slow} -\underset{H}{\overset{|\ \ \ |}{C-\overset{\oplus}{C}}}- + X:^{\ominus} \qquad (8\text{-}87)$$

$$-\underset{H}{\overset{|\ \ \ |}{C-\overset{\oplus}{C}}}- \xrightarrow{fast} -\overset{|}{C}=\overset{|}{C}- + H^{\oplus} \qquad (8\text{-}88)$$

The E_2 mechanism (elimination-bimolecular) is a one-step mechanism where a nucleophilic reagent behaves as a base and removes a β-hydrogen atom from the carbon atom adjacent to the one bearing the substituent. The removal of the β-hydrogen as a proton and the departure of the substituent occur simultaneously. The product obtained by this one step reaction is an alkene.

E2 MECHANISM

$$\underset{\substack{\text{(nucleophile}\\\text{behaving as a base)}}}{B:^{\ominus}} + -\underset{H\ \ X}{\overset{\overset{\beta}{|}\ \ \overset{\alpha}{|}}{C-C}}- \longrightarrow B-H + X:^{\ominus} + \overset{}{\underset{}{C}}=\overset{}{\underset{}{C}} \qquad (8\text{-}89)$$

It can be seen from the E_1 and E_2 mechanisms just outlined that one of the substituents eliminated is usually a proton.

8.14-2 Dehydration of Alcohols Heating alcohols with strong acids, such as sulfuric or phosphoric acid, results in loss of water (dehydration) from adjacent carbon atoms to produce an alkene. The reaction is actually the reverse of the hydration of an alkene.

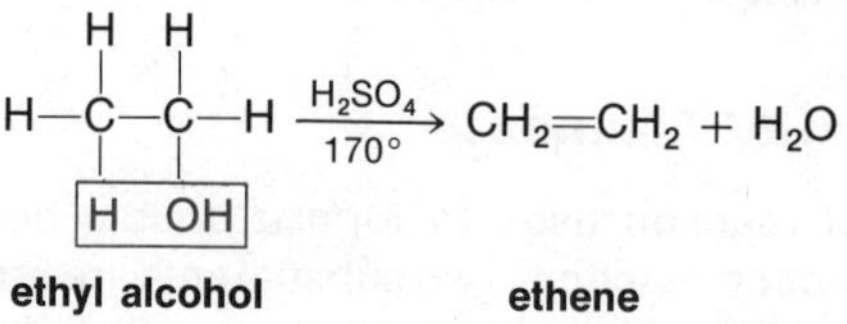

$$\underset{\text{ethyl alcohol}}{H-\underset{H}{\underset{|}{C}}-\underset{\boxed{OH}}{\underset{|}{C}}-H} \xrightarrow[170°]{H_2SO_4} \underset{\text{ethene}}{CH_2=CH_2} + H_2O \qquad (8\text{-}90)$$

The mechanism of the dehydration of alcohols can be represented by the following reaction sequence.

(1) 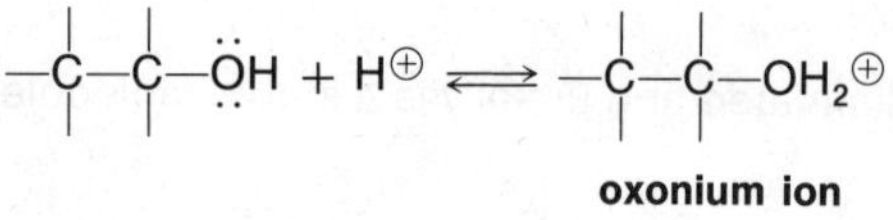 $\quad -\underset{|}{\underset{|}{C}}-\underset{|}{\underset{|}{C}}-\ddot{O}H + H^{\oplus} \rightleftharpoons -\underset{|}{\underset{|}{C}}-\underset{|}{\underset{|}{C}}-OH_2^{\oplus} \qquad (8\text{-}91)$

oxonium ion

(2a) 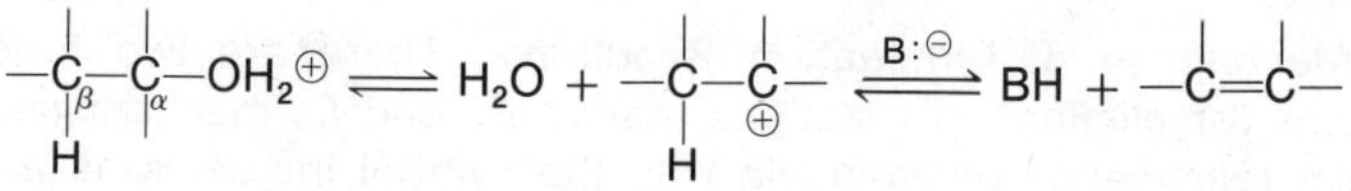 $\quad -\underset{\beta}{\underset{H}{\underset{|}{C}}}-\underset{\alpha}{\underset{|}{\underset{|}{C}}}-OH_2^{\oplus} \rightleftharpoons H_2O + -\underset{H}{\underset{|}{\underset{|}{C}}}-\underset{\oplus}{\underset{|}{\underset{|}{C}}}- \xrightarrow{B:^{\ominus}} BH + -\underset{|}{\underset{|}{C}}=\underset{|}{\underset{|}{C}}- \qquad (8\text{-}92)$

or

(2b) 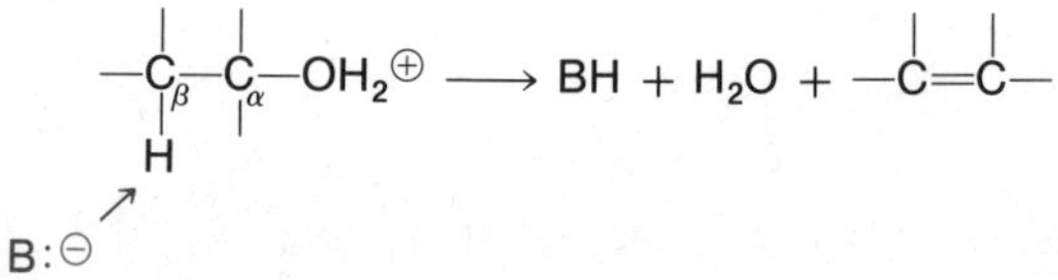 $\quad -\underset{\beta}{\underset{H}{\underset{|}{C}}}-\underset{\alpha}{\underset{|}{\underset{|}{C}}}-OH_2^{\oplus} \longrightarrow BH + H_2O + -\underset{|}{\underset{|}{C}}=\underset{|}{\underset{|}{C}}- \qquad (8\text{-}93)$

$B:^{\ominus} \nearrow$

The function of the sulfuric acid or phosphoric acid used is to convert the alcohol into its conjugate acid, the **oxonium ion,** so that the stable molecule, water, may be easily eliminated later on in the reaction sequence. If an acid catalyst were not used, no alkene would be formed, since the OH group of the alcohol is a poor leaving group; the OH must be protonated to H—OH, which is a good leaving group and departs as a stable water molecule. The oxonium ion can then react by either pathway (2a) or (2b) to form the alkene. The anion of the acid used serves as the base, $B:^{\ominus}$, to remove a β-hydrogen atom in the reaction sequence.

The dehydration of alcohols by the E_1 mechanism occurs most readily with tertiary alcohols, while primary alcohols are dehydrated with difficulty. The order $\underset{3°}{R_3COH} > \underset{2°}{R_2CHOH} > \underset{1°}{RCH_2OH}$ is in accordance with the stability of the intermediate carbonium ions produced in the reaction ($3° > 2° > 1°$). Thus, alcohols that can form stable carbonium ions are dehydrated most easily.

Some alcohols can be dehydrated to produce a mixture of products.

Sec. 8.14 PREPARATION OF ALKENES

$$CH_3-\underset{\underset{CH_3}{|}}{\overset{\overset{OH}{|}}{C}}-CH_2-CH_3 \xrightarrow[\Delta]{conc.\ H_2SO_4} CH_2=\underset{\underset{CH_3}{|}}{C}-CH_2-CH_3 + CH_3-\underset{\underset{CH_3}{|}}{C}=CH-CH_3$$

2-methyl-2-butanol   (17%)   (83%)

(8-94)

The alkene having the most highly substituted double bond is the major product in the reaction (Saytzeff rule).

Dehydration of alcohols by the E_1 mechanism often leads to the formation of unexpected products due to rearrangement of the intermediate carbonium ion. For example, dehydration of n-butyl alcohol results in the formation of *cis* and *trans* 2-butene as the predominant product rather than the expected 1-butene.

$$CH_3-CH_2-CH_2-CH_2OH \xrightarrow[\Delta]{conc.\ H_2SO_4} CH_3-CH=CH-CH_3 \quad (8\text{-}95)$$

n-butyl alcohol   *cis* and *trans* 2-butene
                          (major product)

This can be explained most simply and rationally by the rearrangement of the primary n-butyl carbonium ion to the more stable secondary *sec*-butyl carbonium ion in the course of the reaction sequence. The rearrangement can be accomplished by a migration of a hydrogen atom from one carbon atom to an adjacent carbon atom. This is actually a hydride shift, that is, one in which the hydrogen atom was migrating with its bonding pair of electrons.

(1) $CH_3-CH_2-CH_2-CH_2-OH \xrightarrow{conc.\ H_2SO_4}$
$$H_2O + CH_3-CH_2-CH_2-CH_2^{\oplus} \quad (8\text{-}96)$$
n-butyl carbonium ion (1°)

(2) $CH_3-CH_2-CH_2-CH_2^{\oplus} \xrightarrow{rearr.\ (H\ migration)} CH_3-CH_2-\overset{\oplus}{C}H-CH_3 \quad (8\text{-}97)$
sec-butyl carbonium ion (2°)

(3) $CH_3-CH_2-\overset{\oplus}{C}H-CH_3 \longrightarrow$
$$H^{\oplus} + CH_3-CH=CH-CH_3 + CH_3-CH_2-CH=CH_2 \quad (8\text{-}98)$$
(major product)   (minor product)

An alternative method for dehydrating alcohols is to pass the alcohol vapor over an alumina (Al_2O_3) catalyst at elevated temperatures. The Al_2O_3 is a good dehydrating agent, and by running the reaction in the vapor state one can avoid problems that could result due to rearrangement of intermediate to carbonium ions giving undesired products.

$$CH_3-CH_2-OH \xrightarrow[350°\ C]{Al_2O_3} CH_2=CH_2 + H_2O \quad (8\text{-}99)$$

8.14-3 Dehydrohalogenation of Alkyl Halides

The reaction of alkyl halides with an alcoholic solution of a strong base, such as potassium hydroxide, results in the formation of an alkene. A molecule of a hydrogen halide is eliminated; the term **dehydrohalogenation** describes the reaction.

$$CH_3-CH_2-CH_2Br \xrightarrow{\text{alc. KOH}} H_2O + KBr + CH_3-CH=CH_2 \quad (8-100)$$
1-bromopropane   propene

The reaction can proceed by either the E1 or E2 mechanism, the function of the base being to remove a proton from the carbon atom adjacent to the carbon atom bearing the halogen in the alkyl halide. The order of reactivity for the alkyl halides is the same as that observed for the alcohols (3° > 2° > 1°). When more than one alkene can be formed as a product, the more highly substituted or more stable alkene is the major product (Saytzeff rule).

$$CH_3-CH_2-\underset{\underset{Br}{|}}{CH}-CH_3 \xrightarrow{\text{alc. KOH}} CH_3-CH_2-CH=CH_2 + CH_3-CH=CH-CH_3$$

2-bromobutane   (20 %)   (80 %)

(8-101)

8.15 Synthesis

At this point it is worthwhile to discuss how to determine the proper reaction sequences for synthesizing various organic compounds in the laboratory. We can use the reactions of the alkenes to illustrate the technique of solving synthesis problems. In each case we should select those reactions that will prepare the compound in as pure a form as possible and free from side products.

A typical synthesis problem would be to synthesize the compound 1,2-dibromopropane from n–propyl bromide. First the student must be able to draw the correct structural formulas for the compounds in question and to recognize to which class of compounds each substance belongs.

$$CH_3-CH_2-CH_2Br \longrightarrow CH_3-\underset{\underset{Br}{|}}{CH}-CH_2Br$$

n–propyl bromide   1,2-dibromopropane
(alkyl halide)   (vicinal dihalide)

It is best to work synthesis problems backwards, that is start with the product and work back to the reactant. This may seem difficult at first, but after some practice this will enable you to do synthesis problems more rapidly and efficiently. In doing synthesis it is also advisable to think of the compounds in terms of the class they represent rather than considering them as specific compounds.

Thus, the preceding problem involves the conversion of an alkyl halide to a vicinal dihalide. In solving this problem one must recall all the reactions he has learned. The only one that can be used to prepare a vicinal dihalide is by treating an alkene with a halogen (addition across the carbon–carbon double bond). Since we desire to prepare 1,2-dibromopropane, a three-carbon compound, we need to

Sec. 8.17 CONJUGATE ADDITION

react a three-carbon alkene, propylene, with bromine. The last step in the reaction sequence will be

$$CH_3-CH=CH_2 + Br_2 \longrightarrow CH_3-\underset{\underset{Br}{|}}{CH}-CH_2Br \qquad (8\text{-}102)$$

1,2-dibromopropane

This step requires the use of propylene, an alkene. But, we are not starting with an alkene, and so must prepare an alkene. You will recall that an alkene can be prepared in a number of ways, one of which is by the dehydrohalogenation of an alkyl halide. Treating n-propyl bromide, the starting material, with alcoholic KOH will produce propylene.

$$CH_3-CH_2-CH_2Br \xrightarrow{\text{alc. KOH}} CH_3-CH=CH_2 \qquad (8\text{-}103)$$

n-propyl bromide   propylene

The complete reaction sequence would then be:

$$CH_3-CH_2-CH_2Br \xrightarrow{\text{alc. KOH}} CH_3-CH=CH_2 \xrightarrow{Br_2} CH_3-\underset{\underset{Br}{|}}{CH}-CH_2Br \qquad (8\text{-}104)$$

The synthesis of organic compounds in the laboratory is of prime concern to the organic chemist. We will be using the principles outlined in our discussion to synthesize many different organic compounds. However, as we study more reactions the reaction sequences will contain many more individual steps than in the preceding problem.

8.16 Polyalkenes

The **polyalkenes** or **polyolefins** refer to a class of compounds that contain more than one double bond in the molecule. The chemistry of the polyalkenes is similar to that of the alkenes, except when the molecule contains a series of alternating single and double bonds, as in the compound 1,3-butadiene ($\underset{1}{CH_2}=\underset{2}{CH}-\underset{3}{CH}=\underset{4}{CH_2}$). The arrangement of alternating single and double bonds in a molecule is referred to as a **conjugated system.**

8.17 Conjugate Addition

The addition reactions of the conjugated dienes such as 1,3-butadiene differ somewhat from the electrophilic addition reactions of the simple alkenes. For example, addition of bromine to 1-butene forms 1,2-dibromobutane as the reaction product in which the bromine atoms are attached to adjacent carbon atoms.

$$CH_3-CH_2-CH=CH_2 + Br_2 \longrightarrow CH_3-CH_2-\underset{\underset{Br}{|}}{CH}-CH_2Br \qquad (8\text{-}105)$$

1,2-dibromobutane

But, in conjugated dienes, the addition occurs across the entire conjugated system. Thus, addition of bromine to 1,3-butadiene produces a mixture of *trans*-1,4-dibromo-2-butene and 3,4-dibromo-1-butene.

$$CH_2=CH-CH=CH_2 + Br_2 \longrightarrow$$

$$\underset{\textit{trans}\text{-1,4-dibromo-}\atop\text{2-butene}}{\underset{CH_2Br}{\overset{BrH_2C}{\underset{|}{CH=CH}}}} + \underset{\text{3,4-dibromo-}\atop\text{1-butene}}{CH_2=CH-\underset{|}{\overset{Br}{CH}}-CH_2Br} \quad (8\text{-}106)$$

The formation of 3,4-dibromo-1-butene can be accounted for by simple addition of bromine across one of the double bonds, *1,2-addition* to carbon atoms numbered 1 and 2. The *trans*-1,4-dibromo-2-butene is formed by the addition of bromine to the end carbon atoms of the conjugated system, *1,4-addition*, with a new double bond between carbons 2 and 3 appearing in the product.

A consideration of the mechanism of the reaction will elucidate the formation of the 1,4-addition product. Indeed, it is well to note that addition reactions to conjugated systems will invariably occur at the ends of the conjugated system.

Let us recall the mechanism of halogenation across a double bond by electrophilic addition. Initially, the bromonium ion, :$\overset{\oplus}{Br}$, attacks one of the end carbon atoms (Markovnikov's rule) to produce a substituted allylic carbonium ion

$$CH_2=CH-CH\overset{\frown}{=}CH_2 + :\overset{\oplus}{Br} \longrightarrow CH_2=CH-\overset{\oplus}{CH}-CH_2Br \quad (8\text{-}107)$$

which is a resonance hybrid of the contributing resonance forms.

$$[CH_2=CH\overset{\frown}{-}\overset{\oplus}{CH}-CH_2Br \longleftrightarrow \overset{\oplus}{CH_2}-\overset{\frown}{CH}=CH-CH_2Br]$$

The positive charge is distributed over the conjugated system so that two carbon atoms bearing the positive charge are susceptible to attack by nucleophilic reagents such as bromide ion, $Br^{\ominus}$.

$$CH_2=CH-\overset{\oplus}{CH}-CH_2Br + Br^{\ominus} \longrightarrow CH_2=CH-\underset{\underset{Br}{|}}{CH}-CH_2Br \quad (8\text{-}108)$$

1,2-addition product

$$\overset{\oplus}{CH_2}-CH=CH-CH_2Br + Br^{\ominus} \longrightarrow \underset{\underset{Br}{|}}{CH_2}-CH=CH-CH_2Br \quad (8\text{-}109)$$

1,4-addition product

8.17-1 Kinetic vs. Thermodynamic Control—Rate vs. Equilibrium At ordinary temperatures the 1,4-addition product is the major product of the reaction, whereas at low temperatures the 1,2-addition product is the predominant prod-

Sec. 8.18 ISOPRENE

uct. This is an example of **kinetic control vs. thermodynamic control** in determining the ratio of products formed in a reaction. At low temperatures, the reaction is kinetically controlled, nucleophilic attack occurring at the 2–position at a faster rate than at the 4–position, resulting in formation of more of the 1,2–addition product than the 1,4–addition product. At higher temperatures, the reaction is thermodynamically controlled. The reaction products come to equilibrium with each other, the 1,4–addition product being more stable than the 1,2–addition product. Thus, the 1,2–addition product formed would eventually be converted into the more stable 1,4–isomer, which is the major product under ordinary conditions.

8.18 Isoprene

The conjugated diene **isoprene,** or 2-methyl-1,3-butadiene, is a compound occurring in many important natural products. The heating of natural rubber in the absence of air (pyrolysis) produces, as the major product, isoprene.

$$\text{Natural rubber} \xrightarrow{\Delta} CH_2=\underset{\underset{CH_3}{|}}{C}-CH=CH_2 \qquad (8\text{-}110)$$

isoprene
(2–methyl–1,3–butadiene)

Natural rubber is basically a polymer of many isoprene units joined together by 1,4–addition.

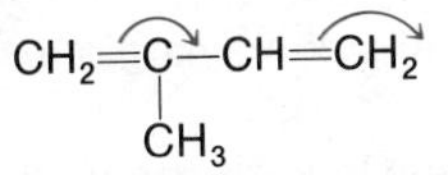

$$CH_2\!=\!\overset{\frown}{C}\!-\!CH\!=\!\overset{\frown}{CH_2} \quad \text{(etc.)}$$
$$\underset{CH_3}{|}$$

(isoprene unit)

($\frown$ indicates electron shift)

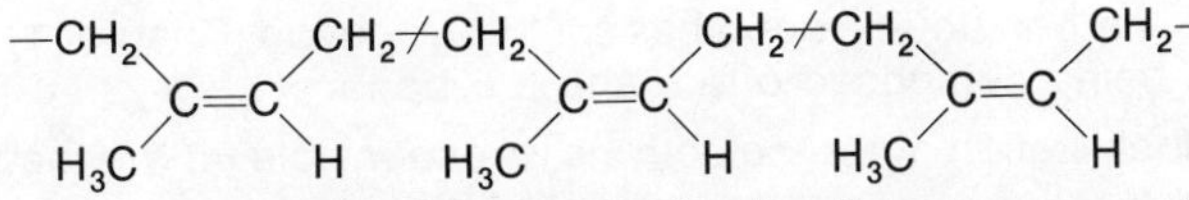

$$\begin{array}{c}-CH_2\\ \diagdown\\ H_3C\end{array}C\!=\!C\begin{array}{c}CH_2\!-\!CH_2\\ \diagup\\ H\end{array}\begin{array}{c}\\ \diagdown\\ H_3C\end{array}C\!=\!C\begin{array}{c}CH_2\!-\!CH_2\\ \diagup\\ H\end{array}\begin{array}{c}\\ \diagdown\\ H_3C\end{array}C\!=\!C\begin{array}{c}CH_2-\\ \diagup\\ H\end{array}$$

(A portion of a natural-rubber molecule showing three isoprene units joined by 1,4–addition. Note that all the double bonds are *cis*.)

The isoprene unit is apparently a very important feature of many biologically active materials. A segment of many molecules can be shown to consist of a series of isoprene units joined together. The yellow plant pigment, **carotene,** which occurs in carrots and tomatoes, has the following structures. Dotted lines indicate the isoprene units in the molecule.

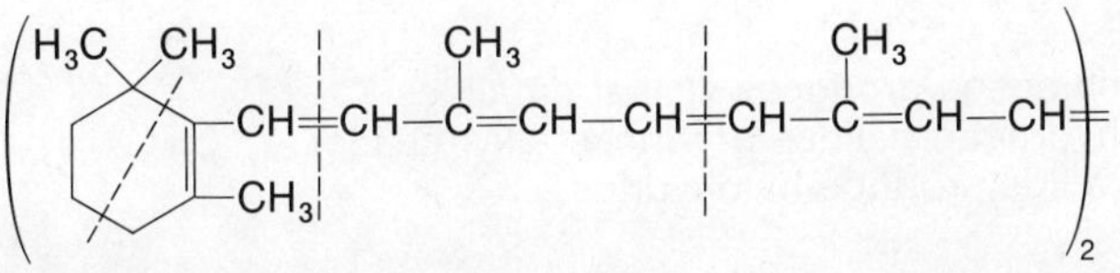

$$\left(\underset{CH_3}{\overset{H_3C\quad CH_3}{\bigcirc}}-CH\!=\!CH\!-\!\underset{}{\overset{CH_3}{C}}\!=\!CH\!-\!CH\!=\!CH\!-\!\underset{}{\overset{CH_3}{C}}\!=\!CH\!-\!CH\!=\right)_2 \quad \beta\text{-carotene}$$

Vitamin A, which is necessary for the synthesis of certain pigments essential for good eyesight, has the structure

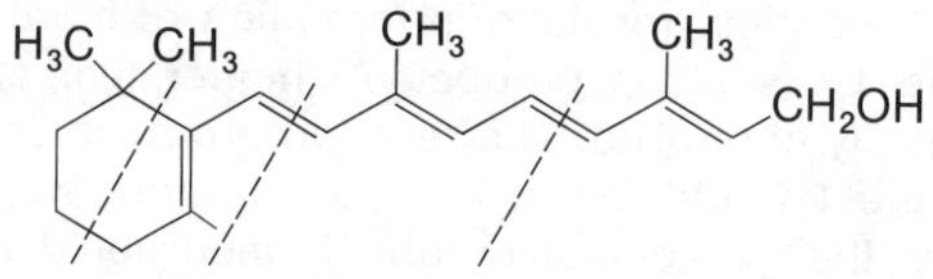

vitamin A

Squalene is made up of six isoprene units with *trans* geometry at the carbon–carbon double bonds. It is a precursor in the biological synthesis of cholesterol and other steroids.

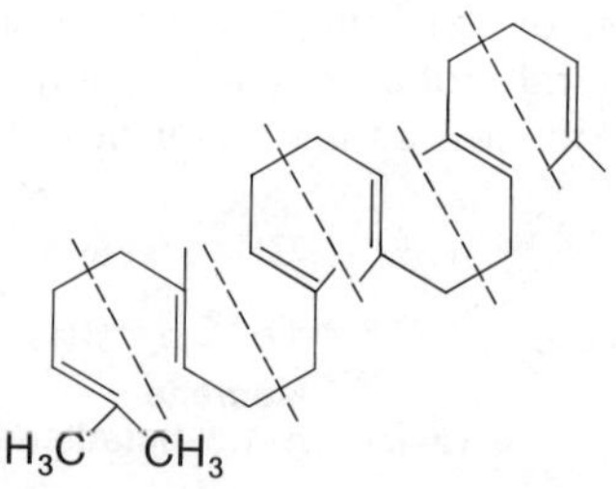

squalene

Summary

- The alkenes or olefins are unsaturated hydrocarbons containing one or more double bonds.
- Alkenes containing one carbon–carbon double bond have the general formula C_nH_{2n}.
- Alkenes can be named by the IUPAC system of nomenclature and are recognized by the *-ene* ending (Section 8.2).
- The carbon–carbon double bond has sp^2 hybridization (Chapter 1) and can be considered as being composed of a σ and a π bond.

    (a) The delocalization of the π electrons is responsible for the reactivity of the alkenes.

    (b) Geometrical isomerism (*cis–trans*) is present in some substituted alkenes and can be attributed to the restricted rotation in the carbon–carbon double bond. The *trans* isomer is usually more stable than the *cis* isomer.

- The characteristic reaction of the alkenes is addition across the double bond, which can occur by either an ionic (electrophilic) or free-radical mechanism.

    Markovnikov's rule predicts the proper orientation of unsymmetrical reagents to unsymmetrical alkenes. The rule can be understood in terms of inductive effects and stability of carbonium ions.

- Addition reactions:

    (a) Addition of halogens produces vicinal dihalides.
    (b) Addition of hydrogen halides produces alkyl halides.
    (c) Hypohalous acids produce halohydrins.

PROBLEMS

   (d) Water produces alcohols.
   (e) H_2SO_4 produces alkyl hydrogen sulfates.
- Free-radical addition of HBr in the presence of peroxide occurs contrary to Marknovnikov's Rule.
- Cycloaddition reactions by *cis* addition:
   (a) Catalytic hydrogenation produces alkanes.
   (b) Hydroboration occurs contrary to Marknovnikov's rule and can produce alcohols.
   (c) Addition of carbenes produces a cyclopropane ring.
   (d) The Diels–Alder reaction produces a new six-membered ring.
- Alkenes can undergo polymerization reactions, forming, as products, polyethylene, polypropylene, etc.
- Substitution of halogen at elevated temperature produces an allyl halide.
- Oxidation reactions:
   (a) Ozonolysis can be used as a tool in structure determination.
   (b) $KMnO_4$ produces *cis*-glycols. The Baeyer test is used in conjunction with this reaction to test for unsaturation.
- Alkenes can be prepared by an elimination reaction. The E1 mechanism can involve a charged species (usually a carbonium ion); the E2 mechanism is a single-step, concerted process.
   (a) Dehydration of alcohols to alkenes—during the reaction it is sometimes possible for the intermediate carbonium ion to rearrange and form a different product because of the rearrangement.
   (b) Dehydrohalogenation of alkyl halides to alkenes.
- Synthesis problems are best solved by working backwards—from product to reactant.
- Polyalkenes contain more than one double bond.
   (a) A conjugated system of alternating single and double bonds is unusually stable due to resonance stabilization.
   (b) Addition to a conjugated system usually occurs at the ends of the system, 1,4-addition in 1,3-butadiene.
   (c) 1,2- vs. 1,4-addition is an example of kinetic vs. thermodynamic control in a reaction.
   (d) Important conjugated dienes include isoprene (polymer is natural rubber), squalene, composed of six isoprene units, etc.

PROBLEMS

1 Draw the correct structural formulas for each of the following compounds.
   (a) propene
   (b) 2-bromo-3-hexene
   (c) tetracyanoethylene
   (d) 3-methyl-2-pentene
   (e) 1,5-hexadiene
   (f) methylethylene
   (g) *cis*-3,4-dimethyl-3-hexene
   (h) 3-chlorocyclohexene
   (i) 5-methyl-1,3-cyclopentadiene
   (j) propylene chlorohydrin
   (k) allyl bromide
   (l) *trans*-2-pentene

2 Give the correct IUPAC name for each of the following compounds.
   (a) $CH(Cl)=CH-CH_2-CH_3$
   (b) $(CH_3)_2-C=CH_2$
   (c) $(CH_3)_3-C-CH=CH-CH_3$
   (d) [cyclopentene with Br and CH_3 substituents]
   (e) [cyclopropane with two CH_3 and two H substituents]
   (f) $CH_2=CH-CH=CH-CH=CH_2$
   (g) $CH_3-CH_2-\overset{Br}{C}=C=CH_2$
   (h) [cyclohexadiene with CH_3 substituent]
   (i) $Br-CH_2-CH=CH_2$
   (j) $\underset{H}{\overset{H_3C}{>}}C=C\underset{H}{\overset{CH_3}{<}}$

3 Explain why each of the following is an incorrect name, and give the correct name of the compound in each case.
   (a) 4-pentene
   (b) 4,4-dimethyl-2-butene
   (c) 1-chloro-3-propene
   (d) cis-3-methyl-1-butene
   (e) 4-i-propyl-2-butene
   (f) 6-bromocyclohexene

4 Draw the structural formulas for all the isomers having the formula C_5H_{10}, and name each isomer by the IUPAC naming system.

5 Which of the following compounds exhibit geometrical isomerism (have cis and trans isomers)?
   (a) 1-butene
   (b) 3-heptene
   (c) 1-chloroethene
   (d) 2,3-dimethyl-1-pentene
   (e) 1,1-dichlorocyclopropane
   (f) 1,3-dimethylcyclohexane
   (g) 3,4-dimethyl-3-hexene
   (h) 2,4-hexadiene

6 Complete each of the following equations by drawing the structural formula(s) of the product(s). If no reaction occurs, write N.R.
   (a) propylene + H_2/Pt $\longrightarrow$
   (b) [methylcyclohexene] + Cl_2 $\xrightarrow{CCl_4}$
   (c) 1-butene + HOCl $\longrightarrow$
   (d) $CH_3-CH=CH_2 + Cl_2$ $\xrightarrow{400°\ C}$
   (e) 1-pentene + HI $\longrightarrow$
   (f) 1,3-butadiene + Br_2 $\longrightarrow$
   (g) 2-methylpropene + HBr + H_2O_2 $\longrightarrow$
   (h) CH_3-CH_2OH + conc. H_2SO_4 $\longrightarrow$
   (i) 2-bromopropane + alc. KOH $\longrightarrow$
   (j) propene + HCN $\longrightarrow$
   (k) cyclohexene + H_2SO_4 $\xrightarrow{H_2O}$
   (l) i-butyl alcohol + conc. H_2SO_4 $\longrightarrow$

PROBLEMS

(m) 1-butene + NBS ⟶
(n) cyclopentene + cold, dil., neutral KMnO$_4$ ⟶
(o) CH$_3$—CH$_2$—CH=CH$_2$ + HBr ⟶

(p) *1,3-butadiene + [p-benzoquinone] ⟶

(q) *2,3-dimethylbutadiene + CH$_3$O$_2$C—C≡C—CO$_2$CH$_3$ ⟶

(r) 1-methylcyclohexene + B$_2$H$_6$ $\xrightarrow[\text{OH}^\ominus]{\text{H}_2\text{O}_2}$

(s) *propene + NOCl ⟶
(t) 1,3-cyclohexadiene + 1 mole HBr ⟶

7 Perform all of the following syntheses from the indicated starting materials.
(a) 1-bromobutane ⟶ 1,2-dibromobutane
(b) n-propyl bromide ⟶ propylene chlorohydrin
(c) 2-propanol ⟶ 1-bromo-3-chloro-2-propanol
(d) 1,2-dibromopropane ⟶ 1-bromopropane
(e) i-butyl alcohol ⟶ i-butyl bromide
(f) cyclohexanol ⟶ trans-1,2-dibromocyclohexane
(g) cyclohexanol ⟶ pure cis-1,2-cyclohexanediol
(h) chlorocyclohexane ⟶ 3-chlorocyclohexene
(i) 2-bromopropane ⟶ 3-bromopropene
(j) t-butyl alcohol ⟶ i-butyl alcohol
(k) propene to 1,1-dichloro-2-methylcyclopropane

8 Name a chemical test or single reagent which will distinguish between the following pairs of compounds.
(a) n-butane and 1-butene
(b) cyclopropane and ethylene
(c) cyclohexane and 2-methylcyclohexene

9 Arrange the following alcohols in order of their increasing ease of dehydration.

CH$_3$CH$_2$OH   CH$_3$—CH(CH$_3$)—OH   CH$_3$—C(CH$_3$)$_2$—OH   CH$_3$OH

CH$_3$—CH$_2$—CH$_2$OH   cyclohexanol

10 Which of the following compounds is more reactive towards electrophilic addition of HBr?
(a) ethylene or propylene
(b) CH$_2$=CH$_2$ or CH$_2$=CH—COOH
(c) CH$_3$—CH=CH$_2$ or 3,3,3-trifluoropropene
(d) 2-butene or 1-chlorocyclohexene

11 Draw the correct structural formulas of each of the following six carbon alkenes that will give the indicated products when treated with ozone followed by hydrolysis, and name each alkene by the IUPAC naming system.

Alkene	Products
A	$CH_3-CH_2-\overset{O}{\overset{\|}{C}}-H + CH_3-\overset{O}{\overset{\|}{C}}-CH_3$
B	$CH_3-\overset{O}{\overset{\|}{C}}-H + CH_3-\overset{CH_3}{\overset{\|}{CH}}-\overset{O}{\overset{\|}{C}}-H$
C	$CH_3-\overset{O}{\overset{\|}{C}}-H + H-\overset{O}{\overset{\|}{C}}-H + H-\overset{O}{\overset{\|}{C}}-CH_2-\overset{O}{\overset{\|}{C}}-H$
D	only $H-\overset{O}{\overset{\|}{C}}-(CH_2)_4-\overset{O}{\overset{\|}{C}}-H$

12 *(a) Trans isomers are usually more stable than the corresponding cis isomers, but *cis*-1,3-dimethylcyclohexane is more stable than the trans isomer. Explain. (HINT: Draw the chair conformations of the compounds.)

(b) Consider the 1,2- and 1,4-dimethylcyclohexanes. Which isomer is the more stable, *cis* or *trans*? Explain.

13 Draw the structural formula for the most stable conformation of the following compounds.
  (a) *cis*-2-methylcyclohexanol
  (b) *trans*-3-methylcyclohexanol
  (c) *cis*-4-methylcyclohexanol

14 What product would you expect to be formed by reacting 1,3,5-hexatriene with one mole of HOCl? Explain.

15 * Saran wrap is actually a polymer of 1,1-dichloroethene. Write all the steps for the free-radical-induced polymerization of 1,1-dichloroethene to form the Saran.

16 * When the compound 1-octene is reacted with chloroform, $CHCl_3$, in the presence of peroxide and sunlight, the product 1,1,1-trichlorononane is obtained. Devise a plausible reaction mechanism, and write all the steps to account for its formation.

17 * Write the steps of the mechanism for the acid-catalyzed dehydration of 3,3-dimethyl-2-butanol. Indicate all the products formed, and the major product formed. Explain your reasoning.

18 * Compound A, $C_6H_{12}O$, is reacted with concentrated sulfuric acid to form compound B, C_6H_{10}. When B is reacted with chlorine at 400° C, compound C, C_6H_9Cl, is obtained. Then C is treated with alcoholic KOH to form D, C_6H_8, which is reacted with one mole of bromine at low temperature to form compound E, $C_6H_8Br_2$. Upon ozonolysis, followed by hydrolysis, E forms only $H-\overset{O}{\overset{\|}{C}}-CHBr-CHBr-CH_2-CH_2-\overset{O}{\overset{\|}{C}}-H$ as a product. From the

PROBLEMS

above information, deduce and write the correct structural formulas of compounds A through E.

19 The addition of HBr by an ionic mechanism to acrylic acid, $CH_2{=}CH{-}COOH$, proceeds in a manner contrary to Markovnikov's rule. Explain.

20 Explain the formation of different products when 2-methyl-1-butene is reacted with HBr in the absence and in the presence of H_2O_2.

9
Unsaturated Hydrocarbons—II: Alkynes

Cutting steel plate with a torch mixing oxygen and the organic compound acetylene, which can achieve temperatures up to 3000°C for cutting and welding metal.

Courtesy of the Linde Division of the Union Carbide Corporation.

9.1 Introduction

The alkynes are hydrocarbons having the general formula C_nH_{2n-2}, and are characterized by the presence of a carbon–carbon triple bond in the molecule. The simplest member of the series has the formula C_2H_2 (H—C≡C—H), and is commonly known as acetylene.

9.2 Nomenclature of Alkynes

The rules governing the nomenclature of the alkynes by the IUPAC naming system are the same as those discussed previously for the alkanes and alkenes, except that the ending *-yne* is used to denote the presence of a carbon–carbon

Sec. 9.4 ADDITION REACTIONS OF ALKYNES

triple bond in the molecule. The following compounds will serve as typical examples.

H—C≡C—H   CH$_3$—C≡C—H   CH$_3$—C≡C—CH$_3$ H—C≡C—CH$_2$—CH—CH$_3$
                                                                                                                                               Cl

ethyne   propyne   2-butyne   4-chloro-1-pentyne

9.3 The Carbon–Carbon Triple Bond

The nature of the carbon–carbon triple bond has been discussed in Chapter 1, and can be considered as being composed of a σ and two π bonds. The alkynes are linear molecules since *sp* hybridization is involved in forming the triple bond. The cylindrical π electron cloud, which virtually surrounds the —C≡C— portion of the molecule, is responsible for most of the chemical properties of this class of compounds. Thus, it will be seen that the addition reactions of the alkynes, or acetylenes as they are sometimes called, are strikingly similar to those of the alkenes. The only difference is that *two moles* of a reagent can add to the triple bond, whereas only *one mole* of a reagent can add across a double bond.

9.4 Addition Reactions of Alkynes

The reactions of the alkynes are similar to those of the alkenes in that the same reagents undergo addition reactions with both classes of compounds. As in the case of the alkenes, addition across the triple bond in the alkynes is in accordance with Markovnikov's rule.

9.4-1 Halogenation For example, the reaction of acetylene (ethyne) with chlorine will occur in two steps, leading to the production of 1,1,2,2-tetrachloroethane as the final product.

$$H-C\equiv C-H + Cl_2 \longrightarrow \underset{\underset{Cl}{|}}{\overset{\overset{Cl}{|}}{H-C=C-H}} \xrightarrow{Cl_2} \underset{\underset{Cl\ Cl}{|\ \ |}}{\overset{\overset{Cl\ Cl}{|\ \ |}}{H-C-C-H}} \quad (9\text{-}1)$$

                                        1,2-dichloroethene   1,1,2,2-tetrachloroethane

9.4-2 Addition of Hydrogen The reduction of alkynes to alkenes by hydrogen in the presence of a catalyst (Pt, Pd, or Ni) results in the formation of a ***cis-alkene*** since the hydrogens add across the triple bond from the same side of the catalyst's surface. A second mole of hydrogen will add to the *cis*-alkene to form an alkane.

$$R-C\equiv C-R' + H_2 \xrightarrow{Pt} \underset{\underset{}{}}{\overset{\overset{H\ H}{|\ \ |}}{R-C=C-R'}} \xrightarrow{H_2,\ Pt} \underset{\underset{H\ H}{|\ \ |}}{\overset{\overset{H\ H}{|\ \ |}}{R-C-C-R'}} \quad (9\text{-}2)$$

                            alkyne   *cis*-alkene   alkane

The reduction of alkynes by the action of active metals, such as sodium or lithium in liquid ammonia, takes place by a different mechanism, and produces **trans-alkenes** rather than the *cis*-isomers.

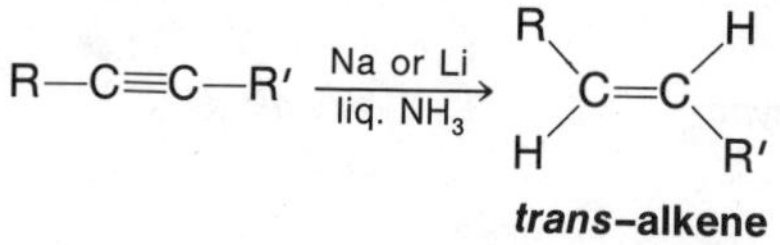

$$R-C\equiv C-R' \xrightarrow{\text{Na or Li} \atop \text{liq. NH}_3} \underset{\text{trans-alkene}}{\overset{R}{\underset{H}{>}}C=C\overset{H}{\underset{R'}{<}}} \quad (9\text{-}3)$$

9.4-3 Addition of Hydrogen Halides The reaction of alkynes with hydrogen halides will produce **gem-dihalides** (*gem*ini = the twins) where the two halogen atoms are on the same carbon atom, consistent with addition by Markovnikov's rule.

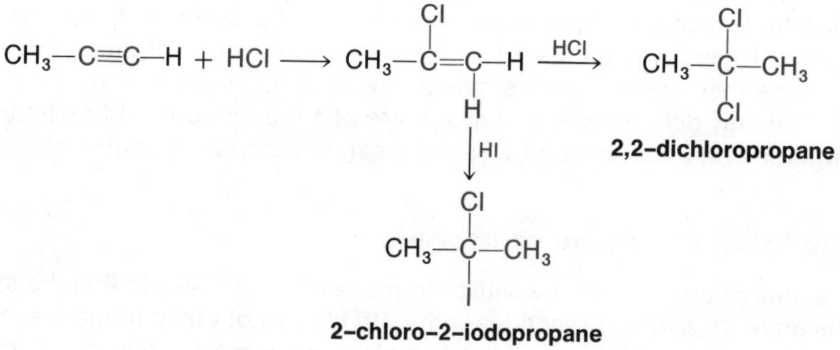

$$CH_3-C\equiv C-H + HCl \longrightarrow CH_3-\underset{H}{\overset{Cl}{C}}=C-H \xrightarrow{HCl} CH_3-\underset{Cl}{\overset{Cl}{C}}-CH_3 \quad (9\text{-}4)$$

2,2-dichloropropane

$$\downarrow HI$$

$$CH_3-\underset{I}{\overset{Cl}{C}}-CH_3 \quad (9\text{-}5)$$

2-chloro-2-iodopropane

9.4-4 Addition of HCN The reaction of acetylene with hydrogen cyanide forms acrylonitrile which can be polymerized to produce the synthetic fabric known as Orlon.

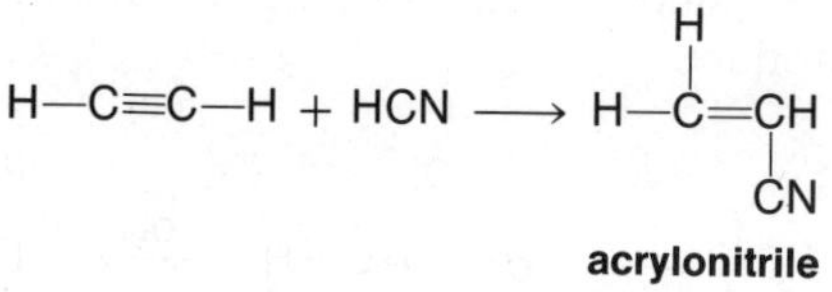

$$H-C\equiv C-H + HCN \longrightarrow H-\overset{H}{\underset{CN}{C}}=CH \quad (9\text{-}6)$$

acrylonitrile

9.4-5 Addition of Water—Tautomerism The addition of water (hydration) to an alkyne requires the presence of mercuric sulfate and sulfuric acid as a catalyst. The reaction is interesting because the final product, which is isolated, is an aldehyde or ketone and *not* an alcohol, as in the hydration of alkenes.

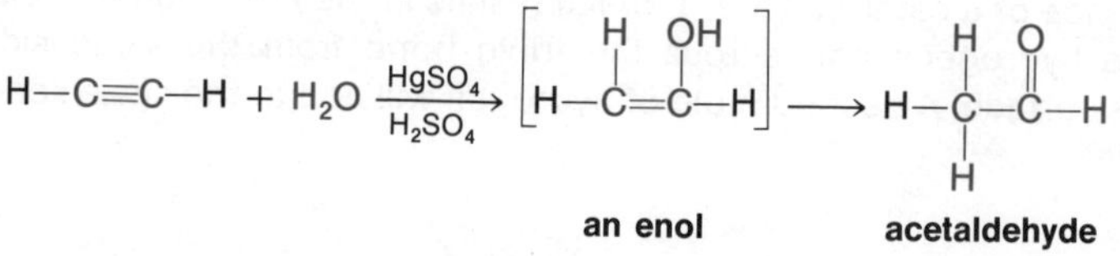

$$H-C\equiv C-H + H_2O \xrightarrow[H_2SO_4]{HgSO_4} \left[H-\overset{H}{C}=\overset{OH}{C}-H \right] \longrightarrow H-\overset{H}{\underset{H}{C}}-\overset{O}{C}-H \quad (9\text{-}7)$$

an enol   acetaldehyde

$$CH_3-C\equiv C-H + H_2O \xrightarrow[H_2SO_4]{HgSO_4} \left[CH_3-\overset{OH}{C}=CH_2 \right] \longrightarrow CH_3-\overset{O}{\overset{\|}{C}}-CH_3 \quad (9\text{-}8)$$

an enol   acetone

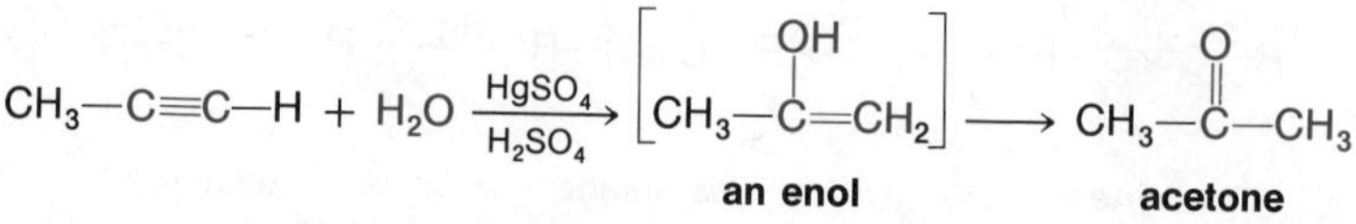

Sec. 9.5 THE ACIDIC HYDROGEN IN ALKYNES

The water adds across the triple bond in accordance with Markovnikov's rule to produce an intermediate known as an **enol** (*ene* = double bond, *-ol* = alcohol or OH group). The enol intermediate is unstable and is in equilibrium by a proton shift with a so-called keto form, containing the $\mathrm{C{=}O}$ or carbonyl group.

$$\underset{\text{enol form}}{\mathrm{H{-}C{=}C{-}H}} \rightleftharpoons \underset{\text{keto form}}{\mathrm{CH_3{-}\overset{O}{\underset{\|}{C}}{-}H}} \qquad (9\text{-}9)$$

Rapid proton shifts of this type are referred to as **tautomerism**. Here we have an example of *enol–keto tautomerism,* where the relatively unstable enol rapidly undergoes a proton shift to form the more stable keto form as either an aldehyde $(\mathrm{R{-}\overset{O}{\underset{\|}{C}}{-}H})$ or a ketone $(\mathrm{R{-}\overset{O}{\underset{\|}{C}}{-}R'})$.

9.5 The Acidic Hydrogen in Alkynes

Hydrogen atoms attached to a triple-bonded carbon atom are acidic. It may be asked why the hydrogen atoms in acetylene, for example, are acidic. The answer can be found in an examination of the acid–base equilibrium between acetylene and its conjugate base, acetylide anion.

$$\underset{\text{acetylene}}{\mathrm{H{-}C{\equiv}C{-}H}} \rightleftharpoons \mathrm{H^{\oplus}} + \underset{\text{acetylide anion}}{\mathrm{H{-}C{\equiv}C{:}^{\ominus}}} \qquad (9\text{-}10)$$

The basicity of the anion is determined by the availability of its electron pair for sharing with acids. In the acetylide anion, the unshared pair of electrons is held tightly, since it is in an *sp* orbital. The *sp* hybridized orbital has some of the character of a pure *s* orbital, so that the unshared electron pair is close to the nucleus and is held tightly. The pair of electrons is not readily available for sharing with acids, making the acetylide anion a weak base, and acetylene a strong conjugate acid, so that the hydrogens in acetylene are somewhat acidic.

As in all acids the hydrogen atom can be replaced by reaction with certain metals. Thus, reaction of an alkyne bearing an acidic hydrogen with a reactive metal, such as sodium, liberates hydrogen gas.

$$\mathrm{2H{-}C{\equiv}C{-}H + 2Na \longrightarrow 2H{-}C{\equiv}C^{\ominus}Na^{\oplus} + H_2} \qquad (9\text{-}11)$$

$$\mathrm{2CH_3{-}C{\equiv}C{-}H + 2Na \longrightarrow 2CH_3{-}C{\equiv}C^{\ominus}Na^{\oplus} + H_2} \qquad (9\text{-}12)$$

A strong base, such as sodamide in liquid ammonia, can also be used to remove the proton.

$$\mathrm{H{-}C{\equiv}C{-}H + NaNH_2} \xrightarrow{\text{liq. NH}_3} \underset{\substack{\text{sodium}\\\text{acetylide}}}{\mathrm{H{-}C{\equiv}C^{\ominus}Na^{\oplus}}} + \mathrm{NH_3} \qquad (9\text{-}13)$$

Alkynes such as 2-butyne ($CH_3-C \equiv C-CH_3$), which have *no* acidic hydrogens will not react with sodium or sodamide.

9.5-1 Use in Synthesis The synthetic utility of the reaction of alkynes with sodium or sodamide is due principally to the high reactivity of the sodium salt of the alkyne formed in the reaction. The sodium acetylides can be used to synthesize more complex alkynes with a longer carbon chain by reaction with a suitable alkyl halide. For example, the reaction of sodium acetylide with *n*-propyl iodide produces 1-pentyne.

$$H-C \equiv \overset{\ominus}{C} Na^{\oplus} + CH_3-CH_2-CH_2I \longrightarrow NaI + \underset{\text{1-pentyne}}{H-C \equiv C-CH_2-CH_2-CH_3} \quad (9-14)$$

The carbon chain of the original alkyne is lengthened by the number of carbon atoms in the alkyl halide. The reaction sequence may be summarized as follows:

$$R-C \equiv C-H \xrightarrow[\text{2. R'X}]{\text{1. Na(or NaNH}_2)} R-C \equiv C-R' + NaX \quad (9-15)$$

9.5-2 Formation of Heavy Metal Acetylides Alkynes with acidic hydrogens (1-alkynes) will also react with ammoniacal solutions of heavy metal ions such as cuprous and silver ion. The *copper and silver acetylides* formed are insoluble in water and *have characteristic colors* which may be used as a test to distinguish 1-alkynes from other alkynes and alkenes.

$$2CH_3-C \equiv C-H + Cu_2Cl_2 + 2NH_4OH \longrightarrow$$
$$2NH_4Cl + 2H_2O + \underset{\text{(red ppt.)}}{(CH_3-C \equiv C)_2Cu_2 \downarrow} \quad (9-16)$$

$$CH_3-C \equiv C-H + AgNO_3 + NH_4OH \longrightarrow$$
$$NH_4NO_3 + H_2O + \underset{\text{(white ppt.)}}{CH_3-C \equiv CAg \downarrow} \quad (9-17)$$

The cuprous and silver acetylides are very explosive when dry. They should be handled with care and kept under solution. Cases have been reported where the dry compounds have exploded at the touch of a feather.

9.6 Preparation of Alkynes

The reactions used for the synthesis of alkynes are very similar to the elimination reactions used to prepare alkenes in the laboratory. Whereas the dehydrohalogenation of an alkyl halide will produce an alkene, the dehydrohalogenation of vicinal dihalides will form alkynes. Alcoholic KOH can be used to remove one molecule of hydrogen halide to form a vinyl halide, or a stronger base, such as sodamide, can remove two molecules of hydrogen halide forming the triple bond.

SUMMARY

$$\underset{\text{vicinal dihalide}}{R-\underset{Br}{\overset{H}{C}}-\underset{H}{\overset{Br}{C}}-R'} \xrightarrow[\text{(or just NaNH}_2\text{)}]{\text{alc. KOH} \atop \text{then NaNH}_2} \underset{\text{alkyne}}{R-C\equiv C-R'} \qquad (9\text{-}18)$$

The elimination reaction proceeds in two steps, first forming a vinyl halide, than an alkyne. The reaction can be stopped at the vinyl halide stage by using alcoholic KOH for the elimination.

$$R-\underset{Br}{\overset{H}{C}}-\underset{H}{\overset{Br}{C}}-R' \xrightarrow{\text{alc. KOH}} \underset{\text{(a vinyl bromide)}}{R-\underset{Br}{C}=CHR'} \qquad (9\text{-}19)$$

$$\underset{Br}{R-C=CHR'} \xrightarrow{\text{NaNH}_2} \underset{\text{alkyne}}{R-C\equiv C-R'} \qquad (9\text{-}20)$$

A specific reaction for the preparation of the simplest Alkyne (Acetylene, $H-C\equiv C-H$), involves treating the inorganic compound calcium carbide, CaC_2, with water.

$$CaC_2 + 2H_2O \longrightarrow H-C\equiv C-H + Ca(OH)_2 \qquad (9\text{-}21)$$

Acetylene is used as a fuel in the oxyacetylene blow torch. When burned with oxygen the flame reaches a temperature as high as 3000° C, which can be used to cut and weld metals.

Summary

• The alkynes are unsaturated hydrocarbons having the general formula C_nH_{2n-2}, and are characterized by the presence of a carbon–carbon triple bond.

• The various alkynes have the *-yne* ending in the IUPAC system of nomenclature.

• The triple bond has *sp* hybridization and can be considered as being composed of one sigma and two pi bonds (Chapter 1 Section 1.6).

• The alkynes undergo addition reactions similar to those of the alkenes.

   (a) Addition of two moles of halogen produces a tetrahalide.

   (b) Addition of one mole of hydrogen can produce either a *cis* or a *trans* alkene.

   (c) Addition of two moles of hydrogen halide produces a *gem*–dihalide.

   (d) Addition of one mole of HCN produces acrylonitrile, which can be polymerized to Orlon.

   (e) Addition of water produces an aldehyde from acetylene, and ketones from other alkynes, because of enol–keto tautomerism.

- The hydrogens in a terminal alkyne are weakly acidic. The resulting sodium acetylides are useful in the synthesis of other alkynes, and in lengthening the carbon chain.

    Terminal alkynes will form colored precipitates with ammoniacal solutions of heavy metal ions such as $Ag^{\oplus}$ and $Cu^{\oplus}$. This may be used as a test to distinguish 1-alkynes from other alkynes.

- Alkynes may be prepared by the dehydrohalogenation of vicinal dihalides, an elimination reaction.

    Acetylene can be prepared by the reaction of calcium carbide and water.

PROBLEMS

1. Draw the correct structural formulas for each of the following compounds.
    (a) 3-bromo-1-butyne
    (b) 2,3-dimethyl-1-hexen-4-yne
    (c) vinylacetylene
    (d) 1,1-dibromo-3-heptyne

2. Give the correct IUPAC name for each of the following compounds.
    (a) $CH_3-C\equiv C-\overset{\overset{Cl}{|}}{C}-(CH_3)_2$
    (b) $CH_2=CH-C\equiv C-H$
    (c) $CH_3-CH_2-C(CH_3)_2-C\equiv C-CH_3$
    (d) dimethylacetylene

3. Complete each of the following equations by drawing the structural formula(s) of the product(s). If no reaction occurs, write N.R.
    (a) 1-butyne + excess HI $\longrightarrow$
    (b) $CH_3-C\equiv C-CH_2-CH_3 + Ag(NH_3)_2^{\oplus} \longrightarrow$
    (c) 2-butyne + 1 mole H_2/Pt $\longrightarrow$
    (d) 2-butyne + H_2O + H_2SO_4 + $HgSO_4 \longrightarrow$
    (e) $C_2H_5-C\equiv C-C_2H_5$ + Na in liq. $NH_3 \longrightarrow$
    (f) propyne + $Cu(NH_3)_2^{\oplus} \longrightarrow$
    (g) 1,2-dibromopropane $\xrightarrow[\text{2. NaNH}_2]{\text{1. alc. KOH}}$

4. Perform all of the following syntheses from the indicated starting materials.
    (a) 2-hydroxypropane to propyne
    (b) calcium carbide and any other inorganic reagents, to 2-chloro-2-iodopropane
    (c) calcium carbide to 1-butanol
    (d) ethylene and propylene to 2-pentyne
    (e) inorganic reagents only to 2-butyne
    (f) *inorganic reagents only to 4-bromoheptane
    (g) *inorganic reagents only to 3,4-dibromohexane
    (h) cis-2-butene to trans-2-butene

5. Name a chemical test, or single reagent, which will distinguish between the following pairs of compounds.
    (a) 2-butene and propyne
    (b) 1-heptyne and 3-heptyne
    (c) 2-methylpentane and ethyne

6. *Compound A, C_6H_{12}, is treated with chlorine in the presence of carbon tetrachloride to form compound B, $C_6H_{12}Cl_2$; B is treated with alcoholic

KOH, followed by NaNH$_2$, resulting in the formation of compound C, C$_6$H$_{10}$; C is treated with hydrogen gas over platinum, and forms a compound identified as 2-methylpentane. Compound C does not react with sodamide or an ammoniacal solution of silver nitrate. When A is treated with potassium permanganate and heated, two acids, D and E, are formed. E is identified as ethanoic acid. From the above information, deduce and write the correct structural formulas of compounds A through E.

10
Aromatic Hydrocarbons

A refinery in Port Arthur, Texas, where benzene, a building block of aromatic compounds, is made 'round-the-clock, seven days a week.

Courtesy of Gulf Oil Chemicals Company.

10.1 Introduction

Because of the great variety in the structural features of organic compounds, it is essential to divide organic compounds into several categories. For example, the compounds discussed in Chapters 7, 8, and 9, the saturated and unsaturated hydrocarbons, are said to be *aliphatic* compounds. **Aliphatic compounds** are saturated or unsaturated open-chain or cyclic (alicyclic) compounds related to methane. A second class of compounds is referred to as *aromatic* because of the characteristic fragrant odors of the compounds. **Aromatic compounds** are related in structure to the parent compound, **benzene,** and will be discussed in this chapter. A third class of organic compounds, **heterocyclic compounds,** will be discussed later. This last class of compounds refers to substances having a ring with at least one element other than carbon (e.g., sulfur, nitrogen, oxygen, phosphorus) as part of the ring.

Benzene is the parent compound of all aromatic compounds. The major source of benzene is as a constituent of coal tar. It was first obtained in the laboratory, from coal gas, by the chemist Michael Faraday in 1825.

10.2 Structure of Benzene

The structure of the benzene molecule has already been discussed in connection with the resonance concept (Section 5.4). The great stability of aromatic compounds is attributed to the resonance stabilization of the conjugated system present in the molecule (see Figure 5.1).

The evidence for the proof of structure of benzene is interesting in that it illustrates the general procedures used to elucidate the structures of organic compounds.

Elemental analysis of benzene indicates the presence of only two elements, carbon and hydrogen; and a molecular formula of C_6H_6, based upon experiments that determine the molecular weight of benzene to be 78. The formula of C_6H_6 indicates that the benzene molecule is unsaturated, and thus it is to be expected that benzene should undergo the characteristic reactions of unsaturated hydrocarbons. However, benzene does not decolorize a solution of bromine in carbon tetrachloride, nor is benzene readily oxidized by potassium permanganate, reactions which are characteristic of the carbon–carbon double bond.

Benzene will react with bromine, but only in the presence of a catalyst, such as iron, the reaction being a *substitution reaction,* and not the addition reaction typical of unsaturated hydrocarbons.

$$C_6H_6 + Br_2 \xrightarrow{Fe} C_6H_5Br + HBr \quad\quad (10\text{-}1)$$
$$\text{bromobenzene}$$

The isolation of only one monobromo derivative suggests that all six of the hydrogens in benzene are equivalent, and replacement of any one hydrogen by bromine will form bromobenzene as a product.

Benzene will react with bromine or chlorine in the presence of sunlight, three moles of the halogen adding to the benzene.

$$C_6H_6 + 3Br_2 \xrightarrow{h\nu} C_6H_6Br_6 \quad\quad (10\text{-}2)$$
$$\text{(hexabromo derivative)}$$

Three moles of hydrogen will add to benzene in the presence of a catalyst to form cyclohexane.

$$C_6H_6 + 3H_2 \xrightarrow{Pt} \bigcirc \; (C_6H_{12}) \quad\quad (10\text{-}3)$$
$$\text{cyclohexane}$$

On the basis of this experimental evidence Kekulé deduced his formulas for benzene (Section 5.4) and led the way to our present day resonance hybrid interpretation of the benzene molecule.

10.3 Other Common Aromatic Compounds—The Hückel $4n + 2$ Rule

Many different types of organic compounds exhibit aromatic character. The theoretical basis for aromaticity comes from quantum mechanics in the form of the $4n + 2$ or Hückel rule. In order for a molecule to be considered aromatic it must contain cyclic clouds of delocalized π electrons above and below the plane of the molecule, and the π clouds must contain a total of $(4n + 2)$ π electrons where *n is a whole integer*.

Some typical aromatic hydrocarbons are

toluene   n-propyl benzene   styrene

In each of these compounds there is a total of six π electrons in the ring, so that $4n + 2 = 6$, or $n = 1$, a whole number. Hence, these compounds would be expected to be aromatic and to exhibit chemical properties similar to benzene.

Some aromatic compounds contain more than one ring; each of which has its own aromatic character.

triphenyl methane

(for each individual ring, $4n + 2 = 6, n = 1$)

Other compounds have condensed-ring systems where two carbon atoms are shared between rings:

naphthalene
$4n + 2 = 10, n = 2$

anthracene
$4n + 2 = 14, n = 3$

phenanthrene
$4n + 2 = 14, n = 3$

Some heterocyclic compounds such as pyridine ($4n + 2 = 6, n = 1$), for example, show aromatic characteristics.

Aromatic character is also shown by the following five- and seven-membered ring species.

Sec. 10.4 NOMENCLATURE

cyclopentadienyl anion
$4n + 2 = 6, n = 1$

cycloheptatrienyl cation
(tropylium ion)
$4n + 2 = 6, n = 1$

10.4 Nomenclature

Many aromatic compounds are named by their common names rather than by a systematic nomenclature, such as the IUPAC system. There are, however, fundamental rules that form a basis for a systematic nomenclature.

The general name for the class of compounds referred to as aromatic hydrocarbons is **arenes**. Removal of a hydrogen from the benzene ring forms the C_6H_5 or **phenyl group**, , analogous to the formation of an alkyl group from an alkane. Groups derived from aromatic compounds are referred to as **aryl groups** and are designated by the symbol Ar-. When the benzene ring contains more than one substituent a numbering system is used as in the IUPAC system. For example, there are three possible disubstituted derivatives of benzene; 1,2 or *ortho* when substituents are on adjacent carbon atoms, 1,3 or *meta* when substituents are separated by one carbon atom in the ring, 1,4 or *para* when substituents are separated by two carbon atoms in the ring.

ortho-nitrotoluene
(1-methyl-2-nitrobenzene)

meta-chloroiodobenzene
(1-chloro-3-iodobenzene)

para-dibromobenzene
(1,4-dibromo-benzene)

The prefixes o- for *ortho*, m- for *meta*, and p- for *para* are used frequently in naming disubstituted derivatives of benzene. A numbering system is used when more than two substituents are present.

2,4-dinitrotoluene

1,3,5-trimethylbenzene
(mesitylene)

Friedrich August Kekulé von Stradonitz (1829–96)—German chemist and professor at the universities of Ghent and Bonn—established the framework for the aromatic theory of organic compounds by his studies of the ring structure of benzene. (From the Dains Collection, courtesy of the Department of Chemistry, The University of Kansas.)

10.5 Electrophilic Aromatic Substitution Reactions

The resonance stabilization of the benzene ring due to the delocalization of the π electrons in the conjugated system makes the benzene ring much less reactive than the alkenes towards electrophilic reagents. The electron cloud in the benzene molecule is spread over the entire ring and so will not aid the attack of a weak electrophilic reagent as the π electrons in the double bond of an alkene do. Thus, only very strong electrophiles would be expected to attack the benzene ring. The weak electrophiles will attack the benzene ring only in the presence of Lewis acid or Brønsted acid catalysts, which serve to increase their electrophilicity by removal of the nucleophilic portion of the electrophilic reagent with an electron pair.

The various electrophilic aromatic substitution reactions differ only in the mode of generation of the reactive electrophilic reagent, $E^{\oplus}$, which then attacks the π electron cloud of the aromatic system to form a resonance stabilized carbonium ion. This carbonium ion can then react with a nucleophilic reagent, resulting in an addition reaction across a carbon–carbon double bond, or by loss of a proton resulting in a substitution reaction and the restoration of the aromatic system. The latter reaction, **electrophilic aromatic substitution,** is characteristic of aromatic compounds. It occurs preferentially to addition since the loss of a proton from the intermediate carbonium ion formed restores the conjugated system to the molecule, and forms a product that is much more stable because of the resonance stabilization associated with a conjugated system. These points can be summarized in the basic mechanism for aromatic electrophilic substitution reactions.

$$\underset{\substack{\text{(weak}\\\text{electrophile)}}}{\text{E—N}} + \text{(acid catalyst)} \longrightarrow \underset{\substack{\text{(strong electrophilic}\\\text{reagent generated)}}}{E^{\oplus}} + \text{A—N}^{\ominus} \quad (10\text{-}4)$$

Sec. 10.5 ELECTROPHILIC AROMATIC SUBSTITUTION REACTIONS

$$E^{\oplus} + C_6H_6 \longrightarrow [C_6H_6E]^{\oplus} \longrightarrow C_6H_5E + H^{\oplus} \qquad (10\text{-}5)$$

(would destroy conjugated system)

Some typical examples of aromatic electrophilic substitution reactions are nitration, sulfonation, halogenation, and Friedel–Crafts alkylation.

10.5-1 Nitration The nitration of aromatic compounds can usually be achieved at relatively low temperatures by using a mixture of concentrated nitric and sulfuric acids. The nitric acid is a source of the electrophilic reagent $NO_2^{\oplus}$, the nitronium ion; the function of the sulfuric acid is to generate the $NO_2^{\oplus}$ from the nitric acid by a Brønsted–Lowry acid–base type reaction. The overall reaction results in the substitution of a NO_2 (nitro) group for a hydrogen atom on the benzene ring.

$$C_6H_6 + HONO_2 \xrightarrow[50°C]{H_2SO_4} C_6H_5NO_2 + H_2O \qquad (10\text{-}6)$$

nitrobenzene

The mechanism for the nitration reaction is believed to be

(1) $\qquad HNO_3 + H_2SO_4 \rightleftharpoons H_2NO_3^{\oplus} + HSO_4^{\ominus} \qquad (10\text{-}7)$
$\qquad\quad$ base $\quad\;$ acid $\qquad\;\;$ conj. $\qquad$ conj.
$\qquad\qquad\qquad\qquad\qquad\qquad$ acid $\qquad\;\;$ base

(2) $\qquad H_2NO_3^{\oplus} \longrightarrow H_2O + NO_2^{\oplus} \qquad (10\text{-}8)$

(3) $\qquad C_6H_6 + NO_2^{\oplus} \longrightarrow [C_6H_6NO_2]^{\oplus} \longrightarrow C_6H_5NO_2 + H^{\oplus} \qquad (10\text{-}9)$

Step (1) is an acid–base reaction in which sulfuric acid acts as an acid and nitric acid (which is a weaker acid than sulfuric) acts as a base relative to the sulfuric acid. At equilibrium the nitric acid is converted to $NO_2^{\oplus}$ because step (2) is driven to completion by the removal of water by H_2SO_4.

$$H_2SO_4 + H_2O \rightleftharpoons H_3O^{\oplus} + HSO_4^{\ominus} \qquad (10\text{-}10)$$

The electrophilic reagent $NO_2^{\oplus}$ can *not* be formed from nitric acid by itself. Nitric acid when ionized will form nitrate ion, $NO_3^{\ominus}$, and concentrated HNO_3 contains little nitronium ion, $NO_2^{\oplus}$ at equilibrium. The sulfuric acid catalyst is needed to form the reactive electrophile, $NO_2^{\oplus}$, which nitrates the benzene ring (step 3).

10.5-2 Sulfonation The substitution of a SO_3H (sulfonic acid) group for an aromatic hydrogen is called **sulfonation.** The reaction can be run using fuming sulfuric acid ($H_2SO_4 \cdot SO_3$) at room temperature, or concentrated sulfuric acid at higher temperatures.

$$C_6H_5\text{-H} + \boxed{HO}SO_3H \xrightarrow{\Delta} C_6H_5\text{-}SO_3H + H_2O \quad (10\text{-}11)$$

benzenesulfonic acid

The electrophilic species in this reaction is believed to be either SO_3 or $SO_3H^{\oplus}$ (each has only six electrons around the sulfur atom), both of which are present in fuming sulfuric acid or hot concentrated sulfuric acid.

The rest of the mechanism is analogous to the general mechanism discussed earlier.

Occasionally aromatic compounds are sulfonated with chlorosulfonic acid, $ClSO_3H$. The sulfonyl chloride formed is hydrolyzed with water to the corresponding sulfonic acid derivative.

$$C_6H_5\text{-H} + \boxed{HO}SO_2Cl \longrightarrow H_2O + C_6H_5\text{-}SO_2Cl \xrightarrow{H_2O} HCl + C_6H_5\text{-}SO_3H$$

benzenesulfonyl chloride   benzenesulfonic acid

$$(10\text{-}12)$$

10.5-3 Halogenation The halogenation of a benzene ring by chlorine or bromine takes place only if a catalyst such as iron or ferric halide is present. This is because the halogens themselves are not very reactive electrophilic reagents, and the catalyst converts them into a more reactive electrophilic species, which can then attack the benzene ring. The actual catalyst for the reaction is the ferric halide, which behaves as a Lewis acid because of the presence of an empty orbital on the iron atom that can accept an electron pair from the halogen molecule. The ferric halide may be added directly to the reaction mixture or may be produced *in situ* by reaction of iron with the halogen used as the reagent. The **halonium ion** formed by reaction of the ferric halide with the halogen is the reactive electrophile that attacks the benzene ring. The reaction sequence may be illustrated by the chlorination of benzene.

$$2Fe + 3Cl_2 \longrightarrow 2FeCl_3 \quad (10\text{-}13)$$

Sec. 10.5 ELECTROPHILIC AROMATIC SUBSTITUTION REACTIONS

$$FeCl_3 + Cl_2 \rightleftharpoons Cl^{\oplus} + FeCl_4^{\ominus} \qquad (10-14)$$
chloronium ion
(strongly electrophilic)

$$\text{C}_6\text{H}_6 + Cl^{\oplus} \longrightarrow [\text{arenium ion}] \longrightarrow \text{C}_6\text{H}_5Cl + H^{\oplus} \qquad (10-15)$$

$$H^{\oplus} + FeCl_4^{\ominus} \longrightarrow HCl + FeCl_3 \qquad (10-16)$$

The intermediate carbonium ion is resonance stabilized and is perhaps best represented by the resonance hybrid structure

(resonance hybrid structure of the carbonium ion)

The resonance stabilization of the carbonium ion, which is present in all aromatic electrophilic substitution reactions, helps the addition of the electrophilic reagent to the aromatic nucleus. The overall reaction may be summarized as

$$\text{C}_6\text{H}_6 + Cl_2 \xrightarrow{Fe} \text{C}_6\text{H}_5Cl + HCl \qquad (10-17)$$
chlorobenzene

The presence of an alkyl side chain on the aromatic nucleus complicates the halogenation reaction. Competing reactions can occur, depending on the selection of reaction conditions. Using toluene, (structure of toluene with CH_3), as an example, halogenation in the presence of an iron catalyst proceeds by the mechanism just discussed, leading to electrophilic substitution of the benzene ring.

$$\text{toluene} + Cl_2 \xrightarrow{Fe\ (cat.)} \text{p-chlorotoluene} + \text{o-chlorotoluene} + \begin{pmatrix}\text{a little of the}\\ \textit{meta } \text{isomer}\end{pmatrix} + HCl \qquad (10-18)$$

However, reaction with chlorine in the presence of sunlight with no catalyst results in free-radical substitution of the alkyl side chain (similar to the free-radical halogenation of alkanes).

$$\text{C}_6\text{H}_5\text{CH}_3 + \text{Cl}_2 \xrightarrow{h\nu} \text{HCl} + \text{C}_6\text{H}_5\text{CH}_2\text{Cl} \xrightarrow[h\nu]{\text{Cl}_2} \text{C}_6\text{H}_5\text{CHCl}_2 \xrightarrow[h\nu]{\text{Cl}_2} \text{C}_6\text{H}_5\text{CCl}_3$$

benzyl chloride   benzal chloride   benzotrichloride

(10-19)

Reaction of benzene with iodine gives very poor yields, and fluorination is too vigorous a reaction to control. The iodination and fluorination of aromatic compounds will be discussed in connection with the reaction of diazonium salts in Section 16.11-1, -2, and -3, and in relation to thallation in Section 10.8.

10.5-4 Friedel-Crafts Alkylation The introduction of alkyl groups into the aromatic nucleus most often makes use of the Friedel-Crafts reaction discovered in 1878 by the French chemist, Charles Friedel, and his American co-worker, James M. Crafts. An alkyl halide is most frequently used as the alkylating agent in the presence of a Lewis acid catalyst such as anhydrous $AlCl_3$. (Other typical Lewis acid catalysts used include $FeCl_3$, $ZnCl_2$, $SnCl_2$, BF_3.) The function of the Lewis acid catalyst is to convert the alkyl halide into a carbonium ion, which is a reactive electrophilic species. The Friedel-Crafts reaction is usually run at or below room temperature since the reaction is very exothermic. During the reaction the catalyst is regenerated as HCl gas is evolved.

The reaction mechanism is typical for an aromatic electrophilic substitution reaction and can be illustrated by the reaction of benzene with ethyl chloride in the presence of anhydrous aluminum chloride as a catalyst.

(1) $CH_3-CH_2Cl + AlCl_3 \rightleftharpoons CH_3-CH_2^{\oplus} + AlCl_4^{\ominus}$ (10-20)

ethyl carbonium ion
(a reactive electrophile)

(2) $C_6H_6 + CH_3-CH_2^{\oplus} + AlCl_4^{\ominus} \longrightarrow [\text{arenium ion intermediate}] + AlCl_4^{\ominus} \longrightarrow C_6H_5CH_2CH_3 + H^{\oplus} + AlCl_4^{\ominus}$ (10-21)

(3) $H^{\oplus} + AlCl_4^{\ominus} \longrightarrow AlCl_3 + HCl$ (10-22)

The overall reaction can be summarized as

Sec. 10.5 ELECTROPHILIC AROMATIC SUBSTITUTION REACTIONS

$$C_6H_5\text{-H} + CH_3\text{-}CH_2\text{-}Cl \xrightarrow{AlCl_3, 0°C} C_6H_5\text{-}CH_2\text{-}CH_3 + HCl \quad (10\text{-}23)$$

ethyl benzene

and the general equation represented as

$$\text{Ar-H} + \text{R-X} \xrightarrow{AlCl_3} \text{Ar-R} + \text{HX} \quad (10\text{-}24)$$

aromatic hydrocarbon | alkyl halide | alkylated aromatic compound

The Friedel–Crafts reaction, despite its wide utility, has its limitations and problems. Polyalkylation is a serious problem as a side reaction, and its product often supercedes the monoalkyl as the predominant derivative.

$$C_6H_6 + 2\,CH_3Br \xrightarrow{AlCl_3} \text{o-xylene} + \text{p-xylene} \quad (10\text{-}25)$$

A serious limitation of the Friedel–Crafts reaction is that an aromatic ring containing electron-withdrawing substituents such as a nitro group (NO_2) cannot be alkylated. The presence of such groups deactivates the ring by withdrawing electrons from the π electron system, and so renders the aromatic ring less susceptible to attack by electrophilic reagents such as the intermediate formed carbonium ion. The Friedel–Crafts reaction will not occur with such compounds as nitrobenzene, for example.

$$C_6H_5NO_2 + \text{R-X} \xrightarrow{AlCl_3} \text{(no reaction)} \quad (10\text{-}26)$$

On the other hand, the presence of strongly electron-releasing substituents such as $-NH_2$, $-NHR$, $-NR_2$, or $-OH$ on the aromatic ring will also result in no alkylation. This is because these substituents have unshared pairs of electrons on nitrogen or oxygen atoms (are typical Lewis bases) and so will undergo a reaction with the Lewis-acid catalyst and prevent the catalyst from performing its proper function of generating the carbonium ion as the reactive electrophile essential for the Friedel–Crafts reaction to occur.

$$-\ddot{N}H_2 + AlCl_3 \rightleftharpoons -\overset{H}{\underset{H}{N^{\oplus}}}-\overset{\ominus}{AlCl_3} \qquad (10\text{-}27)$$

Lewis base   Lewis acid   Lewis acid–base complex

Perhaps the major limitation of the Friedel–Crafts reaction is that it involves carbonium ion intermediates that are prone to rearrangements to more stable carbonium ions. For example, the alkylation of benzene with n-propyl chloride produces i-propyl benzene as a major product rather than the expected n-propyl benzene.

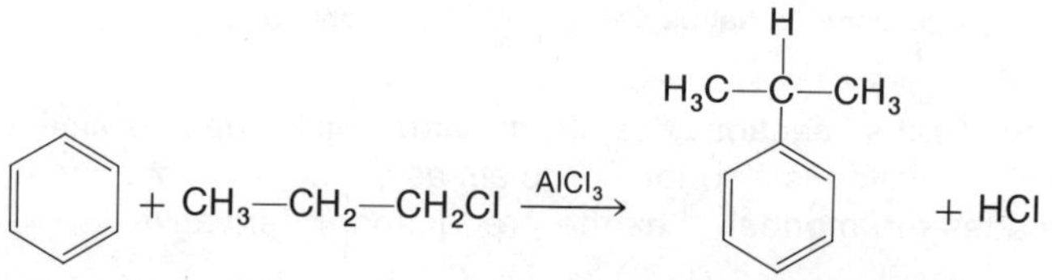

$$\text{C}_6\text{H}_6 + CH_3-CH_2-CH_2Cl \xrightarrow{AlCl_3} \text{C}_6\text{H}_5-C(H)(CH_3)_2 + HCl \qquad (10\text{-}28)$$

i-propyl benzene

This product results from the rearrangement of the n-propyl carbonium ion to the more stable i-propyl carbonium ion.

(1) $CH_3-CH_2-CH_2Cl + AlCl_3 \rightleftharpoons AlCl_4^{\ominus} + CH_3-CH_2-CH_2^{\oplus}$ (10-29)

n-propyl carbonium ion (1°)

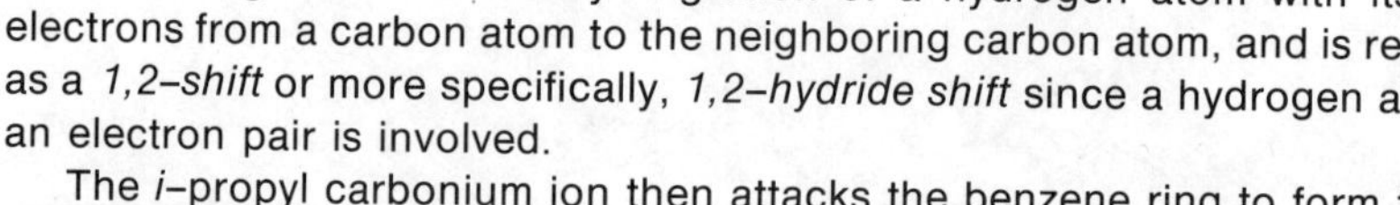

(2) $\underset{3}{CH_3}-\overset{H}{\underset{2}{C}}-\underset{1}{CH_2^{\oplus}}AlCl_4^{\ominus} \longrightarrow \underset{3}{CH_3}-\underset{2}{\overset{\oplus}{CH}}-\underset{1}{CH_3} + AlCl_4^{\ominus}$ (10-30)

(1°)   *i*-propyl carbonium ion (2°)

The rearrangement occurs by migration of a hydrogen atom with its pair of electrons from a carbon atom to the neighboring carbon atom, and is referred to as a *1,2-shift* or more specifically, *1,2-hydride shift* since a hydrogen atom with an electron pair is involved.

The i-propyl carbonium ion then attacks the benzene ring to form the rearranged product.

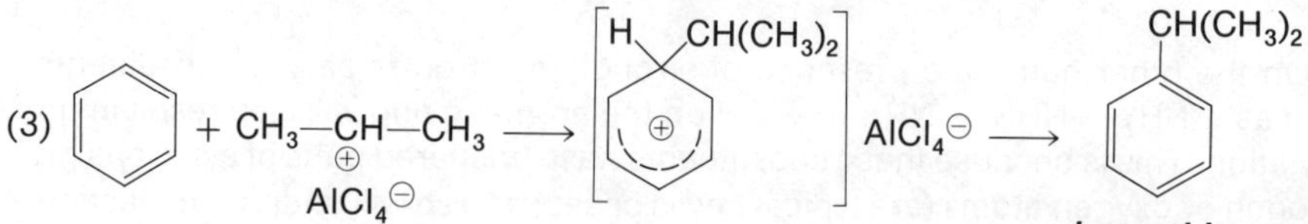

(3) $\text{C}_6\text{H}_6 + CH_3-\overset{\oplus}{CH}-CH_3 \longrightarrow [\text{arenium ion}] \xrightarrow{AlCl_4^{\ominus}} \text{C}_6\text{H}_5-CH(CH_3)_2$

$AlCl_4^{\ominus}$

i-propyl benzene

$+ H^{\oplus} + AlCl_4^{\ominus}$ (10-31)

(4) $H^{\oplus} + AlCl_4^{\ominus} \longrightarrow HCl\uparrow + AlCl_3$ (10-32)

Sec. 10.7 ELECTROPHILIC SUBSTITUTION REACTIONS

Occasionally alkenes are used instead of alkyl halides as alkylating agents.

$$C_6H_6 + CH_2{=}CH_2 \xrightarrow{AlCl_3} C_6H_5{-}CH_2{-}CH_3 \quad \text{(ethyl benzene)} \tag{10-33}$$

A modification of the alkylation of aromatic compounds called the Friedel–Crafts acylation reaction (using acyl halides, $R{-}\overset{\overset{O}{\|}}{C}{-}X$) to form aromatic ketones will be discussed in Section 15.10-2.

10.6 Oxidation of the Alkyl Side Chain

The benzene ring itself is stable to oxidizing agents such as potassium permanganate or potassium dichromate. When alkyl benzenes are treated with strong oxidizing agents it is the alkyl side chain that is oxidized to a carboxyl group, COOH. The length of the alkyl side chain is immaterial; the carbon atom directly attached to the aromatic ring is converted to a carboxyl group, the rest of the side chain is oxidized to carbon dioxide and water.

$$\text{toluene} + KMnO_4 \xrightarrow{H_2SO_4} \text{benzoic acid} \tag{10-34}$$

10.7 Orientation in Aromatic Electrophilic Substitution Reactions

In our discussion on electrophilic substitution reactions the benzene molecule served as an ideal example in that only one monosubstitution product is obtained, since all six positions of the benzene ring are equivalent. The reaction of aromatic compounds already containing a substituent, X, with some electrophilic reagent, $E^{\oplus}$, is more complex in that three possible isomers of the disubstituted aromatic compound can be formed.

ortho   meta   para

On a purely statistical basis (since there are two positions *ortho,* two positions *meta,* and one position *para* to X) one would expect 40% of the *ortho* isomer,

40% of the *meta* isomer, and 20% of the *para* isomer to be formed. However, experimental results rarely coincide with this statistical analysis based on mathematical probability.

As an example, if one nitrates benzene and then brominates the nitrobenzene, the predominant product is the *meta*–nitrobromobenzene, with only insignificant amounts of the *ortho* and *para* isomers formed. But, if the reaction order is reversed, that is if bromination is followed by nitration, a mixture of *ortho* and *para* isomers is formed, with very little of the *meta* isomer.

$$\text{benzene} \xrightarrow[\text{H}_2\text{SO}_4]{\text{HNO}_3} \text{nitrobenzene} \xrightarrow[\text{Br}_2]{\text{Fe}} \textit{m}\text{-nitrobromobenzene} \quad (10\text{-}35)$$

$$\xrightarrow[\text{Br}_2]{\text{Fe}} \text{bromobenzene} \xrightarrow[\text{HNO}_3]{\text{H}_2\text{SO}_4} \textit{o}\text{-nitrobromobenzene} + \textit{p}\text{-nitrobromobenzene} \quad (10\text{-}36)$$

The reaction products isolated from these and numerous other experiments indicate that *any ring substituent (X) that is present in the aromatic ring will determine the position taken by the next entering substituent (E)*. In our illustration the nitro group directed the bromine to the *meta* position to itself, but the bromine directed the incoming nitro group to the *ortho* and *para* positions.

The influence of the substituents (X) in directing the orientation of any in-

Table 10.1 Orientation of Substituents in Electrophilic Aromatic Substitution Reactions

Ortho–para directors	Meta directors
—NH$_2$	—NO$_2$
—OH	—NR$_3^+$
—OR	—CN
—R (alkyl)	—COOH
—C$_6$H$_5$	—SO$_3$H
—F	—CHO
—Cl	—COR
—Br	
—I	

Sec. 10.7 ELECTROPHILIC SUBSTITUTION REACTIONS

coming group (E) appears to be characteristic of the general behavior of these substituents (X), regardless of the nature of the electrophilic substitution reaction which follows. Certain substituents are said to be *ortho–para-directing,* because these substituents direct the entering group *ortho* and *para* to themselves; other substituents are *meta-directing* as they cause the incoming group to go to the *meta* position. Table 10.1 lists some of the more common substituents and places them into one of two classes: *ortho–para*-directing or *meta*-directing.

It is important to know into which class the various substituents belong in order to be able to perform laboratory syntheses and get the desired isomer (*ortho, meta* or *para*) in relatively pure form with as little contamination as possible by the other isomers. The *ortho* and *para* isomers are produced simultaneously, but usually can be separated from each other fairly easily by fractional crystallization or fractional distillation because of differences in their solubility behavior and boiling points. The proper reaction sequence must be chosen, so that the substituents end up in the desired positions in the product. Thus, the preparation of *o*–nitrobenzoic acid for example, would proceed by the reaction sequence

$$\text{benzene} \xrightarrow[\text{AlCl}_3]{\text{CH}_3\text{Cl}} \text{toluene} \xrightarrow[\text{H}_2\text{SO}_4]{\text{HNO}_3} o\text{-nitrotoluene} \xrightarrow[\text{H}_2\text{SO}_4]{\text{K}_2\text{Cr}_2\text{O}_7} o\text{-nitrobenzoic acid} \quad (10\text{-}37)$$

but, in order to prepare *m*–nitrobenzoic acid, the sequence would be

$$\text{benzene} \xrightarrow[\text{AlCl}_3]{\text{CH}_3\text{Cl}} \text{toluene} \xrightarrow[\text{H}_2\text{SO}_4]{\text{K}_2\text{Cr}_2\text{O}_7} \text{benzoic acid} \xrightarrow[\text{H}_2\text{SO}_4]{\text{HNO}_3} m\text{-nitrobenzoic acid} \quad (10\text{-}38)$$

This can be seen from the orientation expected from the various substituents in Table 10.1. *Ortho–para*-directing groups usually donate electrons into the aromatic ring and increase the activity of the aromatic nucleus, whereas *meta*-directing groups tend to withdraw electrons from the aromatic ring (deactivate the ring) making the aromatic π electron cloud less reactive toward electrophilic substitution.

10.7-1 Theory of Reactivity and Orientation

The theoretical explanation of the experimental results can best be seen if one considers the overall mechanism for the aromatic electrophilic substitution reaction. The essential points are that the reaction depends on the reactivity of the aromatic π electron cloud and the stability of the carbonium ion formed in the course of the reaction. Resonance structures can be used to explain why certain substituents are *ortho–para*-directors and others are *meta*-directing groups.

Let us first consider an aromatic nucleus containing a substituent X, which releases electrons into the ring by resonance. The following resonance structures can be drawn.

in which the electron density is distributed primarily to the *ortho* and *para* positions in the ring, which will be favored for attack by an electrophilic reagent, E⊕. The substituent X will then be an *ortho–para*-director. An alternative explanation is to determine the stability of the intermediate carbonium ion formed by examining its resonance structures. Let us cite one example (substitution *para* to X) to illustrate the principles involved. The resonance structures for the intermediate carbonium ion can be represented as

Of the contributing resonance structures, structure IV is the most significant in that the positive charge is stabilized by the unshared electron pairs on substituent X. A similar set of structures could be drawn for *ortho* substitution. *Meta* substitution does not permit the charge to be localized onto X, as can be seen by examination of a resonance structure such as

Thus, the substitution is predominantly *ortho–para* (mostly *para,* possibly because of steric factors in the *ortho* positions).

If the substituent X is an electron-withdrawing group, then electrophilic substitution will occur principally in the *meta* position. Once again, an examination of the contributing resonance structures will be helpful.

It can be seen that the electron density is being taken out of the ring by substituent X and that the depletion of electrons places a positive charge on the *ortho* and

Sec. 10.8 THALLATION

para positions to X. In other words, the *meta* position has a greater share of the electron density than either the *ortho* or *para* positions in these resonance structures, and would be more susceptible to attack by electron-seeking reagents like E⊕. Thus, in the case of removal of electron density from the aromatic ring, electrophilic substitution will occur mainly in the *meta* position, and the substituent X is classified as a *meta*-directing group. The alternative explanation in terms of stability of the carbonium ion formed can be illustrated by the reaction of nitrobenzene (NO_2 group is *meta*-directing) with an electrophile E⊕. Two possible resonance structures for the intermediate carbonium ion are

I
(for *para* substitution)

II
(for *meta* substitution)

Since the nitrogen atom in the NO_2 group is positively charged, resonance structure I, which places a positive charge on the atom adjacent to the nitrogen, is less stable than resonance structure II with the similar positive charge further away from the nitrogen atom. Structure II represents the structure to be expected for *meta* substitution while structure I would be expected if substitution occurred *para* to the nitro group. Structure II is more stable and so the nitro group is a *meta*-directing group. A similar type of reasoning can be used to explain why other groups are *meta*-directing rather than *ortho–para*. The presence of any *meta*-directing group tends to deactivate the aromatic ring by withdrawal of electrons, making the compound less reactive toward electrophilic reagents than the compound would be if the substituent were not present.

10.8 Thallation

From the preceding discussions the reader can see that the major problem in synthesizing aromatic compounds is in the preparation of a pure *ortho,* pure *meta,* or pure *para* isomer; control of orientation of substituents in electrophilic aromatic substitution reactions is extremely difficult.

Recently, Professor Edward C. Taylor at Princeton University and Professor Alexander McKillop at the University of East Anglia, using highly toxic thallium salts, succeeded in preparing arylthallium compounds. These compounds could then be converted by a Lewis acid–base reaction into an *ortho, meta,* or *para* isomer depending on what substituents were present on the aromatic nucleus.

The thallation reaction involves electrophilic attack on the aromatic ring by the Lewis acid thallium in the form of the compound thallium trifluoroacetate, $Tl(OOCCF_3)_3$, in trifluoroacetic acid, CF_3COOH. The arylthallium di-trifluoroacetates formed are stable crystalline compounds.

$$\text{C}_6\text{H}_6 + \text{Tl(OOCCF}_3)_3 \xrightarrow{\text{CF}_3\text{COOH}} \text{CF}_3\text{COOH} + \text{C}_6\text{H}_5\text{Tl(OOCCF}_3)_2 \quad (10\text{-}39)$$

arylthallium di-trifluoroacetate

The usefulness of the thallation reaction is that the orientation of substituents in electrophilic aromatic substitution reactions can be controlled. For example, thallation is almost exclusively *para* to —R, —Cl, and —OCH₃ groups, because of the bulk of the electrophilic reagent, thallium trifluoroacetate, which attacks the uncrowded *para* position.

$$\text{chlorobenzene} + \text{Tl(OOCCF}_3)_3 \xrightarrow{\text{CF}_3\text{COOH}} p\text{-chlorophenyl thallium di-trifluoroacetate} \quad (10\text{-}40)$$

Thallation is *ortho* to substituents such as —COOH and —COOCH₃, even though these groups may be normally *meta*-directing in electrophilic aromatic substitution reactions. This is believed to be due to the formation of a complex between the electrophilic Tl(OOCCF₃)₃ and the substituent on the aromatic ring, the thallium being held at the proper distance for intramolecular bonding to occur in the *ortho* position.

$$\text{methyl benzoate} \xrightarrow{\text{Tl(OOCCF}_3)_3} [\text{complex}] \longrightarrow o\text{-carbomethoxy-phenyl di-trifluoroacetate} \quad (10\text{-}41)$$

The crystalline arylthallium di-trifluoroacetates are useful intermediates in organic syntheses. Although the direct introduction of an iodine in the aromatic ring is difficult, the thallation of benzene, followed by treatment with potassium iodide produces iodobenzene in good yield.

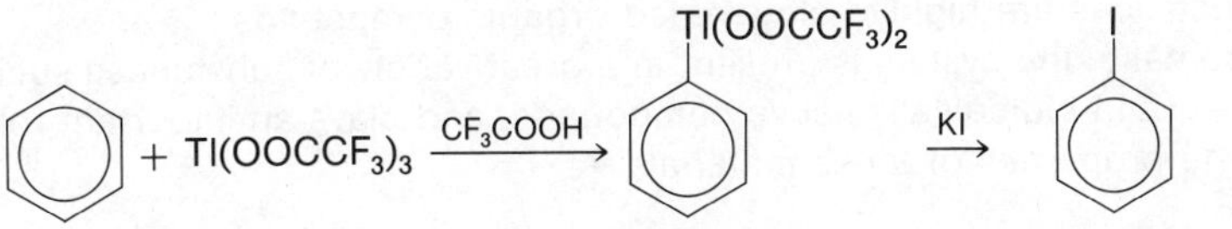

$$\text{C}_6\text{H}_6 + \text{Tl(OOCCF}_3)_3 \xrightarrow{\text{CF}_3\text{COOH}} \text{C}_6\text{H}_5\text{Tl(OOCCF}_3)_2 \xrightarrow{\text{KI}} \text{C}_6\text{H}_5\text{I} \quad (10\text{-}42)$$

iodobenzene

10.9 Some Important Derivatives of Benzene

The syntheses of some very important aromatic compounds can be accomplished by using the reactions and orientation rules discussed in this chapter. Some of the compounds are very familiar to all of us and only a few examples will be presented.

Ethylbenzene, which can be obtained by the Friedel–Crafts reaction from benzene, can be converted to the compound styrene, which can be polymerized to the well-known plastic, polystyrene.

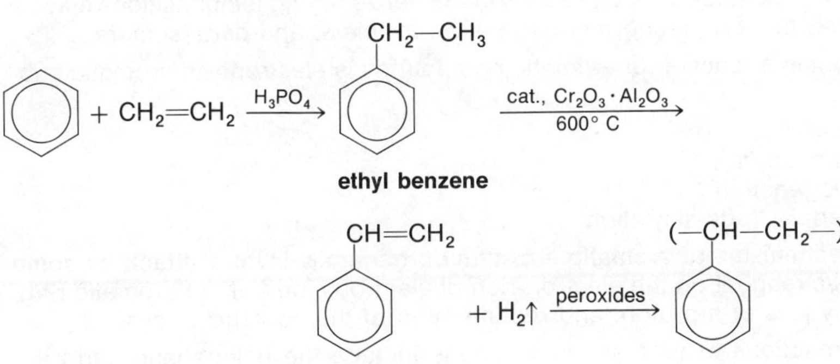

$$\text{C}_6\text{H}_6 + \text{CH}_2{=}\text{CH}_2 \xrightarrow{\text{H}_3\text{PO}_4} \text{C}_6\text{H}_5\text{—CH}_2\text{—CH}_3 \xrightarrow[600°\text{C}]{\text{cat., Cr}_2\text{O}_3 \cdot \text{Al}_2\text{O}_3}$$

ethyl benzene

$$\text{C}_6\text{H}_5\text{—CH}{=}\text{CH}_2 + \text{H}_2\uparrow \xrightarrow{\text{peroxides}} (\text{—CH(C}_6\text{H}_5)\text{—CH}_2\text{—})_n \quad (10\text{-}43)$$

styrene   polystyrene

The stepwise nitration of toluene will produce 2,4,6-trinitrotoluene, commonly known as TNT, an explosive.

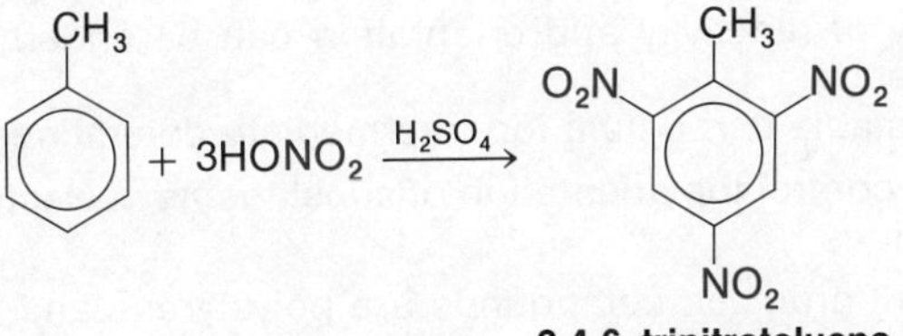

$$\text{C}_6\text{H}_5\text{CH}_3 + 3\text{HONO}_2 \xrightarrow{\text{H}_2\text{SO}_4} \text{C}_6\text{H}_2(\text{CH}_3)(\text{NO}_2)_3 \quad (10\text{-}44)$$

2,4,6-trinitrotoluene
(TNT)

The compound *p*-dichlorobenzene is a moth-repellent and is found in most homes. It is made by chlorination of benzene.

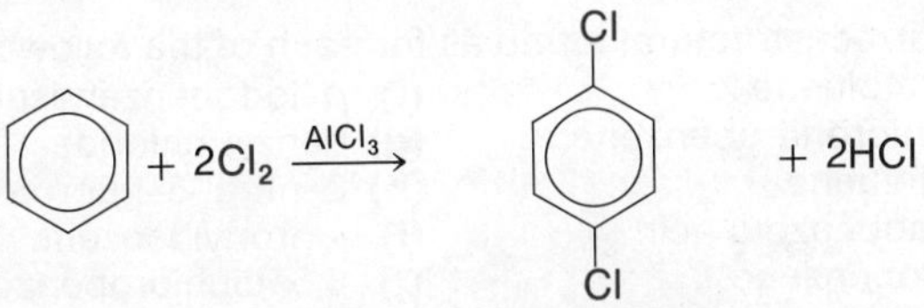

$$\text{C}_6\text{H}_6 + 2\text{Cl}_2 \xrightarrow{\text{AlCl}_3} p\text{-C}_6\text{H}_4\text{Cl}_2 + 2\text{HCl} \quad (10\text{-}45)$$

p-dichlorobenzene

Many insecticides are highly halogenated organic compounds.

The aromatic ring system is present in a great variety of substances such as drugs, dyes, and biologically active compounds, and plays an important role in many of the properties of these materials.

Summary

- Benzene is a typical aromatic compound. It has a cyclic conjugated system stabilized by resonance.

    The resonance hybrid formula for benzene is accepted today over the Kekulé formulas postulated 100 years ago.

- The Hückel $4n + 2$ rule for aromaticity demonstrates that many different types of organic species exhibit aromatic character.

- Aromatic compounds can be named by the IUPAC system; however, many are better known by their common names.

    When two substituents are present on the benzene ring their positions may be indicated by designating them as the *ortho*, *meta*, and *para* isomers.

- The principal reaction of aromatic compounds is electrophilic aromatic substitution.

    (a) Nitration.
    (b) Sulfonation.
    (c) Halogenation.
    (d) Friedel–Crafts alkylation.

- The mechanisms of aromatic substitution generally involve attack by some electrophilic reagent on the electron-rich pi electron cloud of the aromatic ring, followed by loss of a proton, and regeneration of the conjugated system.

- Other reactions of aromatic compounds include the halogenation and the oxidation of alkyl side chains.

- The orientation of a second substituent in electrophilic substitution reactions is influenced by the presence of the first substituent.

    (a) Some substituents are *ortho–para*-directors, others are *meta*-directors.
    (b) The theory of reactivity and orientation can be understood in terms of resonance effects.
    (c) The most stable carbonium ion intermediate determines the orientation.

- Thallation can control the orientation of substituents in electrophilic aromatic substitution reactions.

- Some important aromatic compounds are polystyrene and TNT.

PROBLEMS

1 Draw the correct structural formulas for each of the following compounds.
   (a) 3-chlorotoluene
   (b) 2,4-dichloronitrobenzene
   (c) *o*-nitrotoluene
   (d) *m*-bromobenzoic acid
   (e) *p*-dinitrobenzene
   (f) *p*-iodobenzenesulfonic acid
   (g) benzyl chloride
   (h) 2-nitro-5-fluorotoluene
   (i) *i*-propylbenzene
   (j) 3,5-dichlorobenzoic acid

PROBLEMS

2 Name the following compounds.

(a) 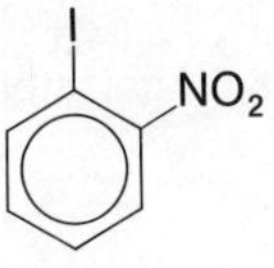 1-iodo-2-nitrobenzene

(b)  1-chloro-4-fluorobenzene

(c) 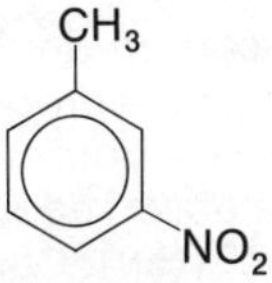 3-nitrotoluene (m-methylnitrobenzene)

(d) 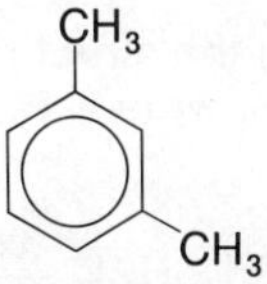 1,3-dimethylbenzene (m-xylene)

(e) 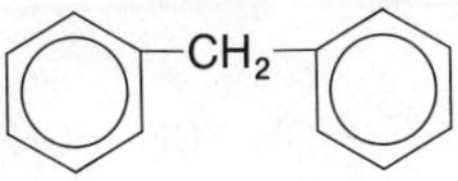 diphenylmethane

(f) 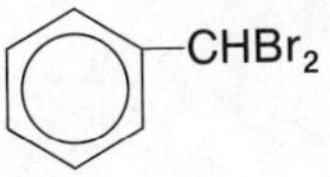 (dibromomethyl)benzene

(g) 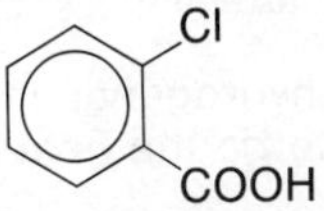 2-chlorobenzoic acid

(h) 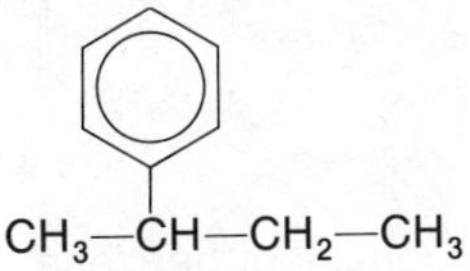 (sec-butyl)benzene

(i) 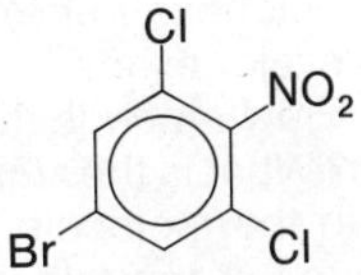 1-bromo-3,5-dichloro-4-nitrobenzene

(j) 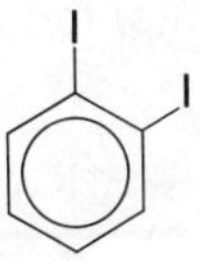 1,2-diiodobenzene

3 (a) Write the correct structural formulas of *all* the mononitrosubstitution products of the three isomeric dibromobenzenes (*ortho, meta,* and *para*) formed by nitrating each of the isomers.

(b) An unknown compound is either *o*-xylene, *m*-xylene, or *p*-xylene. The unknown will form two monochloro-substitution derivatives when reacted with chlorine and iron. Which compound was the unknown?

4 * Considering the Hückel (4*n* + 2) rule, which of the following species would you expect to have aromatic properties?

(a) 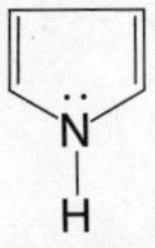 pyrrole

(b) 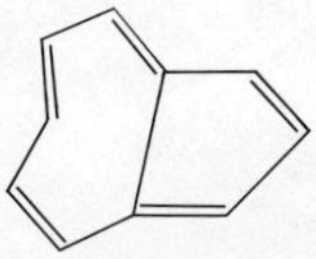 azulene

(c) 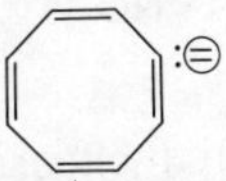 cyclooctatetraenyl dianion

(d)  1,3-cyclohexadiene

5 Consider the following hydrogenation reaction.

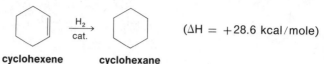

If the heat of hydrogenation is observed to be +28.6 kcal/mole what would you predict to be the heat of hydrogenation for the reaction

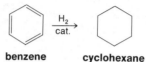

if one considers benzene to contain three carbon–carbon double bonds (like 1,3,5-cyclohexatriene)? The actual observed heat of hydrogenation of benzene to cyclohexane is +49.8 kcal/mole. This differs from your calculated value. How much is the difference between the observed and calculated values? What is the significance of this difference? (HINT: Consider the resonance in the molecule.)

6 Complete each of the following equations by writing the structural formula(s) of the product(s) formed. If no reaction occurs, write N.R.

(a) C$_6$H$_5$COOH + HNO$_3$ $\xrightarrow{H_2SO_4}$

(b) C$_6$H$_5$CH$_3$ + Br$_2$ $\xrightarrow{h\nu}$

(c) C$_6$H$_6$ + CH$_2$=CH$_2$ $\xrightarrow{ZnCl_2}$

(d) (o-C(CH$_3$)$_3$)(NO$_2$)C$_6$H$_4$ $\xrightarrow[H_2SO_4]{K_2Cr_2O_7}$

(e) C$_6$H$_6$ + excess Cl$_2$ $\xrightarrow{h\nu}$

(f) naphthalene + excess H$_2$ $\xrightarrow{Ni}$

(g) C$_6$H$_6$ + CH$_3$CHCl$_2$ $\xrightarrow{AlCl_3}$

(h) C$_6$H$_5$—CH(CH$_3$)—CH$_2$—CH$_3$ + Cl$_2$ $\xrightarrow{h\nu}$

(i) C$_6$H$_5$CH$_3$ + Tl(OOCCF$_3$)$_3$ $\xrightarrow{CF_3COOH}$

(j) C$_6$H$_5$CH$_3$ + H$_2$SO$_4$ $\xrightarrow{\Delta}$

(k) o-xylene + K$_2$Cr$_2$O$_7$ $\xrightarrow{H_2SO_4}$

(l) p-nitrotoluene + Br$_2$ $\xrightarrow{Fe}$

PROBLEMS

7 Perform all of the following syntheses, starting with benzene and any necessary inorganic reagents:
   (a) benzene to *m*-nitrobenzoic acid
   (b) benzene to *m*-chloronitrobenzene
   (c) benzene to styrene
   (d) benzene to phenylacetylene
   (e) benzene to 2-bromo-4-nitrobenzoic acid
   (f) benzene to *p*-toluenesulfonic acid
   (g) *benzene to 1,1-dibromo-1-phenylethane
   (h) benzene to cyclohexyl bromide
   (i) benzene to 1,4-dimethylcyclohexane
   (j) toluene to *p*-iodotoluene

8 Name a chemical test or single reagent which can be used to distinguish between the following pairs of compounds.
   (a) benzene and cyclohexane
   (b) benzene and cyclohexene
   (c) styrene and ethylbenzene
   (d) phenylacetylene and toluene
   (e) benzene and benzenesulfonic acid

9 Arrange the following compounds in order of increasing tendency to undergo electrophilic aromatic substitution:

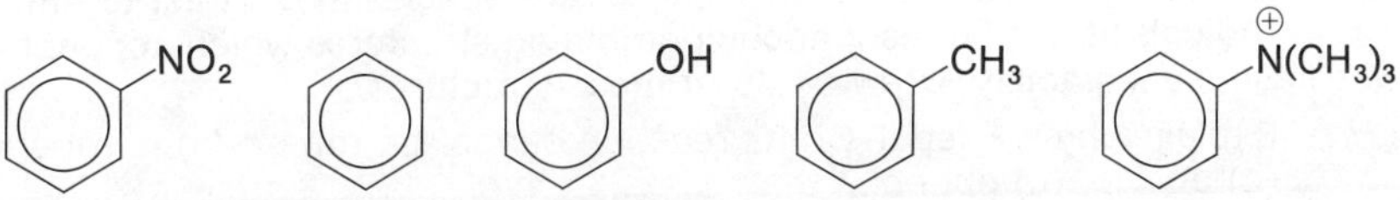

10 Draw appropriate resonance contributing structures for

(a) 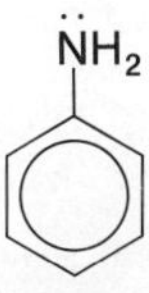 aniline   (b) 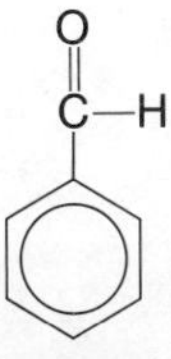 benzaldehyde

to account for their orientation in electrophilic aromatic substitution reactions. Explain the significance of the structures you drew.

11 Consider the reaction of ethylbenzene with bromine (a) in the presence of an iron catalyst, and (b) in the presence of sunlight. What products are formed in each of these reactions? Account for their formation by writing all the steps for the mechanisms involved in (a) and (b).

12 When benzene is reacted with 1-chloro-2-methylpropane in the presence of zinc chloride, the major reaction product obtained is *t*-butylbenzene. Explain. (Write all the steps for the mechanism involved in the reaction).

13 *When *t*-butylbenzene is nitrated, the product formed is almost exclusively *p*-nitro-*t*-butylbenzene, with very little of the *ortho* or *meta* isomer being formed. Explain why this occurs.

14 What would you expect to be the major reaction product in the following reaction? Explain.

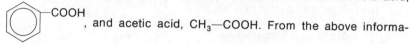

15 Why is 2,4,6-trinitrotoluene weakly acidic?

16 A compound A, C_9H_{10}, is treated with chlorine to form compound B, $C_9H_{10}Cl_2$, which is hydrolyzed by aqueous sodium hydroxide to compound C, $C_9H_{12}O_2$. When C is oxidized, the products formed are benzoic acid, C₆H₅—COOH, and acetic acid, CH_3—COOH. From the above information, deduce and write the correct structural formulas of compounds A through C.

17 The triphenylmethyl free radical, $(C_6H_5)_3C\cdot$, discovered by Moses Gomberg in 1900, is an extremely stable species in contrast to other free radicals. It has 36 resonance contributing structures which account for its unusual stability. Draw all 36 of these structures.

18 Explain why a Friedel–Crafts reaction cannot be run on (a) nitrobenzene, (b) aniline, (c) phenol.

11 Alcohols and Phenols

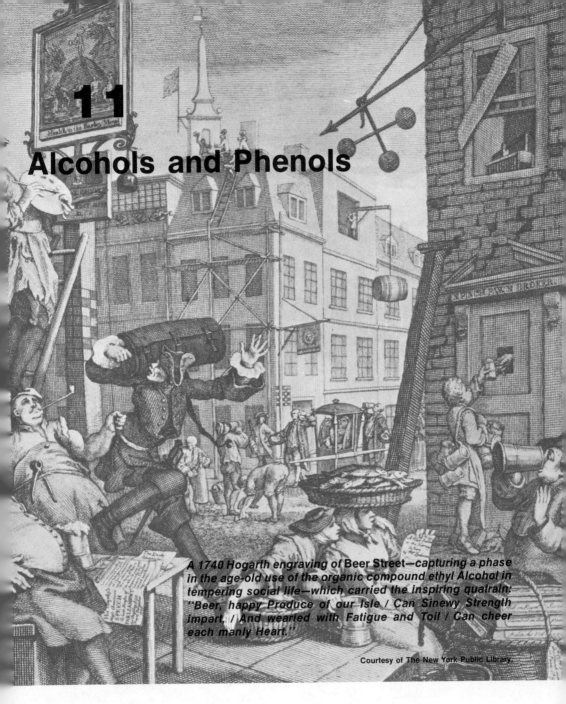

A 1740 Hogarth engraving of Beer Street—capturing a phase in the age-old use of the organic compound ethyl Alcohol in tempering social life—which carried the inspiring quatrain: "Beer, happy Produce of our Isle / Can Sinewy Strength impart, / And wearied with Fatigue and Toil / Can cheer each manly Heart."

Courtesy of The New York Public Library.

11.1 Introduction

Alcohols and phenols contain the hydroxyl, —OH, as the functional group in the molecule. The **alcohols** are derived from aliphatic hydrocarbons where a hydroxyl group has replaced a hydrogen atom. In **phenols** the hydroxyl group has replaced a hydrogen directly attached to an aromatic ring. The chemistry of the alcohols and phenols is due mainly to the characteristics of the hydroxyl group, although the presence of other functional groups in the molecule such as the aromatic ring in phenols exert their influence on reactivity also.

The chemical properties of alcohols and phenols are strikingly similar to those of water in many respects. This can be readily understood by examining the structural formulas of water, alcohols, and phenols.

$$H-OH \qquad R-OH \qquad Ar-OH$$
$$\text{water} \qquad \text{alcohol} \qquad \text{phenol}$$

Alcohols and phenols can be considered to be organic molecules related to water, in which one of the hydrogen atoms has been replaced by an organic group, R or Ar. This accounts for the similarity in chemical reactivity, since all these species contain the characteristic hydroxyl group.

11.2 Classification and Nomenclature

Alcohols may be classified as being primary, secondary, or tertiary, depending on the type of carbon atom to which the OH functional group is attached.

$$R-CH_2OH \qquad R_2-CH-OH \qquad R_3-C-OH$$
$$\text{primary} \qquad \text{secondary} \qquad \text{tertiary}$$

The simpler members of the series of alcohols are often called by their common names (the alkyl group to which the OH is attached followed by the word alcohol).

$$CH_3-CH_2OH \qquad \text{ethyl alcohol}$$
$$CH_3-CH_2-CH_2OH \qquad \textit{n}\text{-propyl alcohol}$$
$$CH_3-\underset{OH}{CH}-CH_3 \qquad \textit{i}\text{-propyl alcohol}$$

The higher members of the series and more complex alcohols are named by the IUPAC system, using the longest continuous chain of carbon atoms containing the hydroxyl group as a basis, and indicating the position of the hydroxyl group by the lowest numbered carbon atom possible. An -*ol* ending is used to show the presence of the hydroxyl group in the molecule.

$$CH_3-CH_2-OH \qquad \text{ethan}\textit{ol}\text{ (ethyl alcohol)}$$
$$\underset{1}{CH_3}-\underset{\underset{OH}{|} \ 2}{CH}-\underset{3}{CH_2}-\underset{4}{CH_3} \qquad \text{2-butan}\textit{ol}\text{ (}\textit{sec}\text{-butyl alcohol)}$$
$$\underset{1}{CH_3}-\underset{\underset{OH}{|} \ 2}{CH}-\underset{3}{CH_2}-\underset{\underset{CH_3}{|} \ 4}{CH}-\underset{5}{CH_3} \qquad \text{4-methyl-2-pentan}\textit{ol}$$

Aromatic alcohols or phenols are usually named as derivatives of phenol or by common names.

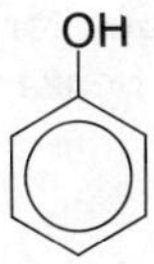

Sec. 11.3 PREPARATION OF ALIPHATIC ALCOHOLS

phenol

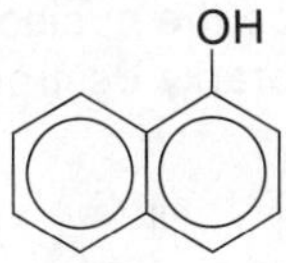

α-naphthol

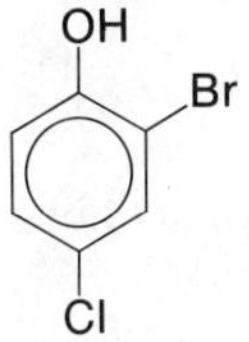

2-bromo-4-chlorophenol

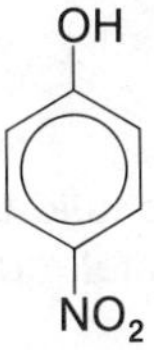

p-nitrophenol

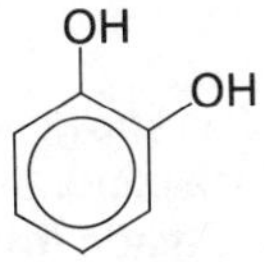

catechol

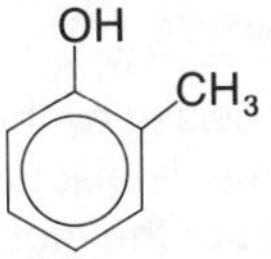

o-cresol

11.3 Preparation of Aliphatic Alcohols

The preparation of alcohols by the hydration of alkenes (Section 8.8–4) and by the hydroboration reaction resulting in anti-Markovnikov addition of water has been discussed in Section 8.10–2. Another general method for preparing alcohols is the hydrolysis of alkyl halides in the presence of aqueous sodium or potassium hydroxide.

$$CH_3-CH_2-Br + OH^\ominus \longrightarrow CH_3-CH_2OH + Br^\ominus \qquad (11-1)$$

When a primary alkyl halide is used, resulting in the preparation of a primary alcohol, the reaction proceeds by the S_N2 mechanism discussed in Section 6.7.

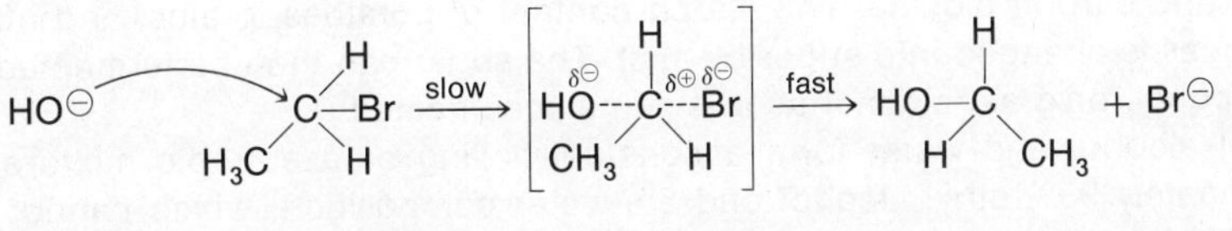

$$HO^\ominus \quad \overset{H}{\underset{H_3C}{\diagup}}C-Br \xrightarrow{slow} \left[HO^{\delta\ominus}\cdots\overset{H}{\underset{CH_3\ H}{C}}\cdots Br^{\delta\ominus} \right] \xrightarrow{fast} HO-\overset{H}{\underset{H}{C}}\diagdown CH_3 + Br^\ominus \qquad (11-2)$$

On the other hand, tertiary halides react by the S_N1 mechanism since the *alkyl groups sterically hinder the approach of the attacking nucleophilic reagent ($OH^\ominus$) to the back-side of the molecule* and at the same time *stabilize the intermediate carbonium ion* formed during the course of the reaction.

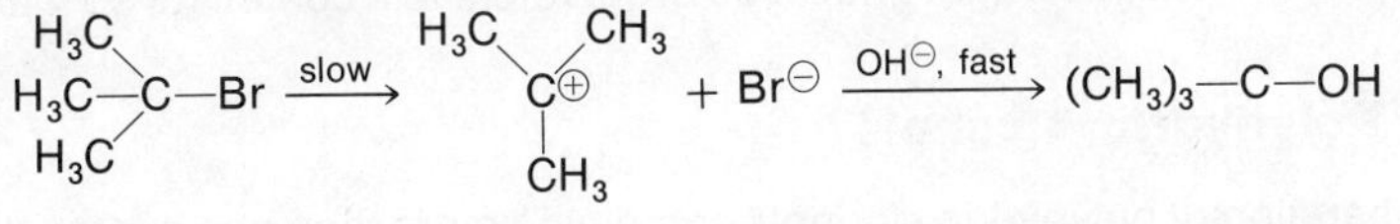

$$\underset{H_3C}{\overset{H_3C}{\diagdown}}\overset{}{\underset{}{C}}-Br \xrightarrow{slow} \underset{CH_3}{\overset{H_3C\ \ CH_3}{\diagdown\diagup}}C^\oplus + Br^\ominus \xrightarrow{OH^\ominus,\ fast} (CH_3)_3-C-OH \qquad (11-3)$$

Secondary halides react by either mechanism, while tertiary halides tend to undergo a competing side reaction, dehydrohalogenation to alkenes (Section 8.14–3) in the presence of strong base. Aryl halides are *not* converted to phenols by base, since the nucleophilic substitution of an hydroxyl group for a halogen would temporarily destroy the resonance stabilization of the benzene ring.

$$C_6H_5Cl + OH^{\ominus} \longrightarrow \text{(no reaction)} \qquad (11\text{-}4)$$

Some of the simpler aliphatic alcohols such as methyl and ethyl alcohol are often prepared by special methods.

11.4 Methyl Alcohol

Methanol can be synthesized by the destructive distillation of wood. It is known as wood alcohol, is extremely poisonous if taken internally and can cause blindness and death. Today, most of the methyl alcohol produced industrially is made from carbon monoxide and hydrogen by heating a mixture of these gases at a temperature of 400° C under a pressure of about 150 atmospheres in the presence of a suitable catalyst (oxides of Cr, Cu, and Zn).

$$CO + 2H_2 \xrightarrow[\text{press. cat.}]{\Delta} CH_3\text{—}OH \qquad (11\text{-}5)$$

Methanol is used as an antifreeze and as a solvent for many organic compounds.

11.5 Ethyl Alcohol

Ethanol can be obtained from the fermentation of cane sugar with yeast. Zymase, an enzyme present in yeast, acts as a catalyst, converting the sugar to ethyl alcohol and carbon dioxide. The starch content of potatoes, grains, and other substances is changed into sugar by malt. The sugar can then be fermented to ethyl alcohol, or grain alcohol as it is sometimes named.

Ethyl alcohol and water form a constant-boiling or azeotropic mixture of approximately 95% ethyl alcohol and 5% water composition, which cannot be separated into the pure components by distillation. Absolute alcohol, or 100% ethanol, can be prepared by other chemical techniques. However, the absolute alcohol is not fit for human consumption, since it contains traces of benzene as an impurity. Alcoholic beverages containing ethanol have been known since ancient times. The term *proof* is often used to indicate twice the percentage of alcohol present in a beverage. Thus, 190 proof refers to a content of 95% alcohol.

11.6 Polyhydric Alcohols

The preparation of polyhydric alcohols containing more than one hydroxyl group can be accomplished by slight modifications of the reactions for preparing aliphatic alcohols. The starting point in the reaction sequence is usually an alkene.

Sec. 11.7 PREPARATION OF PHENOLS

Ethylene glycol, used as an antifreeze in automobiles because of its high boiling point (197° C), can best be prepared by the following reaction sequence.

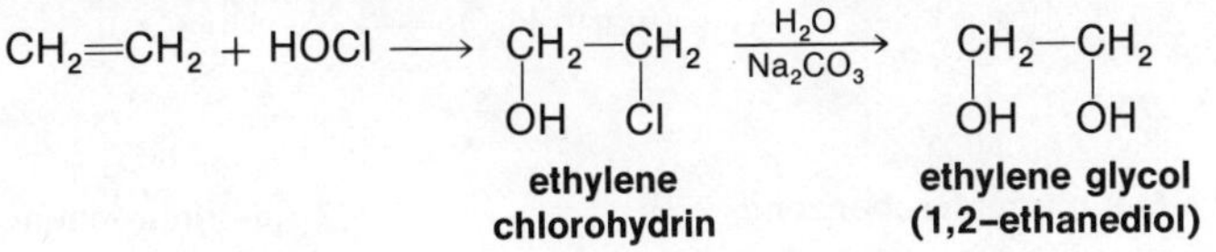

$$CH_2=CH_2 + HOCl \longrightarrow \underset{\underset{OH\ \ Cl}{}}{CH_2-CH_2} \xrightarrow[Na_2CO_3]{H_2O} \underset{\underset{OH\ \ OH}{}}{CH_2-CH_2} \quad (11\text{-}6)$$

ethylene   ethylene glycol
chlorohydrin (1,2-ethanediol)

Glycerol, a trihydric alcohol, is prepared from propylene.

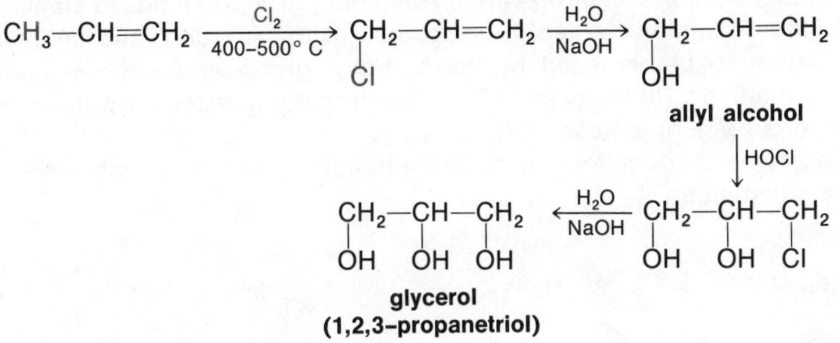

$$CH_3-CH=CH_2 \xrightarrow[400-500°\,C]{Cl_2} \underset{\underset{Cl}{}}{CH_2-CH=CH_2} \xrightarrow[NaOH]{H_2O} \underset{\underset{OH}{}}{CH_2-CH=CH_2}$$

allyl alcohol

$$\downarrow HOCl$$

$$\underset{\underset{OH\ \ OH\ \ OH}{}}{CH_2-CH-CH_2} \xleftarrow[NaOH]{H_2O} \underset{\underset{OH\ \ OH\ \ Cl}{}}{CH_2-CH-CH_2} \quad (11\text{-}7)$$

glycerol
(1,2,3-propanetriol)

11.7 Preparation of Phenols

The preparation of phenols involves reactions quite different from those used in the synthesis of aliphatic alcohols. In our discussion of the hydrolysis of alkyl halides to yield aliphatic alcohols it was mentioned that aryl halides are not readily converted to phenols in the presence of alkali. However, by using more drastic reaction conditions an aryl halide may be converted to a phenol with alkali at high temperatures and under pressure. Phenol is prepared commercially from chlorobenzene in this way (Dow process).

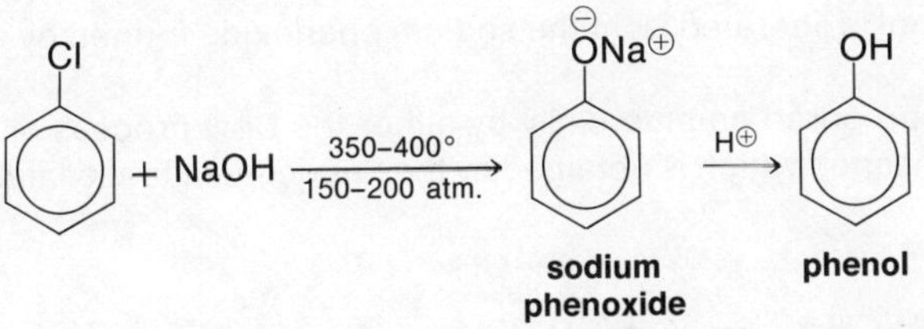

$$\underset{}{\text{C}_6\text{H}_5\text{Cl}} + NaOH \xrightarrow[150-200\,atm.]{350-400°} \underset{\text{sodium phenoxide}}{\text{C}_6\text{H}_5\text{O}^-Na^+} \xrightarrow{H^+} \underset{\text{phenol}}{\text{C}_6\text{H}_5\text{OH}} \quad (11\text{-}8)$$

Activated aryl halides can be converted to phenols under much milder reaction conditions.

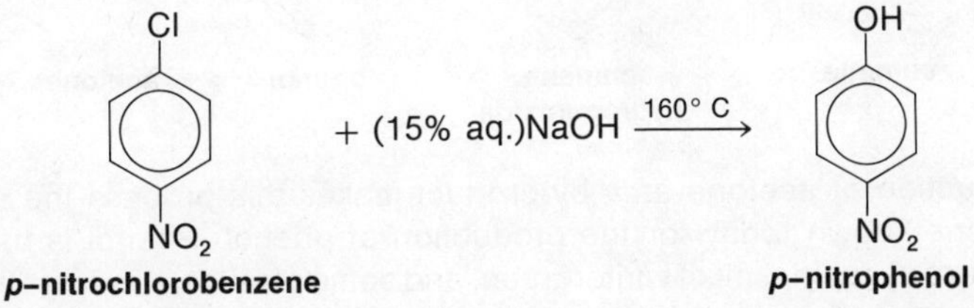

$$\underset{\text{p-nitrochlorobenzene}}{p\text{-}O_2N\text{-}C_6H_4\text{-}Cl} + (15\%\ aq.)NaOH \xrightarrow{160°\,C} \underset{\text{p-nitrophenol}}{p\text{-}O_2N\text{-}C_6H_4\text{-}OH} \quad (11\text{-}9)$$

$$\text{2,4,6-trinitrochlorobenzene} + (\text{warm})H_2O \longrightarrow \text{2,4,6-trinitrophenol (picric acid)} \quad (11\text{-}10)$$

The presence of the electron-withdrawing nitro group(s) tends to stabilize the intermediate carbanion formed in the course of the reaction, and enables the chlorine atom to be replaced by the hydroxyl group under more moderate reaction conditions. (In the case of 2,4,6-trinitrochlorobenzene, warm water can be used as a reagent instead of NaOH.)

The alkali fusion of sulfonates to form phenols is an example of nucleophilic aromatic substitution.

$$\text{benzenesulfonic acid} \xrightarrow{\text{NaOH}} \text{sodium benzene sulfonate} \xrightarrow[300°C]{\text{solid NaOH}} \text{sodium phenoxide} + Na_2SO_3 + H_2O \xrightarrow{H^\oplus} \text{phenol} \quad (11\text{-}11)$$

The free phenol is liberated from the sodium phenoxide formed by treatment with acid.

Phenol is prepared commercially by either the Dow process or from cumene (isopropylbenzene), which is obtained by the Friedel–Crafts alkylation of benzene.

$$\text{cumene} \xrightarrow{O_2} \text{cumene hydroperoxide} \xrightarrow{H^\oplus} \text{phenol} + CH_3-\overset{O}{\underset{\parallel}{C}}-CH_3 \quad (11\text{-}12)$$

The production of acetone as a by-product makes this process the most economical one known today for the production of phenol. Phenol is used in the manufacture of aspirin, oil of wintergreen, and some plastics. It is sometimes used as a mild antiseptic and disinfectant.

11.8 Physical Properties

The physical properties of the alcohols (particularly the lower members of the series) are strikingly similar to water. The lower-molecular-weight alcohols are completely miscible with water since the hydroxyl group comprises a relatively large percentage by weight of the molecule. The solubility in water of the higher members of the series decreases with increasing molecular weight since the larger alkyl groups tend to make the molecule more hydrocarbon and less similar to water in behavior.

A rather interesting property of the alcohols is their abnormally high boiling points with respect to their molecular weights. This has been attributed to the phenomenon of hydrogen bonding, and will be discussed in the following section.

The simplest aromatic alcohol is phenol, a colorless solid with a characteristic odor, somewhat soluble in water. Phenol and some of its derivatives are used as mild disinfectants (e.g., cresols in Lysol).

11.9 The Hydrogen Bond

As has been mentioned, the boiling points of the alcohols are much higher than would be predicted solely on the basis of their molecular weights. For example, ethyl alcohol, CH_3—CH_2OH, is a liquid and boils at 78° C, whereas dimethyl ether, CH_3—O—CH_3, an isomer with the same molecular weight, boils at $-24°$ C and is a gas at room temperature.

The reader will probably recall from his study of inorganic chemistry the rather abnormally high boiling point of water in contrast to the other hydrides (H_2S, H_2Se, H_2Te) of the group VIA elements. Water, H_2O, has the lowest molecular weight (m.w.) of all the group VIA hydrides, but nevertheless has the highest boiling point (i.e., H_2O: m.w. = 18, b.p. = 100° C; H_2S: m.w. = 34, b.p. = $-61.8°$ C). It is believed that water is associated in the liquid state, the water molecules forming a sort of chain network held together by **hydrogen bonds** between adjacent water molecules. The hydrogen bonds are largely electrostatic in nature, the small relatively positive hydrogen atom has some attraction for the unshared electron pairs on the electronegative oxygen atom in a neighboring water molecule.

$$\overset{\delta\oplus}{H}-\overset{\delta\ominus}{O}---\overset{\delta\oplus}{H}-\overset{\delta\ominus}{O}---\overset{\delta\oplus}{H}-\overset{\delta\ominus}{O}---$$
$$|||$$
$$HHH$$

The same type of bonding would be expected to occur in the alcohol series where an R group has replaced one of the hydrogens in the water molecule.

$$\overset{\delta\oplus}{H}-\overset{\delta\ominus}{O}---\overset{\delta\oplus}{H}-\overset{\delta\ominus}{O}---\overset{\delta\oplus}{H}-\overset{\delta\ominus}{O}---$$
$$|||$$
$$RRR$$

Ethers which have the general formula R—O—R' would not be expected to form hydrogen bonds, since they have no appropriate hydrogen atoms available to form the bond.

The presence of hydrogen bonding in a molecule will occur only when the hydrogen is directly attached to a highly electronegative element, such as oxygen, fluorine or nitrogen, that has unshared electrons available to form such a bond. The hydrogen-bonding structure, when present in a molecule, will require additional energy to break these bonds, and this will be reflected in the properties of the molecule, such as in the abnormally high boiling point. This phenomenon accounts for the great difference between the boiling points of ethyl alcohol and dimethyl ether as well as the greater solubility of the lower molecular weight alcohols in water. Evidence for the existence of hydrogen bonds in alcohols is obtained from an examination of their infrared spectra.

Hydrogen bonding is extremely important in connection with the molecular structure and properties of the proteins and nucleic acids, as we will see in later chapters.

11.10 Reactions of the Alcohols

Some of the reactions of the alcohols have already been discussed in earlier chapters, and the reader should refer to the appropriate sections in the text. The majority of the reactions of the alcohols can be subdivided into two general categories: (1) reactions in which the bond between the O and H atoms in the molecule is broken, RO-/-H, resulting in a new substituent's replacing the hydrogen atom originally in the molecule; and (2) reactions involving replacement of the OH group in the molecule by breaking the bond between the R and O atom, R-/-OH.

11.10-1 Acidity The formulas of the alcohols and phenols might imply that these molecules should have basic properties since a hydroxyl group is present. However, it is found experimentally that the alcohols and phenols tend to behave as weak acids. The removal of the weakly acidic hydrogen atom by breaking of the O—H bond in the molecule results in replacement of the hydrogen by some new substituent.

Alcohols are weak acids (essentially neutral) and require strong reagents to remove their proton. For example, even sodium hydroxide will not react with alcohols to any appreciable extent, since the hydroxide ion is too weak a base to remove a proton from the alcohol molecule. The equilibrium lies far towards the left-hand side of the equation, and so we can say that essentially no reaction occurs.

$$\underset{\text{alcohol}}{\text{ROH}} + \text{NaOH} \rightleftharpoons \underset{\text{alkoxide salt}}{\text{R—O}^{\ominus}\text{Na}^{\oplus}} + \text{HOH} \qquad \text{(essentially no reaction)} \qquad (11\text{--}13)$$

Since alcohols are similar structurally to water it should be expected that they have similar chemical properties. Indeed, they do! Although alcohols are much less acidic than water, they will react with very reactive metals, such as sodium or potassium, to form salts (alkoxides) and hydrogen gas, analogous to the reaction of water with these metals.

$$2\text{H—OH} + 2\text{Na} \longrightarrow 2\text{NaOH} + \text{H}_2\uparrow \qquad (11\text{--}14)$$

$$2\text{ROH} + 2\text{Na} \longrightarrow \underset{\text{sodium alkoxide}}{2\text{NaOR}} + \text{H}_2\uparrow \qquad (11\text{--}15)$$

Sec. 11.10 REACTIONS OF THE ALCOHOLS

The salt formed contains a metal atom in place of the hydrogen atom originally present, and is called an alkoxide.

$$2CH_3-CH_2OH + 2Na \longrightarrow 2CH_3-CH_2-O^{\ominus}Na^{\oplus} + H_2\uparrow \quad (11\text{-}16)$$
ethyl alcohol sodium ethoxide

The alkoxides are extremely strong bases, much stronger than the inorganic hydroxides, and are used as basic catalysts in many organic reactions.

In the case of phenols, a base such as sodium hydroxide, or a reactive metal can be used to form a salt.

$$C_6H_5OH + NaOH \longrightarrow C_6H_5O^{\ominus}Na^{\oplus} + H_2O \quad (11\text{-}17)$$
phenol sodium phenoxide

The milder reaction conditions are due to the fact that phenols are far more acidic than the aliphatic alcohols, although phenols are still relatively weak acids. The acidic nature of phenols has been discussed earlier (Section 5.4-2).

11.10-2 Replacement of the Hydroxyl Group The hydroxyl group of aliphatic alcohols can be replaced by various reagents. The hydroxyl group in phenols usually cannot be replaced since it is involved in the resonance stabilization of the aromatic ring in the molecule.

The hydroxyl group is not a good leaving group since it is too basic. However, conversion of the aliphatic alcohols to their conjugate acids, $R-OH_2^{\oplus}$, by protonation, enables the water, H_2O, to be displaced easily as a stable species, and replacement of the hydroxyl group by some nucleophile to occur.

The substitution reaction of alcohols is believed to proceed by the characteristic S_N1 or S_N2 mechanisms, or some modification of them.

$$R-OH + H^{\oplus} \rightleftharpoons ROH_2^{\oplus} \quad (11\text{-}18)$$
$$\text{conj. acid}$$

or

$$N{:}^{\ominus} + R-OH_2^{\oplus} \longrightarrow R-N + H_2O \quad (S_N2) \quad (11\text{-}19)$$

$$ROH_2^{\oplus} \rightleftharpoons R^{\oplus} + H_2O \quad (11\text{-}20)$$
$$R^{\oplus} + N{:} \longrightarrow R-N \quad (S_N1) \quad (11\text{-}21)$$

Alcohols react readily with hydrogen halides to produce alkyl halides.

$$HX + R-OH \longrightarrow R-X + H_2O \quad (11\text{-}22)$$

The reaction proceeds most readily with tertiary alcohols and the rate decreases as one goes to secondary and then to primary alcohols (Rate − $3° > 2° > 1°$). Tertiary butyl alcohol reacts with concentrated hydrochloric acid at room temperature to form t–butyl chloride in several minutes.

$$\underset{\text{t-butyl alcohol}}{\underset{\underset{CH_3}{|}}{\overset{\overset{CH_3}{|}}{CH_3-C-OH}}} + \text{conc. HCl} \xrightarrow{\text{room temp.}} \underset{\text{t-butyl chloride}}{\underset{\underset{CH_3}{|}}{\overset{\overset{CH_3}{|}}{CH_3-C-Cl}}} + H_2O \qquad (11\text{-}23)$$

In contrast, n-butyl alcohol requires the addition of a zinc chloride catalyst and heat for several hours.

$$\underset{\text{n-butyl alcohol}}{CH_3-CH_2-CH_2-CH_2OH} + \underset{\text{(conc.)}}{HCl} \xrightarrow[\Delta]{ZnCl_2} \underset{\text{n-butyl chloride}}{CH_3-CH_2-CH_2-CH_2Cl} + H_2O \qquad (11\text{-}24)$$

Of the hydrogen halides, hydriodic acid is the most reactive, and hydrochloric acid the least. The reaction rate is in the order of the strength of chloride, bromide, and iodide ions as nucleophilic reagents. That is to say, $Cl < Br < I$; chloride reacts the slowest and iodide reacts the fastest.

In the case of tertiary alcohols, which can form a relatively stable carbonium ion, the probable mechanism for the reaction would be the $S_N 1$, whereas the primary alcohols would probably react by the $S_N 2$ mechanism.

The Lucas Test The relative rates of the reaction of alcohols with concentrated hydrochloric acid–anhydrous zinc chloride can be used as a test to differentiate among primary, secondary, and tertiary alcohols. The test is known as the **Lucas test** and is based on the fact that alkyl halides are not soluble in the reaction mixture and produce a turbidity when formed. The time required for the turbidity to be observed will serve as an indication as to the structure of the alcohol. Tertiary alcohols give an almost instantaneous turbidity upon mixing the reagents, and secondary alcohols will turn the solution turbid after several minutes, whereas solutions of primary alcohols will turn turbid after several hours.

Although the reaction of alcohols with hydrogen halides to produce alkyl halides proceeds readily for most alcohols, it has several distinct disadvantages. Perhaps most important is that the reaction is reversible, so that the yields of alkyl halide obtained are usually not very high. Another consideration is the possibility of rearranged products forming by rearrangement of the carbonium ions formed during the course of the reaction under $S_N 1$ reaction conditions. For these reasons we shall discuss other ways to prepare alkyl halides from alcohols which tend to avoid these drawbacks.

Conversion of Alcohols into Alkyl Halides Alcohols can be converted to alkyl halides by treatment with various phosphorus halides.

$$\underset{\text{ethyl alcohol}}{3CH_3-CH_2OH} + PBr_3 \longrightarrow \underset{\text{ethyl bromide}}{3CH_3-CH_2Br} + H_3PO_3 \qquad (11\text{-}25)$$

$$\underset{\text{cyclohexanol}}{\text{C}_6\text{H}_{11}\text{OH}} + PCl_5 \longrightarrow POCl_3 + HCl + \underset{\text{cyclohexyl chloride}}{\text{C}_6\text{H}_{11}\text{Cl}} \qquad (11\text{-}26)$$

Sec. 11.11 INORGANIC ESTERS

The mechanism of this reaction is not completely understood. However, the reaction definitely does not proceed by a carbonium type mechanism, so that the problem of possible rearrangement is eliminated. The reaction goes to completion and so the yields of product are much improved in comparison to the hydrogen halide reaction.

Perhaps the most efficient way of preparing an alkyl chloride or bromide from an alcohol is by treatment with thionyl chloride, $SOCl_2$, or thionyl bromide, $SOBr_2$. The corresponding iodides cannot be made in this manner since the thionyl iodide needed is not commercially available.

Treatment of an alcohol with thionyl chloride proceeds through the formation of an intermediate chlorosulfite ester, which can be isolated if one so desires or can be decomposed to the alkyl halide. The reaction is usually run in the presence of a base such as pyridine, which helps remove the HCl formed and drives the reaction to completion.

$$R\text{—}OH + SOCl_2 \longrightarrow HCl\uparrow + [R\text{—}O\text{—}\overset{\overset{O}{\|}}{S}\text{—}Cl] \longrightarrow SO_2\uparrow + R\text{—}Cl \quad (11\text{-}27)$$

thionyl chloride   chlorosulfite ester

The alkyl halide formed as a product in this reaction has the highest purity of any of the alkyl halides produced by treatment of an alcohol with a hydrogen halide or phosphorus halide. The reason for this is that the other products formed in this reaction are all gases, which can be easily separated from the reaction mixture, resulting in the formation of an extremely pure alkyl halide.

11.11 Inorganic Esters

The reaction of an alcohol with an inorganic acid results in the replacement of the hydrogen atom of the acid by the alkyl group of the alcohol. The compound formed is called an inorganic ester. Thus, the reaction of an alcohol with nitric acid produces esters known as alkyl nitrates.

$$ROH + HONO_2 \longrightarrow RO\text{—}NO_2 + H_2O \quad (11\text{-}28)$$

nitric acid   alkyl nitrate

The reaction of the polyhydric alcohol, glycerol, with nitric acid produces glycerol trinitrate (nitroglycerine), which is the principal explosive component in dynamite. Nitroglycerine is also used as a vasodilator in the treatment of heart disease.

$$\begin{array}{l} CH_2OH \\ | \\ CHOH \\ | \\ CH_2OH \end{array} + 3HONO_2 \xrightarrow{H_2SO_4} \begin{array}{l} CH_2ONO_2 \\ | \\ CHONO_2 \\ | \\ CH_2ONO_2 \end{array} + 3H_2O \quad (11\text{-}29)$$

glycerol   glycerol trinitrate (nitroglycerine)

The reaction of cold sulfuric acid with alcohols produces alkyl hydrogen sulfates.

$$\text{ROH} + \text{HOSO}_3\text{H} \longrightarrow \text{H}_2\text{O} + \text{ROSO}_3\text{H} \quad (11\text{-}30)$$
sulfuric acid   alkyl hydrogen sulfate

The sodium salt of the higher-molecular-weight alkyl hydrogen sulfates ($R \geqq 12$) are used as detergents. Sodium lauryl sulfate, $CH_3-(CH_2)_{10}-CH_2-OSO_3^\ominus Na^\oplus$, for example, is sold commercially as a detergent under the trade name of Dreft.

The esters of the phosphoric acids are extremely important in understanding the biochemical processes in the human body. Phosphates are involved in many enzymatic reactions and are an important constituent in the structure of the nucleic acids. Some typical phosphate esters are represented by the following formulas.

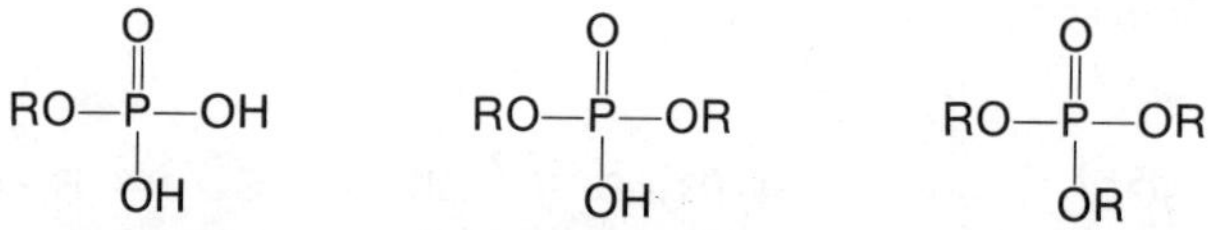

$$\underset{\substack{\text{alkyl dihydrogen}\\\text{phosphate}}}{RO-\underset{\underset{OH}{|}}{\overset{\overset{O}{\|}}{P}}-OH} \quad\quad \underset{\substack{\text{dialkyl hydrogen}\\\text{phosphate}}}{RO-\underset{\underset{OH}{|}}{\overset{\overset{O}{\|}}{P}}-OR} \quad\quad \underset{\text{trialkyl phosphate}}{RO-\underset{\underset{OR}{|}}{\overset{\overset{O}{\|}}{P}}-OR}$$

(from H_3PO_4, orthophosphoric acid)

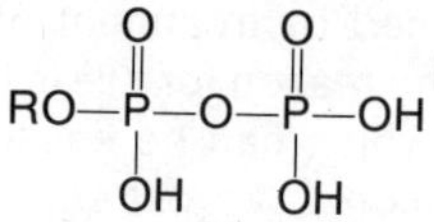

$$\underset{\substack{\text{alkyl diphosphate}}}{RO-\underset{\underset{OH}{|}}{\overset{\overset{O}{\|}}{P}}-O-\underset{\underset{OH}{|}}{\overset{\overset{O}{\|}}{P}}-OH}$$

(from $H_4P_2O_7$, pyrophosphoric acid)

Alcohols also can react with organic acids to form esters.

$$\underset{\text{alcohol}}{\text{ROH}} + \underset{\substack{\text{carboxylic}\\\text{acid}}}{\text{R'COOH}} \overset{H^\oplus}{\rightleftharpoons} \underset{\text{ester}}{R'-\overset{\overset{O}{\|}}{C}-OR} + H_2O \quad (11\text{-}31)$$

This reaction will be discussed in more detail in Section 15.8–2.

11.12 Oxidation of Alcohols

Primary alcohols can be oxidized by an oxidizing agent, such as potassium permanganate or potassium dichromate, to aldehydes, which in turn are oxidized to carboxylic acids. Secondary alcohols can be oxidized to ketones. Tertiary alcohols are not readily oxidized.

$$\underset{\substack{\text{primary}\\\text{alcohol}}}{R-CH_2OH} \xrightarrow[H_2SO_4]{KMnO_4 \text{ or } K_2Cr_2O_7} \underset{\text{aldehyde}}{R-\overset{\overset{O}{\|}}{C}-H} \longrightarrow R-\overset{\overset{O}{\|}}{C}-OH \quad (11\text{-}32)$$

Sec. 11.14 SULFUR ANALOGS OF ALCOHOLS AND PHENOLS

$$R_2CHOH \xrightarrow[H_2SO_4]{KMnO_4 \text{ or } K_2Cr_2O_7} \underset{\text{ketone}}{R-\overset{\overset{O}{\|}}{C}-R} \quad (11\text{-}33)$$

secondary alcohol

$$R_3COH \xrightarrow[H_2SO_4]{KMnO_4 \text{ or } K_2Cr_2O_7} (\text{no reaction}) \quad (11\text{-}34)$$

tertiary alcohol

11.13 Reactions of Phenols

The characteristic reactions of the phenols are quite different in most cases from the reactions we have encountered for the aliphatic alcohols. For example, the replacement of a hydroxyl group by a nucleophilic reagent rarely occurs in the case of phenols. This is due to the involvement of the hydroxyl group in the resonance stabilization of the molecule, which was mentioned earlier.

Thus, most of the reactions of the phenols are electrophilic substitution reactions because of the *ortho–para*-directing and ring-activating influence of the hydroxyl group when directly attached to an aromatic nucleus.

Phenols undergo the characteristic aromatic electrophilic substitution reactions such as nitration and halogenation.

$$\text{C}_6\text{H}_5\text{OH} + \text{HONO}_2 \longrightarrow \underset{o\text{-nitrophenol}}{o\text{-O}_2\text{N-C}_6\text{H}_4\text{-OH}} + \underset{p\text{-nitrophenol}}{p\text{-O}_2\text{N-C}_6\text{H}_4\text{-OH}} + \text{H}_2\text{O} \quad (11\text{-}35)$$

The hydroxyl group activates the benzene ring to such a great extent that no iron catalyst or carrier is needed for halogenation. Indeed, the reaction conditions are quite mild, bromine water producing a **tribromo** derivative instantaneously.

$$\text{C}_6\text{H}_5\text{OH} + 3\text{Br}_2 \xrightarrow{H_2O} \underset{2,4,6\text{-tribromophenol}}{2,4,6\text{-Br}_3\text{C}_6\text{H}_2\text{OH}} + 3\text{HBr} \quad (11\text{-}36)$$

Phenols do not undergo the Friedel–Crafts alkylation reaction with any great degree of success. This is because the unshared electron pairs on the phenolic oxygen (Lewis base) tie up the Lewis-acid catalyst required for the reaction. Other reactions of phenols will be discussed in later chapters.

11.14 Sulfur Analogs of Alcohols and Phenols

Since sulfur and oxygen are both in group VIA of the periodic table, it might be expected that sulfur forms compounds similar to oxygen with similar properties.

This has been found to be the case; the sulfur analogs of alcohols, RSH, are called thiols or mercaptans. The —SH group is called the sulfhydryl group. A characteristic feature of the mercaptans is their unpleasant odor. The *n*-butyl mercaptan or 1-butanethiol, $CH_3-CH_2-CH_2-CH_2-SH$, is the compound responsible for the odor of the skunk's defense mechanism.

Mercaptans are weakly acidic, but more acidic than the ordinary alcohols. They are readily oxidized to disulfides, a reaction of importance in protein chemistry since there are some amino acids (cysteine and methionine) that contain a sulfhydryl group.

$$2R-SH + H_2O_2 \longrightarrow R-S-S-R + 2H_2O \qquad (11-37)$$
mercaptan   disulfide

Summary

- Alcohols and phenols contain the —OH functional group.
- Alcohols can be classified as primary, secondary, or tertiary.
- Alcohols and phenols (where the OH group is directly attached to an aromatic ring) can be named by the IUPAC system of nomenclature. An *-ol* ending is used to show the presence of the hydroxyl group in the molecule.

Some phenols are better known by their common names (catechol, cresol).

- Alcohols can be prepared by hydration of alkenes, hydroboration, or by hydrolysis of alkyl halides by the S_N1 or the S_N2 mechanism.

Methyl and ethyl alcohol can also be prepared by some special methods.

- Ethylene glycol and glycerol are two examples of important polyhydric alcohols.
- Phenols can be prepared by:

    (a) The Dow process.

    (b) Hydrolysis of activated aryl halides.

    (c) Alkali fusion of sulfonates, an example of nucleophilic aromatic substitution.

    (e) From cumene hydroperoxide.

- The abnormally high boiling points of alcohols can be attributed to molecular association through hydrogen bonding.
- The reactions of alcohols and phenols:

    (a) Removal of the weakly acidic hydrogens through breaking the O—H bond by active metals, forming alkoxides or phenoxides.

    (b) Replacement of the OH group by halogen (HX, PX_3, $SOCl_2$) to form alkyl halides.

    (c) The Lucas test can be used to differentiate between primary, secondary, and tertiary alcohols.

    (d) Formation of inorganic esters.

    (e) Primary alcohols can be oxidized to aldehydes, secondary alcohols to ketones.

    (f) Phenols will undergo characteristic aromatic electrophilic substitution reactions.

- The sulfur analogs of alcohols are called thiols or mercaptans.

PROBLEMS

1 Draw the correct structural formulas for each of the following compounds:
   (a) 2-methyl-2-propanol
   (b) potassium isopropoxide
   (c) o-chlorophenol
   (d) 2,3-dinitro-5-bromophenol
   (e) potassium p-nitrophenoxide
   (f) 2-methylcyclopentanol
   (g) allyl alcohol
   (h) p-cresol
   (i) 3-hexyne-1-ol
   (j) 1,3-propanediol

2 Name the following compounds by the IUPAC naming system. Designate each aliphatic alcohol as being either a primary, secondary, or tertiary alcohol.

(a) $CH_3-\underset{\underset{CH_3}{|}}{\overset{\overset{H}{|}}{C}}-CH_2-CH_2OH$

(b) $CH_3-\underset{\underset{OH}{|}}{CH}-\underset{\underset{Br}{|}}{CH}-CH_2-\underset{\underset{CH_3}{|}}{\overset{\overset{CH_3}{|}}{C}}-CH_3$

(c) [benzene ring with OH at top, F at lower-left, Cl at lower-right]

(d) [benzene ring with OH and NO$_2$ ortho]

(e) [benzene ring with CH$_2$OH and Cl ortho]

(f) $CH_3-\underset{\underset{OH}{|}}{\overset{\overset{CH_3}{|}}{C}}-CH_2-CH=CH_2$

(g) $CH_3-\underset{\underset{Cl}{|}}{\overset{\overset{Cl}{|}}{C}}-CH_2-\underset{\underset{OH}{|}}{\overset{\overset{CH_3}{|}}{C}}-CH_3$

(h) [benzene ring with OH, OH, NO$_2$]

(i) [cyclohexane with HO and CH$_3$ on one carbon, OH on another]

(j) [benzene ring with CH$_2$—CH$_2$OH]

3 Explain why each of the following names is incorrect, and give a correct name for each compound.
   (a) isopropanol
   (b) 3,3-dimethyl-4-heptanol
   (c) 5-methylcyclopentanol
   (d) 1,1-dimethyl-1-butanol
   (e) 6-nitrophenol
   (f) 4,5-dibromophenol

4 Arrange the following compounds in order of increasing boiling point.
   ethanol   pentanol
   ethane   ethyl mercaptan
   dimethyl ether   ethyl chloride

5 *Which compound would you expect to have a higher boiling point, o-nitrophenol or p-nitrophenol? Explain.

6 Arrange the following alcohols in increasing order of the rate of their reaction with hydrogen iodide.
benzyl alcohol   t-butyl alcohol
ethanol   3-hexanol
phenol   cyclohexanol

7 Complete each of the following equations by writing the correct structural formula(s) of the product(s). If no reaction occurs, write N.R.

(a) $CH_3-CH_2-OH + PI_3 \longrightarrow$
(b) 2-propanol + $H_2SO_4 \longrightarrow$
(c) cyclohexanol + $SOCl_2 \longrightarrow$
(d) $CH_3-CH_2OH + HNO_3 \longrightarrow$

(e) *p*-nitrophenol + NaOH $\longrightarrow$

(f) phenol + $Br_2(H_2O) \longrightarrow$

(g) 2-butanol + $K_2Cr_2O_7 \xrightarrow{H_2SO_4}$
(h) *p*-bromobenzyl bromide + aq.NaOH $\longrightarrow$

(i) *t*-butyl alcohol + Li $\longrightarrow$
(j) 1-hexanol + $KMnO_4 \xrightarrow{H_2SO_4}$

(k) 1-bromonaphthalene + aq. NaOH $\longrightarrow$

(l) $CH_3-CH_2-CH_2-SH + K \longrightarrow$
(m) β-naphthol + $PCl_5 \longrightarrow$
(n) 3-methyl-3-pentanol + $K_2Cr_2O_7 \xrightarrow{H_2SO_4}$
(o) 2,4,6-trinitrobromobenzene + aq.KOH —

8 Perform all of the following syntheses from the indicated starting materials and any necessary inorganic reagents.

(a) inorganic reagents to ethyl alcohol
(b) ethyl alcohol to 2-bromobutane
(c) ethyl bromide to *n*-butyl alcohol
(d) benzene and carbon tetrachloride to triphenylcarbinol
(e) benzene to resorcinol (*m*-dihydroxybenzene)
(f) benzene to benzyl alcohol
(g) benzene to cyclohexanol
(h) 1-propanol to 2-propanol
(i) benzene to *o*-cresol
(j) cyclopentene to 1,2-dihydroxy-3-bromocyclopentane (structure: cyclopentane with OH, OH, Br)

PROBLEMS

(k) 2-propanol to 1-bromopropane
(l) benzene to 2-phenylethanol

9 In each of the reaction sequences draw the structure of the product represented by the letters A, B, and C. Where a (?) appears, write the formula of the reagent over the arrow which is used for the particular step in the sequence.

(a) $CH_3-\underset{\underset{OH}{|}}{CH}-CH_3 \xrightarrow{SOCl_2}$ (A) $\xrightarrow[\text{dry ether}]{Mg}$ (B) $\xrightarrow[H\oplus]{H_2O}$ (C)

(b) [benzene] $\xrightarrow[AlCl_3]{CH_3Cl}$ (A) $\xrightarrow{(?)}$ [C$_6$H$_5$CH$_2$Br] $\xrightarrow{\text{aq. NaOH}}$ (C)

(c) [benzene] $\xrightarrow[H_2SO_4]{HNO_3}$ (A) $\xrightarrow{H_2SO_4}$ (B) $\xrightarrow{(?)}$ [m-nitrophenol]

(d) $CH_3-CH_2-CH_2I \xrightarrow{\text{alc. KOH}}$ (A) $\xrightarrow[H_2O]{H_2SO_4}$ (B) $\xrightarrow{(?)} CH_3-\underset{\underset{Br}{|}}{CH}-CH_3$

10 Name a chemical test or single reagent that can be used to distinguish between the following pairs of compounds.

(a) *n*–butyl alcohol, and *t*–butyl alcohol (b) phenol, and ethanol
(c) *n*–butyl alcohol, and *n*–butane (d) 1–hexanol, and 3–hexanol

11 Suggest a plausible mechanism to account for the conversion of neopentyl alcohol, $CH_3-\underset{\underset{CH_3}{|}}{\overset{\overset{CH_3}{|}}{C}}-CH_2OH$, into 2–iodo–2–methylbutane, $CH_3-CH_2-\underset{\underset{CH_3}{|}}{\overset{\overset{I}{|}}{C}}-CH_3$, when treated with HI. Write down all the steps of the mechanism in support of your answer.

12 * Compound A, $C_6H_{14}O$, liberates hydrogen gas when reacted with sodium metal; A does not react with sodium hydroxide, and gives a positive Lucas test in several minutes. When A is reacted with PBr$_5$, then compound B, $C_6H_{13}Br$, is formed. When B is treated with alcoholic KOH, compounds C and D, both having the formula C_6H_{12}, are formed; C is the major reaction product, while D is a minor product. When C is treated with ozone, followed by hydrolysis, only a single ketone is obtained. This ketone can be shown to be identical to the compound produced by hydration of propyne in the presence of sulfuric acid and mercuric sulfate. From the above information, deduce and write down the correct structural formulas of compounds A through D.

13 * The reaction of vicinal diols with acid results in a carbonium ion rearrangement commonly known as the *pinacol rearrangement*; the diol is converted into a ketone. Propose a plausible reaction mechanism for the conversion of 2,3–dimethyl–2,3–butandiol (pinacol), in the presence of warm aqueous H_2SO_4, to methyl *t*–butylketone (pinacolone).

12

Ethers

The administration of ethyl ether for an operation in the late 19th century—first used as an anesthetic for surgery over 125 years ago.

Courtesy of The New York Academy of Medicine.

12.1 Introduction

The structural relationship among water, alcohols, and ethers can be seen by their representative formulas, H—O—H, R—O—H, and R—O—R′, where the hydrogen atoms in water are replaced by organic (alkyl or aryl) groups. In alcohols, R—O—H, where one of the hydrogens of water has been replaced by an organic group, the relationship is seen in the great similarity in some of the chemical reactions of alcohols and water. Although ethers are structurally related to water by replacement of the two hydrogens by organic groups, there is *no* similarity in chemical behavior.

Ethers may be aliphatic, aromatic, or mixed compounds of the general formula, R—O—R′. The organic groups, R and R′, may be identical in simple ethers, or different in mixed ethers.

12.1-1 Nomenclature The nomenclature of ethers is such that they are usually called by their common names. When the IUPAC names are used, the usual rules of longest continuous carbon chain and smallest numbers for substituents prevail. (See Table 12.1.)

Table 12.1 Nomenclature of Ethers

Compound	IUPAC name	Common name
CH₃—CH₂—O—CH₂—CH₃ (a simple ether)	ethoxyethane	diethyl ether
CH₃—O—CH₂—CH₂—CH₃ (a mixed ether)	1-methoxypropane	methyl-*n*-propyl ether
C₆H₅—OCH₃ (a mixed ether)	methylphenyl ether methoxybenzene	anisole
Br—CH₂—CH₂—O—CH₃	2-bromoethylmethyl ether	

12.2 Preparation of Ethers

12.2-1 Dehydration of Alcohols When an alcohol is dehydrated (by treatment with concentrated sulfuric or phosphoric acid) in a typical α,β-elimination reaction, the water is split out *intramolecularly* (from one molecule) to form an alkene.

$$\underset{\underset{H}{|}}{\overset{\overset{H}{|}}{H-C}}-\underset{\underset{H}{|}}{\overset{\overset{H}{|}}{C}}-OH \xrightarrow[H_3PO_4, \Delta]{\text{conc. } H_2SO_4 \text{ or}} \underset{\underset{H}{|}}{\overset{\overset{H}{|}}{H-C}}=\underset{\underset{}{}}{\overset{\overset{H}{|}}{C}}-H + H_2O \qquad (12\text{-}1)$$

However, water can also be removed *intermolecularly* (from two molecules of alcohol) to produce an ether.

$$\text{H-CH}_2\text{-CH}_2\text{-OH} + \text{HO-CH}_2\text{-CH}_2\text{-H} \xrightarrow{\text{conc. } H_2SO_4} \text{H-CH}_2\text{-CH}_2\text{-O-CH}_2\text{-CH}_2\text{-H} \qquad (12\text{-}2)$$
diethyl ether

The product obtained depends upon the reaction conditions; at about 140° C the main product of the reaction is the ether; higher temperatures produce more of the alkene.

The ethers produced in this reaction are **simple ethers.** The method can only be conveniently used to synthesize low-molecular-weight ethers that may easily be distilled out of the reaction mixture because of their low boiling points.

If a mixture of two different alcohols is used, dehydration produces three ethers; two simple ethers from each of the alcohols, and a mixed ether resulting from dehydration of the two alcohols and coupling of an alkyl group from each alcohol.

$$\text{ROH} + \text{R'OH} \xrightarrow{\text{conc. } H_2SO_4} \text{R-O-R} + \text{R'-O-R'} + \text{R-O-R'} + H_2O \qquad (12\text{-}3)$$

Due to the production of mixtures of products, this reaction loses its synthetic utility.

12.2-2 The Williamson Ether Synthesis In 1851, Alexander Williamson devised a synthesis for **pure ethers.** Even today, it is the best and by far the most satisfactory method for synthesizing ethers. The reactants used in the Williamson ether synthesis are usually the sodium salt of an alcohol or phenol and an alkyl halide.

$$\text{R-}\overset{\ominus}{\text{O}}\text{-Na}^{\oplus} + \text{R'X} \longrightarrow \text{R-O-R'} + \text{NaX} \qquad (12\text{-}4)$$

The R and R' groups may be identical, in which case a simple ether will be formed; or R and R' may be different, producing a mixed ether.

Alexander William Williamson (1824–1904)—a student of Liebig at Giessen and professor at the University College in London—developed the *Williamson synthesis of ethers*. (From the Dains Collection, courtesy of the Department of Chemistry, The University of Kansas.)

The alkoxide or phenoxide required may be made from the corresponding alcohol or phenol, making use of the fact that the O—H bond is a highly polar one, so that the hydrogen atom is acidic and can be replaced by reactive metals such as sodium.

$$2CH_3OH + 2Na \longrightarrow 2CH_3\text{—}\overset{\ominus}{O}Na^{\oplus} + H_2 \qquad (12\text{–}5)$$

$$2\,C_6H_5OH + 2Na(\text{or NaOH}) \longrightarrow 2\,C_6H_5\overset{\ominus}{O}Na^{\oplus} + H_2(\text{or } H_2O) \qquad (12\text{–}6)$$

The alkoxide or phenoxide ion can then react with an alkyl halide to produce the ether

$$CH_3\text{—}\overset{\ominus}{O}\text{—}Na^{\oplus} + CH_3CH_2Br \longrightarrow \underset{\textbf{methyl ethyl ether}}{CH_3\text{—}O\text{—}CH_2\text{—}CH_3} + NaBr \qquad (12\text{–}7)$$

Aryl halides can *not* be used in place of alkyl halides, unless strong electron-withdrawing groups are present *ortho* or *para* to the halogen.

The mechanism of the Williamson ether synthesis is that of a typical S_N2 reaction, involving nucleophilic attack of the alkoxide (or phenoxide) ion on the alkyl halide and displacing the halogen as a leaving group, producing the ether.

$$R\text{—}O^{\ominus} + R'\text{—}X \longrightarrow [R\text{—}\overset{..}{O}\text{---}R'\text{---}X]^{\ominus} \longrightarrow R\text{—}O\text{—}R' + X^{\ominus} \qquad (12\text{–}8)$$

The Williamson ether synthesis has excellent synthetic utility when the alkyl groups present in the ether are *primary*. However, severe limitations result when

Sec. 12.3 PROPERTIES

one attempts to synthesize an ether containing *secondary* groups and, in particular, *tertiary* groups.

Let us consider the preparation of the ether, methyl–*t*–butyl ether. One envisions two possible reaction sequences.

(a) $CH_3-ONa + CH_3-\underset{\underset{CH_3}{|}}{\overset{\overset{CH_3}{|}}{C}}-Br$
                         tertiary halide

$\longrightarrow NaBr + CH_3-O-\underset{\underset{CH_3}{|}}{\overset{\overset{CH_3}{|}}{C}}-CH_3$ (12-9)

(b) $CH_3-\underset{\underset{CH_3}{|}}{\overset{\overset{CH_3}{|}}{C}}-ONa + CH_3Br$
                         primary halide

Of these two alternatives, (b) is by far the better reaction sequence to follow if a good yield of methyl–*t*–butyl ether as product is desired, since tertiary halides tend to undergo a competitive elimination reaction instead of nucleophilic substitution, when in the presence of strong bases (like alkoxides). Primary halides on the other hand, will undergo nucleophilic substitution reactions. Thus, in order to insure a high yield of ether as product, the primary (or least branched) alkyl group is used as the alkyl halide, and the tertiary (or more branched) alkyl group is incorporated as the alkoxide (reaction b).

Using reaction sequence (a) would have resulted in an elimination reaction (dehydrohalogenation, removal of HBr) and formed an alkene, 2–methylpropene, as the predominant product, rather than the ether.

$$CH_3-\underset{\underset{CH_3}{|}}{\overset{\overset{CH_3}{|}}{C}}-Br \xrightarrow{CH_3ONa} CH_3-\overset{\overset{CH_3}{|}}{C}=CH_2 + NaBr + CH_3OH \quad (12\text{-}10)$$

Alkyl halides are often replaced by alkyl sulfates, especially when preparing ethers of phenols.

$$C_6H_5OH + (CH_3)_2SO_4 \xrightarrow{NaOH} C_6H_5OCH_3 + CH_3-O\overset{\ominus}{S}O_3Na^\oplus \quad (12\text{-}11)$$

                dimethyl sulfate   **anisole**

12.3 Properties

Most of the aliphatic ethers are liquids, except for the lower molecular weight homologs, which are gases (e.g., dimethyl ether). The aromatic ethers are liquids or solids. Ethers are only slightly soluble in water. Perhaps the best known member of the series is ordinary (diethyl) ether which has been used as a general anesthetic. Diethyl ether is an excellent solvent for organic materials, and is easily flammable. On continued exposure to air, ethers form peroxides, which are very explosive materials. Caution must be exercised when distilling ethereal solutions in that the distillation must never be carried to dryness, since explosions may result. It is best to treat ethereal solutions, prior to distillation, with a reducing agent, such as ferrous sulfate, to destroy any peroxides present.

12.4 Reactions of Ethers

The chemical reactivity of ethers is exceedingly limited. The lack of reactivity of ethers is only exceeded by the alkanes as a class of organic compounds. Ethers do *not* react with dilute acids or bases. They are resistant to oxidizing agents such as $KMnO_4$ and to reducing agents. Ethers do *not* react with active metals like sodium and potassium. The use of ethers as solvents in many organic reactions is only because of their lack of reactivity. The reason for the general inertness of ethers is that the oxygen atom in ethers is saturated with alkyl or aryl groups— there is no acidic hydrogen atom present that can be replaced by another group on the oxygen atom (as in alcohols).

The few chemical reactions that ethers do undergo can be attributed to the presence of unshared pairs of electrons on the oxygen atom.

Ethers are soluble in concentrated hydrochloric acid and concentrated sulfuric acid, forming oxonium salts, the proton of the acids being attracted to the unshared electron pairs on the oxygen atom.

$$R-\ddot{O}-R' + HX \longrightarrow R-\overset{\oplus}{\underset{H}{\ddot{O}}}-R' X^{\ominus} \qquad (12\text{-}12)$$

oxonium salt

$$(X = Cl^{\ominus}, \text{ or } HSO_4^{\ominus})$$

12.4-1 Cleavage of Ethers by Acids The principal reaction ethers undergo is a cleavage reaction that occurs when ethers are heated with hydriodic acid. The reaction is a nucleophilic substitution reaction in which the alkyl groups are removed from the oxygen of the ether. HI is used since the iodide ion is a strong nucleophilic reagent. The reaction involves protonation of the ether by strong acid, followed either by an S_N2 reaction by iodide ion, or if the R groups in the ether are tertiary, an S_N1 reaction (ionization) followed with attack by iodide ion. The net result of this reaction with excess HI is the cleavage of the ether and the production of two moles of alkyl halide as product.

$$R-\ddot{O}-R' + H^{\oplus} \rightleftharpoons \left[R-\overset{H}{\underset{}{\ddot{O}}}-R' \right]^{\oplus} \xrightarrow[S_N2]{I^{\ominus}} RI + R'OH \qquad (12\text{-}13)$$

(R is tertiary) \qquad\qquad\qquad (R is primary or secondary)

$$\downarrow S_N1$$

$$R^{\oplus} + R'OH \xrightarrow{I^{\ominus}} RI \qquad (12\text{-}14)$$

The alcohol formed is converted in the acid solution to its conjugate acid, which then undergoes a second nucleophilic displacement reaction with iodide ion to form the second mole of alkyl halide product.

Sec. 12.5 SULFUR ANALOGS

$$R'-\overset{..}{\underset{..}{O}}-H + H^{\oplus} \longrightarrow \left[R'-\overset{H}{\underset{..}{\overset{|}{O}}}-H\right]^{\oplus} \xrightarrow{I^{\ominus}} R'-I + H_2O \quad (12\text{-}15)$$

The reaction can be summarized by the following equation.

$$R_2O + 2HI \longrightarrow 2RI + H_2O \quad (12\text{-}16)$$

(i.e., $CH_3-O-CH_2-CH_3 + 2HI \longrightarrow CH_3I + CH_3-CH_2I + H_2O$) (12-17)

The cleavage of *aromatic* ethers produces *phenols* rather than aryl halides. This is because of the great reluctance of the aromatic systems (phenols) to undergo nucleophilic substitution reactions readily.

$$\underset{\text{anisole}}{C_6H_5-O-CH_3} + HI \longrightarrow \underset{\text{phenol}}{C_6H_5-OH} + \underset{\text{methyl iodide}}{CH_3-I} \quad (12\text{-}18)$$

Aromatic ethers undergo the usual aromatic electrophilic substitution reactions. The alkoxyl group (R—O—), being an electron-donating group and activating the benzene ring, directs the entering group to the *ortho* and *para* positions.

$$\underset{\text{phenetole}}{C_6H_5-O-CH_2-CH_3} + Cl_2 \xrightarrow{Fe} \text{o-Cl-}C_6H_4\text{-O-CH}_2\text{-CH}_3 + \text{p-Cl-}C_6H_4\text{-O-CH}_2\text{-CH}_3 + HCl \quad (12\text{-}19)$$

12.5 Sulfur Analogs

The sulfur analogs of the ethers, the thioethers or sulfides, can be prepared from the mercaptans (thio alcohols) by the Williamson ether synthesis sequence.

$$\underset{\substack{\text{sodium salt} \\ \text{of mercaptan}}}{R-\overset{\ominus}{S}-Na^{\oplus}} + R'X \longrightarrow NaX^{\ominus} + \underset{\text{thioether (sulfide)}}{R-S-R'} \quad (12\text{-}20)$$

Oxidation of sulfides produces sulfoxides or sulfones.

$$R-\overset{..}{\underset{..}{S}}-R' \xrightarrow{H_2O_2} \underset{\text{sulfoxide}}{R-\overset{\overset{\displaystyle :\overset{..}{O}:^{\ominus}}{|}}{\underset{\oplus}{S}}-R'} \xrightarrow{HNO_3} \underset{\text{sulfone}}{R-\overset{\overset{\displaystyle :\overset{..}{O}:^{\ominus}}{|}}{\underset{\underset{\displaystyle :\overset{..}{O}:^{\ominus}}{|}}{\overset{\oplus\oplus}{S}}}-R'} \quad (12\text{-}21)$$

Dimethyl sulfoxide, $CH_3-\overset{\overset{O}{\|}}{S}-CH_3$, is an interesting compound. It has excellent solvent properties, even to the extent of being miscible with water, no doubt due to the polar S—O bond in the molecule. Dimethyl sulfoxide also possesses the rather unique property of relieving the pain in some cases of arthritis or bursitis, although the mechanism of its action is not known.

12.6 Epoxides

The reaction of alkenes with certain oxidizing agents, notably peroxy acids, produces a class of compounds in which an oxygen atom has been inserted between two carbon atoms to form a three-membered ring called an epoxide (*cyclic ether*). The reaction of styrene with peracetic acid produces the cyclic ether styrene oxide.

$$\text{styrene} \xrightarrow{CH_3-\overset{\overset{O}{\|}}{C}-OOH} \text{styrene oxide (phenylepoxyethane)} \quad (12\text{-}22)$$

Whereas ethers are relatively inert, the epoxides (oxiranes) are extremely reactive compounds, as are their sulfur analogs (thiiranes). This great difference in reactivity between ordinary ethers and cyclic ethers can be best explained in terms of structural differences between the two classes of compounds (e.g., ring strain).

The normal bond angle around a saturated carbon atom is the tetrahedral angle of 109° 28'. When the atoms form a three-membered ring (as in an epoxide), the bond angle is only about 60°. The difference between this bond angle of about 60° in epoxides and the normal tetrahedral angle (109° 28') creates a *strain* in the molecule (Baeyer ring strain theory). This strain in the small ring compounds is reflected in the reactions of epoxides, where the ring tends to open up to form open-chain compounds and thus relieve the strain, by increasing the bond angles from 60° to 109° 28' (the normal tetrahedral angle). Ring opening reactions occur readily in epoxides to relieve the ring strain, and so epoxides are far more reactive than ordinary open-chain ethers.

12.6-1 Ethylene Oxide Among the epoxides, the compound ethylene oxide is the one most frequently encountered in organic synthesis. Ethylene oxide can be prepared by treating ethylene with peracetic or perbenzoic acid. The commercial preparation involves oxidation of ethylene with air in the presence of a silver catalyst.

$$2CH_2=CH_2 + O_2 \xrightarrow[350°\text{ C, press.}]{Ag} 2 \underset{\text{ethylene oxide}}{CH_2-CH_2} \quad (12\text{-}23)$$

Sec. 12.6 EPOXIDES

Ethylene oxide can also be synthesized by the following reaction sequence involving addition of HOCl across the carbon–carbon double bond, followed by an elimination reaction.

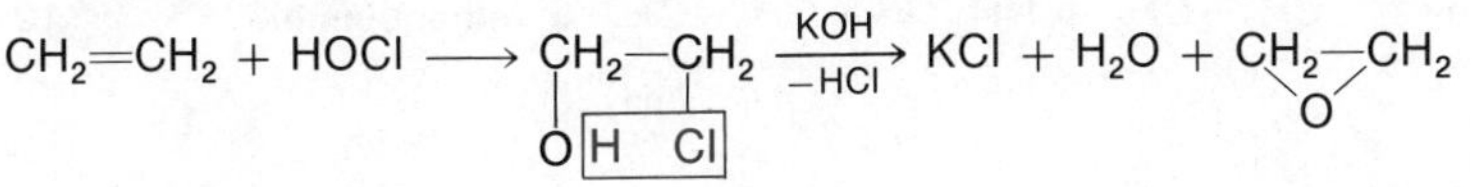

$$CH_2{=}CH_2 + HOCl \longrightarrow \underset{\underset{O\;\;\boxed{H\;\;Cl}}{|\;\;\;\;\;|}}{CH_2{-}CH_2} \xrightarrow[-HCl]{KOH} KCl + H_2O + \underset{O}{CH_2{-}CH_2} \quad (12\text{-}24)$$

The ring strain present in the ethylene oxide molecule enables the C—O bond to be broken much more easily than in an open-chain ether. The cleavage of the cyclic ether involves opening the ring, relieving the strain. Ethylene oxide and other epoxides will react with nucleophilic reagents in the presence of acid catalysts. The overall reaction resulting from the ring opening appears to be an addition reaction producing 1,2-disubstituted derivatives as products. One of the C—O bonds breaks and the positive part of the attacking reagent becomes attached to the more electronegative oxygen atom, as is shown in the general equation

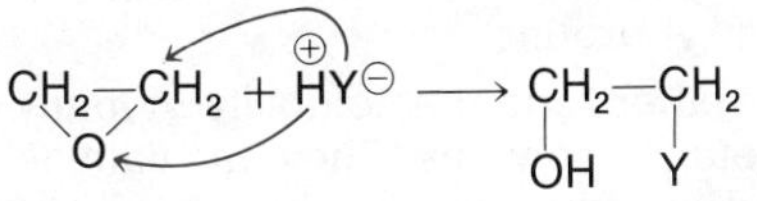

$$\underset{O}{CH_2{-}CH_2} + \overset{\oplus}{H}Y^{\ominus} \longrightarrow \underset{\underset{OH\;\;\;Y}{|\;\;\;\;\;|}}{CH_2{-}CH_2} \quad (12\text{-}25)$$

The Y can be any group that has a reactive hydrogen attached to it. Some typical reagents are H_2O, HCl, NH_3, CH_3OH.

Reaction of ethylene oxide with some typical reagents is illustrated in the reactions

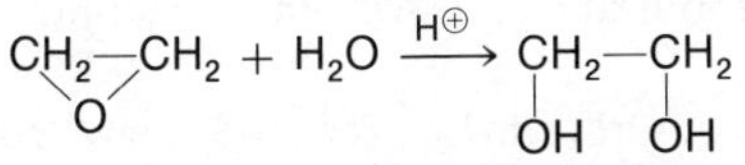

$$\underset{O}{CH_2{-}CH_2} + H_2O \xrightarrow{H^{\oplus}} \underset{\underset{OH\;\;\;OH}{|\;\;\;\;\;|}}{CH_2{-}CH_2} \qquad \textbf{ethylene glycol} \quad (12\text{-}26)$$
(used in Prestone antifreeze)

$$\underset{O}{CH_2{-}CH_2} + \underset{\underset{OH\;\;\;OH}{|\;\;\;\;\;|}}{CH_2{-}CH_2} \xrightarrow{H^{\oplus}} \underset{\underset{OH\;\;\;O{-}CH_2{-}CH_2{-}OH}{|\;\;\;\;\;\;\;\;\;\;\;\;\;|}}{CH_2{-}CH_2} \quad (12\text{-}27)$$

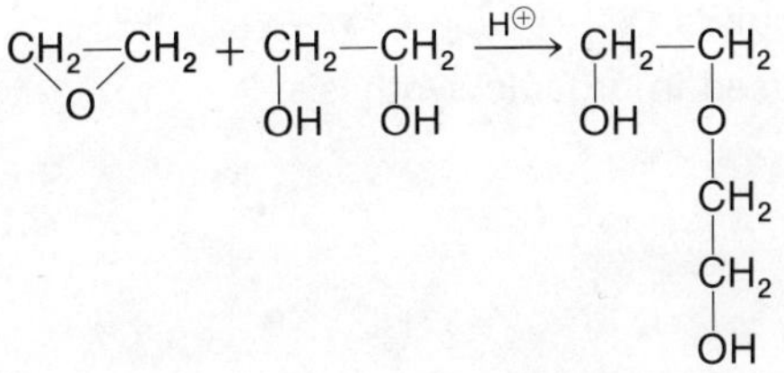

diethylene glycol (an excellent solvent)

$$\underset{O}{CH_2{-}CH_2} + HCl \longrightarrow \underset{\underset{OH\;\;\;Cl}{|\;\;\;\;\;|}}{CH_2{-}CH_2} \qquad \textbf{ethylene chlorohydrin} \quad (12\text{-}28)$$

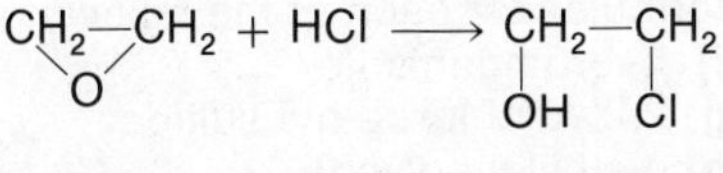

Reaction of alcohols with ethylene oxide produces as products a class of compounds that contains *both* the *ether* and *alcohol* functional groups (Cellosolves, which are excellent solvents).

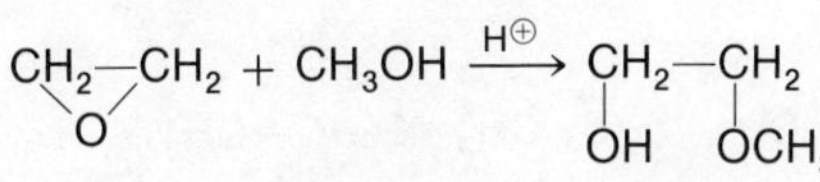

$$\underset{O}{CH_2{-}CH_2} + CH_3OH \xrightarrow{H^{\oplus}} \underset{\underset{OH\;\;\;OCH_3}{|\;\;\;\;\;|}}{CH_2{-}CH_2} \qquad \textbf{methyl cellosolve} \quad (12\text{-}29)$$

Reaction of ethylene oxide with ammonia produces ethanolamine. Using an excess of ethylene oxide leads to the formation of diethanolamine and triethanolamine as products.

$$CH_2\overset{O}{\underset{}{-}}CH_2 + NH_3 \longrightarrow \underset{OH}{CH_2}-\underset{NH_2}{CH_2} \quad \text{ethanolamine} \quad (12\text{-}30)$$

Summary

- The structure of ethers can be considered to be related to water and alcohols.
- The ethers may be simple or mixed.
- Ethers may be prepared by the dehydration of alcohols or by the Williamson synthesis.

  (a) The Williamson ether synthesis, which is the best method for the preparation of ethers, involves the displacement of an alkyl halide by an alkoxide. Alkyl sulfates are sometimes used instead of alkyl halides when preparing alkyl aryl ethers from phenols.

  (b) The Williamson synthesis has some limitations when preparing ethers containing tertiary alkyl groups.

- The low reactivity of ethers and the solubility of many organic compounds in them makes ethers useful as solvents. They are flammable and form explosive peroxides on continued exposure to air. They must be handled with care.
- Ethers are quite unreactive. However, they may be cleaved by hot concentrated HI or HBr.
- Aromatic ethers undergo the usual aromatic electrophilic substitution reactions; the —OR group is *ortho–para*-directing.
- Sulfur analogs of the ethers can be prepared by the Williamson synthesis from thiols.
- The epoxides (three-membered cyclic ethers) can be prepared easily. Because of Baeyer ring strain the epoxides are unusually reactive and can be used to prepare quite a variety of compounds.
- Ethylene oxide is frequently used in organic synthesis.

PROBLEMS

1 Draw the correct structural formulas for each of the following compounds.
  (a) *n*–propyl ether
  (b) ethoxyethane
  (c) ethyl *n*–butyl ether
  (d) 2,3–epoxybutane
  (e) phenylvinyl ether
  (f) 4–bromoanisole
  (g) 2,2′–dichloroethyl sulfide
  (h) diethyleneglycol
  (i) methyl cellosolve
  (j) propylene oxide

2 Name the following compounds.

(a) $CH_3-\underset{\underset{}{OCH_3}}{CH}-CH_2-CH_2-CH_3$ (b) $CH_3-O-\underset{\underset{CH_3}{|}}{\overset{\overset{H}{|}}{C}}-CH_3$

PROBLEMS

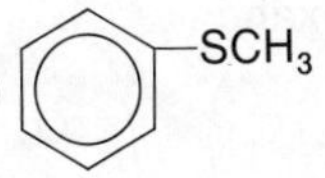

(c) 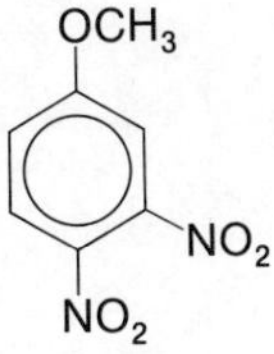 1-methoxy-3,4-dinitrobenzene (OCH₃, NO₂, NO₂ on benzene ring)

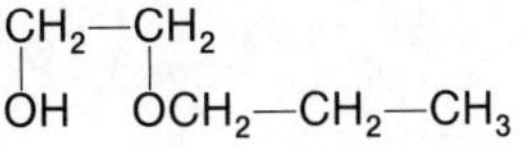

(d) 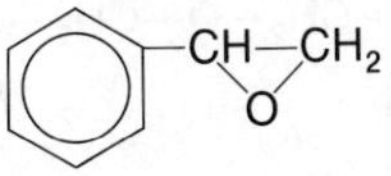 styrene oxide (C₆H₅—CH—CH₂ with O bridge)

(e) thioanisole (C₆H₅—SCH₃)

(f) CH₂—CH₂
     | |
     OH   OCH₂—CH₂—CH₃

3 Name a test or single reagent which can be used to distinguish between the following pairs of compounds.
  (a) ethyl ether, and n-butyl alcohol
  (b) 1-methoxynonane, and nonane
  (c) diphenyl ether, and ethyl ether
  (d) $CH_3-CH_2-CH_2-O-CH_2-CH_3$, and $CH_2=CH-CH_2-O-CH_2CH_3$

4 Complete each of the following equations by writing the correct structural formula(s) of the product(s) formed. If no reaction occurs, write N.R.

  (a) $CH_3CH_2Br + CH_3O^\ominus K^\oplus \longrightarrow$
  (b) $CH_3-O-CH_3 + Na \longrightarrow$
  (c) $(CH_3)_3-C-Br + CH_3CH_2O^\ominus Na^\oplus \longrightarrow$
  (d) $CH_3CH_2SH \xrightarrow[Na]{CH_3Cl}$
  (e) 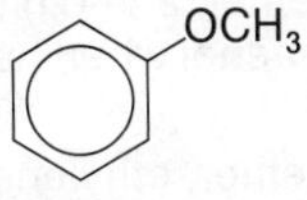 anisole (C₆H₅—OCH₃) + conc. HBr $\xrightarrow{\Delta}$
  (f) 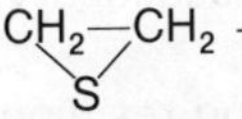 $\underset{S}{CH_2-CH_2}$ + $NH_3 \longrightarrow$ (ethylene sulfide)
  (g) 2,3-epoxybutane + $CH_3CH_2OH \longrightarrow$
  (h) ethylene oxide + HBr $\longrightarrow$
  (i) α-naphthol + $(CH_3)_2SO_4 \xrightarrow{KOH}$
  (j) methoxypropane + HCl $\longrightarrow$
  (k) 2-propanol + conc.$H_2SO_4 \xrightarrow{\Delta}$
  (l) anisole + $Br_2 \xrightarrow{Fe}$

5 Perform all of the following syntheses starting with the indicated starting materials and any necessary inorganic reagents.

  (a) ethylene to ethyl n-propyl ether
  (b) ethyl ether to ethylene glycol
  (c) any four carbon alkenes to n-butyl t-butyl ether
  (d) benzene to anisole
  (e) ethyl mercaptan to diethylsulfone
  (f) inorganic reagents to ethyl cellosolve
  (g) propylene to allyl i-propyl ether
  (h) *ethylene to divinyl ether
  (i) *benzene to phenyl p-bromobenzyl ether

(j) *ethylene to 1,4-dioxane

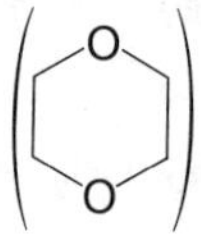

6 *Diglyme, diethyleneglycol methyl ether, $CH_2-CH_2-O-CH_2-CH_2$ is an
$\quad\quad\quad\quad\quad\quad\quad\quad\quad\quad\quad\quad\quad\quad\quad\quad\quad\quad\;\;\;\;|\quad\quad\quad\quad\quad\quad\quad\quad\quad\quad\;\;|$
$\quad\quad\quad\quad\quad\quad\quad\quad\quad\quad\quad\quad\quad\quad\quad\quad\quad\quad\;OCH_3\quad\quad\quad\quad\quad\quad\;OCH_3$
excellent solvent. Can you suggest a reason to account for the fine solvent properties of diglyme? How would you synthesize diglyme starting with inorganic reagents?

7 * (a) Why can't t-butyl ether be prepared by the Williamson ether synthesis?

(b) Could you prepare t-butyl ether by reacting t-butyl alcohol with concentrated H_2SO_4?

(c) When 2-methylpropene and t-butyl alcohol are reacted with concentrated H_2SO_4 under pressure, a good yield of t-butyl ether is obtained. Suggest a detailed mechanism for the reaction which occurred.

8 Can you use dimethylsulfate, $(CH_3)_2SO_4$, instead of a methyl halide in the Williamson ether synthesis? Explain.

9 *Why can ethers be cleaved by hot, concentrated HI or HBr, but *not* by concentrated HCl?

10 *What would you expect to happen when diphenyl ether is treated with hot, concentrated hydriodic acid? Explain.

11 *Most ethers are inert towards bases. However, the ether, 2,4-dinitroanisole is readily cleaved to methyl alcohol and 2,4-dinitrophenol when heated with dilute, aqueous sodium hydroxide. Explain.

12 Although most ethers are relatively inert, the cyclic ether, ethylene oxide, is extremely reactive towards nucleophilic displacement reactions. Would you expect all cyclic ethers to be very reactive? Explain.

13 Suggest detailed mechanisms for each of the following reactions.

(a) $CH_3-CH_2-CH_2I + CH_3CH_2\overset{\ominus}{O}Na^{\oplus} \longrightarrow$ (?)

(b) $(CH_3)_3C-O-CH_3$ + hot conc. HI $\longrightarrow$ (?)

(c) CH_2-CH_2 + HCN $\longrightarrow$ (?)
$\quad\;\;\diagdown\;\diagup$
$\quad\quad\;O$

14 Compound A, C_8H_9OBr, is reacted with hot alkaline $KMnO_4$ to form compound B, $C_8H_8O_3$. When B is treated with hot, concentrated HI, compound C, $C_7H_6O_3$, is formed. Compound A is soluble in cold, concentrated sulfuric acid, and does *not* decolorize a dilute, neutral $KMnO_4$ solution or bromine in carbon tetrachloride. Compound C is a hydroxybenzoic acid about 100 times stronger than benzoic acid. From this information suggest a set of structures for compounds A, B, and C.

13
Organic Halogen Compounds

In this new ultra-swift immersion freezer, using liquid Freon, 10,000 pounds of corn on the cob can be quick-frozen in an hour.

Courtesy of General Foods Corporation.

13.1 Introduction

Organic halogen compounds are an extremely important class of compounds in that they are perhaps the most useful compounds in organic synthesis. It is from many of these compounds, particularly the monohalogen compounds, that many other compounds are made. For this reason, it is essential to know the preparations and reactions of these compounds, which serve as invaluable tools to the synthetic organic chemist.

The chemistry of the aliphatic and aromatic halides has been discussed throughout the previous chapters in the text. The reader should review this material if he does not recall the reactions of these compounds. We shall briefly refer back to the most important aspects of this material and extend our knowledge of organic halogen compounds in this chapter.

13.2 Preparation of Organic Monohalogen Compounds

Alkyl halides can be prepared by the free-radical halogenation of alkanes (Section 7.7-1), the reaction of alkenes with hydrogen halides (Sections 8.8-2 and 8.9), the free-radical substitution reaction of an alkene with a halogen (Section 8.12), and by treating alcohols with hydrogen halides, phosphorus halides, or thionyl chloride (Section 11.10-2).

The halogenation of aromatic compounds to produce aryl halides has been discussed in Section 10.5-3. The free-radical halogenation of alkylbenzenes produces side-chain halogenation, the halogen attacking the carbon atom of the alkyl side chain directly attached to the benzene ring. This is due to the greater stability of benzylic type free radicals in contrast to free radicals formed at other positions in the side chain.

$$\text{C}_6\text{H}_5\text{CH}_2\text{—CH}_3 + \text{Cl}_2 \xrightarrow{h\nu} \text{HCl} + \text{C}_6\text{H}_5\text{CHCl—CH}_3 \qquad (13\text{–}1)$$

1–chloro–1–phenylethane

13.3 Preparation of Iodides and Fluorides

The reader will undoubtedly have noticed that in our discussions of organic halogen compounds we limited ourselves to chlorides and bromides, and virtually no mention was made of iodides and fluorides. This is because the reactions we discussed were not feasible for the preparation of iodides and fluorides.

Alkyl iodides and fluorides are usually prepared from the corresponding chloride or bromide by a **halide exchange reaction.** Alkyl iodides are prepared by treating an alkyl chloride or bromide with sodium iodide in acetone. The sodium chloride or bromide formed is less soluble than the sodium iodide in acetone and precipitates out of the solution, driving the equilibrium to the right.

$$R-Cl + Na^{\oplus}I^{\ominus} \xrightleftharpoons{\text{acetone}} R-I + NaCl \downarrow \qquad (13-2)$$

Alkyl fluorides can also be prepared by an exchange reaction using mercurous fluoride or antimony trifluoride.

$$2RCl + Hg_2F_2 \longrightarrow 2R-F + Hg_2Cl_2 \qquad (13-3)$$

$$3CCl_4 + 2SbF_3 \xrightarrow[\text{cat.}]{SbCl_5} 3CCl_2F_2 + 2SbCl_3 \qquad (13-4)$$
dichlorodifluoromethane
(Freon-12) (a refrigerant)

Direct fluorination or iodination of organic compounds is not feasible, so that the halogen exchange reaction is extremely important in the preparation of fluorides and iodides. Aryl iodides and fluorides can be prepared by using diazonium salts (Sections 16.11-1, -2, and -3).

The polyfluorides or fluorocarbons are becoming increasingly important compounds. They are best prepared by replacement of hydrogen with a strong fluorinating agent such as cobalt(III) fluoride, CoF_3.

$$n\text{-}C_7H_{16} + 32CoF_3 \longrightarrow C_7F_{16} + 16HF + 32CoF_2 \qquad (13-5)$$
n-heptane   perfluoroheptane
(per- = fully substituted)

13.4 Displacement Reactions of Alkyl Halides

The displacement reactions of alkyl halides (S_N1 and S_N2) have been discussed in various sections of the preceding chapters in the text. The substitution of many different functional groups for halogen occurs easily because the halides ($Cl^{\ominus}$, $Br^{\ominus}$, $I^{\ominus}$) are good leaving groups. The reader should review Sections 6.7, 6.7-1, 6.7-2, and 6.7-3 dealing with S_N1 and S_N2 reactions. The overall reaction can be simply written:

$$N{:}^{\ominus} + R-X \longrightarrow R-N + X^{\ominus} \qquad (13-6)$$

where N is a nucleophilic reagent.

The yields obtained from displacement reactions will of course depend on the structure of the alkyl halide used, the nature of the nucleophilic reagent, polarity

Sec. 13.5 THE GRIGNARD REAGENT

of solvent, and other experimental conditions. (In some cases [tertiary halides] a competing elimination [dehydrohalogenation] reaction will occur to a greater extent than substitution.) The reaction rate also depends on the halide ion being displaced as a leaving group. Since anions of strong acids are good leaving groups, iodides react faster than bromides which in turn react more rapidly than chlorides.

Some typical nucleophilic substitution (displacement) reactions of the alkyl halides will be illustrated by the following equations. Many of these reactions have already been discussed in earlier chapters. In each case the reaction is an attack of a nucleophilic reagent on an alkyl halide.

$$OH^\ominus + RX \xrightarrow{H_2O} \underset{\text{alcohol}}{R-OH} + X^\ominus \qquad (13\text{-}7)$$

$$\underset{\text{alkoxide}}{OR'^\ominus} + RX \longrightarrow \underset{\text{ether}}{R'-O-R} + X^\ominus \qquad (13\text{-}8)$$
$$\text{(Williamson ether synthesis)}$$

$$Na^\oplus I^\ominus + RX \underset{}{\overset{\text{acetone}}{\rightleftarrows}} \underset{\text{iodide}}{R-I} + NaX\downarrow \qquad (13\text{-}9)$$
$$\text{(halide exchange)}$$

$$CN^\ominus + RX \longrightarrow \underset{\text{nitrile (cyanide)}}{R-CN} + X^\ominus \qquad (13\text{-}10)$$

$$SH^\ominus + RX \longrightarrow \underset{\text{thio alcohol (mercaptan)}}{R-SH} + X^\ominus \qquad (13\text{-}11)$$

$$NH_2^\ominus + RX \xrightarrow{NH_3} \underset{\text{amine}}{R-NH_2} + X^\ominus \qquad (13\text{-}12)$$

$$Na-C\equiv C-R' + RX \longrightarrow \underset{\text{alkyne}}{R-C\equiv C-R'} + X^\ominus \qquad (13\text{-}13)$$

$$\underset{\text{malonate}}{^\ominus CH(COOC_2H_5)_2} + RX \longrightarrow \underset{\substack{\text{substituted malonic}\\ \text{ester}}}{R-CH(COOC_2H_5)_2} + X^\ominus \qquad (13\text{-}14)$$

One can see the great variety of organic compounds that can be prepared from alkyl halides by even these few reactions. Aryl halides usually do not undergo nucleophilic substitution reactions unless electron-withdrawing substituents are located *ortho* or *para* to the halogen on the aromatic ring.

13.5 Preparation and Use of the Grignard Reagent to Prepare Primary Alcohols

The Grignard reagent is perhaps the most useful reagent in synthetic organic chemistry. The reactions of Grignard reagents are so numerous that entire volumes of books have been written and devoted exclusively to their reactions. We will touch upon a few of the many uses of the Grignard reagent.

A French chemist, Victor Grignard (1871–1935) discovered that alkyl or aryl halides and magnesium metal turnings will react in anhydrous diethyl ether as a solvent to produce a solution known as a Grignard reagent.

$$R\!-\!X + Mg \xrightarrow{\text{anhy. ether}} \overset{\ominus}{R}\!-\!\overset{\oplus}{Mg}X \qquad (13\text{-}15)$$
$$\textbf{Grignard reagent}$$
$$(R = \text{alkyl or aryl})$$

For his work in the field of synthetic organic chemistry Victor Grignard received the Nobel prize in 1912.

The exact structure of the Grignard reagent is not known. It is believed that some solvent ether molecules are incorporated into its structure, but as to the exact nature of this interaction no one is absolutely certain. For this reason, the structural formula is written simply as RMgX. The Grignard reagent can be more appropriately represented as $R^{\ominus}Mg^{\oplus}X$, in order to understand its chemical behavior in many reactions.

$$CH_3\!-\!CH_2\!-\!I + Mg \xrightarrow{\text{anhy. ether}} CH_3\!-\!CH_2^{\ominus}\!-\!\overset{\oplus}{Mg}I \qquad (13\text{-}16)$$
$$\textbf{ethyl magnesium iodide}$$

Grignard reagents are extremely reactive. Usually the ethereal solutions of Grignard reagents are treated immediately upon their preparation with some reagent to form other products. Grignard reagents are not stored for any length of time since they tend to react with moisture and decompose to form a hydrocarbon. This reaction is between the Grignard reagent and an acidic hydrogen. For this reason, *anhydrous* ether is used as a solvent in the preparation of Grignard reagents. If *water* or any substance containing an acidic hydrogen is added to a Grignard reagent, decomposition of the reagent occurs, producing a hydrocarbon.

$$\overset{\ominus\ \oplus}{RMgX} + \overset{\oplus\ \ominus}{H\!-\!OH} \longrightarrow R\!-\!H + \overset{\oplus\oplus\ \ominus}{Mg(OH)(X)^{\ominus}} \qquad (13\text{-}17)$$

$$\overset{\ominus\ \oplus}{RMgX} + \overset{\oplus\ \ominus}{H\!-\!OR} \longrightarrow R\!-\!H + \overset{\oplus\oplus\ \ominus}{Mg(OR)(X)^{\ominus}} \qquad (13\text{-}18)$$
$$\textbf{alcohol}$$

Occasionally ethers other than diethyl ether (which is extremely flammable and tends to form explosive peroxides) are used as solvents to prepare Grignard reagents. One such ether that is used for this purpose is tetrahydrofuran, ⌷.

There are numerous organic reactions involving Grignard reagents. Let us discuss one. In Section 12.6-1, you will recall, it was mentioned that the cyclic ether ethylene oxide was extremely reactive, due to the ring strain associated with three-membered rings. The ring opens readily and ethylene oxide will react with Grignard reagents to produce primary alcohols after hydrolysis of the intermediate reaction product.

PROBLEMS

$$RMgX + \underset{O}{CH_2-CH_2} \longrightarrow R-CH_2-CH_2-OMgX \xrightarrow{HOH}$$

$$R-CH_2-CH_2-OH + Mg(OH)(X) \quad (13\text{-}19)$$
primary alcohol

The primary alcohol formed contains two more carbon atoms than the original Grignard reagent used. This is an excellent way of increasing the chain length by two carbon atoms. Oil of roses, 2-phenylethanol can be prepared in this way.

$$C_6H_5-MgBr + \underset{O}{CH_2-CH_2} \longrightarrow C_6H_5-CH_2-CH_2-OMgBr \xrightarrow{H_2O}$$

$$C_6H_5-CH_2-CH_2OH + Mg(OH)Br \quad (13\text{-}20)$$
2-phenylethanol

Summary

- Alkyl and aryl halides can be prepared by various methods.
    - (a) Free radical halogenation of alkanes.
    - (b) Halogenation of aromatic compounds.
    - (c) Replacement of the —OH group in alcohols.
    - (d) Addition of hydrogen halides to alkenes.
- Alkyl iodides and fluorides are usually prepared by a halide exchange reaction from the corresponding chloride or bromide.
- The principal use of alkyl halides is in synthesis by displacement reactions (S_N1 and S_N2).
- Grignard reagents can be prepared from alkyl and aryl halides, and are very useful in synthesis (of alcohols, etc.).

PROBLEMS

1 Draw the correct structural formulas for each of the following compounds.

- (a) allyl bromide
- (b) 3-chlorohexane
- (c) *p*-nitrobenzyl chloride
- (d) cyclopentyl iodide
- (e) 2,4-dichlorotoluene
- (f) perfluorobutane

2 Name the following compounds.

(a) $CH_3-\underset{\underset{Cl}{|}}{\overset{\overset{Cl}{|}}{C}}-\underset{\underset{H}{|}}{\overset{\overset{Cl}{|}}{C}}-CH=CH_2$

(b) 3-iodotoluene (benzene ring with CH_3 and I in meta positions)

(c) 2,6-dichloro-4-chlorobenzoic acid (benzene with COOH, two Cl ortho to COOH, and Cl para)

(d) 1-bromo-1-methyl-3-bromocyclohexane (cyclohexane with Br and CH_3 on C1 and Br on C3)

(e) $CH_3-(CH_2)_4-\underset{\underset{CH_3}{|}}{\overset{\overset{CH_3}{|}}{C}}-Cl$

(f) 4-nitrophenyl group with $-CH_2-CH_2-CH_2Br$ (benzene with NO_2 para to $CH_2CH_2CH_2Br$)

3 Complete each of the following equations by writing the correct structural formula(s) of the product(s) formed. If no reaction occurs, write N.R.

(a) $CH_3-CH_2-OH + PCl_5 \longrightarrow$

(b) $CH_3-CH_3 \xrightarrow[h\nu]{Cl_2}$

(c) $CH_3-CH_2-CH=CH_2 + HI \longrightarrow$

(d) propene + HBr + $H_2O_2 \longrightarrow$

(e) toluene + $Cl_2 \xrightarrow{Fe}$

(f) $CH_3-CH_2-Cl + NaI \xrightarrow{acetone}$

(g) 2,3-dimethylpentane + $CoF_3 \longrightarrow$

(h) H_3C-⟨benzene ring⟩$-Br$ + Mg $\xrightarrow{anhy.\ ether}$ (?) $\xrightarrow{H_2O}{H^\oplus}$

(i) $CH_2=CH-CH_2Cl + SbF_3 \xrightarrow[cat.]{SbCl_5}$

(j) ⟨phenyl⟩$-\underset{\underset{CH_3}{|}}{CH}-CH_2-CH_3 \xrightarrow[h\nu]{Cl_2}$

(k) ethyl iodide + $Na^\oplus \overset{\ominus}{O}CH_2CH_3 \longrightarrow$

(l) cyclohexyl chloride + NaCN $\longrightarrow$

(m) $CH_3Br + Na^\oplus-\overset{\ominus}{C}\equiv C-H \longrightarrow$

(n) CH_3-CH_2Br + aq. NaOH $\longrightarrow$

(o) $CH_3-CH_2Br + EtOOC-CH_2-COOEt \xrightarrow{NaOEt}$

(p) bromobenzene + aq. NaOH $\longrightarrow$

PROBLEMS

(q) 1-chlorobutane + $CH_3-\overset{O}{\underset{\|}{C}}-CH_2-\overset{O}{\underset{\|}{C}}-OC_2H_5 \xrightarrow{NaOEt}$

(r) $CH_3Br + :P(C_6H_5)_3 \longrightarrow$

4 Perform all of the following syntheses from the indicated starting materials and any necessary inorganic reagents.
  (a) 1-bromobutane to 2-bromo-2-chlorobutane
  (b) ethyl bromide to 1-butanol
  (c) benzene to 1,1-dibromo-1-phenylethane
  (d) benzene to p-bromobenzyl bromide
  (e) i-butyl bromide to 4-methyl-1-pentyne
  (f) benzene to p-chlorobenzoic acid
  (g) *1-propanol to 1,1-dichloro-2-methylcyclopropane
  (h) n-propyl chloride to isopropyl iodide
  (i) m-bromotoluene to toluene
  (j) inorganic reagents to allyl chloride.

5 (a) Draw all the isomers having the molecular formula C_4H_9Cl.

  (b) *Which of the isomers would you expect to form a precipitate readily upon addition of an alcoholic silver nitrate solution? Explain.

  (c) *Would you expect chlorobenzene to form a precipitate with alcoholic silver nitrate? Explain.

6 *The compound 2,4-dinitrofluorobenzene reacts readily with ammonia or primary amino groups. This reaction is used to determine the sequence of amino acids in protein molecules. Fluorobenzene will not undergo any similar type reaction; but, when 2,4-dinitrofluorobenzene is treated with ammonia (a nucleophilic aromatic substitution reaction), 2,4-dinitroaniline,

[structure: benzene ring with NH$_2$, NO$_2$ (ortho), NO$_2$ (para)], is formed. Explain why the reaction occurs with the 2,4-dinitro compound and not with the fluorobenzene.

7 Neopentyl halides react very slowly in nucleophilic substitution reactions whatever the experimental conditions. How do you explain this fact?

8 *Explain in terms of resonance why allyl chloride is extremely reactive, and vinyl chloride is very unreactive in nucleophilic substitution reactions.

9 Arrange the following compounds in order of increasing activity towards bromide ion under S_N1 reaction conditions.
CH_3-CH_2-Cl
$CH_2=CH-CH_2Cl$
CH_3Cl
$CH_2=CHCl$
$[(CH_3)_3-C-CH_2]_2-C(CH_3)Cl$

10 Arrange the following compounds in order of increasing activity towards bromide ion under S_N2 reaction conditions.

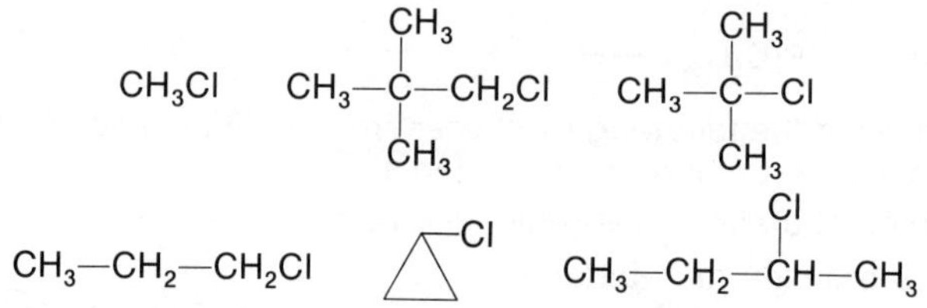

CH_3Cl $CH_3-\underset{\underset{CH_3}{|}}{\overset{\overset{CH_3}{|}}{C}}-CH_2Cl$ $CH_3-\underset{\underset{CH_3}{|}}{\overset{\overset{CH_3}{|}}{C}}-Cl$

$CH_3-CH_2-CH_2Cl$ △―Cl $CH_3-CH_2-\underset{\underset{}{}}{\overset{\overset{Cl}{|}}{C}H}-CH_3$

11 *In each of the following cases, indicate which reaction of the pair occurs most readily. Explain why.

(a) $NaOH + CH_3CH_2I \longrightarrow CH_3-CH_2OH + NaI$
$NaF + CH_3-CH_2I \longrightarrow CH_3-CH_2F + NaI$

(b) $KI + CH_3-CH_2Cl \longrightarrow CH_3-CH_2I + KCl$
$KI + CH_3-CH_2Br \longrightarrow CH_3-CH_2-I + KBr$

(c) $NaOCH_2CH_3 + CH_3(CH_2)_2-CH_2Br \longrightarrow$
$CH_3-CH_2-O-(CH_2)_3CH_3 + NaBr$

$NaOCH_2CH_3 + (CH_3)_3CBr \longrightarrow CH_3-\overset{\overset{CH_3}{|}}{C}=CH_2 + NaBr + CH_3CH_2OH$

(d) $CH_2=CHCl + AgNO_3 \longrightarrow CH_2=CHNO_3 + AgCl\downarrow$
$CH_2=CH-CH_2Cl + AgNO_3 \longrightarrow CH_2=CH-CH_2NO_3 + AgCl\downarrow$

12 The major product of the reaction of $CH_3-CH(CH_3)-CH_2Br$ with aqueous NaOH is $(CH_3)_3-C-OH$. Explain.

13 Why does benzhydryl bromide, $(C_6H_5)_2-CH-Br$, hydrolyze slowly $[(C_6H_5)_2CHBr + H_2O \longrightarrow (C_6H_5)_2CHOH]$ in the presence of some LiBr? (HINT: The benzhydryl bromide reacts by the S_N1 mechanism.)

14 Suggest a detailed mechanism for each of the following reactions.

(a) $(C_6H_5)_3-C-Cl + aq. NaOH \longrightarrow (C_6H_5)_3-C-OH + Na^\oplus + Cl^\ominus$
(b) $CH_3-CH_2Br + aq. NaOH \longrightarrow CH_3-CH_2OH + Na^\oplus + Br^\ominus$
(c) *$(CH_3)_3-C-Cl + alc. KOH \longrightarrow (CH_3)_2-C=CH_2 + KCl + H_2O$

14
Aldehydes and Ketones

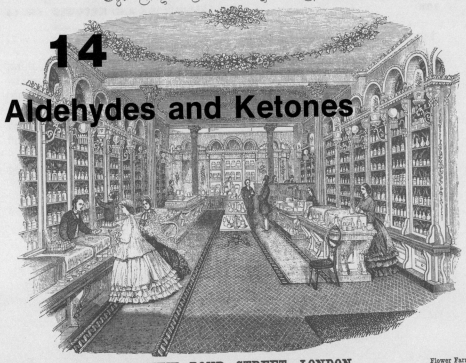

In 1865, as today, certain flowers containing aldehydes and ketones in their essences were the source of the sweet scents of perfumes.

Courtesy of The New York Public Library.

14.1 Introduction

Aldehydes and ketones are characterized as classes of organic compounds by the presence of the carbonyl group, $\diagdown\!\!\text{C}\!\!=\!\!\text{O}$, as the functional group. Aldehydes have the carbonyl group directly attached to at least one hydrogen atom, $\text{R}\!-\!\underset{\underset{\text{O}}{\|}}{\text{C}}\!-\!\text{H}$. (R may be H, alkyl, or aryl.) In ketones the carbonyl group is bound to two organic hydrocarbon (alkyl or aryl) groups, $\text{R}\!-\!\underset{\underset{\text{O}}{\|}}{\text{C}}\!-\!\text{R}'$.

In most cases the chemical reactivity of the aldehydes and ketones is very similar. It is therefore convenient and advantageous to discuss these two classes of compounds at the same time. We will see that any differences in the reactivity of aldehydes and ketones is due mainly to steric factors caused by the bulkier substituents (R vs. H) attached to the carbonyl group in ketones.

14.2 Nomenclature

Aldehydes and ketones can be named by the IUPAC system of nomenclature, using the principles outlined in earlier chapters. *Al*dehydes are characterized by an *-al* ending in their nomenclature; ket*ones* have names ending in *-one*.

Some of the simpler aldehydes and ketones are often given common names. In the case of aldehydes the common names are usually related to the acid obtained when the aldehyde undergoes oxidation (e.g., acetaldehyde produces acetic acid). The common names of the ketones are obtained by simply naming the groups attached to the carbonyl group. The following examples illustrate these points.

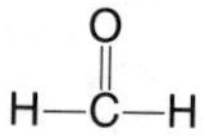

methan*al*
(formaldehyde)

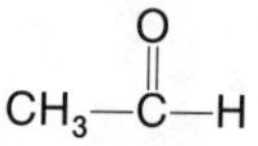

ethan*al*
(acetaldehyde)

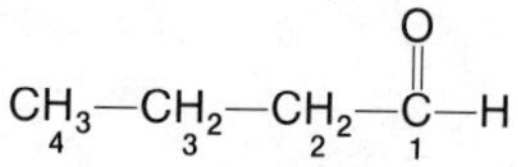

butan*al*
(*n*-butyraldehyde)

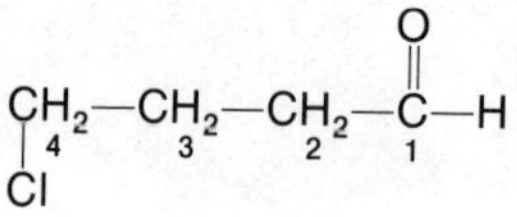

4-chlorobutan*al*

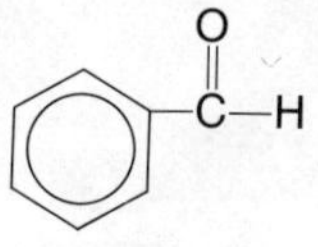

benzaldehyde
(found in almonds)

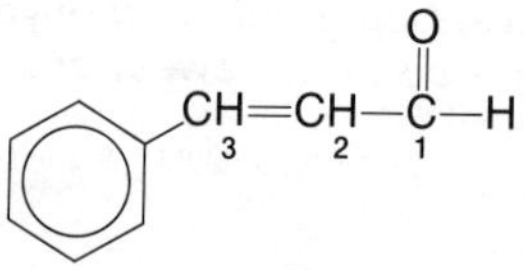

3-phenylpropen*al*
(cinnamaldehyde, found in cinnamon)

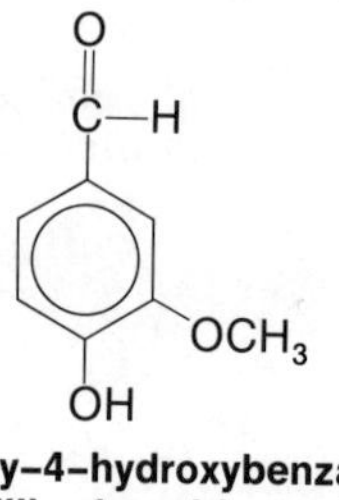

3-methoxy-4-hydroxybenzaldehyde
(vanillin, found in vanilla)

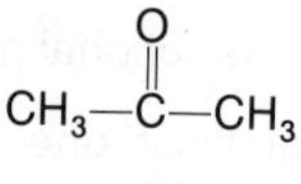

acet*one*
(dimethylketone)

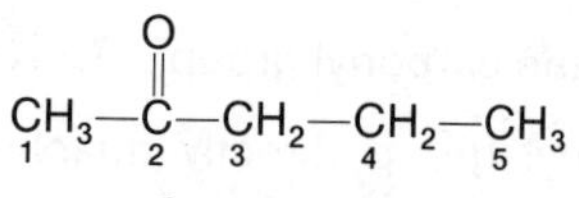

2-pentan*one*
(methyl *n*-propylketone)

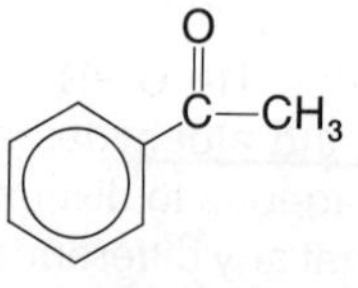

acetophen*one*
(methyl phenyl ketone)

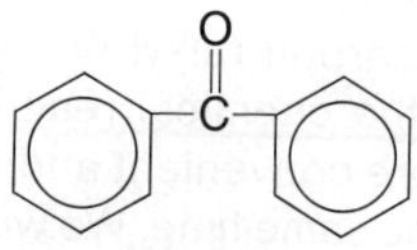

benzophen*one*
(diphenyl ketone)

14.3 Preparation of Aldehydes and Ketones

The preparation of aldehydes and ketones may be accomplished from quite a variety of compounds. The most common procedures used in their synthesis are the oxidation of alcohols, hydrolysis of *gem*-dihalides, and the Reimer–Tiemann reaction. The Rosenmund reduction of acyl halides to aldehydes and the Friedel–Crafts ketone synthesis will be discussed in Sections 15.10-1 and 15.10-2.

14.3-1 Oxidation of Alcohols As was mentioned in Chapter 11, aldehydes can be prepared by oxidizing primary alcohols, and ketones, by oxidation of secondary alcohols. In order to understand the oxidation and reduction of organic compounds it is important to remember that **oxidation** refers to *gain of oxygen or loss of hydrogen* in a compound, whereas **reduction** involves *gain of hydrogen or loss of oxygen* in a compound. It can be seen that the carbon atom becomes more oxidized as one goes from an alkane, R—CH$_3$, to an alcohol, R—CH$_2$OH, to an aldehyde, R—$\overset{\overset{O}{\|}}{C}$—H, to a carboxylic acid, R—$\overset{\overset{O}{\|}}{C}$—OH. Partial oxidation of a carbon atom produces compounds that are more susceptible to oxidation. For example, alcohols and aldehydes are oxidized much more easily than alkanes, in which the carbon atom is in its most reduced state.

The treatment of a primary alcohol with a strong oxidizing agent such as potassium dichromate or potassium permanganate in sulfuric acid solution produces an aldehyde.

$$R-\underset{\underset{H}{|}}{\overset{\overset{H}{|}}{C}}-OH \xrightarrow[\text{[O]}]{K_2Cr_2O_7,\ H_2SO_4} \left[R-\underset{\underset{H}{|}}{\overset{\overset{\text{(OH)}}{|}}{C}}-O\text{\textcircled{H}} \right] \xrightarrow{-H_2O} R-\overset{\overset{O}{\|}}{C}-H \quad (14\text{-}1)$$

primary alcohol   **aldehyde**

The intermediate formed during the course of the oxidation reaction (by the addition of an oxygen atom) is unstable and cannot be isolated. The presence of two hydroxyl groups (or two very electronegative groups) attached to the same carbon atom forms an unstable species that loses water to form a more stable product (Erlenmeyer's rule). The aldehydes formed are more susceptible to oxidation than the original alcohols used in the synthesis. The aldehydes must therefore be removed out of the reaction mixture as fast as they are formed to prevent their oxidation to carboxylic acids. Since the aldehydes (no hydrogen bonding present) have lower boiling points than the alcohols from which they are derived they can usually be distilled out of the reaction mixture easily.

$$CH_3-CH_2-CH_2-CH_2OH \xrightarrow[\Delta]{K_2Cr_2O_7,\ H_2SO_4} CH_3-CH_2-CH_2-\overset{\overset{O}{\|}}{C}-H \quad (14\text{-}2)$$

1-butanol   **butanal**
(b.p. = 118° C)   **(b.p. = 76° C)**

Ketones can be prepared by a similar oxidation reaction of secondary alcohols. They are less prone to further oxidation than are the aldehydes.

$$\underset{\substack{\text{secondary} \\ \text{alcohol}}}{R-\overset{H}{\underset{R'}{C}}-OH} \xrightarrow[\text{[O]}]{K_2Cr_2O_7,\ H_2SO_4} \left[R-\overset{\text{\textcircled{OH}}}{\underset{R'}{C}}-O\text{\textcircled{H}} \right] \xrightarrow{-H_2O} \underset{\text{ketone}}{R-\overset{O}{\overset{\|}{C}}-R'} \quad (14\text{-}3)$$

$$\underset{\text{2-butanol}}{CH_3-\overset{H}{\underset{OH}{C}}-CH_2-CH_3} \xrightarrow[\Delta]{K_2Cr_2O_7,\ H_2SO_4} \underset{\text{butanone}}{CH_3-\overset{O}{\overset{\|}{C}}-CH_2-CH_3} \quad (14\text{-}4)$$

The treatment of alcohols with milder oxidizing agents, such as copper gauze or powder, usually avoids the possibility of further oxidation of the aldehyde formed. The reaction is called **dehydrogenation** since hydrogen gas is produced along with the aldehyde or ketone.

$$\underset{\text{ethyl alcohol}}{CH_3-\overset{H}{\underset{H}{C}}-OH} \xrightarrow[300°\,C]{Cu} \underset{\text{acetaldehyde}}{CH_3-\overset{O}{\overset{\|}{C}}-H} + H_2 \uparrow \quad (14\text{-}5)$$

14.3-2 Hydrolysis of *gem*-Dihalides

The preparation of alcohols by the hydrolysis of alkyl halides has been discussed in Section 11.3. An analogous reaction with *gem*-dihalides can be used to prepare aldehydes and ketones. The hydrolysis of 1,1-dihalides, with the halogens at the end of an alkyl chain, produces aldehydes. The *gem*-dihalides situated at carbon atoms other than at the end of the chain will form ketones. The intermediate dihydroxy compound formed by hydrolysis of the dihalide loses water to form the more stable carbonyl compound.

$$\underset{\textit{gem}\text{-dihalide}}{R-\overset{H}{\underset{X}{C}}-X} + H_2O \xrightarrow{NaOH} \left[R-\overset{H}{\underset{\text{\textcircled{OH}}}{C}}-O\text{\textcircled{H}} \right] \xrightarrow{-H_2O} \underset{\text{aldehyde}}{R-\overset{O}{\overset{\|}{C}}-H} \quad (14\text{-}6)$$

Benzaldehyde is prepared commercially by the chlorination of toluene to benzal chloride, followed by hydrolysis.

$$\underset{\text{toluene}}{C_6H_5-CH_3} \xrightarrow[h\nu]{2Cl_2} \underset{\text{benzal chloride}}{C_6H_5-CHCl_2} \xrightarrow[NaOH]{H_2O} \underset{\text{benzaldehyde}}{C_6H_5-CH{=}O} \quad (14\text{-}7)$$

The overall reaction can be pictured as the replacement of two halogen atoms by a doubly bonded oxygen atom.

14.3-3 Reimer–Tiemann Reaction

The Reimer–Tiemann reaction for the synthesis of phenolic aldehydes involves treatment of phenol with chloroform in the presence of a strong base such as aqueous sodium hydroxide. The reaction is believed to proceed by the attack of the electrophilic reagent, dichlorocarbene ($:CCl_2$), on the highly reactive phenoxide ring. The dichlorocarbene is generated from chloroform by the action of a strong base (Section 8.10–3).

$$OH^\ominus + CHCl_3 \rightleftharpoons H_2O + :CCl_3^\ominus \longrightarrow Cl^\ominus + \;:CCl_2 \qquad (14\text{-}8)$$
dichlorocarbene

phenoxide (from phenol + base) + $:CCl_2$ → [substituted benzal chloride] → salicylaldehyde (14-9)

The substituted benzal chloride formed in the course of the reaction is hydrolyzed by the reaction medium to the aldehyde. Although para isomer is also obtained, the major product in the Reimer–Tiemann reaction is with the aldehyde group *ortho* to the phenolic hydroxyl group.

phenol + $CHCl_3$ + aq. NaOH $\xrightarrow[2.\; H^\oplus]{1.\; \Delta,\; 70°\text{C}}$ salicylaldehyde (major product) (14-10)

14.4 Preparation of Formaldehyde, Acetaldehyde, and Acetone

From an industrial standpoint the most important carbonyl compounds are the aldehydes formaldehyde and acetaldehyde, and the ketone, acetone. These three compounds can be prepared by the synthetic procedures discussed earlier. In addition, other methods specific to these compounds are also available.

Formaldehyde can be prepared by the reaction of methyl alcohol in the vapor state with air over a copper or silver catalyst at a temperature of 250–300° C.

$$2CH_3OH + O_2 \xrightarrow[250°-300°\,C]{\text{Ag or Cu}} 2H-\overset{\overset{O}{\|}}{C}-H + 2H_2O \qquad (14\text{-}11)$$
formaldehyde

Formaldehyde (b.p. = −21° C), a gas at room temperature, is very soluble in water. Formaldehyde has many uses such as the silvering of mirrors, as a disinfectant, as a preservative for biological materials, and in the manufacture of some plastics.

Acetaldehyde can be prepared by the analogous oxidation of ethyl alcohol.

$$2CH_3-CH_2OH + O_2 \xrightarrow[250°-300°\,C]{\text{Ag or Cu}} 2CH_3-\overset{\overset{O}{\|}}{C}-H + 2H_2O \qquad (14\text{-}12)$$
acetaldehyde

However, commercially acetaldehyde is usually prepared by the hydration of acetylene (Section 9.4–5).

$$H-C\equiv C-H + H_2O \xrightarrow[H_2SO_4]{\text{HgSO}_4} \left[H-\overset{\overset{H}{|}}{C}=\overset{\overset{OH}{|}}{C}-H\right] \longrightarrow CH_3-\overset{\overset{O}{\|}}{C}-H \qquad (14\text{-}13)$$
acetylene   enol   acetaldehyde

Acetone, which is perhaps most useful as a solvent for many materials, is familiar to most everyone as a constituent of nail polish remover. The very good solvent properties of acetone and the fact that it is miscible with water make acetone a most desirable material. The commercial preparation of acetone involves dehydrogenation of isopropyl alcohol or as a by-product formed in the synthesis of phenol from cumene (Section 11.7). Acetone can also be prepared by the hydration of propyne (Section 9.4–5) or by the pyrolysis of acetic acid in the presence of a manganous or thorium oxide catalyst.

$$2CH_3-\overset{\overset{O}{\|}}{C}-OH \xrightarrow[ThO_2,\,\Delta]{\text{MnO or}} CH_3-\overset{\overset{O}{\|}}{C}-CH_3 + CO_2\uparrow + H_2O \qquad (14\text{-}14)$$
acetic acid   acetone

Large quantities of acetone are also made by the fermentation of sugar by the bacterium *Clostridium acetobutylicum,* a process discovered by the chemist and first president of Israel, Chaim Weizmann.

14.5 The Nature of the Carbonyl Group

Since the chemistry of the aldehydes and ketones is determined mainly by the presence of the carbonyl, $\text{C}=\text{O}$, functional group, let us consider the molecular geometry and electronic distribution of the carbonyl group.

Since the carbonyl carbon is attached to three other substituents (double bonded to an oxygen atom and single bonded to two other atoms), it is trigonal in nature and three of its orbitals are of the sp^2 hybridization (Chapter 1). The remaining p orbital of the carbon atom overlaps with a p orbital of the oxygen atom to form a π bond above and below the plane of the molecule (see Figures 14.1 and 14.2).

Sec. 14.6 BONDING OF THE CARBONYL GROUP   203

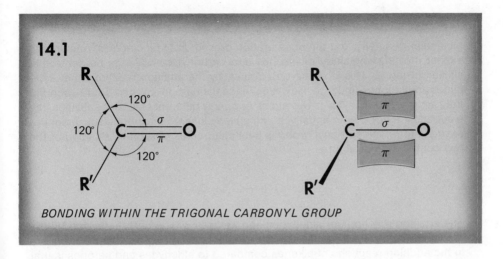

14.1

BONDING WITHIN THE TRIGONAL CARBONYL GROUP

14.6 Bonding of the Carbonyl Group

The bond angles in the carbonyl group are the 120° angles characteristic of sp^2 hybridization, and the carbonyl portion of the molecule is flat and coplanar with

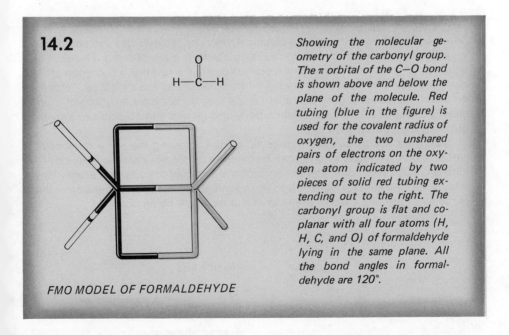

14.2

FMO MODEL OF FORMALDEHYDE

Showing the molecular geometry of the carbonyl group. The π orbital of the C–O bond is shown above and below the plane of the molecule. Red tubing (blue in the figure) is used for the covalent radius of oxygen, the two unshared pairs of electrons on the oxygen atom indicated by two pieces of solid red tubing extending out to the right. The carbonyl group is flat and coplanar with all four atoms (H, H, C, and O) of formaldehyde lying in the same plane. All the bond angles in formaldehyde are 120°.

the four atoms (R, R', C, and O) all lying in the same plane. The molecular geometry of the carbonyl group $\left(\diagup\!\!\!\diagdown C=O\right)$ is similar to that of the carbon–carbon double bond $\left(\begin{array}{c}\diagdown\\C=C\\\diagup\end{array}\begin{array}{c}\diagup\\\\\diagdown\end{array}\right)$ present in the alkenes. Because of the similarity in molecular geometry and the unsaturation present in the molecules, one should expect that the chemistry of alkenes and carbonyl compounds be somewhat similar in nature. This is borne out, in fact, by the addition reactions that aldehydes and ketones undergo. However, since the carbon–oxygen double bond of the carbonyl group involves two atoms of greatly different electronegativities, the electrons are not shared equally, being pulled closer to the more electronegative oxygen atom. The carbonyl group is best pictured as a resonance hybrid of the two contributing structures.

$$\left[\begin{array}{c}\diagdown\\C=\ddot{O}:\\\diagup\end{array} \longleftrightarrow \begin{array}{c}\diagdown\\\overset{\oplus}{C}-\ddot{\underset{..}{O}}:^{\ominus}\\\diagup\end{array} \right]$$

The addition reactions of aldehydes and ketones will be best understood if the reader keeps in mind the polar nature of the carbonyl group. The major difference in the addition reactions of alkenes compared to aldehydes and ketones is that the carbon–carbon double bond is attacked by electron-seeking (electrophilic reagents) and serves as a source of electrons, whereas the carbonyl group reacts mainly with nucleophilic reagents, which can supply electrons to the relatively positive carbonyl carbon atom. (See resonance structures of the carbonyl group.)

14.7 Reactions of Aldehydes and Ketones

14.7-1 Oxidation

Although the chemical reactivity of aldehydes and ketones is very similar, their behavior toward oxidizing agents is quite different. The ease of oxidation of aldehydes to carboxylic acids is due to the fact that a hydrogen atom can be removed without breaking the carbon skeleton of the molecule. Ketones are much more resistant to oxidation, since they do not have a hydrogen atom directly attached to the functional carbonyl group, and can be oxidized only by breaking the carbon skeleton.

Aldehydes are oxidized to carboxylic acids containing the same number of carbon atoms, whereas ketones can be oxidized under more drastic reactions to form two carboxylic acids with fewer carbon atoms than the original ketone. This is because the ketone is cleaved into two fragments by attack on either side of the carbonyl group, the carbon skeleton of the molecule being broken.

$$\underset{\text{aldehyde}}{R-\overset{\overset{\displaystyle O}{\|}}{C}-H} \xrightarrow{[O]} \underset{\text{carboxylic acid}}{R-\overset{\overset{\displaystyle O}{\|}}{C}-OH} \tag{14-15}$$

$$\underset{\substack{\uparrow \ \ \ \uparrow \\ \left[\begin{array}{c}\text{attack at either}\\\text{of these positions}\end{array}\right]\\ \text{ketone}}}{R-\overset{\overset{\displaystyle O}{\|}}{C}-R'} \xrightarrow{[O]} \underset{\text{carboxylic acids}}{R-\overset{\overset{\displaystyle O}{\|}}{C}-OH + R'-\overset{\overset{\displaystyle O}{\|}}{C}-OH} \tag{14-16}$$

Sec. 14.7 REACTIONS OF ALDEHYDES AND KETONES

The ease of oxidation of aldehydes, as compared to ketones, is often used as a basis for distinguishing between the two classes of compounds. Strong oxidizing agents such as alkaline potassium permanganate or potassium dichromate in sulfuric acid can be used, the time required for the appearance of a brown precipitate of MnO_2 or the green color of the chromium $+3$ ion formed by reduction of the $MnO_4^{\ominus}$ and $Cr_2O_7^{\ominus}$ ions, respectively, distinguishing between the presence of an aldehyde or ketone. Aldehydes will produce a color change in a much shorter period of time.

Aldehydes also can be oxidized by much milder oxidizing agents, such as complex cupric or silver ions in alkaline solution. Ketones will *not* be oxidized under these conditions, so that these reagents can be used to clearly differentiate between aldehydes and ketones. The tests involving the complex cupric ion are known as the **Benedict's** and **Fehling's tests,** and the **Tollens' test** refers to an alkaline ammoniacal solution of complex silver ion.

The reagents using cupric ion as an oxidizing agent are known as **Benedict's solution** (cupric citrate complex) and **Fehling's solution** (cupric tartrate complex).

$$Cu^{+2} + \text{citric acid} + \text{base} \longrightarrow [Cu(citrate)_2]^{-6}$$
Benedict's solution

$$Cu^{+2} + \text{tartaric acid} + \text{base} \longrightarrow [Cu(tartrate)_2]^{-6}$$
Fehling's solution

Benedict's and Fehling's solutions oxidize aliphatic aldehydes and are reduced in the process to a red precipitate of cuprous oxide, Cu_2O. The presence of the red precipitate of cuprous oxide serves as an indication of an aldehyde group in a molecule. The presence of glucose (which contains an aldehydic group) in the urine is often detected in this way.

$$\underset{\substack{\text{aliphatic}\\\text{aldehyde}}}{R-\overset{\overset{\displaystyle O}{\|}}{C}-H} \xrightarrow{\text{Cu(II) citrate or tartrate}} R-\overset{\overset{\displaystyle O}{\|}}{C}-O^{\ominus} + \underset{(red)}{Cu_2O \downarrow} \qquad (14\text{-}17)$$

The reagent that uses an ammoniacal solution of silver nitrate as an oxidizing agent is known as **Tollens' solution.** It is prepared by carefully adding just enough ammonia to dissolve the silver oxide precipitate present in a solution of silver nitrate and sodium hydroxide. The silver diamino complex formed serves as the oxidizing agent.

$$Ag^{\oplus} + 2NH_3 \longrightarrow Ag(NH_3)_2^{\oplus}$$
silver diamino complex

Tollens' reagent is a slightly more powerful oxidizing agent than Benedict's or Fehling's solution. It can oxidize aromatic aldehydes in addition to aliphatic aldehydes. The silver complex is reduced to metallic silver in a few minutes at room temperature. When a thoroughly clean glass surface is used with the Tollens' test, the metallic silver formed will deposit on the glass surface in the form of a mirror. The **silver mirror** formed indicates the presence of an aldehyde grouping in a molecule. The silvering of mirrors is done commercially using formaldehyde in this way.

$$R-\overset{O}{\underset{\parallel}{C}}-H + 2Ag(NH_3)_2^{\oplus} + 3OH^{\ominus} \longrightarrow R-\overset{O}{\underset{\parallel}{C}}-O^{\ominus} + 2Ag\downarrow + 4NH_3\uparrow + 2H_2O \quad (14\text{–}18)$$

aldehyde   silver mirror

$$C_6H_5-CHO \xrightarrow[OH^{\ominus}]{Ag(NH_3)_2^{\oplus}} C_6H_5-COO^{\ominus} + Ag\downarrow \quad (14\text{–}19)$$

14.7–2 Reduction Aldehydes and ketones can be reduced to alcohols or hydrocarbons by a variety of reducing agents. Catalytic hydrogenation of aldehydes forms primary alcohols; ketones are reduced to secondary alcohols.

$$\overset{}{\underset{}{>}}C=O + H_2 \xrightarrow[Ni]{\Delta,\text{ press.}} \overset{H}{\underset{OH}{>C<}} \quad (14\text{–}20)$$

The reaction may be pictured as the addition of hydrogen across the carbon–oxygen double bond of the carbonyl group to form the corresponding alcohol.

Complex metal hydrides reduce aldehydes and ketones to alcohols. Much of the knowledge concerning the use of complex metal hydrides as reducing agents is due principally to the research work of Professor Herbert C. Brown at Purdue University. The most common of these hydrides is lithium aluminum hydride, $LiAlH_4$; but sodium borohydride, $NaBH_4$, is often used instead, because it is cheaper and safer (not flammable) than lithium aluminum hydride.

$$R-\overset{O}{\underset{\parallel}{C}}-H \xrightarrow{LiAlH_4 \text{ or } NaBH_4} R-CH_2OH \quad (14\text{–}21)$$

aldehyde   primary alcohol

$$R-\overset{O}{\underset{\parallel}{C}}-R' \xrightarrow{LiAlH_4 \text{ or } NaBH_4} R-\underset{OH}{\underset{|}{CH}}-R' \quad (14\text{–}22)$$

ketone   secondary alcohol

The most interesting application of these complex metal hydrides is that they will reduce a carbonyl group to an alcohol without reducing a carbon–carbon double bond present in the same molecule. Thus, 2-butenal (crotonaldehyde) can be converted to 2-buten-1-ol (crotyl alcohol) in one step.

$$CH_3-CH=CH-\overset{O}{\underset{\parallel}{C}}-H \xrightarrow{LiAlH_4 \text{ or } NaBH_4} CH_3-CH=CH-CH_2OH \quad (14\text{–}23)$$

crotonaldehyde   crotyl alcohol

Sec. 14.8 NUCLEOPHILIC ADDITION REACTIONS

The reduction of a carbonyl group to a methylene (CH_2) group is usually accomplished by either the **Clemmensen reduction** or the **Wolff–Kishner reaction**.

Aldehydes or ketones can be reduced to hydrocarbons ($C{=}O$ to CH_2) by using a zinc–mercury amalgam in concentrated hydrochloric acid as a reducing agent. The reaction is known as a Clemmensen reduction.

$$R-\underset{\text{(usually aromatic)}}{\overset{\overset{O}{\|}}{C}}-Ar \xrightarrow[\text{HCl}]{\text{Zn–Hg}} R-CH_2-Ar \qquad (14\text{-}24)$$

The reduction of aliphatic aldehydes and ketones to methylene groups by a basic solution of hydrazine ($NH_2{-}NH_2$) is known as the Wolff–Kishner reaction.

$$R-\underset{\text{(usually aliphatic)}}{\overset{\overset{O}{\|}}{C}}-R' + H_2N-NH_2 \xrightarrow[\text{diethylene glycol, }\Delta]{\text{NaOH}} R-CH_2-R' + N_2 \uparrow \qquad (14\text{-}25)$$

14.8 Nucleophilic Addition Reactions

The polar nature and unsaturation of the carbonyl group in aldehydes and ketones would lead one to predict that these classes of compounds should undergo a large variety of addition reactions. This is indeed the case; the carbonyl carbon atom is susceptible to attack by many nucleophilic reagents. The vast majority of the reactions of aldehydes and ketones can be classified as **nucleophilic addition reactions**.

A general mechanism for nucleophilic addition to a carbonyl group involves attack by a nucleophilic species on the carbonyl carbon atom, followed by a proton or other electrophilic species attacking the carbonyl oxygen atom.

$$N{:} + \overset{}{\underset{}{C}}{=}O \rightleftharpoons N-\underset{|}{\overset{|}{C}}-O^{\ominus} \qquad (14\text{-}26)$$

$$N-\underset{|}{\overset{|}{C}}-O^{\ominus} + H^{\oplus} \rightleftharpoons N-\underset{|}{\overset{|}{C}}-OH \qquad (14\text{-}27)$$

The net result of the reaction is the addition of HN across the carbonyl group of the aldehyde or ketone.

The order of reactivity for addition reactions decreases as one goes from aldehydes to ketones.

$$H_2C{=}O > R-\overset{\overset{O}{\|}}{C}-H > Ar-\overset{\overset{O}{\|}}{C}-H > R_2-C{=}O > Ar_2-C{=}O$$

The lesser reactivity of ketones in comparison with aldehydes can be attributed to a combination of electronic and steric factors. The two alkyl or aryl groups around the carbonyl carbon in ketones are bulkier in size than the one alkyl or aryl group and the *small hydrogen atom* attached to the carbonyl group in aldehydes. The steric factor of the two bulky substituents present in ketones somewhat hinders the approach of a nucleophilic reagent seeking to attack the carbonyl carbon. This accounts to some extent for the greater reactivity of aldehydes in contrast to ketones in addition reactions. However, steric hindrance is not the entire story. Electronic effects also exert their influence in the addition reactions of aldehydes and ketones. The inductive effect of alkyl groups (in aldehydes and ketones) is electron-releasing in contrast to the hydrogen (in aldehydes). An aromatic ring, if present, tends to release electrons to the electron-withdrawing carbonyl group by resonance. These electronic effects, which *release electrons towards the carbonyl group* in the molecule, tend to somewhat *neutralize the positive charge on the carbonyl carbon atom* (due to the polar nature of the carbonyl group) to a greater extent in ketones than in aldehydes. This is simply because ketones have *two* R or Ar groups present whereas aldehydes have only *one* R or Ar group and a H atom attached to the carbonyl group; thus, the electronic effects are more strongly felt in ketones. The increase in the electron density on the carbonyl carbon in ketones *lessens the positive character of the carbon atom* and makes it less susceptible to attack by nucleophilic (electron-seeking) reagents. Ketones generally react more slowly than aldehydes in addition reactions. Some ketones containing bulky structures do not react at all because of the tremendous steric and electronic effects present in the molecule.

14.8-1 Addition of HCN (Formation of Cyanohydrins) The addition of HCN across the carbonyl group in aldehydes or ketones results in the formation of a class of organic compounds known as **cyanohydrins** (since the adduct contains both cyano and hydroxyl groups).

$$\underset{\text{aldehyde (or ketone)}}{R-\overset{O}{\underset{\|}{C}}-H \text{ (or } R-\overset{O}{\underset{\|}{C}}-R')} + HCN \xrightarrow[(pH\,=\,8)]{} \underset{\text{cyanohydrin}}{R-\underset{CN}{\overset{OH}{\underset{|}{C}}}-H \text{ (or } R-\underset{CN}{\overset{OH}{\underset{|}{C}}}-R')} \qquad (14\text{--}28)$$

The reaction mechanism is typical of the nucleophilic addition across the carbonyl group—attack by the nucleophilic cyanide ion, followed by addition of a proton.

$$-\overset{O}{\underset{\|}{C}}- + CN^{\ominus} \longrightarrow -\underset{CN}{\overset{O^{\ominus}}{\underset{|}{C}}}- \underset{}{\overset{H^{\oplus}}{\rightleftharpoons}} -\underset{CN}{\overset{OH}{\underset{|}{C}}}- \qquad (14\text{--}29)$$

The cyanohydrin reaction is usually run by treating the aldehyde or ketone with an aqueous solution of sodium or potassium cyanide; and then the acid is added slowly, with the pH of the solution adjusted to pH = 8. These reaction conditions are chosen because the reaction is reversible, and the equilibrium must be shifted to the right in order to get a good yield of product. For most aldehydes and small ketones the equilibrium is favorable, so that the yields of cyanohydrin formed are quite high.

Sec. 14.8 NUCLEOPHILIC ADDITION REACTIONS

$$CH_3-\overset{O}{\overset{\|}{C}}-H + HCN \longrightarrow CH_3-\underset{CN}{\overset{OH}{\underset{|}{C}}}-H \qquad (14\text{-}30)$$

acetaldehyde   acetaldehyde cyanohydrin

However, the yields of cyanohydrin obtained from ketones containing bulky substituents are very low because of the steric factors discussed earlier. (The steric hindrance will be even greater in the cyanohydrin product than in the original ketone.)

$$CH_3-\underset{H}{\overset{CH_3}{\underset{|}{\overset{|}{C}}}}-\overset{O}{\overset{\|}{C}}-\underset{H}{\overset{CH_3}{\underset{|}{\overset{|}{C}}}}-CH_3 + HCN \rightleftharpoons CH_3-\underset{H}{\overset{CH_3}{\underset{|}{\overset{|}{C}}}}-\underset{CN}{\overset{OH}{\underset{|}{\overset{|}{C}}}}-\underset{H}{\overset{CH_3}{\underset{|}{\overset{|}{C}}}}-CH_3 \qquad (14\text{-}31)$$

diisopropyl ketone   (less than 5 % yield at equilibrium)

The most important use of the cyanohydrins is as intermediates in the synthesis of α-hydroxy and α-amino acids. The hydrolysis of the cyano (nitrile) group in acid solution to a carboxyl group (COOH) results in the formation of an α-hydroxy acid.

$$CH_3-\overset{O}{\overset{\|}{C}}-H \xrightarrow{HCN} CH_3-\underset{H}{\overset{OH}{\underset{|}{\overset{|}{C}}}}-CN \xrightarrow[H_2O]{HCl} CH_3-\underset{H}{\overset{OH}{\underset{|}{\overset{|}{C}}}}-COOH \qquad (14\text{-}32)$$

acetaldehyde α-hydroxypropionic acid (lactic acid)

An important modification of this reaction using ammonia and hydrogen cyanide as nucleophilic reagents, converts aldehydes or ketones into α-amino acids, which are the building blocks of proteins. The reaction is known as the **Strecker amino acid synthesis.**

$$R-\overset{O}{\overset{\|}{C}}-H \xrightarrow[HCN]{NH_3} R-\underset{H}{\overset{NH_2}{\underset{|}{\overset{|}{C}}}}-CN \xrightarrow[H_2O, \Delta]{HCl} R-\underset{H}{\overset{NH_2}{\underset{|}{\overset{|}{C}}}}-COOH \qquad (14\text{-}33)$$

α-amino acid

14.8-2 Addition of Water Since water is a nucleophilic reagent one might expect water to add across the carbonyl group of an aldehyde or ketone to form hydrates having a *gem*-diol structure.

$$R-\overset{O}{\overset{\|}{C}}-R' + H_2O \rightleftharpoons R-\underset{OH}{\overset{OH}{\underset{|}{\overset{|}{C}}}}-R' \qquad (14\text{-}34)$$

aldehyde or ketone   (hydrate)

The reaction is reversible and for most aldehydes and ketones the equilibrium lies far to the left in favor of the reactants, since most of the hydrates formed are not stable.

Some carbonyl compounds do form stable hydrates that can be isolated out of the reaction mixture as such. This usually occurs when the aldehyde or ketone has certain structural features, such as the presence of a powerfully electron-withdrawing substituent directly attached to the carbonyl group. This increases the positive charge on the carbonyl carbon atom, and the ease of nucleophilic attack by water, as well as stabilizing the hydrate formed.

Chloral (trichloroacetaldehyde) forms a stable crystalline hydrate because the dipole–dipole repulsions in the aldehyde are reduced in the *gem*–diol, and this factor tends to make the equilibrium more favorable for the formation of the hydrate.

$$Cl_3\overset{\delta^-}{-}C\overset{\delta^+}{-}\underset{\delta^+}{\overset{O^{\delta^-}}{\overset{\|}{C}}}-H + H_2O \longrightarrow Cl_3-C-\underset{OH}{\overset{OH}{\overset{|}{C}}}-H \quad (14\text{-}35)$$

chloral   chloral hydrate

Chloral hydrate is one of the very few stable compounds known with two hydroxyl groups on the same carbon atom in the molecule. It is used as a hypnotic and narcotic, and is perhaps most familiar to the reader as the active ingredient of the "knock-out drops" used in the drink known as the "Mickey Finn."

14.8–3 Addition of Alcohols   Alcohols can undergo nucleophilic addition reactions with the carbonyl group in aldehydes and ketones to form unstable addition products known as **hemiacetals** and **hemiketals.** The reaction is catalyzed by mineral acids, which make the carbonyl carbon more electron deficient and reactive towards the weakly nucleophilic reagent (alcohols).

$$\searrow\!\!C=O + H^\oplus \rightleftharpoons \left[\searrow\!\!C\overset{\oplus}{=}\ddot{O}H \longleftrightarrow \searrow\!\!\overset{\oplus}{C}-\ddot{O}H \right] \overset{N:}{\longrightarrow} -\underset{|}{\overset{N}{\overset{|}{C}}}-OH \quad (14\text{-}36)$$

In the presence of catalytic amounts of acid, aldehydes and ketones can react with excess alcohol to form compounds known as **acetals** or **ketals,** which can be isolated.

$$R-\overset{O}{\overset{\|}{C}}-H + R'OH \rightleftharpoons R-\underset{OR'}{\overset{OH}{\overset{|}{C}}}-H \overset{H^\oplus, R'OH}{\rightleftharpoons} R-\underset{OR'}{\overset{OR'}{\overset{|}{C}}}-H + H_2O \quad (14\text{-}37)$$

aldehyde   alcohol   hemiacetal   acetal

$$R-\overset{O}{\overset{\|}{C}}-R' + R''OH \rightleftharpoons R-\underset{OR''}{\overset{OH}{\overset{|}{C}}}-R' \overset{H^\oplus, R''OH}{\rightleftharpoons} R-\underset{OR''}{\overset{OR''}{\overset{|}{C}}}-R' + H_2O \quad (14\text{-}38)$$

ketone   alcohol   hemiketal   ketal

Sec. 14.8 NUCLEOPHILIC ADDITION REACTIONS

The intermediate in the reaction is referred to as a *hemiacetal* or *hemiketal* (half converted; *hemi-* = half) and contains a hydroxyl and alkoxyl (ether) group on the same carbon atom. The final product containing two alkoxy (ether) groups on the same carbon atom is called an *acetal* or *ketal*.

The overall reaction sequence is reversible and is catalyzed by acid. A large excess of alcohol is used to shift the equilibrium in favor of acetal formation. The acidic solution must be neutralized and the solvent should be anhydrous in order to isolate the acetal. In acid solution the reverse reaction occurs rapidly, that is reconversion of the acetal to the original aldehyde (and ketal to ketone) by hydrolysis.

The reaction of acetaldehyde with ethyl alcohol results in the formation of acetaldehyde diethyl acetal, in a typical reaction.

$$CH_3-\overset{O}{\underset{\|}{C}}-H + CH_3-CH_2OH \rightleftharpoons CH_3-\underset{\underset{OCH_2-CH_3}{|}}{\overset{\overset{OH}{|}}{C}}-H \quad \underset{CH_3-CH_2OH}{\overset{dry\ HCl,}{\rightleftharpoons}} \quad CH_3-\underset{\underset{OCH_2-CH_3}{|}}{\overset{\overset{OCH_2CH_3}{|}}{C}}-H$$

hemiacetal   acetaldehyde diethyl acetal

(14–39)

The great similarity in structure between acetals and ketals as compared to ethers is borne out by the unreactivity of these compounds, except for the relative ease of hydrolysis under acidic conditions. Acetals and ketals do not react with alkaline solutions, but are readily cleaved by acid to the aldehyde or ketone and alcohol. Hemiacetals and acetals are extremely important in understanding the chemistry of the carbohydrates.

The relative inertness of acetals and ketals makes these compounds extremely valuable in synthetic techniques where an aldehyde or ketone functional group must be *protected* in a molecule. (See Problem 14.9 at the end of this chapter.)

14.8–4 Addition of Ammonia and Its Derivatives (Addition Reactions Followed by Dehydration)

Ammonia and some of its derivatives can behave as nucleophilic reagents since they have an unshared electron pair on the nitrogen atom, and can undergo addition reactions with the carbonyl group in aldehydes and ketones. These reactions are characterized by the loss of water (dehydration) after addition of the nucleophile to the carbonyl group.

$$-\overset{O}{\underset{\|}{C}}- + NH_2-Z \rightleftharpoons -\underset{\underset{Z}{\overset{|}{N-\textcircled{H}}}}{\overset{\overset{\textcircled{OH}}{|}}{C}}- \xrightarrow{-H_2O} -\overset{N-Z}{\underset{\|}{C}}- + H_2O \quad (14\text{–}40)$$

$$(Z = H,\ OH,\ NH_2,\ \bigcirc\!\!\!-NH-,\ H_2N-\overset{O}{\underset{\|}{C}}-\underset{|}{N}-H,\ etc.)$$

The addition of ammonia itself (Z=H) to aldehydes and ketones is a reversible reaction and produces compounds having an N=C or an N—C—N bonding in the molecule. The reaction of ammonia with formaldehyde is perhaps most important in this type of addition reaction. The product obtained, hexamethylenetetramine, is both useful and interesting because of its unusual structure.

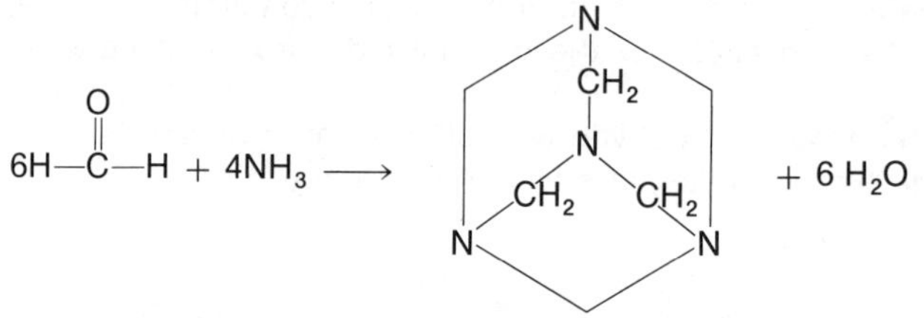

$$6H-\underset{\underset{H}{\|}}{C}-H + 4NH_3 \longrightarrow \text{hexamethylenetetramine} + 6H_2O \quad (14\text{-}41)$$

Hexamethylenetetramine has a cage-like structure in which every nitrogen atom is attached to three other carbon atoms, and every carbon atom is attached to two nitrogen atoms. The name *hexamethylenetetramine* can be seen as arising from the presence of six CH_2 (methylene) groups and four nitrogen atoms (amine) in the molecule. The product is perhaps better known by the name *urotropine*, a medicine used as a urinary antiseptic. Nitration of the product produces a high explosive, cyclonite (RDX).

The addition of various derivatives of ammonia to aldehydes and ketones is extremely useful as a means of identifying specific aldehydes or ketones. The reaction products are usually crystalline solids whose melting points can be used to identify aldehydes and ketones, most of which are liquids.

Addition of Hydroxylamine to Form Oximes. Hydroxylamine, NH_2OH, reacts with aldehydes and ketones to form a crystalline derivative known as an *oxime*.

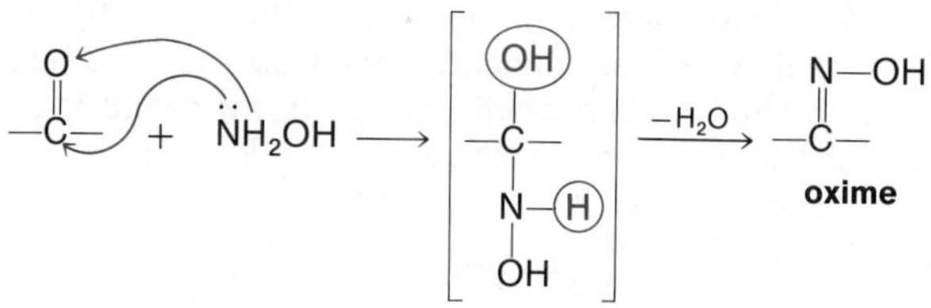

$$-\underset{\|}{\overset{O}{C}}- + NH_2OH \longrightarrow \left[-\underset{\underset{\underset{OH}{|}}{N-H}}{\overset{OH}{|}}C- \right] \xrightarrow{-H_2O} -\underset{\|}{\overset{N-OH}{C}}- \quad (14\text{-}42)$$

oxime

$$CH_3-\underset{\|}{\overset{O}{C}}-CH_3 + NH_2OH \longrightarrow \left[CH_3-\underset{\underset{\underset{OH}{|}}{N-H}}{\overset{OH}{|}}C-CH_3 \right] \xrightarrow{-H_2O} CH_3-\underset{\|}{\overset{N-OH}{C}}-CH_3 \quad (14\text{-}43)$$

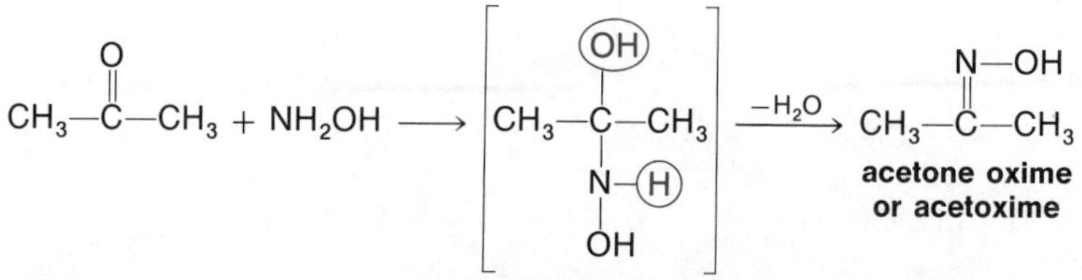

acetone oxime
or acetoxime

Sec. 14.8 NUCLEOPHILIC ADDITION REACTIONS

Addition of Hydrazine, Phenylhydrazine, and 2,4-Dinitrophenylhydrazine A similar type reaction of aldehydes and ketones with hydrazine, phenylhydrazine, or 2,4-dinitrophenylhydrazine produces hydrazones, phenylhydrazones, and 2,4-dinitrophenylhydrazones as useful classes of derivatives.

$$CH_3-\underset{\underset{\text{acetaldehyde}}{}}{\overset{O}{\underset{\|}{C}}}-H + \underset{\text{hydrazine}}{NH_2-NH_2} \xrightarrow{-H_2O} CH_3-\underset{\underset{\text{acetaldehyde hydrazone}}{}}{\overset{N-NH_2}{\underset{\|}{C}}}-H \quad (14\text{-}44)$$

benzaldehyde + phenylhydrazine $\xrightarrow{-H_2O}$ benzaldehyde phenylhydrazone (14-45)

acetone + 2,4-dinitrophenylhydrazine $\xrightarrow{-H_2O}$ acetone 2,4-dinitrophenylhydrazone (14-46)

The 2,4-dinitrophenylhydrazones are usually yellow, orange, or red in color and the formation of a 2,4-DNP derivative from an unknown is excellent evidence for the presence of a carbonyl group in the molecule.

Addition of Semicarbazide The reaction of aldehydes or ketones with semicarbazide results in the formation of solid semicarbazone derivatives.

cyclohexanone + semicarbazide $\xrightarrow{-H_2O}$ cyclohexanone semicarbazone (14-47)

14.8-5 Addition of Sodium Bisulfite The nucleophilic attack of the sulfur atom in sodium bisulfite on the carbonyl group of aldehydes and some simple ketones produces a white solid crystalline bisulfite addition product. The reaction is reversible and the equilibrium is shifted to the right by using an excess of sodium bisulfite.

$$R-\overset{O}{\underset{\|}{C}}-R' + Na\overset{\oplus}{H}SO_3{}^{\ominus} \rightleftharpoons R-\underset{\underset{SO_3Na^{\oplus}}{|}}{\overset{OH}{\underset{|}{C}}}-R' \quad (14\text{-}48)$$

aldehyde or small ketone   bisulfite addition product

(R and R' = H, or CH_3)

Ketones that have bulky groups adjacent to the carbonyl group will not form a bisulfite addition product because of the steric and electronic effects that make the equilibrium unfavorable for the reaction.

The bisulfite addition compounds can be reconverted into the original aldehyde or ketone by treatment with acid or base. This provides a means of separating aldehydes and small ketones from a mixture by converting them to bisulfite addition products that are water soluble, and then regenerating the aldehyde or ketone by treatment with acid or base.

14.8-6 Addition of Grignard Reagents The reaction of Grignard reagents with the carbonyl group in aldehydes and ketones results in the formation of an addition product which can be hydrolyzed by aqueous acid to an alcohol. The Grignard reagent undergoes nucleophilic addition as if it were an ionic compound composed of $R^{\ominus}$ and $MgX^{\oplus}$.

$$R'-\underset{\|}{C}-R'' + R Mg X^{\oplus} \longrightarrow R'-\underset{R}{\overset{OMgX}{C}}-R'' \xrightarrow[HCl]{H_2O} R'-\underset{R}{\overset{OH}{C}}-R'' + MgXCl \qquad (14\text{-}49)$$

aldehyde   addition   alcohol
or ketone   product

The reaction is extremely useful in organic syntheses and provides a convenient way of preparing alcohols that contain a longer carbon chain than the starting materials. *Grignard reagents will react with formaldehyde* to form *primary alcohols* upon hydrolysis of the addition product formed; *other aldehydes* will result in the formation of *secondary alcohols;* and *ketones* will produce *tertiary alcohols.*

$$H-\underset{\|}{\overset{O}{C}}-H + R-MgX \longrightarrow H-\underset{R}{\overset{OMgX}{C}}-H \xrightarrow[HCl]{H_2O} H-\underset{R}{\overset{OH}{C}}-H \text{ (or } R-CH_2-OH) \qquad (14\text{-}50)$$

formaldehyde   a primary alcohol

$$R'-\underset{\|}{\overset{O}{C}}-H + R-MgX \longrightarrow R'-\underset{R}{\overset{OMgX}{C}}-H \xrightarrow[HCl]{H_2O} R'-\underset{R}{\overset{OH}{C}}-H \qquad (14\text{-}51)$$

aldehyde other   a secondary alcohol
than formaldehyde

$$R'-\underset{\|}{\overset{O}{C}}-R'' + R-MgX \longrightarrow R'-\underset{R}{\overset{OMgX}{C}}-R'' \xrightarrow[HCl]{H_2O} R'-\underset{R}{\overset{OH}{C}}-R'' \qquad (14\text{-}52)$$

ketone   a tertiary alcohol

14.9 Enol–Keto Tautomerism

Enol–keto tautomerism has been described in connection with the addition of water to an alkyne (Section 9.4–5). The reactions of aldehydes and ketones involving the active hydrogens attached to the carbon atom adjacent to the carbonyl group (the α-carbon atom) can best be understood by referring to the enol–keto tautomerism in the molecule.

The hydrogen atoms attached to the α–carbon atom in an aldehyde or ketone are somewhat acidic and can be removed by a basic reagent. The acidity of the α–hydrogens to the carbonyl group can be attributed to the electron-withdrawing inductive effect of the carbonyl group (facilitating removal of a proton) and the resonance stabilization of the enolate anion formed when a proton has been removed.

$$\underset{\underset{H}{\overset{|}{\underset{\alpha}{C}}}-\overset{O}{\overset{||}{C}}-}{} \;\underset{base}{\rightleftharpoons}\; base{:}H^{\oplus} + \underset{\underset{\alpha}{\overset{|}{\underset{\ominus}{C}}}-\overset{O}{\overset{||}{C}}-}{} \qquad (14\text{-}53)$$

enolate anion

$$\left[\begin{array}{c} \overset{\ddot{O}:}{\underset{\ominus}{\overset{|}{C}}}\!\!-\!\!\overset{|}{C}- \end{array} \longleftrightarrow \begin{array}{c} :\!\overset{\ddot{O}:^{\ominus}}{}\\ \overset{}{C}\!=\!\overset{}{C}- \end{array} \right]$$

The reaction of an aldehyde or ketone having hydrogens attached to the α-carbon with a base produces a small amount of enolate anions at equilibrium. The enolate anions are good nucleophilic reagents and can attack a carbonyl group as we will see in our discussion of the aldol condensation in Section 14.13. The enolate anions are capable of accepting a proton from a donor to form the so-called enol product (*en(e)-* = double, *-ol* = OH group).

$$\underset{\text{enolate anion}}{\overset{:\ddot{O}{:}^{\ominus}}{\underset{}{\overset{|}{C}}\!=\!\overset{|}{C}-}} + H^{\oplus} \longrightarrow \underset{\text{enol}}{\overset{OH}{\underset{}{\overset{|}{C}}\!=\!\overset{|}{C}-}} \qquad (14\text{-}54)$$

The equilibrium that exists between carbonyl compounds (such as aldehydes and ketones) in the keto and enol forms is known as **enol–keto tautomerism.**

$$\underset{\text{keto form}}{\underset{\underset{H}{\overset{|}{C}}}{\overset{O}{\overset{||}{\underset{}{C}}}}\!-\!\overset{|}{C}-} \;\rightleftharpoons\; \underset{\text{enol form}}{\overset{OH}{\underset{}{\overset{|}{C}}\!=\!\overset{|}{C}-}} \qquad (14\text{-}55)$$

The concept of tautomerism is different from the phenomenon of resonance. Tautomerism is an actual equilibrium between two structures, whereas the resonance hybrid is the intermediate structure resulting from a combination of a series of resonance contributing forms; and, furthermore, resonance involves only the movement of electrons from one region to another, whereas in tautomerism not only electrons, but atoms (e.g., —H atom) also are being moved to a different location.

Most aldehydes and ketones exist predominantly in their keto form, which is the more stable form for these compounds. However, some compounds do exist mainly in the enol form. For example, the compound acetylacetone, $CH_3-\overset{O}{\underset{\|}{C}}-CH_2-\overset{O}{\underset{\|}{C}}-CH_3$, has a stable enol, and the pure liquid is about 80 % in the enol form and only 20 % in the keto form. Acetoacetaldehyde, $CH_3-\overset{O}{\underset{\|}{C}}-CH_2-\overset{O}{\underset{\|}{C}}-H$, as pure liquid, is 98 % in the enol form. In contrast the simple ketone acetone, $CH_3-\overset{O}{\underset{\|}{C}}-CH_3$, exists mainly in the keto form, the pure liquid being only 0.00025 % in the enol form.

14.10 Halogenation (Substitution of α-Hydrogens)

The α-hydrogen atoms in an aldehyde or a ketone can be substituted by halogens (Cl_2, Br_2, or I_2) in alkaline solution. This reaction probably proceeds through the formation of an enolate anion formed by the action of base on the weakly acidic α-hydrogen atoms in the aldehyde or ketone, followed by electrophilic attack of the halogen at the α-carbon atom. The overall result is the substitution of a halogen for an α-hydrogen. The reaction can be illustrated for acetaldehyde with bromine in the presence of base.

$$\underset{\text{acetaldehyde}}{CH_3-\overset{O}{\underset{\|}{C}}-H} + OH^\ominus \rightleftharpoons H_2O + \underset{\text{enolate anion}}{^\ominus{:}CH_2-\overset{O}{\underset{\|}{C}}-H} \qquad (14\text{-}56)$$

$$^\ominus{:}CH_2-\overset{O}{\underset{\|}{C}}-H + Br_2 \longrightarrow \underset{\text{bromoacetaldehyde}}{Br-CH_2-\overset{O}{\underset{\|}{C}}-H} + Br^\ominus \qquad (14\text{-}57)$$

It is extremely difficult to stop the reaction at the stage where only one α-hydrogen atom has been replaced by halogen. This is due to the introduction of the first halogen atom into the molecule, which by its electron-withdrawing inductive effect makes the remaining α-hydrogen atoms more acidic than usual and enables them to be replaced more easily. The halogenation reaction will continue until all the α-hydrogen atoms have been replaced by halogen.

$$\underset{\text{acetaldehyde}}{CH_3-\overset{O}{\underset{\|}{C}}-H} + 3Br_2 + 3OH^\ominus \longrightarrow \underset{\substack{\text{tribromo-}\\\text{acetaldehyde}}}{CBr_3-\overset{O}{\underset{\|}{C}}-H} + 3H_2O + 3Br^\ominus \qquad (14\text{-}58)$$

14.11 The Haloform Reaction—Iodoform Test

As we discussed in the preceding section, the final product obtained by the halogenation is an aldehyde or ketone that is fully halogenated in the α-position.

Sec. 14.11 THE HALOFORM REACTION

For example, a methyl ketone would form an α,α,α-trihaloketone as the product.

$$R-\overset{O}{\underset{\|}{C}}-CH_3 + X_2 \xrightarrow{OH^\ominus} R-\overset{O}{\underset{\|}{C}}-\underset{\underset{X}{|}}{\overset{\overset{X}{|}}{\underset{\alpha}{C}}}-X \quad (14\text{-}59)$$

methyl ketone   **α,α,α-trihaloketone**

In the basic reaction medium the α,α,α-trihaloketone undergoes a cleavage reaction between the carbonyl carbon and α-carbon atom.

$$R-\overset{\overset{\ddot{O}:}{\|}}{C}\!\!\!\!\!+\!\!\!\!\!\underset{\underset{X}{|}}{\overset{\overset{X}{|}}{C}}-X + :\ddot{O}H^\ominus \longrightarrow R-\underset{\underset{OH}{|}}{\overset{\overset{:\ddot{O}:^\ominus}{|}}{C}}-\underset{\underset{X}{|}}{\overset{\overset{X}{|}}{C}}-X \longrightarrow X-\underset{\underset{X}{|}}{\overset{\overset{X}{|}}{C}}:^\ominus + R-\overset{O}{\underset{\|}{C}}-OH \quad (14\text{-}60)$$

                                                                             trihalo-   **carboxylic**
                                                                           methyl anion   **acid**

$$R-\overset{O}{\underset{\|}{C}}-OH + X-\underset{\underset{X}{|}}{\overset{\overset{X}{|}}{C}}:^\ominus \longrightarrow R-\overset{O}{\underset{\|}{C}}-O^\ominus + X-\underset{\underset{X}{|}}{\overset{\overset{X}{|}}{C}}-H \quad (14\text{-}61)$$

                                       salt of   **haloform**
                                    carboxylic
                                      acid

The cleavage reaction occurs because the bond between the α-carbon and carbonyl carbon atom is weakened, since both these carbon atoms are relatively positive due to the presence of the electron-withdrawing halogens on the α-carbon atom.

$$R-\underset{\delta^+}{\overset{\overset{\delta^\ominus O}{\|}}{C}}\!\!\!\!\!+\!\!\!\!\!\underset{\underset{X^{\delta\ominus}}{|}}{\overset{\overset{X^{\delta\ominus}}{|}}{\underset{\delta^+}{C}}}-X^{\delta\ominus}$$

The cleavage by hydroxide ion results in the formation of a trihalogenated derivative of methane known as a *haloform* (hence the name, **haloform reaction**) and the salt of a carboxylic acid. When chlorine is used, $CHCl_3$ or chloroform is formed, bromine produces $CHBr_3$ (bromoform), and iodine yields CHI_3, iodoform, which is an insoluble yellow solid with a medicinal odor. Iodoform can be easily identified, and since the cleavage reaction to form it is only given by methyl ketones $\left(R-\overset{O}{\underset{\|}{C}}-CH_3\right)$ and acetaldehyde $\left(CH_3-\overset{O}{\underset{\|}{C}}-H\right)$, the **iodoform test** serves as a convenient way to distinguish methyl ketones from other ketones. The test is actually for the presence of a $CH_3-\overset{O}{\underset{\|}{C}}-$ group in a molecule. Acetaldehyde is the only aldehyde that has this structure. However, since the alkaline solutions of halogens are good oxidizing agents, the hypoiodite ($OI^\ominus$) present can oxidize ethanol (CH_3-CH_2OH) to acetaldehyde $\left(CH_3-\overset{\overset{O}{\|}}{C}-H\right)$, and a secondary alcohol like 2-propanol $\left(CH_3-\overset{\overset{OH}{|}}{\underset{\underset{H}{|}}{C}}-CH_3\right)$ to acetone $\left(CH_3-\overset{\overset{O}{\|}}{C}-CH_3\right)$. Thus, it should be noted that ethanol and any secondary alcohol of the general structure

$CH_3-\overset{\overset{OH}{|}}{\underset{\underset{H}{|}}{C}}-R$, which can be oxidized to a $CH_3-\overset{\overset{O}{\|}}{C}-$ grouping under the reaction

conditions, will also give a positive iodoform test. The reaction can be used to distinguish between two ketones such as 2-pentanone and 3-pentanone.

$$CH_3-\overset{\overset{O}{\|}}{C}-CH_2-CH_2-CH_3 \xrightarrow[OH^\ominus]{I_2} CHI_3 \downarrow \qquad (14\text{-}62)$$
$$\text{2-pentanone} \qquad\qquad \text{iodoform}$$
$$\text{(yellow)}$$

$$CH_3-CH_2-\overset{\overset{O}{\|}}{C}-CH_2-CH_3 \xrightarrow[OH^\ominus]{I_2} \text{(no ppt. formed)} \qquad (14\text{-}63)$$
$$\text{3-pentanone}$$
$$\text{(get } CH_3-\overset{\overset{H}{|}}{C}l_2$$
$$\text{instead of } CHI_3\text{)}$$

14.12 Halogenation of Ketones in Acid Medium

Ketones can also be halogenated in acid solution. The value of the acid-catalyzed reaction is that the reaction can be stopped after only one halogen atom has replaced an α-hydrogen atom. This reaction is useful in synthesis where it is desirable to introduce only one halogen atom in the α-position.

$$CH_3-\overset{\overset{O}{\|}}{C}-CH_3 + Cl_2 \xrightarrow{H^\oplus} CH_3-\overset{\overset{O}{\|}}{C}-CH_2Cl + HCl \qquad (14\text{-}64)$$
$$\text{acetone} \qquad\qquad \alpha\text{-chloroacetone}$$

14.13 The Aldol Condensation

There are many known condensation reactions involving aldehydes and ketones. All of these reactions have one common step in the reaction sequence, namely, the formation of an enolate anion followed by its addition to a carbonyl group. One of the best known of these reactions is the aldol condensation, in which aldehydes and ketones having at least one α-hydrogen undergo a self-condensation reaction in the presence of base.

The reaction of acetaldehyde in the presence of dilute sodium hydroxide to form 3-hydroxybutanal is a typical aldol condensation. The base removes an

Sec. 14.13 THE ALDOL CONDENSATION

α–hydrogen atom from a small, but significant, number of acetaldehyde molecules to form an enolate anion, which then attacks the carbonyl group of another molecule of acetaldehyde to form the aldol (aldehyde and alcohol functional groups present), 3–hydroxybutanal. The mechanism for the reaction is believed to be as follows.

$$CH_3-\overset{O}{\overset{\|}{C}}-H + OH^{\ominus} \rightleftharpoons H_2O + \left[:\overset{\ominus}{C}H_2-\overset{O}{\overset{\|}{C}}-H \longleftrightarrow CH_2=\overset{\overset{\ominus}{O}}{\overset{|}{C}}-H\right] \quad (14\text{-}65)$$

$$CH_3-\overset{O}{\overset{\|}{C}}-H + :\overset{\ominus}{C}H_2-\overset{O}{\overset{\|}{C}}-H \rightleftharpoons CH_3-\overset{O^{\ominus}}{\overset{|}{\underset{H}{C}}}-CH_2-\overset{O}{\overset{\|}{C}}-H \quad (14\text{-}66)$$

$$CH_3-\overset{O^{\ominus}}{\overset{|}{\underset{H}{C}}}-CH_2-\overset{O}{\overset{\|}{C}}-H + H_2O \rightleftharpoons CH_3-\overset{OH}{\overset{|}{\underset{H}{C}}}-CH_2-\overset{O}{\overset{\|}{C}}-H + OH^{\ominus} \quad (14\text{-}67)$$

3–hydroxybutanal

The reaction is reversible, and the basic catalyst, $OH^{\ominus}$, is regenerated at the end of the reaction. The product obtained can be classified as a **β–hydroxyaldehyde**, with the hydroxyl group located two carbon atoms down the chain from the carbonyl group. Small ketones that can undergo the reaction condense to produce **β–hydroxy ketones**. For example, acetone undergoes an aldol condensation to form 4–hydroxy–4–methyl–2–pentanone.

$$2CH_3-\overset{O}{\overset{\|}{C}}-CH_3 \xrightarrow{OH^{\ominus}} \underset{5}{CH_3}-\underset{4}{\overset{OH}{\overset{|}{\underset{CH_3}{\overset{|}{C}}}}}-\underset{3}{CH_2}-\underset{2}{\overset{O}{\overset{\|}{C}}}-\underset{1}{CH_3} \quad (14\text{-}68)$$

4–hydroxy–4–methyl–2–pentanone

The synthetic utility of the aldol condensation is that it is an excellent means of lengthening the carbon chain. The aldols (and ketols) formed can be easily dehydrated by concentrated sulfuric acid or distilling with I_2 to form a carbon–carbon double bond. The resulting products are either an **α,β–unsaturated aldehyde** or **α,β–unsaturated ketone**. The carbon–carbon double bond in the product is located between the α and β carbon atoms, rather than between the β and γ carbon atoms, because in the α,β position the double bond extends the conjugated system from the carbonyl group.

$$CH_3-\overset{OH}{\overset{|}{\underset{H}{C}}}-CH_2-\overset{O}{\overset{\|}{C}}-H \xrightarrow{\text{conc. } H_2SO_4} H_2O + CH_3-\underset{\gamma}{CH}=\underset{\beta}{CH}-\underset{\alpha}{\overset{O}{\overset{\|}{C}}}-H \quad (14\text{-}69)$$

3–hydroxybutanal   **crotonaldehyde**

14.13-1 Mixed or Crossed Aldol Condensations The reaction of two different aldehydes (at least one of which has an α–hydrogen atom) in the presence of base is known as a **mixed aldol condensation.** If both of the compounds used have α-hydrogen atoms, we obtain a mixture of all four possible condensation products formed by removal of an α-hydrogen atom from a molecule and combination with either one of the two original compounds.

$$CH_3-\underset{O}{\underset{\|}{C}}-H + CH_3-CH_2-\underset{O}{\underset{\|}{C}}-H \xrightarrow{OH^\ominus} CH_3-\underset{H}{\underset{|}{C}}(\overset{OH}{|})-\underset{CH_3}{\underset{|}{C}}(\overset{H}{|})-\underset{O}{\underset{\|}{C}}-H + CH_3-\underset{H}{\underset{|}{C}}(\overset{OH}{|})-CH_2-\underset{O}{\underset{\|}{C}}-H +$$

$$CH_3-CH_2-\underset{H}{\underset{|}{C}}(\overset{OH}{|})-CH_2-\underset{O}{\underset{\|}{C}}-H + CH_3-CH_2-\underset{H}{\underset{|}{C}}(\overset{OH}{|})-\underset{CH_3}{\underset{|}{C}}(\overset{H}{|})-\underset{O}{\underset{\|}{C}}-H \quad (14\text{-}70)$$

On the other hand, if one of the compounds used does *not* have an α-hydrogen atom it cannot undergo an aldol condensation, but it can serve as an acceptor for the carbanion (enolate anion) formed from another carbonyl compound having α-hydrogens in the presence of base. A mixed aldol condensation run on these compounds will result in the formation of a major product formed from the addition of the enolate anion across the carbonyl group of the other compound, whose sole function in the reaction is to permit the enolate anion to add across its carbonyl group. For example, the reaction of benzaldehyde with acetaldehyde forms cinnamaldehyde by the addition of the enolate anion derived from acetaldehyde across the carbonyl group of benzaldehyde. Since benzaldehyde has no α-hydrogen atoms it cannot undergo an aldol condensation with itself. Acetaldehyde (although it possesses α-hydrogen atoms and can condense with itself) reacts mainly with the benzaldehyde to form cinnamaldehyde. This occurs because one usually starts the reaction with an excess of benzaldehyde, and once the enolate anion of acetaldehyde has been formed from some of the acetaldehyde molecules, a competition occurs between acetaldehyde and benzaldehyde for the carbanion. The enolate anions will, however, have a greater probability of reacting with the substance present in the greater concentration, namely benzaldehyde, thus forming cinnamaldehyde as the major reaction product.

$$CH_3-\underset{O}{\underset{\|}{C}}-H + \underset{\text{benzaldehyde}}{\underset{\text{(has no }\alpha\text{-}}{\underset{\text{hydrogens)}}{C_6H_5-\underset{O}{\underset{\|}{C}}-H}}} \xrightarrow{OH^\ominus} C_6H_5-\underset{H}{\underset{|}{C}}(\overset{OH}{|})-CH_2-\underset{O}{\underset{\|}{C}}-H$$

acetaldehyde (has α-hydrogens)

$$\downarrow \Delta,\text{ conc. } H_2SO_4$$

$$H_2O + C_6H_5-CH=CH-\underset{O}{\underset{\|}{C}}-H \quad (14\text{-}71)$$

cinnamaldehyde (main constituent in cinnamon)

14.14 The Perkin Reaction

A condensation reaction closely related to the aldol condensation is the Perkin reaction. The mechanism of the Perkin reaction is very similar to that of the aldol condensation, and involves the condensation of an acid anhydride with an aromatic aldehyde.

As a typical example of a Perkin reaction let us consider the reaction of acetic anhydride with benzaldehyde. The α-hydrogen atoms of the anhydride are weakly acidic due to the electron-withdrawing effect of the adjacent carbonyl groups.

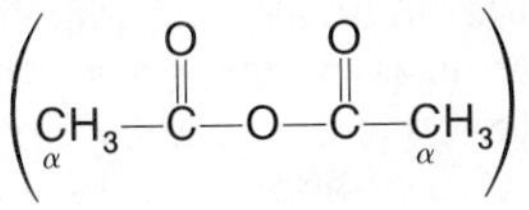

$$\left(\underset{\alpha}{CH_3}-\overset{O}{\underset{\|}{C}}-O-\overset{O}{\underset{\|}{C}}-\underset{\alpha}{CH_3} \right)$$

acetic anhydride

When treated with a weak base such as sodium acetate the anhydride will lose an α-hydrogen atom to form a carbanion, which will then condense with the aromatic aldehyde (i.e., benzaldehyde) by adding across the carbonyl group in a manner analogous to the aldol condensation. The intermediate compound formed, which is a mixed anhydride, is usually hydrolyzed to form cinnamic acids.

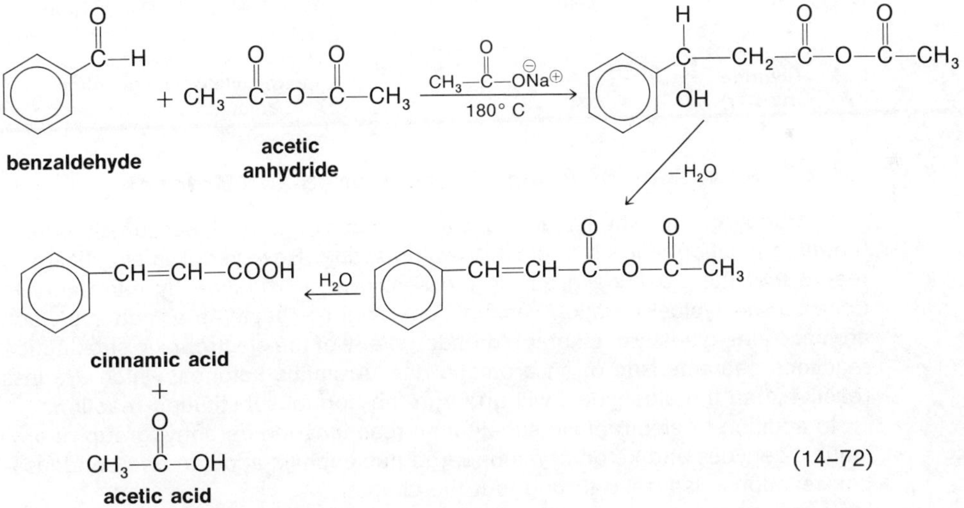

benzaldehyde + acetic anhydride $\xrightarrow{CH_3-C(O)-ONa, 180°C}$ intermediate $\xrightarrow{-H_2O}$

Ph—CH=CH—COOH (cinnamic acid) + CH$_3$—COOH (acetic acid) $\xleftarrow{H_2O}$ Ph—CH=CH—C(O)—O—C(O)—CH$_3$

(14-72)

14.15 The Cannizzaro Reaction

We have previously stated that aldehydes that have no α-hydrogen atoms cannot undergo an aldol condensation with themselves. In our discussion of the Perkin reaction we used a weak base, sodium acetate, to remove an α-hydrogen from the anhydride. A strong base, such as sodium hydroxide, could *not* be used in the Perkin reaction. The reason for this is that in the presence of strong bases, such as concentrated sodium hydroxide, aldehydes that have no α-hydrogen atoms undergo a self oxidation–reduction reaction called the **Cannizzaro reaction**.

Benzaldehyde, for example, undergoes a Cannizzaro reaction to form benzyl alcohol and sodium benzoate when treated with concentrated sodium hydroxide.

$$\text{benzaldehyde} \xrightarrow{\text{conc. NaOH}} \text{benzyl alcohol} + \text{sodium benzoate} \quad (14\text{-}73)$$

The reaction is limited to aldehydes that have no α-hydrogens; the aldehydes that do have α-hydrogens will undergo an aldol condensation in the presence of base.

The mechanism of the Cannizzaro reaction is interesting in that the key step in the reaction sequence is believed to proceed by the transfer of a hydride ion from an anion to the carbonyl group of another aldehyde molecule.

$$\underset{\substack{(R \text{ has no} \\ \alpha\text{-hydrogens})}}{R-\overset{O}{\underset{\|}{C}}-H} + OH^{\ominus} \rightleftharpoons R-\underset{\underset{O^{\ominus}}{|}}{\overset{\overset{OH}{|}}{C}}-H \quad (14\text{-}74)$$

$$\underset{\text{(hydride transfer)}}{R-\underset{\underset{O^{\ominus}}{|}}{\overset{\overset{OH}{|}}{C}}-H} + \overset{H}{\underset{\underset{O}{\|}}{C}}-R \longrightarrow R-\underset{\underset{O}{\|}}{\overset{\overset{OH}{|}}{C}} + H-\underset{\underset{O^{\ominus}}{|}}{\overset{\overset{H}{|}}{C}}-R \longrightarrow \underset{\text{carboxylate anion}}{R-\overset{O}{\underset{\|}{C}}-O^{\ominus}} + \underset{\text{alcohol}}{R-CH_2OH}$$

$$(14\text{-}75)$$

14.16 Reactions of Aromatic Aldehydes and Ketones

The carbonyl group, when attached to an aromatic ring, will direct substituents to the *meta* position since it is electron-withdrawing. However, it is significant to realize that the aldehyde group is too sensitive to oxidation by nitric acid to undergo the typical nitration reaction one might expect. As a matter of fact, aromatic aldehydes are reluctant to undergo any of the electrophilic substitution reactions characteristic of an aromatic ring. Aromatic ketones, which are less reactive than the aldehydes, will undergo the normal substitution reactions.

In addition to electrophilic substitution reactions the carbonyl group of aromatic aldehydes and ketones will undergo nucleophilic addition and condensation reactions discussed throughout the chapter.

Summary

- Aldehydes and ketones are characterized by the presence of the carbonyl, $\text{\Large>}C=O$, functional group.

- The nomenclature of aldehydes (IUPAC system) is indicated by an *-al* ending, and ketones end in *-one*. Some aldehydes and ketones are better known by their common names.

- Aldehydes and ketones can be prepared in various ways:

SUMMARY

  (a) Oxidation of primary alcohols to aldehydes—secondary alcohols to ketones.
  (b) Hydrolysis of *gem*-dihalides.
  (c) The Reimer–Tiemann Reaction.

- Special methods can be used to prepare formaldehyde, acetaldehyde, and acetone.
- The nature of the carbonyl group, and its sp^2 hybridization, molecular geometry, and polarization, are helpful in explaining the chemical reactivity of aldehydes and ketones.
  The principal reaction of aldehydes and ketones is nucleophilic addition.
- Reactions of aldehydes and ketones:

  (a) Oxidation—aldehydes are more easily oxidized (Tollens' test, and Benedict's or Fehling's test).
  (b) Catalytic reduction by hydrogen or by complex metallic hydrides to alcohols.
  (c) Clemmensen reduction or Wolff–Kishner reduction to hydrocarbons.

- Nucleophilic addition reactions—ketones react more slowly than aldehydes mainly because of the steric and electronic effects present in the molecule.

  (a) HCN produces cyanohydrins, which can be converted into α-hydroxy acids and α-amino acids.
  (b) $H_2O \longrightarrow$ hydrates.
  (c) Alcohols $\longrightarrow$ hemiacetals or hemiketals and acetals or ketals.
  (d) Hydroxylamine $\longrightarrow$ oxime.
  (e) Hydrazine $\longrightarrow$ hydrazone.
  (f) Phenylhydrazine $\longrightarrow$ phenylhydrazone.
  (g) 2,4-DNP $\longrightarrow$ 2,4-DNPH.
  (h) Semicarbazide $\longrightarrow$ semicarbazones.
  (i) Ammonia + HCHO $\longrightarrow$ hexamethylenetetramine (urotropine), used as a urinary antiseptic.
  (j) $NaHSO_3 \longrightarrow$ addition product.
  (k) Addition of Grignard reagent + HCHO $\longrightarrow$ 1° alcohol. Addition of Grignard reagent + aldehyde $\longrightarrow$ 2° alcohol. Addition of Grignard reagent + ketone $\longrightarrow$ 3° alcohol.

- Ionization of the α-hydrogen atom in aldehydes and ketones results in enol-keto tautomerism. This is important in the base-catalyzed reactions of aldehydes and ketones, such as the aldol condensation, haloform reaction (iodoform test), Perkin reaction, etc., all of which involve substitution for a hydrogen on the α-carbon atom of an adehyde or ketone.
  Methyl ketones can be distinguished from other ketones since they give a positive iodoform test.
- The Cannizzaro reaction takes place on aldehydes with no α-hydrogens, when they are treated with concentrated base.
- The carbonyl group is *meta*-directing in electrophilic aromatic substitution.

PROBLEMS

1 Draw the correct structural formulas for each of the following compounds.
   (a) 2-methyl-5,6-dibromooctanal
   (b) 3-hexanone
   (c) *p*-nitrobenzaldehyde
   (d) γ,γ,γ-trichlorobutyraldehyde
   (e) 2-bromoacetophenone
   (f) 4-methyl-3-pentene-2-one
   (g) acetone cyanohydrin
   (h) benzophenone oxime
   (i) methylethylketone
   (j) 3-hydroxybutanal
   (k) *n*-heptaldehyde diethyl acetal
   (l) 1,3-cyclohexadione

2 Name the following compounds.

(a) 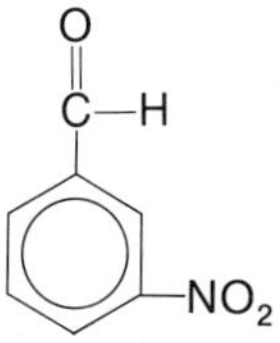 3-nitrobenzaldehyde structure (benzene with C(=O)–H and NO$_2$)

(b) $CH_3-CH(CH_3)-CH_2-CH(CH_3)-C(=O)-H$

(c) diphenyl ketone with 3,5-dimethyl on one ring

(d) cyclopentanone with two CH$_3$ groups on α-carbon

(e) $Br-CH_2-CH=CH-C(=O)-CH_2-CH_3$

(f) $CH_3-C(OCH_3)(OCH_3)-H$

(g) $CH_3-C(=N-OH)-$phenyl

(h) $CH_3-C(=N-NH-Ar)-CH_3$ where Ar = 2,4-dinitrophenyl

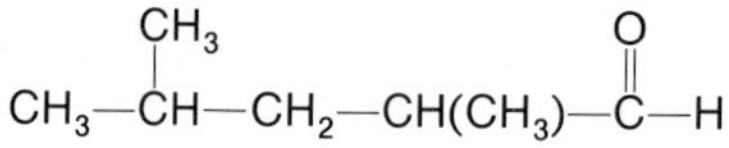

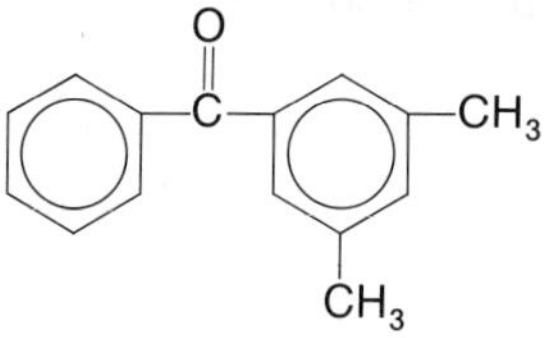

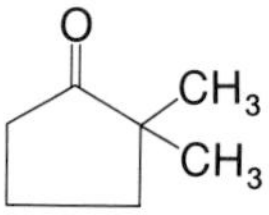

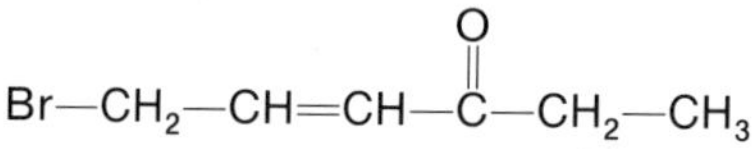

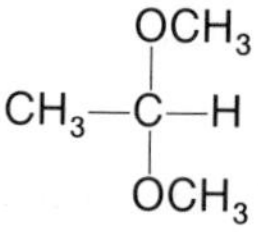

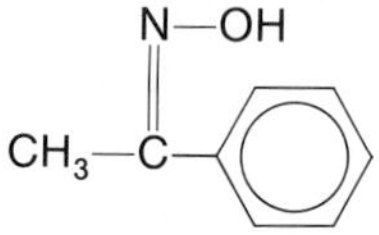

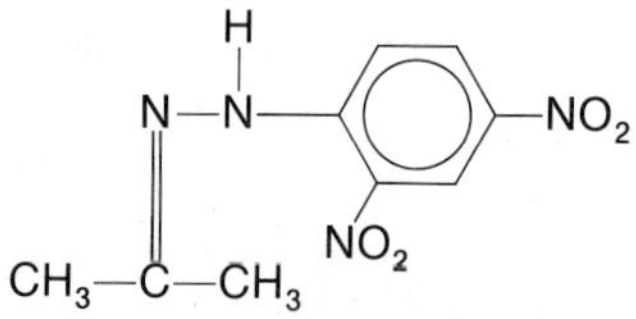

(i) 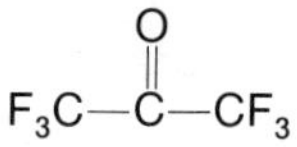 $F_3C-C(=O)-CF_3$

(j) 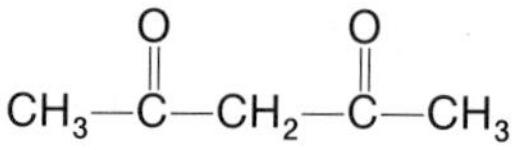 $CH_3-C(=O)-CH_2-C(=O)-CH_3$

(k) 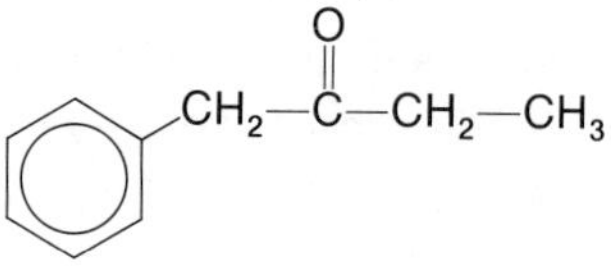 phenyl-$CH_2-C(=O)-CH_2-CH_3$

(l) 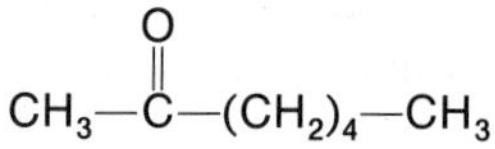 $CH_3-C(=O)-(CH_2)_4-CH_3$

3 Complete each of the following equations by writing the formula(s) of the product(s) formed. If no reaction occurs, write N.R.

(a) 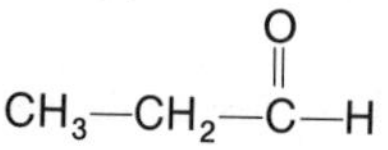 $CH_3-CH_2-C(=O)-H$ + HCN ⟶

(b) 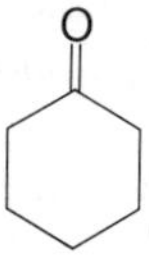 cyclohexanone + 2,4-DNPH ⟶

PROBLEMS

(c) 2-butanol + $K_2Cr_2O_7$ $\xrightarrow{H_2SO_4}$

(d) 1-pentanol + Cu $\xrightarrow{\Delta}$

(e) benzaldehyde + NH_2OH $\xrightarrow{OH^\ominus}$

(f) $H-\underset{\underset{O}{\|}}{C}-H$ $\xrightarrow{NaOH}$

(g) trimethylacetaldehyde $\xrightarrow{50\% \text{ NaOH}}$

(h) 2-hexanone + $NaBH_4$ $\longrightarrow$

(i) benzophenone $\xrightarrow{Zn(Hg) / HCl}$

(j) 4-bromo-(dibromomethyl)benzene $\xrightarrow{H_2O / OH^\ominus}$

(k) n-butyraldehyde + excess CH_3CH_2OH $\xrightarrow{H^\oplus}$

(l) cyclopentanone + Br_2 $\xrightarrow{OH^\ominus}$

(m) benzaldehyde + $CH_3-\underset{\underset{O}{\|}}{C}-CH_3$ $\xrightarrow{OH^\ominus / \Delta}$

(n) benzophenone + $CH_3-\underset{\underset{O}{\|}}{C}-H$ $\xrightarrow{OH^\ominus / \Delta}$

(o) benzaldehyde ethylidene acetal + $H^\oplus$ $\longrightarrow$

(p) 3-nitrobenzaldehyde + Ac_2O $\xrightarrow{\text{sodium acetate}}$

(q) $CH_3-CH_2-CH=CH-\underset{\underset{O}{\|}}{C}-H$ + $LiAlH_4$ $\xrightarrow{\text{then } H_2O}$

(r) di-t-butyl ketone + $NaHSO_3$ $\longrightarrow$

(s) $CH_3-\underset{\underset{O}{\|}}{C}-CH_3$ + CH_3MgBr $\xrightarrow{H_2O / H^\oplus}$

(t) diphenyl ketone + $Ag(NH_3)_2^\oplus$ $\xrightarrow{OH^\ominus}$

4 Perform all of the following syntheses from the indicated starting materials and any necessary inorganic reagents.
  (a) benzene to *p*-bromobenzaldehyde
  (b) propylene to 2-methyl-3-hydroxypentanal
  (c) *inorganic reagents to 4-methyl-2-pentanol
  (d) benzaldehyde to 2-hydroxy-2-phenylethanoic acid
  (e) *inorganic reagents to 2-ethyl-1-hexanol
  (f) benzaldehyde and acetic anhydride to $C_6H_5-C\equiv C-COOH$
  (g) inorganic reagents to 1,3-dihydroxybutane
  (h) propyne to mesityl oxide (4-methyl-3-pentene-2-one)
  (i) benzene to 3-ortho-chlorophenylpropenal
  (j) benzene to 2-phenyl-2-propanol
  (k) inorganic reagents to 1,1-diethoxyethane
  (l) bromobenzene to cyclohexanone oxime

5 Starting with any aliphatic alcohols of five carbon atoms or less as your only source of organic compounds, synthesize the following compounds.
  (a) 3-pentanol
  (b) 4-ethyl-4-heptanol
  (c) 3-methyl-3-hexanol
  (d) 3-methyl-4-heptanol
  (e) 3-methyl-3-pentanol
  (f) 3-ethyl-1-pentanol

6 Name a chemical test or single reagent which can be used to distinguish between the following pairs of compounds.
  (a) benzaldehyde and acetophenone
  (b) 2-hexanone and 3-hexanone
  (c) 2-propanol and acetone
  (d) acetaldehyde and acetone
  (e) methyl alcohol and ethyl alcohol

7 Arrange the following compounds in increasing order towards nucleophilic addition reactions.
acetone
acetaldehyde
3-methylpropanal
3,3-dimethylbutanone
2-cyanobutanal

8 Trifluoroacetaldehyde is more reactive than acetaldehyde towards nucleophilic addition. Explain.

9 Why can't crotonaldehyde, $CH_3-CH=CH-\overset{\overset{O}{\|}}{C}-H$, be directly reacted with cold alkaline $KMnO_4$, to produce 2,3-dihydroxybutanal? How would you prepare 2,3-dihydroxybutanal from crotonaldehyde?

10 What is the function of the base used in the Reimer–Tiemann reaction?

11 Which hydrogens in the compound acetylacetone, $CH_3-\overset{\overset{O}{\|}}{C}-CH_2-\overset{\overset{O}{\|}}{C}-CH_3$, would you expect to be somewhat acidic? Why is this compound much more acidic than acetone?

12 Suggest a detailed mechanism for the mixed aldol condensation between benzaldehyde and propionaldehyde in the presence of base to form 2-methyl-3-phenylpropenal.

13 Aldol condensations can be catalyzed by base or by acid. Propose a mechanism for the acid-catalyzed aldol condensation of acetaldehyde.

14 Compound A, $C_6H_{13}Br$, forms a Grignard reagent with magnesium in anhydrous ether. The Grignard reagent can be hydrolyzed by water and acid to 2-methylpentane. When A is reacted with aqueous sodium hydroxide,

PROBLEMS

compound B, $C_6H_{14}O$, is formed, which is heated with copper metal to form C, $C_6H_{12}O$. Compound C gives a negative Fehling's test, and a negative Tollens' test, but a positive iodoform test. From this information, deduce and write the correct structural formulas of compounds A, B, and C.

15 * Compound A, C_8H_9Cl, is hydrolyzed by dilute acid to form compound B, $C_8H_{10}O$. Compound B can be oxidized under mild reaction conditions to compound C, C_8H_8O. Compound C gives a positive haloform (iodoform) test. Deduce and write the correct structural formulas of compounds A, B, and C.

16 * Compound A, $C_5H_{11}Br$, forms a Grignard reagent which is hydrolyzed to form *n*-pentane. When A is treated with aqueous potassium hydroxide, compound B, $C_5H_{12}O$, is formed. Compound B gives a positive Lucas test. When B is treated with potassium dichromate in sulfuric acid, C is formed. Compound C gives a negative iodoform test. From this information, deduce and write the correct structural formulas of compounds A, B, and C.

17 * Compound A, $C_6H_{12}O$, forms an oxime, but gives a negative Fehling's test. When A is reduced with hydrogen over a platinum catalyst, compound B, $C_6H_{14}O$, is formed. Compound B is heated with concentrated sulfuric acid to form C, C_6H_{12}. Compound C is oxidized to form two compounds, D and E. Compound D gives a negative Tollens' test and a positive iodoform test. Compound E gives a positive Tollens' test and a negative iodoform test. From this information, deduce and write the correct structural formulas of compounds A through E.

18 * Compound A, $C_{10}H_{12}$, is insoluble in water, but soluble in concentrated sulfuric acid. Compound A reacts with excess hydrogen gas under pressure to form compound B, $C_{10}H_{20}$; and A also reacts with ozone, followed by zinc and water to form compounds C and D. Compound C, C_7H_6O, reacts with a solution of $Ag(NH_3)_2^{\oplus}$ to form compound E, which is soluble in NaOH, and reacts with $NaHCO_3$ to evolve a gas. Compound E forms three different monobromo-substitution products. Compound D reacts with $NaHSO_3$ and phenylhydrazine, but gives a negative Tollens' test. From this information, deduce and write the correct structural formulas of compounds A through E.

15 Carboxylic Acids and Their Derivatives

Carboxylic acids are important components of fruits—with their low pH—such as the grapes of the Italian vineyard shown under spray irrigation.

Courtesy of Massey-Ferguson Ltd.

15.1 Introduction

The group of organic compounds characterized by the presence of a **carboxyl group**, $-\overset{\overset{\text{O}}{\|}}{\text{C}}-\text{OH}$, in the molecule are known as **carboxylic acids.** The acidity of the carboxylic acids has already been discussed in terms of the electronic effects present in the molecule and the resonance-stabilization of the carboxylate anion (Chapters 4 and 5). The general formula of a carboxylic acid may be represented as $\text{R}-\overset{\overset{\text{O}}{\|}}{\text{C}}-\text{OH}$, where R may be either aliphatic or aromatic. The derivatives of the carboxylic acids are related to the parent compound by the replacement of the hydroxyl group of the carboxyl group by another atom or group. The derivatives to be discussed in this chapter include salts, acyl halides, anhydrides, esters, and amides, formed by substitution of an oxide ion, halogen, acyloxy group, alkoxy group, or amino group, for the carboxyl hydroxyl group, respectively.

15.2 Nomenclature of the Carboxylic Acids

Many of the carboxylic acids are better known by their common names than by the IUPAC system nomenclature. The common names usually are derived from the Latin or Greek word that indicates the original source of the acid. For example, the simplest carboxylic acid, $\text{H}-\overset{\overset{\text{O}}{\|}}{\text{C}}-\text{OH}$, (R = H), is commonly called *formic acid* and may be obtained in the laboratory by the distillation of ants. The

Sec. 15.2 NOMENCLATURE OF THE CARBOXYLIC ACIDS

common name is derived from the Latin word *formica* (which means ant). The carboxylic acid, $CH_3-\overset{O}{\underset{\|}{C}}-OH$ (R = CH_3) is called *acetic acid* from the Latin word *acetum* (meaning vinegar), while the acid $CH_3-CH_2-CH_2-\overset{O}{\underset{\|}{C}}-OH$ is called *butyric acid* from the Latin word *butyrum* (butter). Indeed, butyric acid can be easily recognized by its characteristic and rather disagreeable odor of rancid butter. Other carboxylic acids are often referred to by their common names, as are related compounds, so it is essential that the reader learn both the common names and the IUPAC system nomenclature for the carboxylic acids.

The IUPAC naming of the carboxylic acids applies all the rules discussed previously, with the additional use of the *-oic* ending to indicate that the compound is an *acid*. The carboxyl group is located at the end of the chain and the carboxyl carbon atom is designated as number 1 in numbering the carbon atoms in the longest chain. When naming a compound by its common name, the Greek letters, α, β, γ, etc., are used, rather than numbers, to indicate the presence of substituents on the chain, the carbon atom adjacent to the carboxyl group being designated as the alpha(α)-carbon atom, etc.

$\overset{\beta}{\underset{3}{C}}H_3-\overset{\alpha}{\underset{2}{C}}H-\overset{O}{\underset{\underset{1}{\|}}{C}}-OH$
$|$
Cl

Common name: α-chloropropionic acid
IUPAC name: 2-chloropropanoic acid

$\overset{\gamma}{\underset{4}{C}}H_3-\overset{\beta}{\underset{3}{C}}H-\overset{\alpha}{\underset{2}{C}}H_2-\overset{O}{\underset{\underset{1}{\|}}{C}}-OH$
$|$
NH_2

Common name: β-aminobutyric acid
IUPAC name: 3-aminobutanoic acid

Compounds that contain two carboxyl groups in the molecule are called *dicarboxylic acids.* The aliphatic dicarboxylic acids are known by their common names, the IUPAC names being used very rarely. Aromatic carboxylic acids are named as derivatives of the parent compound benzoic acid, [benzene ring]—COOH, or by a common name if the compound has one. Table 15.1 illustrates these points with examples of representative compounds.

Most of the reactions of the carboxylic acids that are related to the carboxyl functional group involve the replacement or substitution of the OH portion of the carboxyl group by another substituent. A carboxylic acid, $R-\overset{O}{\underset{\|}{C}}-OH$, from which the OH has been removed leaves an $R-\overset{O}{\underset{\|}{C}}-$ fragment that is important in the numerous substitution reactions that the carboxylic acids and their derivatives undergo. The $R-\overset{O}{\underset{\|}{C}}-$ group is referred to as an **acyl group,** analogous to the R or alkyl groups derived from the alkanes by removal of a hydrogen atom. The acyl

Table 15.1 Nomenclature of Carboxylic Acids

Formula	Common name	IUPAC name
HCOOH	Formic acid	Methanoic acid
CH_3COOH	Acetic acid	Ethanoic acid
CH_3CH_2COOH	Propionic acid	Propanoic acid
$CH_3CH_2CH_2COOH$	n-Butyric acid	Butanoic acid
CH_3CHCH_2COOH | Br	β-Bromobutyric acid	3-Bromobutanoic acid
$CH_3(CH_2)_{10}COOH$	Lauric acid	Dodecanoic acid
$CH_3(CH_2)_{14}COOH$	Palmitic acid	Hexadecanoic acid
$CH_3(CH_2)_{16}COOH$	Stearic acid	Octadecanoic acid
HOOC—COOH	Oxalic acid	
HOOC—CH_2—COOH	Malonic acid	
HOOC—CH_2—CH_2—COOH	Succinic acid	
HOOC—$(CH_2)_3$—COOH	Glutaric acid	
HOOC—$(CH_2)_4$—COOH	Adipic acid	
⌬(COOH)(COOH)	Phthalic acid	
⌬—COOH	Benzoic acid	
O_2N—⌬—COOH	p-Nitrobenzoic acid	4-Nitrobenzoic acid
⌬(COOH)(OH)	Salicylic acid	2-Hydroxybenzoic acid
⌬(COOH)(CH_3)	m-Toluic acid	3-Methylbenzoic acid

Sec. 15.3 ACID STRENGTH

groups are named in relation to the acids from which they are derived, using an -yl ending instead of the -ic ending used in the common name of the acid.

$$R-\overset{O}{\underset{\|}{C}}-\quad \text{acyl group}$$

$$H-\overset{O}{\underset{\|}{C}}-\quad \text{formyl group} \qquad CH_3-\overset{O}{\underset{\|}{C}}-\quad \text{acetyl group}$$

$$CH_3-CH_2-\overset{O}{\underset{\|}{C}}-\quad \text{propionyl group} \qquad C_6H_5-\overset{O}{\underset{\|}{C}}-\quad \text{benzoyl group}$$

The derivatives of the carboxylic acids can be represented by the general formula $R-\overset{O}{\underset{\|}{C}}-N$, where N is the substituent that has replaced the hydroxyl portion of the carboxyl group.

$$R-\overset{O}{\underset{\|}{C}}-O^{\ominus}M^{\oplus} \qquad R-\overset{O}{\underset{\|}{C}}-X \qquad R-\overset{O}{\underset{\|}{C}}-O-\overset{O}{\underset{\|}{C}}-R$$
$$\text{salt} \qquad\qquad \text{acyl halide} \qquad\qquad \text{anhydride}$$

$$R-\overset{O}{\underset{\|}{C}}-OR' \qquad R-\overset{O}{\underset{\|}{C}}-NH_2$$
$$\text{ester} \qquad\qquad \text{amide}$$

15.3 Acid Strength of the Carboxylic Acids

The carboxylic acids are acidic since they can donate a proton to a more basic species. Generally, the carboxylic acids are weak acids in comparison with hydrochloric, nitric, or sulfuric acid. They have about the same acid strength regardless of the length of the chain, but the acid strength can be increased or decreased by the presence of electron-withdrawing or electron-donating substituents in the molecule.

The ionization of carboxylic acids can be represented by the following acid–base equilibrium equation in water solution:

$$R-\overset{O}{\underset{\|}{C}}-OH + H_2O \rightleftharpoons R-\overset{O}{\underset{\|}{C}}-O^{\ominus} + H_3O^{\oplus} \qquad (15\text{--}1)$$
$$\text{acid} \qquad \text{base} \qquad \text{carboxylate anion}$$

The resonance stabilization of the carboxylate anion is probably the most important factor in the acidity of the carboxyl group to give up its proton to a basic species. The concept of acid strength has already been discussed in detail in Chapters 4 and 5.

15.4 Physical Properties of the Carboxylic Acids

The lower molecular weight aliphatic carboxylic acids are all liquids with characteristic odors. The aliphatic carboxylic acids of ten or more carbon atoms are waxlike solids. Stearic acid, $CH_3-(CH_2)_{16}-COOH$, is a constituent in the manufacture of wax candles. The dicarboxylic acids and aromatic carboxylic acids are usually crystalline solids. The rather high boiling points of the carboxylic acids are attributed to the presence of hydrogen bonding in the molecule. For example, formic and acetic acid exist in the form of a cyclic dimer in the vapor state by the formation of hydrogen bonds between the molecules.

$$CH_3-C\begin{matrix}O\cdots H-O\\ \diagup \quad \diagdown \\ O-H\cdots O\end{matrix}C-CH_3 \qquad \text{acetic acid dimer}$$

The slight solubility of the lower molecular weight carboxylic acids in water is also attributed to hydrogen bonding between the acid and the water molecules.

15.5 Preparation of Carboxylic Acids

Carboxylic acids are usually prepared by one of three synthetic routes: (1) oxidation; (2) hydrolysis; or (3) use of a Grignard reagent.

15.5-1 Oxidation Reactions The oxidation of primary alcohols or aldehydes produces a carboxylic acid containing the same number of carbon atoms as were present in the original compound.

$$\underset{\text{primary alcohol}}{R-CH_2OH} \xrightarrow[H_2SO_4]{K_2Cr_2O_7 \text{ or } KMnO_4,} \underset{\text{carboxylic acid}}{R-COOH} \qquad (15\text{-}2)$$

$$\underset{\text{aldehyde}}{R-\overset{\overset{\displaystyle O}{\|}}{C}-H} \xrightarrow[H_2SO_4]{K_2Cr_2O_7} \underset{\text{carboxylic acid}}{R-\overset{\overset{\displaystyle O}{\|}}{C}-OH} \qquad (15\text{-}3)$$

The oxidation of alkanes to carboxylic acids is not a good method to prepare carboxylic acids, due to difficulties in the experimental procedures. However, the oxidation of an alkyl side chain, when directly attached to an aromatic nucleus, is an excellent way to prepare aromatic carboxylic acids. The oxidizing agent used may be potassium dichromate, potassium permanganate, or nitric acid. This reaction has been discussed in Section 10.6 in connection with the reactions of aromatic compounds.

$$\underset{o\text{-xylene}}{\begin{matrix}\text{CH}_3\\ \bigcirc\\ \text{CH}_3\end{matrix}} \xrightarrow[H_2SO_4]{KMnO_4} \underset{\text{phthalic acid}}{\begin{matrix}\text{COOH}\\ \bigcirc\\ \text{COOH}\end{matrix}} \qquad (15\text{-}4)$$

Sec. 15.5 PREPARATION OF CARBOXYLIC ACIDS

15.5-2 Hydrolysis Reactions Carboxylic acids can also be prepared by the hydrolysis of *gem*-trihalides or nitriles. The hydrolysis of *gem*-trihalides to form carboxylic acids is an extremely limited reaction, primarily because of the great difficulty in preparing pure 1,1,1-trihalides in good yield. However, the preparation of benzoic acid by this method can be accomplished readily.

$$\text{toluene} \xrightarrow[h\nu]{3Cl_2} \text{benzotrichloride (a } gem\text{-trihalide)} \xrightarrow[H^{\oplus}]{H_2O} \text{benzoic acid} \qquad (15\text{-}5)$$

The hydrolysis of an organic cyanide (nitrile) leads to the formation of a carboxylic acid. The reaction is run in the presence of an acidic or basic catalyst, and is believed to proceed through the intermediate formation of an amide, which is actually hydrolyzed to the final product, the carboxylic acid. When the reaction is run in base the nitrogen of the cyanide is converted to ammonia and the salt of the acid is isolated; in acid solution the nitrogen of the cyanide is converted to an ammonium salt and the carboxylic acid is isolated as the product.

$$R-C\equiv N \xrightarrow{HOH} \left[R-\underset{\text{enol form}}{\overset{OH}{\underset{|}{C}}=NH} \right] \xrightarrow{\text{rearr.}} \underset{\text{amide}}{R-\overset{O}{\overset{\|}{C}}-NH_2} \xrightarrow[H^{\oplus}]{HOH} R-\overset{O}{\overset{\|}{C}}-OH + NH_4^{\oplus} \qquad (15\text{-}6)$$

$$\xrightarrow[OH^{\ominus}]{HOH} R-\overset{O}{\overset{\|}{C}}-O^{\ominus} + NH_3 \qquad (15\text{-}7)$$

If the reaction sequence is started with an alkyl halide, for example, which is first converted to a cyanide and then hydrolyzed to a carboxylic acid, the product will contain one more carbon atom than the starting material. This serves as a convenient way of synthesizing carboxylic acids containing one more carbon atom in the molecule.

$$\underset{\text{ethyl chloride}}{CH_3-CH_2Cl} \xrightarrow{NaCN} \underset{\substack{\text{ethyl cyanide} \\ \text{(propiono-nitrile)}}}{CH_3-CH_2-C\equiv N} \xrightarrow[H^{\oplus}]{H_2O} \underset{\text{propionic acid}}{CH_3-CH_2-\overset{O}{\overset{\|}{C}}-OH} \qquad (15\text{-}8)$$

15.5-3 Use of Grignard Reagents to Prepare Carboxylic Acids Grignard reagents react with carbon dioxide to form addition products that can be hydrolyzed to carboxylic acids. The acids formed contain one more carbon atom than the Grignard reagent used, the extra carbon coming from the CO_2. This is still another way of lengthening the carbon chain in a molecule by one carbon atom.

$$CH_3-Br + Mg \xrightarrow{\text{anhy. ether}} CH_3^{\ominus}\overset{\oplus}{M}gBr + \overset{\overset{\overset{\delta\ominus}{O}}{\|}}{\underset{\delta\oplus}{C}}=O \longrightarrow CH_3-\overset{\overset{OMgBr}{|}}{\underset{|}{C}}=O \xrightarrow[H_2O]{H^{\oplus}} CH_3-\overset{\overset{O}{\|}}{C}-OH$$

methyl
bromide
acetic acid

(15-9)

o-bromotoluene → (with Mg, anhy. ether or tetrahydrofuran) → o-tolyl-MgBr → (1. CO$_2$; 2. H$_2$O, H$^{\oplus}$) → o-toluic acid

(15-10)

15.6 Reactions of the Carboxylic Acids and Their Derivatives

The preparation of derivatives from carboxylic acids and the reactions of these derivatives can best be understood by considering the carboxyl functional group,

$-\overset{\overset{O}{\|}}{C}-OH$, as being composed of a carbonyl, $-\overset{\overset{O}{\|}}{C}-$, and a hydroxyl, $-OH$, group and possessing some of the properties and characteristics of these components.

The preparation of salts from carboxylic acids involves the replacement of the hydrogen atom of the —OH group by a metal. Most of the reactions of the carboxylic acids and acid derivatives can be classified as nucleophilic substitution reactions. These reactions involve attack by a nucleophilic reagent on the carbonyl carbon atom (of the carboxyl group) similar to the reactions of aldehydes and ketones. But, in contrast to the reactions of aldehydes and ketones, the net result of these reactions is the subsequent replacement of the hydroxyl group (or other group present in the acid derivatives) by the nucleophilic species. The reason why the carboxylic acids and their derivatives tend to undergo nucleophilic substitution reactions is that the substituents attached to the carbonyl portion of the carboxyl group represented by Y (R—$\overset{\overset{O}{\|}}{C}$—Y) are good leaving groups, and can be replaced by some nucleophilic reagent. Whereas the Y would represent a hydrogen atom or an alkyl or aryl group for aldehydes and ketones, for carboxylic acids and acid derivatives the Y must be a hydroxyl group, halogen, amino, or alkoxy group, etc. Hydrogen atoms, and alkyl and aryl groups, are not good leaving groups, and since they are not easily displaced, aldehydes and ketones tend to undergo nucleophilic addition across the carbonyl group rather than nucleophilic substitution. The groups displaced from the carboxylic acids and acid derivatives (i.e., —Cl$^{\ominus}$, OR$^{\ominus}$, etc.) form relatively stable anions or stable neutral species (H$_2$O, in the cases of carboxylic acids); they can be displaced easily by nucleophilic reagents; and therefore, in these compounds nucleophilic substitution is favored over nucleophilic addition.

Sec. 15.7 SALT FORMATION

From the previous discussion it should be apparent that the fundamental mechanism for nucleophilic substitution reactions of the carboxylic acids and their derivatives can be represented by the following sequence.

$$R-\overset{O}{\underset{\|}{C}}-Y + N:^{\ominus} \rightleftharpoons R-\overset{O^{\ominus}}{\underset{Y}{\overset{|}{C}}}-N \longrightarrow R-\overset{O}{\underset{\|}{C}}-N + :Y^{\ominus} \qquad (15\text{-}11)$$

(nucleo- (tetrahedral (good leaving group)
phile) intermediate)

The success of these substitution reactions depends to some degree on the ease of displacement of the leaving group. Other significant factors in determining the feasibility of the reaction are the reactivity of the carbonyl group (of the carboxyl group) to undergo reaction with the nucleophile and form a tetrahedral reaction intermediate, and the strength of the attacking nucleophilic reagent.

As we saw in the reactions of the aldehydes and ketones, a protonated carbonyl group, $>C=\overset{\oplus}{O}H$, is much more reactive than the carbonyl group, $>C=O$, toward nucleophilic attack. For this reason, occasionally the nucleophilic substitution reactions of the carboxylic acids and their derivatives are catalyzed by acid, where a proton adds to the carbonyl oxygen and thereby forms a reaction intermediate represented by $R-\overset{OH}{\underset{N}{\overset{|}{C}}}-Y$, which helps facilitate the removal of Y as YH.

The reactions may also be catalyzed by a base that can remove a proton from a relatively weak nucleophilic reagent, as in the removal of a proton from water to form hydroxide ion, which is a stronger nucleophilic reagent.

15.7 Salt Formation

Since the carboxyl hydrogen atom of the carboxylic acids is acidic it can be replaced by a metal to form a salt. The acids can be reacted with the metal directly, but usually are treated with an aqueous solution of a base to form a salt and water, as in a typical acid–base neutralization reaction.

$$\underset{\text{acid}}{R-\overset{O}{\underset{\|}{C}}-OH} + \underset{\text{base}}{MOH} \rightleftharpoons \underset{\text{salt}}{R-\overset{O}{\underset{\|}{C}}-O^{\ominus}M^{\oplus}} + H_2O \qquad (15\text{-}12)$$

$$\underset{\text{acetic acid}}{CH_3-\overset{O}{\underset{\|}{C}}-OH} + NaOH \rightleftharpoons \underset{\text{sodium acetate}}{CH_3-\overset{O}{\underset{\|}{C}}-O^{\ominus}Na^{\oplus}} + H_2O \qquad (15\text{-}13)$$

The salts are ionic compounds, usually water soluble, that conduct an electric current. The salts are named by the name of the metal followed by the name of the acid from which the salt is derived, but the -ic ending of the acid is changed to -ate.

$$\underset{\text{lithium benzoate}}{C_6H_5-\overset{O}{\underset{\|}{C}}-O^\ominus Li^\oplus}$$

$$\underset{\text{barium butyrate}}{\left(CH_3-CH_2-CH_2-\overset{O}{\underset{\|}{C}}-O^\ominus\right)_2 Ba^{\oplus\oplus}}$$

The sodium and potassium salts of the long chain acids (or fatty acids) are known as **soaps** and are used in the preparation of commercial **detergents**.

$$\underset{\text{sodium stearate (a soap)}}{CH_3-(CH_2)_{16}-CO\overset{\ominus}{O}Na^\oplus}$$

15.8 Formation of Other Derivatives of the Carboxylic Acids (Acyl Halides, Esters, Anhydrides, Amides)

As we just saw in the preceding section salt formation involves the replacement of a hydrogen atom by a metal. The formation of the other derivatives of the carboxylic acids involves the replacement of the hydroxyl group of the carboxylic acid by some nucleophile. The formation of acyl halides, esters, anhydrides, and amides from carboxylic acids is similar in this respect, and it is convenient to discuss the preparation of these derivatives at the same time. The only difference in their preparation is in the nature of the nucleophilic reagent used.

15.8–1 Formation of Acyl Halides The formation of acyl halides from carboxylic acids involves the replacement of the hydroxyl portion of the carboxyl group by a halogen. The reaction is similar in many respects to the preparation of alkyl halides from alcohols (Section 11.10–2) and indeed, the same reagents (phosphorus tri- and penta-halides and thionyl chloride) may be used.

$$\underset{\substack{\text{carboxylic}\\\text{acid}}}{R-\overset{O}{\underset{\|}{C}}-OH} + PCl_5 \longrightarrow \underset{\substack{\text{acyl}\\\text{chloride}}}{R-\overset{O}{\underset{\|}{C}}-Cl} + HCl\uparrow + POCl_3 \qquad (15\text{-}14)$$

$$\underset{\text{acetic acid}}{CH_3-\overset{O}{\underset{\|}{C}}-OH} + PCl_5 \longrightarrow \underset{\substack{\text{acetyl}\\\text{chloride}}}{CH_3-\overset{O}{\underset{\|}{C}}-Cl} + HCl\uparrow + POCl_3 \qquad (15\text{-}15)$$

$$\underset{\text{carboxylic acid}}{R-\overset{O}{\underset{\|}{C}}-OH} + SOCl_2 \longrightarrow \underset{\text{acyl chloride}}{R-\overset{O}{\underset{\|}{C}}-Cl} + SO_2\uparrow + HCl\uparrow \qquad (15\text{-}16)$$

Sec. 15.8 FORMATION OF OTHER DERIVATIVES

The corresponding acyl bromides and iodides are prepared by treating the carboxylic acids with the phosphorus tri- or penta-bromide or iodide.

$$\text{C}_6\text{H}_5\text{COOH} + \text{PBr}_3 \longrightarrow \text{C}_6\text{H}_5\text{COBr} + \text{H}_3\text{PO}_3 \quad (15\text{-}17)$$

benzoic acid → benzoyl bromide

Acyl halides cannot be prepared by reaction of the carboxylic acids with the hydrogen halides, HX, since the reaction is reversible, with the equilibrium lying far to the left in favor of the reactants. The acyl halides formed are hydrolyzed, by the water formed, to the corresponding carboxylic acid and hydrogen halide.

$$R-\underset{\underset{O}{\|}}{C}-OH + HX \rightleftharpoons R-\underset{\underset{O}{\|}}{C}-X + H_2O \quad (15\text{-}18)$$

15.8-2 Formation of Esters When a carboxylic acid is heated with an alcohol in the presence of a strong acid, such as H_2SO_4 or HCl, as a catalyst, an equilibrium is established favoring the ester side, resulting in the formation of an ester and water.

$$R-\underset{\underset{O}{\|}}{C}-OH + R'OH \underset{}{\overset{H^\oplus}{\rightleftharpoons}} \left[R-\underset{OR'}{\overset{OH}{\underset{|}{C}}}-OH \right] \overset{H^\oplus}{\rightleftharpoons} R-\underset{\underset{O}{\|}}{C}-O-R' + H_2O \quad (15\text{-}19)$$

carboxylic acid   alcohol   →   ester

A typical example of an esterification reaction is the reaction between acetic acid and ethyl alcohol to form the ester ethyl acetate.

$$CH_3-\underset{\underset{O}{\|}}{C}-OH + CH_3-CH_2-OH \overset{H^\oplus}{\rightleftharpoons} CH_3-\underset{\underset{O}{\|}}{C}-O-CH_2-CH_3 + H_2O \quad (15\text{-}20)$$

acetic acid   ethyl alcohol   →   ethyl acetate

The equilibrium may be shifted to the right by using an excess amount of one of the reactants or by removal of water as rapidly as it is formed (by distillation from the reaction mixture).

The nomenclature of the esters is similar to that used in naming the salts of the carboxylic acids, except that the name of the group from the alcohol replaces the name of the metal in the salt.

$$CH_3-\underset{\underset{O}{\|}}{C}-O^\ominus Na^\oplus \qquad CH_3-\underset{\underset{O}{\|}}{C}-O-CH_2-CH_3$$

sodium acetate (a salt)   ethyl acetate (an ester)

$$CH_3-CH_2-\overset{\overset{O}{\|}}{C}-O-CH_3$$

methyl propionate

$$C_6H_5-\overset{\overset{O}{\|}}{C}-O-CH_2-CH_3$$

ethyl benzoate

The general formula for an ester is $R-\overset{\overset{O}{\|}}{C}-OR'$; the group R' attached to the oxygen atom originated in the alcohol used in the formation of the ester, and the $R-\overset{\overset{O}{\|}}{C}-$ portion of the ester came from the carboxylic acid used. Esters usually have pleasant odors or aromas. Many esters are the constituents responsible for the characteristic odors of certain fruits (e.g., *n*–amyl acetate, bananas; ethyl buryrate, pineapples; methyl anthranilate, oranges).

The mechanism of the esterification reaction is very well understood, and illustrates many of the theoretical concepts discussed in earlier chapters. Much of the experimental evidence is based on kinetic data dealing with rates of reaction and the use of isotopic tracer experiments using the radioactive isotope, O^{18}. Let us now discuss the general mechanism for esterification in some detail.

First, the function of a strong-acid catalyst in esterification is to protonate some of the carboxylic-acid molecules to render the carbonyl groups more susceptible to nucleophilic attack by alcohol molecules.

$$R-\overset{\overset{O}{\|}}{C}-OH + H^{\oplus} \rightleftharpoons R-\overset{\overset{\oplus OH}{|}}{C}-OH \qquad (15\text{-}21)$$

If a series of esterification experiments is run on *a given acid* under identical reaction conditions using a series of *different alcohols*, a comparison of the relative rates of reaction will reveal that the rate of the reaction decreases as the bulk of the R' group of the alcohol increases. This can be interpreted as meaning that steric factors are extremely important in the esterification reaction and large bulky groups in the alcohol will hinder the formation of the transition state required for esterification (since the carbonyl carbon atom of the acid goes from sp^2 to sp^3 hybridization, resulting in more crowding in the transition state). (Thus, primary alcohols are usually esterified more easily than secondary alcohols, which in turn are more easily esterified than tertiary alcohols.) If an alcohol enriched in O^{18} ($RO^{18}H$) is used in the esterification reaction, the product obtained will be an ester containing O^{18} ($R-\overset{\overset{O}{\|}}{C}-O^{18}R'$); but no O^{18} is found in the water formed.

The *mechanism for esterification,* which can account for and explain all these experimental facts, can be represented by the following series of equations.

$$R-\overset{\overset{O}{\|}}{C}-OH + H^{\oplus} \rightleftharpoons R-\overset{\overset{\oplus OH}{|}}{C}-OH \quad \text{(protonation to increase reactivity of the carbonyl group)} \qquad (15\text{-}22)$$

Sec. 15.8 FORMATION OF OTHER DERIVATIVES

$$R-\overset{\overset{\oplus}{O}H}{\underset{}{C}}-OH + R'O^{18}H \rightleftharpoons \left[R-\overset{OH}{\underset{\overset{18}{O}\overset{\oplus}{\underset{H}{-}}R'}{C}}-OH \right] \quad \text{(nucleophilic attack by alcohol to form a transition state)} \quad (15\text{-}23)$$

$$\left[R-\overset{OH}{\underset{\overset{18}{O}\overset{\oplus}{\underset{H}{-}}R'}{C}}-OH \right] \rightleftharpoons \left[R-\overset{OH_2^{\oplus}}{\underset{{}^{18}OR'}{C}}-OH \right] \quad \text{(proton shift to form } OH_2^{\oplus}, \text{ which is a better leaving group than OH)} \quad (15\text{-}24)$$

$$\left[R-\overset{OH_2^{\oplus}}{\underset{{}^{18}OR'}{C}}-OH \right] \rightleftharpoons R-\overset{\overset{\oplus}{O}H}{\underset{}{C}}-O^{18}R' + H_2O \quad \text{(H}_2\text{O is a good leaving group)} \quad (15\text{-}25)$$

$$R-\overset{\overset{\oplus}{O}H}{\underset{\|}{C}}-O^{18}R' \rightleftharpoons R-\overset{O}{\underset{\|}{C}}-O^{18}R' + H^{\oplus} \quad (15\text{-}26)$$

This mechanism applies to most esterification reactions and to the reverse reaction, the acid-catalyzed hydrolysis of esters.

15.8-3 Formation of Anhydrides The anhydrides of the carboxylic acids are represented by the general formula, $R-\overset{O}{\underset{\|}{C}}-O-\overset{O}{\underset{\|}{C}}-R$, and can be pictured as being derived from two molecules of a carboxylic acid by removal of water, or by replacement of the hydroxyl group of the carboxyl group by the $-O-\overset{O}{\underset{\|}{C}}-R$ group. Anhydrides are usually prepared by the reaction between an acyl halide and the sodium salt of an acid. The anhydrides can not usually be prepared from the corresponding carboxylic acids themselves.

$$\underset{\substack{\text{acyl}\\\text{halide}}}{R-\overset{O}{\underset{\|}{C}}-X} + \underset{\text{salt}}{R-\overset{O}{\underset{\|}{C}}-O^{\ominus}Na^{\oplus}} \xrightarrow{\Delta} \underset{\text{acid anhydride}}{R-\overset{O}{\underset{\|}{C}}-O-\overset{O}{\underset{\|}{C}}-R} + Na\overset{\oplus}{X}{}^{\ominus} \quad (15\text{-}27)$$

$$\underset{\substack{\text{acetyl}\\\text{chloride}}}{CH_3-\overset{O}{\underset{\|}{C}}-Cl} + \underset{\text{sodium acetate}}{CH_3-\overset{O}{\underset{\|}{C}}-O^{\ominus}Na^{\oplus}} \xrightarrow{\Delta} \underset{\substack{\text{acetic}\\\text{anhydride}}}{CH_3-\overset{O}{\underset{\|}{C}}-O-\overset{O}{\underset{\|}{C}}-CH_3} + NaCl \quad (15\text{-}28)$$

Phthalic anhydride, which is used to make certain plasticizers, can be prepared by the oxidation of naphthalene or o-xylene.

$$\text{o-xylene} \xrightarrow{\Delta, \text{HNO}_3, \text{V}_2\text{O}_5} \text{phthalic acid} \xrightarrow{\Delta, -\text{H}_2\text{O}} \text{phthalic anhydride} \quad (15\text{-}29)$$

naphthalene

15.8-4 Preparation of Amides

Amides can be represented by the general formula $R-\overset{\overset{O}{\|}}{C}-NH_2$ and are named as derivatives from the related acids (e.g., $CH_3-\overset{\overset{O}{\|}}{C}-NH_2$, acetamide).

The preparation of amides can be accomplished by treating various derivatives (acyl halides, esters, or anhydrides) with ammonia, or by converting a carboxylic acid into its ammonium salt, which when heated loses water to form the amide.

$$R-\overset{\overset{O}{\|}}{C}-OH + NH_3 \longrightarrow R-\overset{\overset{O}{\|}}{C}-O^{\ominus}NH_4^{\oplus} \xrightarrow{\Delta, -H_2O} R-\overset{\overset{O}{\|}}{C}-NH_2 \quad (15\text{-}30)$$

carboxylic acid   ammonium salt   amide

The actual formula of amides can be considered to be a resonance hybrid of several contributing structures, of which

$$\left[R-\overset{\overset{:\ddot{O}:}{\|}}{C}-\ddot{N}H_2 \longleftrightarrow R-\overset{\overset{:\ddot{O}:^{\ominus}}{|}}{C}=\underset{\oplus}{N}H_2 \right]$$

are important in that these structures show that the carbon–nitrogen bond has some double bond character. As a consequence of the resonance phenomenon present in amides, all of the atoms attached to the carbon and nitrogen atoms of the amide functional group ($-\overset{\overset{O}{\|}}{C}-NH_2$) must lie in the same plane. The peptide linkage present in the proteins is actually an amide grouping ($-\overset{\overset{H}{|}}{N}-\overset{\overset{O}{\|}}{C}-$) and therefore the geometry of the amides is important when one considers the characteristics and properties of the proteins.

Sec. 15.9 REACTIONS OF THE DERIVATIVES

15.9 Reactions of the Derivatives of Carboxylic Acids

15.9-1 Reactions of Carboxylate Salts The salts of the carboxylic acids can be converted into the acids by treatment with a strong acid, such as HCl.

$$CH_3-CH_2-\underset{\underset{\text{sodium propionate}}{}}{\overset{\overset{O}{\|}}{C}}-O^{\ominus}Na^{\oplus} + HCl \longrightarrow CH_3-CH_2-\underset{\underset{\text{propionic acid}}{}}{\overset{\overset{O}{\|}}{C}}-OH + NaCl \quad (15\text{-}31)$$

This reaction is analogous to the preparation of some inorganic acids (such as nitric acid, by treating sodium nitrate with concentrated H_2SO_4) where reaction of a salt with an acid produces a new salt and new acid.

$$(salt_1) + (acid_1) \longrightarrow (salt_2) + (acid_2)$$

The net reaction involves replacement of the metal ion in the salt by a proton.

Heating the calcium or barium salt of a carboxylic acid results in the loss of carbon dioxide as a carbonate and the formation of a ketone.

$$\underset{\text{calcium acetate}}{\left[\begin{array}{c}CH_3-\overset{\overset{O}{\|}}{C}-O^{\ominus}\\ CH_3-\overset{\overset{O}{\|}}{C}-O^{\ominus}\end{array}\right]Ca^{\oplus\oplus}} \overset{\Delta}{\longrightarrow} CaCO_3 + \underset{\text{acetone}}{CH_3-\overset{\overset{O}{\|}}{C}-CH_3} \quad (15\text{-}32)$$

Cyclic ketones can be prepared by heating the calcium or barium salt of a dicarboxylic acid.

$$\underset{\text{barium adipate}}{\left[^{\ominus}O-\overset{\overset{O}{\|}}{C}-(CH_2)_4-\overset{\overset{O}{\|}}{C}-O^{\ominus}\right]Ba^{\oplus\oplus}} \overset{\Delta}{\longrightarrow} BaCO_3 + \underset{\text{cyclopentanone}}{\bigcirc\!\!=\!\!O} \quad (15\text{-}33)$$

15.9-2 Hydrolysis, Alcoholysis, and Ammonolysis of Acyl halides, Esters, Anhydrides, and Amides The most important reaction of the carboxylic acid derivatives is nucleophilic substitution. This type of reaction occurs readily because of the presence of good leaving groups (Y) in the molecule, $R-\overset{\overset{O}{\|}}{C}-Y$, and the high reactivity of the carbonyl group towards nucleophilic reagents.

It is convenient to discuss the hydrolysis (reaction with water), alcoholysis (reaction with alcohol), and ammonolysis (reaction with ammonia) reactions of the carboxylic acid derivatives at the same time, since the various derivatives give similar reaction products. The product formed depends on the nucleophilic reagent used.

Hydrolysis.

The acyl halides, $R-\overset{O}{\overset{\|}{C}}-X$, are the most reactive of the carboxylic acid derivatives. This is due to the presence of the halogen, which is a good leaving group and, by its electron-withdrawing inductive effect, greatly increases the reactivity of the carbonyl group.

The acyl halides can be hydrolyzed easily to the corresponding carboxylic acid. Many acyl halides even fume in air since they react violently with the moisture present in the atmosphere.

$$CH_3-\overset{O}{\overset{\|}{C}}-Cl + H_2O \longrightarrow CH_3-\overset{O}{\overset{\|}{C}}-OH + HCl \qquad (15\text{-}34)$$
acetyl chloride   acetic acid

The hydrolysis of esters is the reverse of the esterification process, the ester being converted into the acid and alcohol from which it was originally formed.

$$CH_3-\overset{O}{\overset{\|}{C}}-O-C_6H_5 + H_2O \underset{}{\overset{H^\oplus}{\rightleftharpoons}} CH_3-\overset{O}{\overset{\|}{C}}-OH + C_6H_5OH \qquad (15\text{-}35)$$
phenyl acetate   acetic acid   phenol

Esters may also be hydrolyzed in alkaline solution by heating with NaOH. The hydrolysis of esters in basic solution is referred to as **saponification,** since soaps are made in this fashion from fats (esters of glycerol).

$$R-\overset{O}{\overset{\|}{C}}-OR' + NaOH \overset{\Delta}{\longrightarrow} R-\overset{O}{\overset{\|}{C}}-\overset{\ominus}{O}Na^\oplus + R'OH \qquad (15\text{-}36)$$
ester   sodium salt of acid   alcohol

The reaction of the acid anhydrides with water is less vigorous, and easier to control than the corresponding reaction with acyl halides. The hydrolysis of a symmetrical acid anhydride produces two moles of a carboxylic acid.

$$CH_3-\overset{O}{\overset{\|}{C}}-O-\overset{O}{\overset{\|}{C}}-CH_3 + HOH \longrightarrow 2CH_3-\overset{O}{\overset{\|}{C}}-OH \qquad (15\text{-}37)$$
acetic anhydride   acetic acid

Amides are perhaps the least reactive of the carboxylic acid derivatives. This is shown by the need to use more drastic reaction conditions, in that amides are hydrolyzed slowly even in the presence of acid or base.

$$C_6H_5-\overset{O}{\overset{\|}{C}}-NH_2 + HOH \overset{H^\oplus}{\underset{\Delta}{\longrightarrow}} C_6H_5-\overset{O}{\overset{\|}{C}}-OH + NH_4^\oplus \qquad (15\text{-}38)$$
benzamide   benzoic acid

Sec. 15.9 REACTIONS OF THE DERIVATIVES

Alcoholysis. The alcoholysis of the carboxylic acid derivatives is very similar to the hydrolysis reaction. The only essential difference is that the nucleophilic reagent is alcohol instead of water, and the product formed is an ester instead of a carboxylic acid. The general term "solvolysis" is used to indicate a reaction in which the solvent (i.e., H_2O, ROH, etc.) is a reactant.

$$\text{Ph-CO-Br} + CH_3\text{-OH} \longrightarrow \text{Ph-CO-OCH}_3 + HBr \quad (15\text{-}39)$$

benzoyl bromide + methyl alcohol → methyl benzoate

Esters will undergo an alcoholysis reaction in which the alcohol portion of the ester is replaced by the alcohol used in the reaction. A new ester is formed in this exchange type reaction, which is catalyzed by base. The process is referred to as **transesterification.**

$$\underset{\text{ester}}{R\text{-}\overset{O}{\underset{\|}{C}}\text{-}OR'} + \underset{\text{alcohol}}{R''OH} \underset{}{\overset{\text{base}}{\rightleftharpoons}} \underset{\text{new ester}}{R\text{-}\overset{O}{\underset{\|}{C}}\text{-}OR''} + \underset{\text{new alcohol}}{R'OH} \quad (15\text{-}40)$$

$$\underset{\substack{\text{methyl} \\ \text{acetate}}}{CH_3\text{-}\overset{O}{\underset{\|}{C}}\text{-}OCH_3} + \underset{\substack{\text{ethyl} \\ \text{alcohol}}}{CH_3CH_2OH} \overset{\text{base}}{\rightleftharpoons} \underset{\text{ethyl acetate}}{CH_3\text{-}\overset{O}{\underset{\|}{C}}\text{-}OCH_2CH_3} + \underset{\substack{\text{methyl} \\ \text{alcohol}}}{CH_3OH} \quad (15\text{-}41)$$

The alcoholysis of anhydrides results in the formation of an ester and a carboxylic acid.

$$\underset{\text{acetic anhydride}}{CH_3\text{-}\overset{O}{\underset{\|}{C}}\text{-}O\text{-}\overset{O}{\underset{\|}{C}}\text{-}CH_3} + \underset{\substack{\text{methyl} \\ \text{alcohol}}}{CH_3\text{-}OH} \longrightarrow \underset{\substack{\text{methyl} \\ \text{acetate}}}{CH_3\text{-}\overset{O}{\underset{\|}{C}}\text{-}O\text{-}CH_3} + \underset{\text{acetic acid}}{CH_3\text{-}\overset{O}{\underset{\|}{C}}\text{-}OH}$$

$$(15\text{-}42)$$

Amides react slowly with alcohols to form esters.

$$\underset{\text{acetamide}}{CH_3\text{-}\overset{O}{\underset{\|}{C}}\text{-}NH_2} + \underset{\substack{\text{ethyl} \\ \text{alcohol}}}{CH_3\text{-}CH_2OH} \longrightarrow \underset{\text{ethyl acetate}}{CH_3\text{-}\overset{O}{\underset{\|}{C}}\text{-}OCH_2\text{-}CH_3} + NH_3 \quad (15\text{-}43)$$

Ammonolysis. The ammonolysis of the carboxylic acid derivatives leads to the formation of amides. For example, acyl halides react rapidly with ammonia to form the corresponding amide.

$$\text{CH}_3\text{—}\underset{\underset{\text{acetyl}}{}}{\overset{\overset{\text{O}}{\|}}{\text{C}}}\text{—Cl} + \text{NH}_3 \longrightarrow \text{CH}_3\text{—}\underset{\underset{\text{acetamide}}{}}{\overset{\overset{\text{O}}{\|}}{\text{C}}}\text{—NH}_2 + \text{HCl} \qquad (15\text{–}44)$$

chloride

The reaction of esters and acid anhydrides with ammonia can be illustrated by the following equations.

$$\underset{\text{ester}}{\text{R—}\overset{\overset{\text{O}}{\|}}{\text{C}}\text{—OR}'} + \text{NH}_3 \longrightarrow \underset{\text{amide}}{\text{R—}\overset{\overset{\text{O}}{\|}}{\text{C}}\text{—NH}_2} + \underset{\text{alcohol}}{\text{R}'\text{OH}} \qquad (15\text{–}45)$$

$$\underset{\text{acid anhydride}}{\text{R—}\overset{\overset{\text{O}}{\|}}{\text{C}}\text{—O—}\overset{\overset{\text{O}}{\|}}{\text{C}}\text{—R}} + \text{NH}_3 \longrightarrow \underset{\text{amide}}{\text{R—}\overset{\overset{\text{O}}{\|}}{\text{C}}\text{—NH}_2} + \underset{\text{carboxylic acid}}{\text{R—}\overset{\overset{\text{O}}{\|}}{\text{C}}\text{—OH}} \qquad (15\text{–}46)$$

The hydrolysis, alcoholysis, and ammonolysis reactions of the carboxylic acid derivatives are summarized in Table 15.2.

Table 15.2 Hydrolysis, Alcoholysis, and Ammonolysis of Carboxylic Acid Derivatives

Reactant	Acyl halide	Ester	Anhydride	Amide
H_2O	$R\text{—}\overset{\overset{O}{\|}}{C}\text{—OH}$	$R\text{—}\overset{\overset{O}{\|}}{C}\text{—OH} + R'\text{OH}$	$R\text{—}\overset{\overset{O}{\|}}{C}\text{—OH}$	$R\text{—}\overset{\overset{O}{\|}}{C}\text{—OH}$
$R'OH$	$R\text{—}\overset{\overset{O}{\|}}{C}\text{—OR}'$	$R\text{—}\overset{\overset{O}{\|}}{C}\text{—OR}'$	$R\text{—}\overset{\overset{O}{\|}}{C}\text{—OR}' + R\text{—}\overset{\overset{O}{\|}}{C}\text{—OH}$	$R\text{—}\overset{\overset{O}{\|}}{C}\text{—OR}'$
NH_3	$R\text{—}\overset{\overset{O}{\|}}{C}\text{—NH}_2$	$R\text{—}\overset{\overset{O}{\|}}{C}\text{—NH}_2 + R'\text{OH}$	$R\text{—}\overset{\overset{O}{\|}}{C}\text{—NH}_2 + R\text{—}\overset{\overset{O}{\|}}{C}\text{—OH}$	—

15.10 Formation of Aldehydes and Ketones from Acyl Halides and Acid Anhydrides

In addition to the nucleophilic substitution reactions discussed above, the acyl halides and acid anhydrides undergo some extremely important reactions used in the synthesis of aldehydes and ketones.

A common problem the synthetic organic chemist encounters in the laboratory is to reduce a carboxylic acid directly to an aldehyde. The reduction of a carboxylic acid by hydrogen gas over a platinum catalyst occurs only with great difficulty and the use of $LiAlH_4$ or other reducing agents usually leads to the formation of the alcohol as the reaction product rather than the aldehyde. The aldehyde stage is a difficult one to attain as a final step in oxidation or reduction reactions.

Sec. 15.10 FORMATION OF ALDEHYDES AND KETONES

15.10-1 The Rosenmund Reduction Thus, it is significant that, although a carboxylic acid cannot be directly converted into an aldehyde, if the acid is first converted into an acyl halide the acyl halide can undergo a rather unique reaction, the **Rosenmund reduction,** and be reduced to the aldehyde. The Rosenmund reduction merely involves reducing an acyl halide with hydrogen gas passed over a Pd–BaSO$_4$ catalyst containing a catalyst poison or moderator. Usually the organic compound quinoline is used as the poison whose function is to cut down the activity of the catalyst so the reduction is stopped at the aldehyde stage.

$$R-\overset{O}{\underset{}{C}}-X \xrightarrow[\text{quinoline}]{H_2,\ Pd-BaSO_4} R-\overset{O}{\underset{}{C}}-H \qquad (15\text{-}47)$$

(R can be aliphatic or aromatic)

$$\text{Ph}-\overset{O}{\underset{}{C}}-Cl \xrightarrow[\text{quinoline}]{H_2,\ Pd-BaSO_4} \text{Ph}-\overset{O}{\underset{}{C}}-H \qquad (15\text{-}48)$$

benzoyl chloride   **benzaldehyde**

15.10-2 The Friedel–Crafts Ketone Synthesis Acyl halides or acid anhydrides react with aromatic compounds in the presence of Lewis acid catalysts to form aromatic ketones. This reaction acylation is known as the **Friedel–Crafts ketone synthesis** or Friedel–Crafts acylation (analogous to the ordinary Friedel–Crafts alkylation reaction).

$$R-\overset{O}{\underset{}{C}}-Cl + C_6H_6 \xrightarrow{AlCl_3} R-\overset{O}{\underset{}{C}}-C_6H_5 + HCl\uparrow \qquad (15\text{-}49)$$

acyl halide   **aromatic**
(or anhydride)   **ketone**

The reaction is an example of an electrophilic aromatic substitution reaction, the electrophilic species being the acyl cation (R—$\overset{O}{\underset{}{C}}{}^{\oplus}$), whereas in the ordinary Friedel–Crafts reaction the reactive species is the carbonium ion (R$^{\oplus}$). The mechanism is illustrated by the following series of equations.

$$R-\overset{O}{\underset{}{C}}-Cl + AlCl_3 \rightleftharpoons R-\overset{O}{\underset{}{C}}{}^{\oplus} + AlCl_4^{\ominus} \qquad (15\text{-}50)$$

$$C_6H_6 + R-\overset{O}{\underset{}{C}}{}^{\oplus} + AlCl_4^{\ominus} \longrightarrow \left[\overset{\oplus}{\underset{}{C_6H_6}}\overset{H}{\underset{C-R}{}}\overset{O}{}\right] + AlCl_4^{\ominus} \qquad (15\text{-}51)$$

$$\left[\bigcirc\!\!\!\!\!\!\!\!\!\!\!{}^{\oplus}\!\!\!\!\!\!{}^{H}\!\!\!\!\!\!\underset{C-R}{\overset{O}{\|}}\right] + AlCl_4^{\ominus} \longrightarrow HAlCl_4 + \bigcirc\!\!\!\!\!\!\!\!\!\!\!\!\overset{O}{\underset{\|}{C}}-R \qquad (15\text{-}52)$$

$$HAlCl_4 \longrightarrow HCl \uparrow + AlCl_3 \qquad (15\text{-}53)$$

Acid anhydrides can be used instead of the acyl halides in the reaction.

$$CH_3-\overset{O}{\underset{\|}{C}}-Cl \quad + \quad \bigcirc \quad \xrightarrow{AlCl_3} \quad \bigcirc\!\!\!\!\!\!\!\!\!\!\!\!\overset{O}{\underset{\|}{C}}-CH_3 \qquad (15\text{-}54)$$

$$\left(\text{or } CH_3-\overset{O}{\underset{\|}{C}}-O-\overset{O}{\underset{\|}{C}}-CH_3\right) \qquad \text{acetophenone}$$

15.11 Other Reactions of Carboxylic Acid Derivatives

The Claisen Condensation. Esters that have hydrogen atoms on the alpha carbon atom adjacent to the carbonyl group can undergo condensation reactions similar in mechanism to the Aldol Condensation in aldehydes and ketones. The alpha hydrogen atoms are weakly acidic (as in aldehydes and ketones), can be removed by a strong base to form a carbanion which can then attack the carbonyl group of another ester molecule. The nucleophilic addition reaction that occurs in esters is similar in many respects to the Aldol Condensation and is referred to as the Claisen condensation.

Ethyl acetate can condense with itself in the presence of sodium ethoxide (why can't NaOH be used?) to form ethyl acetoacetate commonly known as acetoacetic ester.

$$2CH_3-\overset{O}{\underset{\|}{C}}-OCH_2-CH_3 \xrightarrow{Na^{\oplus},\ C_2H_5O^{\ominus}} CH_3-\overset{O}{\underset{\|}{C}}-CH_2-\overset{O}{\underset{\|}{C}}-O-CH_2-CH_3 \qquad (15\text{-}55)$$
$$\text{ethyl acetoacetate}$$

15.12 Substitution Reactions of Carboxylic Acids

15.12-1 Electrophilic Aromatic Substitution The aromatic carboxylic acids will undergo the usual electrophilic aromatic substitution reactions. Since the carboxyl group is an electron-withdrawing group it deactivates the aromatic ring and is a *meta*-directing group.

$$\bigcirc\!\!\!\!\!\!\!\!\!\!\!\!\overset{COOH}{} \quad \xrightarrow{HONO_2}{H_2SO_4} \quad \bigcirc\!\!\!\!\!\!\!\!\!\!\!\!\overset{COOH}{\underset{NO_2}{}} \qquad (15\text{-}56)$$
$$\text{\textit{m}-nitrobenzoic acid}$$

Sec. 15.13 SUBSTITUTED CARBOXYLIC ACIDS

15.12-2 Preparation of Alpha-Halogenated Carboxylic Acids The substitution of the α-hydrogen atoms in a carboxylic acid (alpha to the carboxyl group) by halogen can be accomplished easily. The influence of the adjacent carboxyl group is such that by the nature of its electron-withdrawing inductive effect, it renders the α-hydrogen atoms susceptible to being displaced by reaction with a halogen and phosphorus.

$$R-\underset{\alpha}{CH_2}-COOH \xrightarrow[X_2]{P} R-CH_2-\overset{O}{\underset{\|}{C}}-X \xrightarrow[X_2]{P} R-\underset{X}{\underset{|}{CH}}-\overset{O}{\underset{\|}{C}}-X \xrightleftharpoons{R-CH_2-COOH}$$

$$R-\underset{X}{\underset{|}{CH}}-COOH + R-CH_2-\overset{O}{\underset{\|}{C}}-X \quad (15\text{-}57)$$

α-halogenated
acid

The reaction is known as the **Hell-Volhard-Zelinsky** reaction and results in the substitution of a halogen atom for an α-hydrogen atom.

$$CH_3-\overset{O}{\underset{\|}{C}}-OH \xrightarrow[Cl_2]{P} ClCH_2-\overset{O}{\underset{\|}{C}}-OH \quad (15\text{-}58)$$

acetic acid   chloroacetic acid

$$CH_3-CH_2-CH_2-\overset{O}{\underset{\|}{C}}-OH \xrightarrow[Br_2]{P} CH_3-CH_2-\underset{Br}{\underset{|}{CH}}-\overset{O}{\underset{\|}{C}}-OH \quad (15\text{-}59)$$

butyric acid   α-bromobutyric acid

Only the α-hydrogens are affected by the neighboring electron-withdrawing carboxyl group, since the inductive effect decreases with increasing distance down the carbon chain. Thus, the carboxyl group has little or no effect on β-, γ-, δ-, etc., carbon atoms further down the chain in the molecule. The α-halogenated acid formed contains two functional groups, the carboxyl group and the halogen, which behaves as a typical alkyl halide. The molecule will undergo the reactions characteristic of each functional group. For example, the α-halogen undergoes the nucleophilic substitution reactions characteristic of the alkyl halides. The α-halogenated acids are extremely useful compounds in that they can be used in the synthesis of various α-substituted acids such as α-hydroxy, α-amino, and even α,β-unsaturated acids.

15.13 Preparation of Various Substituted Carboxylic Acids

The α-halogenated carboxylic acids, which can be prepared by the Hell-Volhard-Zelinsky reaction, are extremely valuable compounds in that they can be converted into many other substituted carboxylic acids. For example, α-bromopropionic acid can be converted into β-bromopropionic acid by a dehydrohalogenation reaction, followed by addition of HBr across the carbon–carbon double bond of the α,β-unsaturated acid formed in the reaction sequence.

$$CH_3-\underset{Br}{CH}-COOH \xrightarrow{alc.\ KOH} CH_2=CH-COOH \xrightarrow{HBr} \underset{Br}{CH_2}-CH_2-COOH \quad (15\text{-}60)$$

α-bromopropionic acid   acrylic acid   β-bromopropionic acid

The addition of HBr to acrylic acid appears to be contrary to Markovnikov's rule. However, the presence of the carboxyl group which is electron-withdrawing pulls the electron cloud towards itself and makes the β-carbon end of the molecule more positive than the α-carbon atom.

$$Br^{\ominus}{}^{\delta\oplus}\ \ \underset{}{CH_2=CH}-\overset{O}{\underset{}{C}}-OH\ \ {}^{\delta\ominus}H^{\oplus}$$
$$\xrightarrow{\text{electron density}}$$

This explains why the addition is in an opposite direction from what was predicted by Markovnikov's rule. An alternative explanation to account for the addition is that the acrylic acid molecule has a conjugated system, and that the 1,4-addition that can result leads to the formation of a resonance-stabilized allylic carbonium ion.

$$\underset{4\ \ \ 3\ \ \ 2}{CH_2=CH-\overset{O^1}{\overset{\|}{C}}-OH} + H^{\oplus} \rightleftharpoons CH_2=CH-\overset{OH}{\underset{\oplus}{\overset{|}{C}}}-OH \longleftrightarrow \overset{\oplus}{CH_2}-CH=C\overset{OH}{\underset{OH}{}}$$

$$\downarrow Br^{\ominus}$$

$$Br-CH_2-CH_2-\overset{O}{\overset{\|}{C}}-OH \rightleftharpoons Br-CH_2-CH=C\overset{OH}{\underset{OH}{}}$$

more stable keto form of product   enol form of product

(15-61)

15.13-1 Preparation of Alpha-Hydroxy Acids Alpha hydroxy acids can be made by reaction of an α-halogenated acid with dilute base (a typical nucleophilic substitution reaction) followed by acidification of the solution. Lactic acid, $CH_3-\underset{OH}{CH}-COOH$, which is a constituent in sour milk, can be formed in this manner.

Sec. 15.14 PHENOLIC ACIDS

$$CH_3-\underset{Br}{CH}-COOH \xrightarrow{\text{dil. NaOH}^{\ominus}} CH_3-\underset{OH}{CH}-\overset{O}{\overset{\|}{C}}-O^{\ominus}Na^{\oplus} \xrightarrow{H^{\oplus}} CH_3-\underset{OH}{CH}-COOH$$

lactic acid
(α-hydroxypropionic acid)

(15-62)

Lactic acid is an important compound biologically. It is formed by enzymatic reactions during the course of carbohydrate metabolism in the human body. It is also produced in muscle tissue by enzymatic reduction of pyruvic acid ($CH_3-\overset{O}{\overset{\|}{C}}-\overset{O}{\overset{\|}{C}}-OH$), and its accumulation is believed to be responsible for slow reflexes due to "tired" muscles.

Alpha hydroxy acids can also be prepared by hydrolysis of cyanohydrins as in Section 14.8-1. Oxidation of α-hydroxy acids produces the corresponding α-keto acid.

$$CH_3-\underset{OH}{CH}-COOH \xrightarrow{\text{oxid.}} CH_3-\overset{O}{\overset{\|}{C}}-COOH \quad (15\text{-}63)$$

lactic acid   pyruvic acid
(α-ketopropionic acid)

15.13-2 Preparation of Alpha-Aminoacids Alpha amino acids can be prepared by nucleophilic attack of ammonia on α-halogenated acids.

$$CH_3-\underset{Br}{CH}-COOH \xrightarrow{NH_3} CH_3-\underset{NH_2}{CH}-COOH \quad (15\text{-}64)$$

α-bromopropionic acid α-aminopropionic acid (alanine)

15.13-3 Preparation of Dicarboxylic Acids The reaction of α-halogenated acids with cyanide ion produces an α-cyano acid, which can be hydrolyzed to form a dicarboxylic acid.

$$\underset{Cl}{CH_2}-COOH \xrightarrow{CN^{\ominus}} \underset{CN}{CH_2}-COOH \xrightarrow[H^{\oplus}]{H_2O} HOOC-CH_2-COOH \quad (15\text{-}65)$$

chloroacetic acid   cyanoacetic acid   malonic acid

15.14 Preparation of Phenolic Acids

The most important member of this series of compounds is salicylic acid prepared by heating sodium phenoxide with carbon dioxide under a pressure of 4-7 atmospheres. The reaction is known as the **Kolbe reaction**.

250   CARBOXYLIC ACIDS AND THEIR DERIVATIVES   Ch. 15

$$\text{sodium phenoxide} \xrightarrow[\substack{\text{NaOH,} \\ 125°\text{C,} \\ 4-7 \text{ atm}}]{CO_2} \text{(sodium salicylate intermediate)} \xrightarrow{H^\oplus} \text{salicylic acid (o-hydroxybenzoic acid)} \quad (15\text{-}66)$$

The phenolic acids undergo all the characteristic reactions of the carboxylic acids and phenols. Methyl salicylate or oil of wintergreen can be prepared by simply esterifying salicylic acid with methyl alcohol.

$$\text{salicylic acid} + CH_3OH \xrightleftharpoons{H^\oplus} \text{methyl salicylate (oil of wintergreen)} + H_2O \quad (15\text{-}67)$$

Treatment of salicylic acid with acetyl chloride or acetic anhydride acetylates the phenolic hydroxyl group to form acetylsalicylic acid, the sodium salt of which is **aspirin**.

$$\text{salicylic acid} + CH_3\text{-}\underset{\underset{O}{\|}}{C}\text{-}Cl \longrightarrow \text{acetylsalicylic acid} + HCl\uparrow \xrightarrow{Na^\oplus} \text{sodium acetylsalicylate (aspirin)}$$

$$(15\text{-}68)$$

15.15   Maleic and Fumaric Acids

Two interesting organic compounds are the α,β-unsaturated acids, maleic and fumaric acid. These compounds have the same molecular formula, $C_4H_4O_4$. Both are dibasic acids and are obtained by heating malic acid, a naturally occurring constituent in many fruits.

$$\underset{\text{malic acid}}{\begin{array}{c} \text{(HO)-CH-COOH} \\ \text{(H)-CH-COOH} \end{array}} \xrightarrow[-H_2O]{160°\text{C}} \underset{\substack{\text{maleic acid} \\ (cis)}}{\begin{array}{c} \text{H-C-COOH} \\ \| \\ \text{H-C-COOH} \end{array}} + \underset{\substack{\text{fumaric acid} \\ (trans)}}{\begin{array}{c} \text{H-C-COOH} \\ \| \\ \text{HOOC-C-H} \end{array}} \quad (15\text{-}69)$$

SUMMARY

Maleic and fumaric acid are related to each other as geometrical isomers due to the restricted rotation about the carbon–carbon double bond in the molecule. Maleic acid is the *cis* isomer with both the carboxyl groups being located on the same side of the C=C plane in the molecule. Fumaric acid is the *trans* acid. The *cis* structure was assigned to maleic acid, mainly on experimental evidence showing that this acid could form an anhydride easily by loss of water from the two carboxyl groups.

$$\begin{array}{c}\text{H-C-COOH}\\ \parallel \\ \text{H-C-COOH}\end{array} \xrightarrow[\text{vacuum}]{140°\text{C}} \text{H}_2\text{O} + \begin{array}{c}\text{H-C-C}\overset{\text{O}}{\parallel}\\ \parallel \quad\quad \text{O}\\ \text{H-C-C}\underset{\text{O}}{\parallel}\end{array} \qquad (15\text{-}70)$$

maleic acid   maleic anhydride

In the *trans* isomer (fumaric acid) the carboxyl groups were too far from each other (on opposite sides of the molecule) to lose water and form a stable anhydride. Of the two isomers, the *trans* (fumaric acid) is the more stable. The *cis* and *trans* isomers also have different biological activity, fumaric acid being a metabolic reaction intermediate, whereas maleic acid is found to be toxic.

Summary

- The carboxylic acids are characterized by the presence of a carboxyl group,
$-\overset{\overset{\displaystyle O}{\parallel}}{C}-\text{OH}$, in the molecule.
- The nomenclature of the carboxylic acids is discussed in Section 15.2.
  - (a) The IUPAC ending is *-oic*.
  - (b) Many acids are better known by their common names.
  - (c) Can be named by indicating presence of substituents with Greek letters, α, β, etc., along with common name.
- The acidity of the carboxylic acids has been discussed in Chapters 4 and 5. Both inductive and resonance effects play a vital role in determining the acidity of a molecule. The resonance stabilization of the carboxylate anion is extremely important in accounting for the acidity.
- The rather high boiling points of the carboxylic acids may be attributed to hydrogen bonding.
- The carboxylic acids may be prepared by:
  - (a) Oxidation of a primary alcohol, aldehyde, or aromatic side chain.
  - (b) Hydrolysis of nitriles or *gem*-trihalides.
  - (c) Reaction of a Grignard reagent with carbon dioxide.
- The principal reactions of the carboxylic acids and their derivatives are nucleophilic substitution reactions. These include formation of acyl halides, esters, anhydrides, and amides from carboxylic acids.

- Esterification and hydrolysis are the reverse reactions of one another. Isotopic labeling has been used experimentally to elucidate this and many other mechanisms.
- The reactions of the carboxylate salts include:

    (a) Heating ammonium salts to form amides.

    (b) Thermal decomposition of calcium or barium salt of dibasic acid to form a cyclic ketone.

- The acyl halides, esters, anhydrides, and amides can undergo hydrolysis, alcoholysis, and ammonolysis reactions (see Section 15.9-2).
- Acyl halides can be used to prepare aldehydes (Rosenmund reduction) and ketones via the Friedel–Crafts ketone synthesis.
- Esters with α-hydrogen atoms can undergo Claisen condensations.
- Electrophilic aromatic substitution occurs on aromatic carboxylic acids. The COOH group is a *meta*-director.
- The HVZ reaction can be used to prepare α-halogenated acids which can be converted into other types of substituted carboxylic acids.
- The Kolbe reaction can be used to prepare phenolic acids.
- Maleic and fumaric acids are unsaturated acids that illustrate the phenomenon of geometrical isomerism.

PROBLEMS

1 Draw the correct structural formulas for each of the following compounds.

(a) *i*-butyric acid

(b) 2,3-dibromohexanoic acid

(c) *p*-nitrobenzamide

(d) γ-hydroxybutyric acid

(e) barium acetate

(f) phenyl benzoate

(g) butanoic anhydride

(h) β-naphthoic acid

(i) *n*-butyl isobutyrate

(j) ethyl α-chloroacetate

(k) butanoyl fluoride

(l) 3,5-dinitrobenzoyl chloride

2 Name the following compounds.

(a) 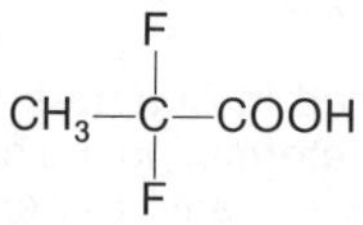 CH_3-CF_2-COOH

(b) 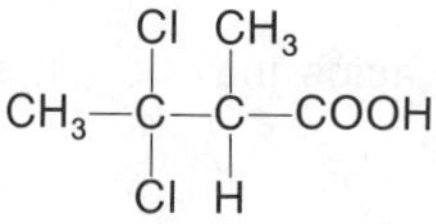 $CH_3-CCl_2-C(Cl)(CH_3)-COOH$

(c) cyclopropyl-COOH

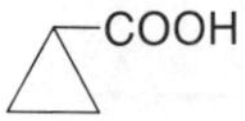

(d) $CH_3-CH=CH-COOH$

(e) 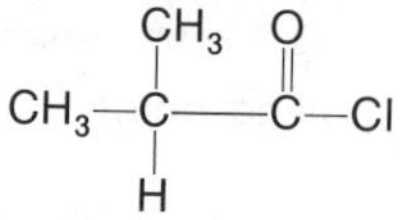 $CH_3-CH(CH_3)-C(=O)-Cl$

(f) 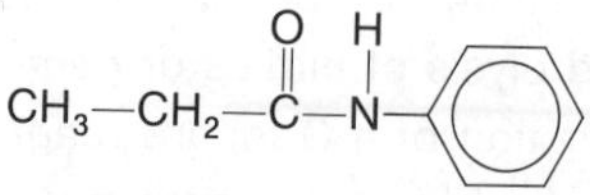 $CH_3-CH_2-C(=O)-NH-C_6H_5$

(g) $CH_3-CH_2-C\equiv N$

PROBLEMS

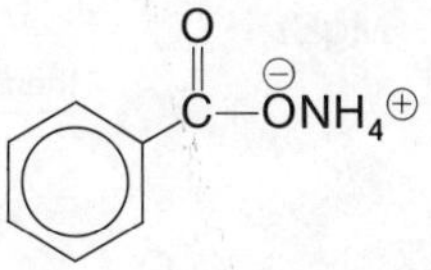

(h) 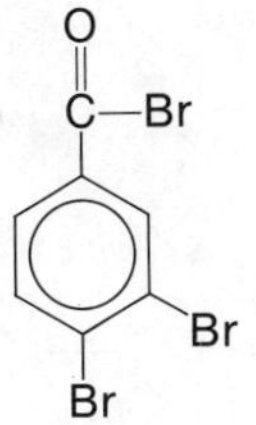 3,4-dibromobenzoyl bromide (benzene ring with C(=O)Br and two Br substituents)

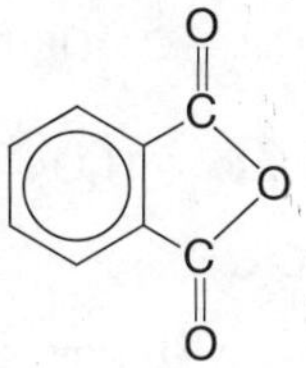

(i) $CH_3-\underset{\underset{O}{\|}}{C}-CH_2-COOH$

(j) 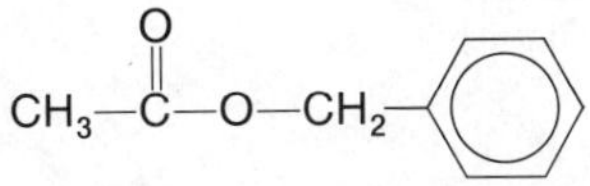 $CH_3-\underset{\underset{O}{\|}}{C}-O-CH_2-C_6H_5$

(k) $C_6H_5-\underset{\underset{O}{\|}}{C}-O^{-}NH_4^{+}$

(l) phthalic anhydride

3 Complete each of the following equations by writing the correct structural formula(s) of the product(s) formed. If no reaction occurs, write N.R.

(a) $CH_3-CH_2-\underset{\underset{H}{|}}{\overset{\overset{CH_3}{|}}{C}}-CH_2OH \xrightarrow[H_2SO_4]{K_2Cr_2O_7}$

(b) propanoic acid + KOH ⟶

(c) $CH_3-CH_2-CN \xrightarrow[H^{\oplus}]{H_2O}$

(d) 2-methylbutanoic acid + ethyl alcohol $\xrightarrow{H^{\oplus}}$

(e) benzoyl chloride + $NH_3 \xrightarrow{\Delta}$

(f) 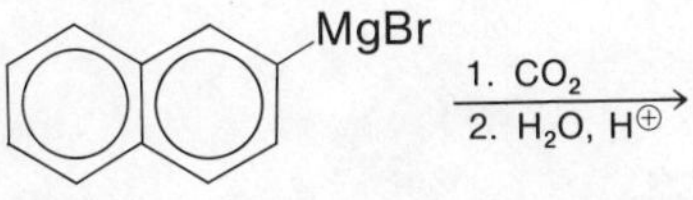 2-naphthyl-MgBr $\xrightarrow[\text{2. }H_2O, H^{\oplus}]{\text{1. }CO_2}$

(g) ethanoic acid + PI_3 ⟶

(h) 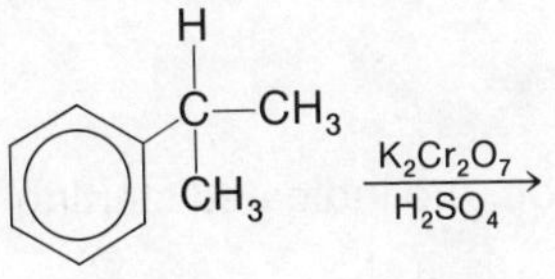 $C_6H_5-\underset{\underset{CH_3}{|}}{\overset{\overset{H}{|}}{C}}-CH_3 \xrightarrow[H_2SO_4]{K_2Cr_2O_7}$

(i) 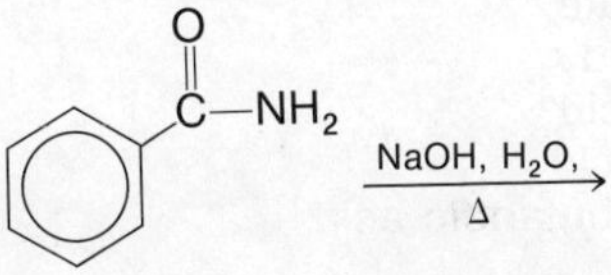 $C_6H_5-\underset{\underset{O}{\|}}{C}-NH_2 \xrightarrow[\Delta]{NaOH, H_2O,}$

(j) propionic acid $\xrightarrow[Cl_2]{P}$

(k) C₆H₅—MgBr + SO₂ $\xrightarrow{\text{then } H_2O, H^{\oplus}}$

(l) $CH_3-CH_2-CH_2-\overset{O}{\underset{\|}{C}}-O^{\ominus}K^{\oplus}$ + $CH_3-CH_2-CH_2-\overset{O}{\underset{\|}{C}}-Br \longrightarrow$

(m) methyl benzoate + H₂O $\xrightarrow{H^{\oplus}}$

(n) $CH_3-\underset{\underset{CH_3}{|}}{\overset{\overset{CH_3}{|}}{C}}-\overset{O}{\underset{\|}{C}}-OH \xrightarrow{\underset{Br_2}{P}}$

(o) $CH_3-\underset{\underset{Cl}{|}}{CH}-\overset{O}{\underset{\|}{C}}-OH + CN^{\ominus} \longrightarrow$

(p) C₆H₆ + $CH_3-\overset{O}{\underset{\|}{C}}-Cl \xrightarrow{AlCl_3}$

(q) p-cresol (HO—C₆H₄—CH₃) + CO₂ $\xrightarrow[\Delta, \text{press.}]{NaOH}$

(r) C₆H₅—COOH $\xrightarrow[Br_2]{Fe}$

(s) $CH_3-CH_2-\overset{O}{\underset{\|}{C}}-Br \xrightarrow[S, \text{quinoline}]{H_2, Pd-BaSO_4}$

(t) $CH_3-\overset{O}{\underset{\|}{C}}-OCH_3 + CH_3CH_2OH \xrightarrow{OH^{\ominus}}$

4 Perform all of the following syntheses from the indicated starting materials and any necessary inorganic reagents.

(a) inorganic reagents to ethyl acetate
(b) *i*-propyl bromide to butanoic acid
(c) inorganic reagents to malonic acid
(d) inorganic reagents to succinic acid
(e) inorganic reagents to 2-hydroxybutanoic acid
(f) benzene to *n*-propyl benzene
(g) *benzene to ethyl cinnamate
(h) benzene to benzophenone oxime

PROBLEMS

(i) ethyl bromide to acetic anhydride
(j) n-propyl alcohol to 2-butynoic acid
(k) phenylacetic acid to phenylmalonic acid
(l) benzene to 4-chlorobenzamide

5 Which of the following pairs of compounds is the more acidic?
  (a) α-chlorobutyric acid or α-iodobutyric acid
  (b) acetic acid or trimethylacetic acid
  (c) 2-iodopropionic acid or 5-fluoroheptanoic acid
  (d) 2-nitrobenzoic acid or o-toluic acid (2-methylbenzoic acid)

6 *Which is more acidic, benzoic acid or benzamide? Explain.

7 Arrange the following compounds in increasing order towards nucleophilic substitution. Explain the basis for your answer.

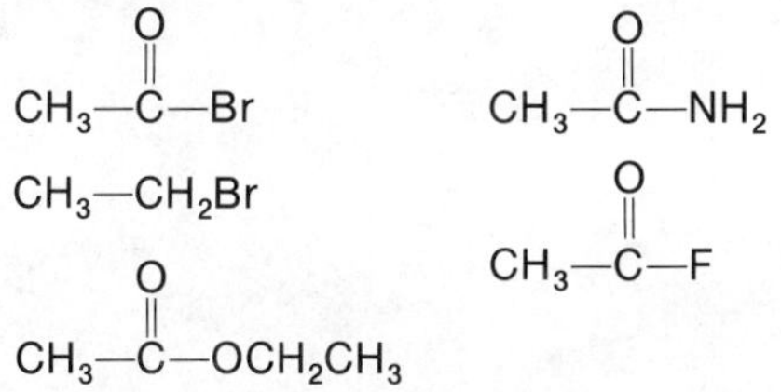

$$CH_3-\overset{O}{\underset{\|}{C}}-Br \qquad CH_3-\overset{O}{\underset{\|}{C}}-NH_2$$

$$CH_3-CH_2Br$$

$$CH_3-\overset{O}{\underset{\|}{C}}-F$$

$$CH_3-\overset{O}{\underset{\|}{C}}-OCH_2CH_3$$

8 Arrange the following compounds in increasing order towards the rate of esterification with a given alcohol or acid.

(a) 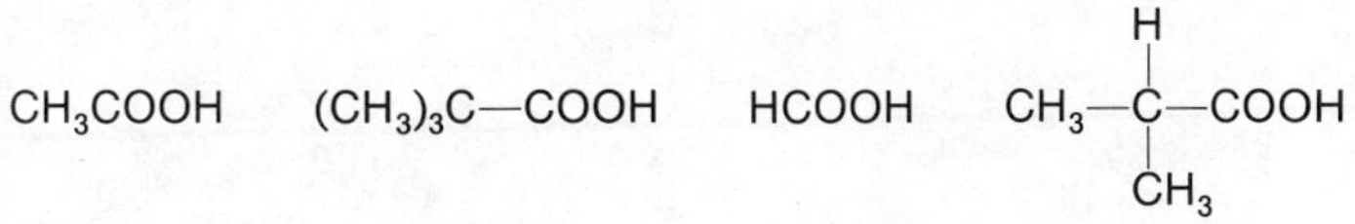 CH_3COOH $(CH_3)_3C-COOH$ $HCOOH$ $CH_3-\underset{CH_3}{\overset{H}{\underset{|}{C}}}-COOH$

(b) 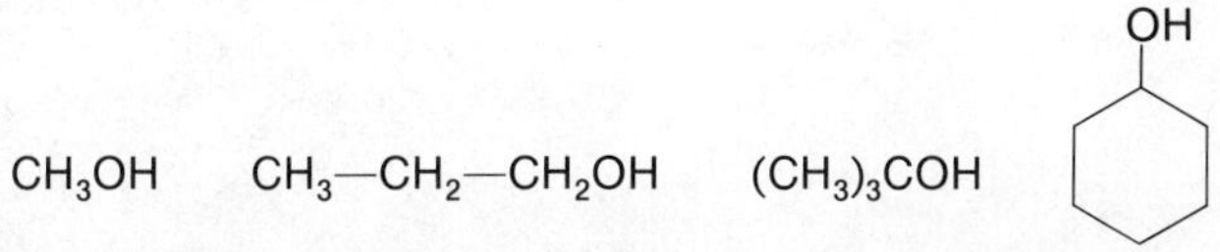 CH_3OH $CH_3-CH_2-CH_2OH$ $(CH_3)_3COH$ cyclohexanol

9 Propose a detailed mechanism for the esterification of α-bromoacetic acid with ethyl alcohol to form the ester ethyl α-bromoacetate.

10 *Suggest detailed reaction mechanisms for the following reactions.

(a) 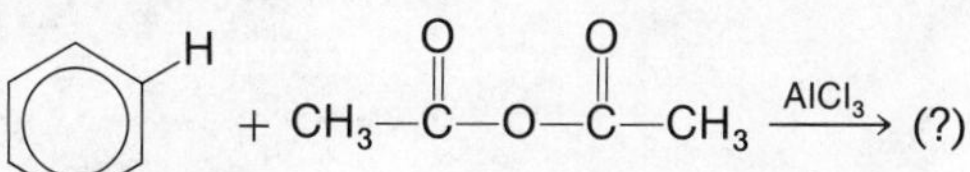 benzene + $CH_3-\overset{O}{\underset{\|}{C}}-O-\overset{O}{\underset{\|}{C}}-CH_3 \xrightarrow{AlCl_3}$ (?)

(b) Acid catalyzed hydrolysis of propionamide
(c) Ammonolysis of ethyl benzoate

11 An ester has a saponification equivalent (equivalent weight) of 150. Hydrolysis of the ester produces an acid and alcohol. The acid has a neutralization equivalent of 122. What is the structural formula of the original ester?

12 *Compound A has a neutralization equivalent of 116. Compound A forms a semicarbazone or phenylhydrazone, and gives a positive iodoform test. When A is treated with PCl_5, compound B is formed, which can undergo a Rosenmund reduction to compound C. A Clemmensen reduction of C forms n-pentane as the reaction product. From this information, deduce and write the correct structural formulas of compounds A, B, and C.

16

Amines, Diazonium Salts, and Dyes

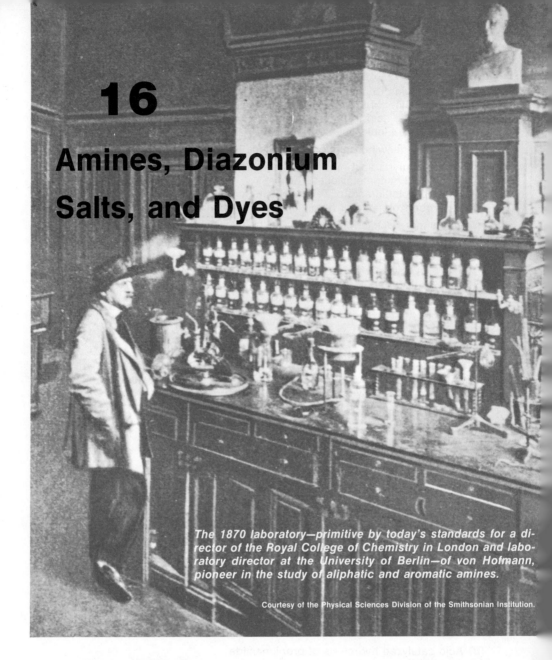

The 1870 laboratory—primitive by today's standards for a director of the Royal College of Chemistry in London and laboratory director at the University of Berlin—of von Hofmann, pioneer in the study of aliphatic and aromatic amines.

Courtesy of the Physical Sciences Division of the Smithsonian Institution.

16.1 Introduction

The organic chemistry of nitrogen is extremely interesting and complex. Numerous classes of organic compounds contain nitrogen, and some of these classes such as the amides and nitriles have already been discussed. In this chapter we will concentrate on several other types of organic nitrogen compounds, paying particular attention to the **amines.**

Amines are found in nature in many plants and animals. They are believed responsible for the odor of decaying fish, and the characteristic odor of olive oil is due to amines. They are used to manufacture many drugs, medicinal products,

Sec. 16.2 NOMENCLATURE OF AMINES

and dyes. Amines represent the family of organic compounds with the most basic properties and their basicity has already been referred to in Chapter 5.

16.2 Nomenclature and Classification of Amines

The amines are related structurally to, and are considered to be derivatives of ammonia, NH_3, where the hydrogen atoms have been replaced by organic groups. Amines are classified as primary, secondary, or tertiary, according to the number of groups attached to the nitrogen atom. The groups may be aliphatic, aromatic, or the nitrogen atom may be found in the form of a heterocyclic ring system.

Some of the properties of amines, such as basicity, are a characteristic of all the different types of amines. However, many of the properties of amines depend on the number of hydrogen atoms attached to the nitrogen atom, that is to say whether an amine is primary, secondary, or tertiary.

$$\begin{array}{ccc}
\text{H} & \text{H} & \text{R}'' \\
| & | & | \\
\text{R}-\text{N}-\text{H} & \text{R}-\text{N}-\text{R}' & \text{R}-\text{N}-\text{R}' \\
\text{primary (1°)} & \text{secondary (2°)} & \text{tertiary (3°)}
\end{array}$$

The chemistry of the amines is also somewhat dependent on whether the amine is aliphatic, aromatic, or heterocyclic.

Aliphatic amines are named by naming the alkyl groups attached to the nitrogen atom, followed by the suffix *amine*. More complex amines are named by using the prefix *amino-*, with a capital *N* to indicate the groups attached to the nitrogen atom. The following examples illustrate these points.

$$\begin{array}{ccc}
\text{H} & \text{H} & \text{CH}_3 \\
| & | & | \\
\text{CH}_3-\text{N}-\text{H} & \text{CH}_3-\text{N}-\text{CH}_2-\text{CH}_3 & \text{CH}_3-\text{CH}_2-\text{N}-\text{CH}_3 \\
\text{methylamine (1°)} & \text{methylethylamine (2°)} & \text{dimethylethylamine (3°)}
\end{array}$$

$$\begin{array}{cc}
 & \text{H} \\
 & | \\
\text{H}_2\text{N}-\text{CH}_2-\text{CH}_2\text{OH} & \text{CH}_3-\text{N}-\text{CH}-\text{CH}_2-\text{CH}_2-\text{CH}_3 \\
 & \quad\quad\quad | \\
 & \quad\quad\quad \text{CH}_3 \\
\text{2-aminoethanol} & \text{2-(N-methylamino)pentane (2°)} \\
\text{(ethanolamine) (1°)}
\end{array}$$

Aromatic amines, where the nitrogen atom is directly attached to an aromatic nucleus, are usually named as derivatives of the simplest aromatic amine, **aniline**. Some aromatic amines have special names, such as the aminotoluenes, which are called **toluidines**.

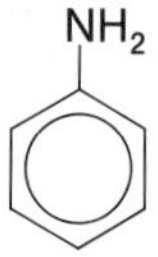

258   **AMINES, DIAZONIUM SALTS, AND DYES   Ch. 16**

aniline
(phenylamine) (1°)

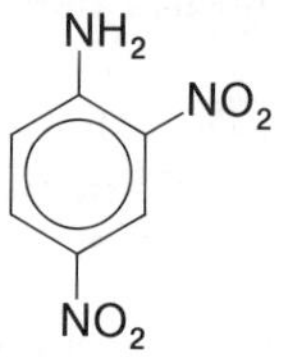

2,4-dinitroaniline (1°)

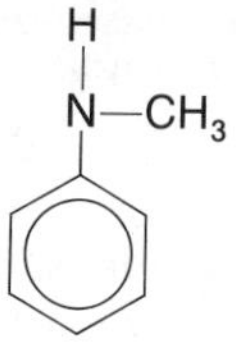

N-methylaniline (2°)

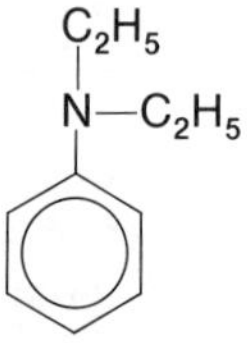

N,N-diethylaniline (3°)

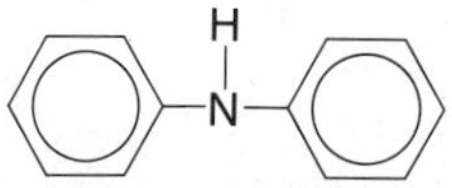

diphenylamine (2°)

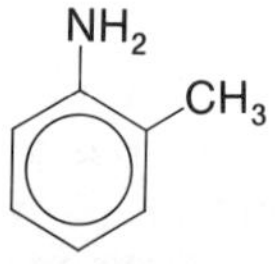

o-toluidine (1°)

Heterocyclic amines are usually named by their common names. Since there are many heterocyclic ring systems, each ring system has its own individual name, and in the numbering system used for substituents the hetero atom is usually the number one position. The numbering continues around the ring, usually counterclockwise.

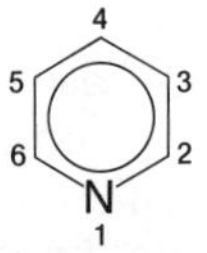

pyridine (3°)

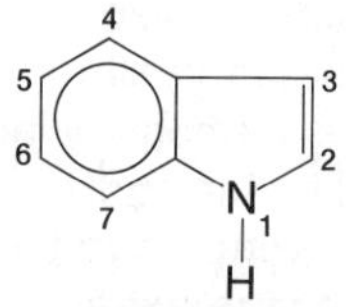

indole (2°)

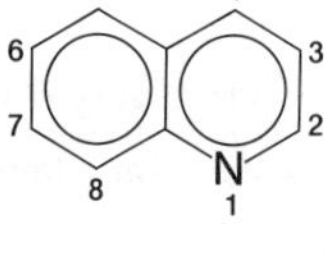

quinoline (3°)

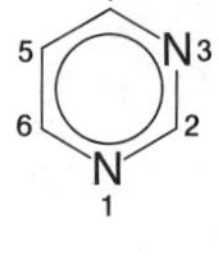

pyrimidine (3°)

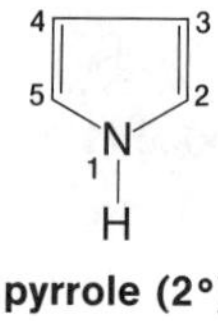

pyrrole (2°)

16.3 Salts of Amines—Nomenclature

Since all amines are basic one would expect that they react with acids to form salts, a reaction that will be discussed later in this chapter. The salts can be considered as being related to the ammonium ($NH_4^{\oplus}$) salts where the hydrogens of the ammonium ion have been replaced by organic groups. Thus, they are named by using the prefix ammonium in place of amine (or anilinium in place of aniline, etc.) and then adding the name of the anion present in the salt.

Sec. 16.4 PHYSICAL PROPERTIES OF AMINES

$$CH_3-\overset{\overset{CH_3}{|}}{\underset{H}{N}}\overset{\oplus}{-}H \; I^{\ominus}$$
dimethylammonium iodide
(dimethylamine hydroiodide)

$$CH_3-\overset{\overset{CH_3}{|}}{\underset{H}{N}}\overset{\oplus}{-}C_2H_5 \; OH^{\ominus}$$
dimethylethylammonium
hydroxide

$$\underset{}{\text{anilinium chloride}}$$ $NH_3Cl^{\ominus}$ on benzene ring
anilinium chloride
(aniline hydrochloride)

N-methylpyridinium iodide

16.4 Physical Properties of Amines

Since the amines are structurally similar to ammonia it is not too surprising to find that the amines and ammonia have many similar properties and characteristics. The lower members of the aliphatic amines are all gases and are very soluble in water (like ammonia). They form basic solutions in water, and the lower molecular weight amines (methylamine and ethylamine in particular) smell very much like ammonia. With increasing molecular weight the amines' solubility in water decreases and their odor becomes less pungent and more "fish-like."

The amines are polar compounds and have a tendency to form intermolecular hydrogen bonds. Although the hydrogen bonding is weaker in amines than in the alcohols or carboxylic acids (N—H-----N vs. O—H-----O), since nitrogen is less electronegative than oxygen the boiling points of the amines are higher than one might expect if hydrogen bonding were not present. Thus, the amines have higher boiling points than non-polar compounds of the same or similar molecular weights, but lower boiling points than the alcohols or carboxylic acids ($CH_3-CH_2-CH_3$, b.p. = $-42°C$; $CH_3-CH_2-NH_2$, b.p. = $17°C$;

CH_3-CH_2-OH, b.p. = $78°C$; $H-\overset{\overset{O}{\|}}{C}-OH$, b.p. = $100.5°C$). The solubility of the amines in water can also be attributed to the phenomenon of hydrogen bonding between the amine and the solvent. Amines are soluble in less polar solvents than water, such as ether, and benzene.

Some aromatic amines are extremely toxic. They are readily absorbed through the skin and can be fatal. The β-naphthylamine, (naphthalene-NH$_2$ structure), is one of the most potent chemical carcinogens known to man.

16.5 Basicity of Amines

The basic strength of the amines is due to the availability of the unshared electron pair on the nitrogen atom for bonding with a proton or other Lewis acid. As bases, amines will react with acids such as hydrochloric acid to form salts.

$$R_3N: + H^{\oplus}Cl^{\ominus} \longrightarrow R_3NH^{\oplus} + Cl^{\ominus} \qquad (16\text{-}1)$$
Base   acid   salt

This reaction is analogous to the reaction of ammonia with acids to form ammonium salts. Like ammonia, the amines will undergo an acid–base reaction with water which can be expressed by

$$R\ddot{N}H_2 + HOH \rightleftharpoons RNH_3^{\oplus} + OH^{\ominus} \qquad (16\text{-}2)$$
base   acid   conj.   conj.
                           acid   base

The ionization constant for this equilibrium can be expressed as

$$K_b = \frac{[RNH_3^{\oplus}][OH^{\ominus}]}{[RNH_2]}$$

the larger the numerical value of K, the stronger the base. In contrast to the strength of acids, the presence of electron-donating groups in a molecule tends to increase the basic strength. For example, dimethylamine ($K_b = 5.1 \times 10^{-4}$) is a stronger base than methylamine ($K_b = 4.4 \times 10^{-4}$) or ammonia ($K_b = 1.8 \times 10^{-5}$) due to the presence of the electron-releasing alkyl groups in the molecule. (Trimethylamine has a $K_b = 5.9 \times 10^{-5}$, which makes it a *weaker* base than either methylamine or dimethylamine. How can you explain this fact, which is not in agreement with predictions based on the inductive effects of the alkyl groups? HINT: Consider the steric factors present in the molecule.)

Aliphatic amines are much stronger bases than the aromatic amines, due to the resonance stabilization present in aromatic compounds. This has been discussed earlier (Section 5.4-2).

16.6 Preparation of Amines

There are numerous synthetic procedures used to prepare amines. However, there is no one general type of reaction that can be used to prepare all classes of amines. Generally, it is far more difficult to prepare a secondary or tertiary amine than to prepare a primary amine. The following discussion will be mainly concerned with the synthesis of primary amines.

The most important synthetic methods used to prepare amines are

1. Reaction of alkyl halides with ammonia.
2. Reduction reactions.
3. Reductive amination.
4. Gabriel phthalimide synthesis.
5. Hofmann degradation of amides.
6. Reaction of aryl halides with sodamide in liquid ammonia.

16.6-1 Reaction of Alkyl Halides with Ammonia Ammonia will react with primary or secondary alkyl halides in a nucleophilic substitution (S_N2) reaction to form amines. The reaction, however, is of little synthetic use, since a mixture of products is obtained. The reaction does not stop at the primary amine stage, but instead, the amine, having an unshared electron pair on the nitrogen atom,

Sec. 16.6 PREPARATION OF AMINES

behaves as a nucleophile and reacts further with the alkyl halide to form a secondary amine. The secondary amine in turn reacts to form a tertiary amine, which finally forms a quaternary ammonium salt.

$$RX + \overset{..}{N}H_3 \xrightarrow{OH^\ominus} R\overset{..}{N}H_2 \xrightarrow[OH^\ominus]{RX} R_2\overset{..}{N}H \xrightarrow[OH^\ominus]{RX} R_3\overset{..}{N} \xrightarrow[OH^\ominus]{RX} R_4\overset{\oplus}{N}X^\ominus$$

| alkyl halide | primary amine | secondary amine | tertiary amine | quaternary ammonium salt |

(16-3)

A mixture of all these products is obtained in which they are often difficult to separate, and for this reason a *pure* primary aliphatic amine usually cannot be prepared in this way. Using a large excess of ammonia favors the production of the primary amine, but nevertheless some of the other products are still formed. Tertiary alkyl halides cannot be used in this reaction, since in the presence of the base ammonia they tend to undergo a competing elimination reaction rather than nucleophilic substitution. Because of all these difficulties and limitations, the alkylation of ammonia to amines has little synthetic value.

Aryl halides do *not* react with ammonia or amines in a similar reaction *unless* electron-withdrawing substituents are present on the ring.

picryl chloride (2,4,6-trinitrochlorobenzene) → 2,4,6-trinitroaniline (16-4)

Aryl halides containing no other substituents, such as chlorobenzene, will *not* undergo this nucleophilic substitution reaction since the halogen is involved in the resonance stabilization of the molecule, which would temporarily be destroyed during the course of the reaction.

However, aniline will react with alkyl halides to form *N*-alkyl derivatives.

aniline + $CH_3Cl \xrightarrow{NaOH}$ *N*-methylaniline + $NaCl + H_2O \xrightarrow[NaOH]{CH_3Cl}$ (16-5)

N,N-dimethylaniline + $NaCl + H_2O$ (16-6)

16.6-2 Reduction Reactions to Prepare Amines

Many organic compounds containing nitrogen can be reduced to form amines by various reducing agents. These compounds include the nitro compounds, amides, oximes, and nitriles. The reductions vary from catalytic hydrogenation over a metal catalyst to chemical hydrogenation, using such reducing agents as iron or tin and hydrochloric acid, ammonium bisulfide, and lithium aluminum hydride.

The reduction of nitro compounds is by far the most useful method for preparing aromatic amines. The major reason for this is that aromatic nitro compounds are readily available by nitration of the aromatic ring, and reduction of these nitro compounds forms primary aromatic amines that are themselves extremely important compounds, in that they can be converted into many different types of aromatic compounds by suitable synthetic routes.

$$ArNO_2 \xrightarrow{\text{red. agent}} ArNH_2 \quad (16\text{-}7)$$
(or R—NO$_2$)   (or R—NH$_2$)
nitro compound   primary amine

$$\text{nitrobenzene} \xrightarrow{\text{Sn}/\text{HCl}} \text{aniline} \quad (16\text{-}8)$$

The reducing agent ammonium bisulfide has the rather unique ability to selectively reduce one nitro group in compounds where more than one nitro group is present in the molecule.

$$m\text{-dinitrobenzene} \xrightarrow{NH_4SH} m\text{-nitroaniline} \quad (16\text{-}9)$$

The reduction of nitro groups to amines is a most essential step in one of the most important reaction sequences in organic chemistry. Aromatic nitro compounds can be reduced to primary aromatic amines, which, as we shall see later in this chapter, can be converted to diazonium salts, which in turn can be converted into many different types of aromatic compounds not readily available by other synthetic routes. The reduction of aliphatic nitro compounds is not as important since the starting materials are not easily obtained.

Amides can be readily reduced to amines. Unsubstituted amides yield primary amines, whereas N-substituted amides form secondary amines, and N,N-disubstituted amides form tertiary amines.

$$CH_3-\underset{\underset{O}{\|}}{C}-NH_2 \xrightarrow{LiAlH_4} CH_3-CH_2-NH_2 \quad (16\text{-}10)$$
acetamide   ethylamine
                    (1°)

Sec. 16.6 PREPARATION OF AMINES

$$\text{acetanilide} \xrightarrow{\text{LiAlH}_4} \text{N-ethylaniline (2°)} \quad (16\text{-}11)$$

$$\text{N-ethylacetanilide} \xrightarrow{\text{LiAlH}_4} \text{N,N-diethylaniline (3°)} \quad (16\text{-}12)$$

The overall reaction involves the reduction of the carbonyl group ($\overset{\text{O}}{\underset{\|}{\text{C}}}$) of the amide to a methylene group (CH$_2$).

Oximes can also be reduced to primary amines by lithium aluminum hydride.

$$\underset{\text{acetaldoxime}}{CH_3-\overset{H}{\underset{|}{C}}=N-OH} \xrightarrow{\text{LiAlH}_4} \underset{\text{ethylamine}}{CH_3-CH_2-NH_2} \quad (16\text{-}13)$$

The reduction of nitriles also produces primary amines. Catalytic hydrogenation or chemical reducing reagents, such as sodium metal in alcohol, can be used to afford this reduction.

$$\underset{\substack{\text{propiononitrile}\\ \text{(ethyl cyanide)}}}{CH_3-CH_2-C\equiv N} \xrightarrow[\substack{\text{Ni, }\Delta,\\ \text{press.}}]{2H_2} \underset{\textit{n}\text{-propylamine}}{CH_3-CH_2-CH_2-NH_2} \quad (16\text{-}14)$$

If a reaction sequence is started with an alkyl halide that is converted into a nitrile, and the nitrile is then reduced to an amine, the amine will contain one more carbon atom in the chain than the original alkyl halide. This is a convenient way of synthesizing amines containing one more carbon atom in the chain than the starting material.

$$\underset{\substack{\text{alkyl}\\ \text{halide}}}{R-X} + CN^\ominus \longrightarrow \underset{\text{nitrile}}{R-C\equiv N} \xrightarrow{\text{redn.}} \underset{\substack{\text{primary amine}\\ \text{(has one more carbon atom)}}}{R-CH_2-NH_2} \quad (16\text{-}15)$$

$$\underset{\substack{\textit{i}\text{-propyl}\\ \text{bromide}}}{CH_3-\overset{CH_3}{\underset{|}{CH}}-Br} \xrightarrow{CN^\ominus} \underset{\substack{\textit{i}\text{-butyronitrile}\\ (\textit{i}\text{-propyl}\\ \text{cyanide})}}{CH_3-\overset{CH_3}{\underset{|}{CH}}-C\equiv N} \xrightarrow[\text{ROH}]{\text{Na}} \underset{\textit{i}\text{-butylamine}}{CH_3-\overset{CH_3}{\underset{|}{CH}}-CH_2-NH_2} \quad (16\text{-}16)$$

16.6-3 Reductive Amination

The catalytic reduction of aldehydes and ketones in the presence of ammonia or an amine is known as **reductive amination.** The reaction can be used to prepare any class of amine, and the reaction has the advantage of controlling the formation of mixtures more easily than in the reaction of alkyl halides with ammonia. The mechanism for the reaction is not clearly understood, but the reaction sequence probably goes through the formation of an **imine** (R—CH=NH, or R—C(R')=NH) as an intermediate.

$$R\text{—CHO} + NH_3 \longrightarrow [R\text{—CH=N—H}] + H_2O \xrightarrow{H_2, Ni} R\text{—CH}_2\text{—NH}_2 \quad (16\text{-}17)$$

aldehyde (or ketone) → imine → primary amine

The aldehyde or ketone can react with the primary amine formed in the reaction to yield a secondary amine as a side-product.

$$R\text{—CHO} + R'\text{—CH}_2\text{—NH}_2 \longrightarrow [R\text{—CH=N—CH}_2R'] + H_2O \xrightarrow{H_2, Ni} R\text{—CH}_2\text{—NH—CH}_2R'$$

aldehyde + primary amine → an imine → secondary amine

$$(16\text{-}18)$$

The tendency for the primary or secondary amine formed to react further with the aldehyde or ketone used can be readily controlled by the relative amounts of the reactants used in the reaction. The overall reaction involves addition of ammonia to the carbonyl group in the aldehyde or ketone, loss of water to form an imine, and the subsequent addition of hydrogen across the carbon–nitrogen double bond in the imine to form the amine.

16.6-4 Gabriel Phthalimide Synthesis of Pure Primary Aliphatic Amines

Reaction of phthalic anhydride with ammonia produces an amide, which when heated loses water, and the product obtained is an **imide** (phthalimide) containing two acyl groups attached to a nitrogen atom in the form of a ring.

$$\text{phthalic anhydride} + NH_3 \xrightarrow{\Delta} H_2O + \text{phthalimide} \quad (16\text{-}19)$$

The hydrogen attached to the nitrogen atom in phthalimide is weakly acidic due to the close proximity of the two electron-withdrawing acyl groups. Phthalimide can be easily converted to its potassium salt by treatment with a base like alcoholic KOH or by reaction with metallic potassium. The salt is then treated with an alkyl halide to form an N–substituted imide, which can be hydrolyzed to form a

Sec. 16.6 PREPARATION OF AMINES

primary aliphatic amine. This reaction sequence is known as the **Gabriel phthalimide synthesis** for pure primary aliphatic amines.

phthalimide → (K, Δ, or alc. KOH) → potassium phthalimide → (RX, Δ) → N-substituted phthalimide → (H$_2$O, OH$^⊖$, Δ) →

R—NH$_2$ + phthalate salt (with two COO$^⊖$ groups on benzene ring)   (16-20)

pure primary amine

The great value of the Gabriel phthalimide synthesis is that it produces *pure* primary aliphatic amines, uncontaminated by secondary or tertiary amines as side products.

16.6-5 The Hofmann Degradation of Amides

The Hofmann degradation of amides to form amines was discovered by the German chemist, A. W. Hofmann in 1881. The reaction involves reacting an amide with an alkaline solution of chlorine or bromine to produce a primary amine. The most important feature of the reaction is that the amine formed has one *less* carbon atom in the molecule than the original amide, and the reaction is a convenient way of *decreasing* the length of a carbon chain by one carbon atom. As we shall see, the reaction mechanism involves a rearrangement, and for this reason the reaction is often referred to as the **Hofmann rearrangement.** Because of this rearrangement the reaction is of considerable theoretical as well as synthetic interest.

$$R-\underset{\text{amide}}{\overset{\overset{O}{\|}}{C}}-NH_2 \xrightarrow[Br_2]{NaOH} \underset{\substack{\text{primary amine} \\ \text{(has one less carbon atom)}}}{R-NH_2} + NaBr + Na_2CO_3 + H_2O \qquad (16-21)$$

$$CH_3-\underset{\text{acetamide}}{\overset{\overset{O}{\|}}{C}}-NH_2 \xrightarrow[Br_2]{NaOH} \underset{\text{methylamine}}{CH_3-NH_2} \qquad (16-22)$$

The mechanism of the Hofmann rearrangement is important since a group migrates from carbon to an adjacent nitrogen atom, and a number of other organic reactions have been found to proceed by a similar mechanism. The reaction is believed to proceed by the following sequence of steps.

(1) $\text{R}-\overset{\text{O}}{\underset{}{\text{C}}}-\ddot{\text{N}}\text{H}_2 + \text{OBr}^{\ominus} \longrightarrow \text{R}-\overset{\text{O}}{\underset{}{\text{C}}}-\ddot{\text{N}}\text{HBr} + \text{OH}^{\ominus}$ (16-23)
    amide   N-haloamide

(2) $\text{R}-\overset{\text{O}}{\underset{}{\text{C}}}-\ddot{\text{N}}\text{HBr} + \text{OH}^{\ominus} \longrightarrow \text{R}-\overset{\text{O}}{\underset{}{\text{C}}}-\overset{\ominus}{\underset{}{\text{N}}}-\text{Br} + \text{H}_2\text{O}$ (16-24)

(3) $\text{R}-\overset{\text{O}}{\underset{}{\text{C}}}-\overset{\ominus}{\underset{}{\text{N}}}-\text{Br} \longrightarrow \left[\text{R}-\overset{\text{O}}{\underset{}{\text{C}}}-\ddot{\text{N}}\right] + \text{Br}^{\ominus}$ (16-25)

(4) $\text{R}-\overset{\text{O}}{\underset{}{\text{C}}}-\ddot{\text{N}} \longrightarrow \text{R}-\ddot{\text{N}}=\text{C}=\text{O}$ (16-26)
                                alkyl isocyanate

(5) $\text{R}-\ddot{\text{N}}=\text{C}=\text{O} + \text{HOH} \longrightarrow \left[\text{R}-\text{NH}-\overset{\text{O}}{\underset{}{\text{C}}}-\text{OH}\right] \longrightarrow$
                              carbamic acid

$\quad\quad\quad\quad\quad\quad\quad\quad\quad\quad\quad\quad\text{CO}_2 \uparrow + \quad \text{RNH}_2$ (16-27)
$\quad\quad\quad\quad\quad\quad\quad\quad\quad\quad\quad\quad\quad\quad\quad\quad$ primary amine

(6) $\quad 2\text{NaOH} + \text{CO}_2 \longrightarrow \text{Na}_2\text{CO}_3 + \text{H}_2\text{O}$ (16-28)

The first step of the proposed mechanism is the base-catalyzed halogenation of an amide. This is a known reaction. The hypobromite ($\text{OBr}^{\ominus}$) ion is formed by reaction of the sodium hydroxide with the bromine.

$$2\text{NaOH} + \text{Br}_2 \longrightarrow \text{NaBr} + \text{NaOBr} + \text{H}_2\text{O} \quad (16\text{-}29)$$

The second step in the reaction involves removal of the weakly acidic hydrogen (due to the presence of the strongly electron-withdrawing bromine) from the N-haloamide by the basic hydroxide ion. Step (3) is merely the departure of bromide ion, a good leaving group, to form a hypothetical species with a nitrogen atom having only six electrons around it. The electron-deficient nitrogen atom needing two more electrons to complete its outer shell is extremely reactive and is the driving force for the rearrangement that follows. In step (4) the rearrangement occurs, the electron-deficient nitrogen atom stabilizes itself by migration of the R group with its electrons from carbon to nitrogen. Steps (3) and (4) are believed to occur simultaneously. The isocyanate formed can actually be isolated out of the reaction mixture at this point, but usually is hydrolyzed (step 5) to form a carbamic acid. The carbamic acids are extremely unstable substances. They cannot be isolated, and they lose carbon dioxide readily to form a primary amine. The amine contains one less carbon atom in its skeleton due to the loss of CO_2. The CO_2 is converted to carbonate ion and water in the basic reaction medium (step 6).

The mechanism is supported by much experimental evidence, and especially because the rearrangement step is similar to the rearrangement of carbonium ions where a group with its electrons migrates to an electron-deficient atom as in a 1,2-shift.

Sec. 16.6 PREPARATION OF AMINES

16.6-6 Reaction of Aryl Halides with Sodamide in Liquid Ammonia—Benzyne

Aryl halides are comparatively unreactive towards nucleophilic substitution reactions. This has been discussed earlier in the text. Chlorobenzene, for example, *will not react under ordinary conditions* with sodium hydroxide or ammonia to form phenol or aniline respectively.

$$C_6H_5Cl \xrightarrow{NH_3} \!\!\!\!\!\!/\!\!\!\!\!\!\!\!\!\!\!\!\!\! \to C_6H_5NH_2 \qquad (16\text{-}30)$$

However, in the presence of a very strong base, such as sodamide in liquid ammonia, a rather special reaction occurs in that the halogen is *easily* replaced by the nucleophilic amide ion. This reaction was first studied and reported by Professor John D. Roberts at the California Institute of Technology in 1953. The reaction is rather unique in that the experimental evidence suggests that the reaction proceeds through the formation of a hypothetical species known as *benzyne, a benzene ring containing a triple bond*. Although benzyne is too reactive and unstable to be isolated, there is much experimental evidence and reason to substantiate its rather short-lived existence (a few millionths of a second).

$$C_6H_5X \xrightarrow[\text{liq. NH}_3,\ -33°C]{\text{NaNH}_2} [\text{benzyne}] \xrightarrow[\text{liq. NH}_3]{NH_2^{\ominus}} C_6H_5NH_2 \qquad (16\text{-}31)$$

The overall reaction is actually an example of a nucleophilic aromatic substitution reaction leading to the formation of an aromatic amine. Aniline can be made very easily from chlorobenzene in this manner.

$$C_6H_5Cl \xrightarrow[\text{liq. NH}_3,\ -33°C]{\text{NaNH}_2} C_6H_5NH_2 \qquad (16\text{-}32)$$

Reactions which involve the formation of a benzyne-type reaction transition state are said to proceed by an **elimination–addition mechanism.** Can you suggest a reason for this descriptive phrase for the reaction?

It is worthwhile to mention that the benzyne structure is that of a symmetrical intermediate in which the amine function ($NH_2^{\ominus}$) can attack (add to) either end of the triple bond. The experimental evidence for this was obtained by the use of radioactive isotopes in labeling experiments performed by Roberts.

Labeled chlorobenzene in which C^{14} (a radioactive isotope of carbon) was strategically placed at the ring carbon atom to which the chlorine atom was attached, was reacted with sodamide in liquid ammonia. The product, which was aniline, was examined and it was found that in approximately half the aniline the amino group was attached to the C^{14} ring carbon atom, and in the other half the NH_2 group was held by an adjacent ring carbon atom.

268   AMINES, DIAZONIUM SALTS, AND DYES   Ch. 16

$$\text{C}_6\text{H}_4\text{Cl} \xrightarrow[\text{liq. NH}_3, -33°\text{C}]{\text{NaNH}_2} \text{C}_6\text{H}_5\text{NH}_2 + \text{C}_6\text{H}_5\text{NH}_2 \qquad (\text{* designates C}^{14}) \qquad (16\text{--}33)$$

(47%)   (53%)

The interpretation of the above experimental result could only be that in benzyne the labeled carbon and the ones directly adjacent to it become equivalent (in other words, benzyne is a symmetrical species), and the $\text{NH}_2^\ominus$ adds randomly to either one of the carbon atoms. The result of the isotopic tracer experiment is seen in the ratio of the products obtained in the previous equation.

16.7 Reactions of Amines

The chemical reactivity of the amines is mainly due to the tendency for the nitrogen to share its pair of electrons with various electron-seeking reagents. The basicity of amines, their behavior as nucleophilic reagents, and the extremely high reactivity of aromatic rings with amino substituents, can be attributed to the unshared electron pair on the nitrogen atom.

16.7-1 Salt Formation (Basicity)

Salt formation involves the donation of the unshared electron pair on the nitrogen of the amine to a proton from an acid.

$$\underset{\text{amine}}{\text{R}\ddot{\text{N}}\text{H}_2} + \text{HX} \longrightarrow \underset{\text{salt}}{\text{RNH}_3^\oplus} + \text{X}^\ominus \qquad (16\text{--}34)$$

The reaction is related to the basicity of the amine. Most higher molecular weight amines are insoluble in water, whereas the ionic salts are water soluble. Since the amines are basic they can be extracted with an acid from a reaction mixture. The resultant salt formed is water soluble, and the free amine can be regenerated simply by making the solution basic. The amines can then be extracted out of the solution with ether and can be separated from neutral or acidic components in a mixture in this way.

$$\underset{\text{salt}}{\text{RNH}_3^\oplus \text{X}^\ominus} + \text{Na}^\oplus \text{OH}^\ominus \longrightarrow \text{NaX} + \text{H}_2\text{O} + \underset{\text{amine}}{\text{RNH}_2} \qquad (16\text{--}35)$$

All classes of amines, whether primary, secondary, or tertiary, will form salts when treated with an acid solution. An exception is triphenylamine which is nearly neutral and does not form salts even with hot concentrated acid solutions.

16.7-2 Quaternary Ammonium Salts

The salts formed by reaction of amines with acid can be considered as being substituted ammonium compounds, where organic groups have replaced one or more of the hydrogens of the ammonium ion, $\text{NH}_4^\oplus$.

As we have seen in our discussion of the preparation of amines, all amines have a tendency to undergo a nucleophilic substitution reaction with alkyl halides. The reaction can be considered to be an alkylation of the amine by the alkyl halide where the hydrogen atoms attached to the amine nitrogen can be replaced by alkyl groups from the alkyl halide to form a substituted ammonium salt. Tertiary

Sec. 16.7 REACTIONS OF AMINES

amines react with alkyl halides in this manner to form a compound in which all four of the hydrogens of the ammonium ion have been replaced by organic groups. These compounds are called **quaternary ammonium salts.**

$$R_3N: + R'X \longrightarrow R_3N-R'^{\oplus} + X^{\ominus} \quad (16\text{-}36)$$

tertiary amine | alkyl halide | quaternary ammonium salt

The quaternary ammonium salts have properties very similar to the typical inorganic salts. They are ionic, water-soluble and their aqueous solutions conduct an electric current.

Quaternary ammonium salts can be converted into quaternary ammonium hydroxides, $R_4N^{\oplus}OH^{\ominus}$, by treatment with silver oxide. The precipitate of the silver halide that is formed helps drive the reaction to completion.

$$R_4\overset{\oplus}{N}X^{\ominus} + Ag_2O \longrightarrow AgX\downarrow + R_4\overset{\oplus}{N}OH^{\ominus} \quad (16\text{-}37)$$

quaternary ammonium salt | quaternary ammonium hydroxide

The silver halide precipitate can be removed by filtration, and upon evaporation of the filtrate a white solid is obtained. The solid is the quaternary ammonium hydroxide, which when in aqueous solution is a strong base, comparable in strength to sodium or potassium hydroxide.

Several of the quaternary ammonium hydroxides are important biochemical substances and have physiological activity. For example, trimethyl-β-hydroxyethylammonium hydroxide, commonly known as **choline,** is a constituent part of substances known as phospholipids, which constitute part of the brain and spinal cord. These phospholipids are better known by the name **lecithins,** some of which can be isolated from egg yolk.

$$\underset{\text{trimethyl-}\beta\text{-hydroxyethylammonium hydroxide (choline)}}{CH_3-\underset{\underset{CH_3OH^{\ominus}}{|}}{\overset{\overset{CH_3}{|}}{\underset{\oplus}{N}}}-\underset{\alpha}{CH_2}-\underset{\beta}{CH_2}OH}$$

a typical lecithin:

$$\begin{array}{l} CH_2-O-\overset{O}{\overset{\|}{C}}-R \\ CH-O-\overset{O}{\overset{\|}{C}}-R \\ CH_2-O-\underset{\underset{O^{\ominus}}{|}}{\overset{\overset{O}{\|}}{P}}-O-CH_2-CH_2-\overset{\oplus}{N}(CH_3)_3 \end{array}$$

Choline has biological activity. It is a depressant and has been shown to cause a decrease in blood pressure. When choline is reacted with acetic anhydride the acetate ester, better known as the compound **acetylcholine,** is formed.

$$CH_3-\underset{\underset{CH_3}{|}}{\overset{\overset{CH_3}{|}}{\underset{\oplus}{N}}}-CH_2-CH_2-O-\overset{O}{\overset{\|}{C}}-CH_3OH^{\ominus}$$

acetylcholine

Acetylcholine is an important substance involved in the transmission of nerve impulses in the body. Medical research has shown that acetylcholine has some connection with certain mental disorders related to schizophrenia. This may be expected since both choline and acetylcholine have been shown to be involved in the body's attempt to cope with stress situations.

16.7-3 Acylation of Amines The reaction of ammonia with acyl halides or acid anhydrides to prepare amides as an example of nucleophilic substitution has been discussed in Section 15.9-2. Primary and secondary amines can also behave as nucleophilic reagents because of the unshared pair of electrons on the nitrogen atom of the amine. These amines will react with acyl halides or acid anhydrides, attacking the carbonyl carbon atom, in a manner analogous to ammonia to yield substituted amides.

$$R\ddot{N}H_2 + R'-\overset{O}{\underset{\|}{C}}-Cl \longrightarrow R-\underset{H}{\overset{|}{N}}-\overset{O}{\underset{\|}{C}}-R' + HCl \qquad (16\text{-}38)$$

primary amine   acyl chloride   N-substituted amide

$$R'-\underset{\underset{H}{|}}{\overset{\overset{R''}{|}}{N}}-H + R-\overset{O}{\underset{\|}{C}}-O-\overset{O}{\underset{\|}{C}}-R \longrightarrow R'-\underset{\underset{}{}}{\overset{\overset{R''}{|}}{N}}-\overset{O}{\underset{\|}{C}}-R + R-\overset{O}{\underset{\|}{C}}-OH \quad (16\text{-}39)$$

secondary amine   acid anhydride   N,N-disubstituted amide

The compound acetanilide can be made from aniline and acetic anhydride. It is used as a medicine to treat neuralgia and as a preservative in aqueous solutions of hydrogen peroxide.

$$\underset{\text{aniline}}{\phi\text{-}NH_2} + \underset{\substack{\text{acetic} \\ \text{anhydride}}}{CH_3\text{-}\overset{O}{\underset{\|}{C}}\text{-}O\text{-}\overset{O}{\underset{\|}{C}}\text{-}CH_3} \longrightarrow \underset{\text{acetanilide}}{\phi\text{-}\underset{H}{\overset{|}{N}}\text{-}\overset{O}{\underset{\|}{C}}\text{-}CH_3} + CH_3COOH \quad (16\text{-}40)$$

Tertiary amines cannot be acylated since they do not have any hydrogen atoms directly attached to the amine nitrogen.

16.7-4 Preparation of Sulfonamides Primary and secondary amines, as well as ammonia, undergo a very similar reaction to acylation to produce compounds known as **sulfonamides.** Sulfonyl halides (RSO_2Cl) are used in place of the acyl halides, the unshared electron pair on the nitrogen of the amine attacking the sulfur atom of the sulfonyl halide in a manner analogous to the acylation of amines. Usually aromatic sulfonyl halides are used in the reaction.

Sec. 16.7 REACTIONS OF AMINES

$$RNH_2 + ArSO_2Cl \xrightarrow{NaOH} ArSO_2NHR + NaCl + H_2O \quad (16\text{-}41)$$
primary   sulfonyl   *N*–substituted
amine   halide   sulfonamide

$$R_2NH + ArSO_2Cl \xrightarrow{NaOH} ArSO_2NR_2 + NaCl + H_2O \quad (16\text{-}42)$$
secondary   sulfonyl   *N,N*–disubstituted
amine   halide   sulfonamide

$$R_3N + ArSO_2Cl \longrightarrow \text{(no reaction)} \quad (16\text{-}43)$$
tertiary   sulfonyl
amine   halide

The preparation of sulfonamides is used as the basis of a test to distinguish primary, secondary, and tertiary amines. The test is known as the **Hinsberg test** and can be used to differentiate the three classes of amines. An unknown amine is reacted with benzenesulfonyl chloride (C$_6$H$_5$—SO$_2$Cl) in the presence of base. Primary and secondary amines will produce a white solid, the sulfonamide derivative. Tertiary amines will not react and usually give an oil, or two immiscible layers of liquid. The sulfonamide prepared from a primary amine still has an acidic hydrogen left on the nitrogen atom; it will therefore dissolve in base to give a clear solution containing a salt.

$$ArSO_2\ddot{N}HR + NaOH \longrightarrow ArSO_2\ddot{N}RNa^{\oplus} + H_2O \quad (16\text{-}44)$$
N–substituted   soluble salt
sulfonamide
(from primary amine)

The *N,N*–disubstituted sulfonamide formed from a secondary amine has *no* hydrogen atoms attached to the nitrogen. It will therefore not dissolve in base and will remain as an insoluble precipitate. Thus, if an amine does not react with the benzenesulfonyl chloride, it must be a tertiary amine. If an amine reacts to form a solid sulfonamide derivative that is soluble in base, the amine must be primary. If the sulfonamide derivative is not soluble in base, the amine must be secondary.

16.7-5 Reaction of Amines with Nitrous Acid

Occasionally the reaction of an amine with nitrous acid (HONO or HNO_2) can also be used as a test to differentiate various classes of amines. However, this means of identification is not nearly as reliable and conclusive as the Hinsberg test.

Nitrous acid, HONO, is an unstable substance and is therefore prepared *in situ* by reaction of sodium nitrite with an acid such as hydrochloric or sulfuric acid at a temperature of 0° C.

Primary amines react with nitrous acid to form diazonium salts, which are usually unstable and decompose on standing. In the case of the primary aliphatic

$$R\!-\!NH_2 + HONO \longrightarrow H_2O + N_2\uparrow + ROH \quad (16\text{-}45)$$
primary   alcohol
aliphatic
amine

amines the diazonium salt formed is so unstable that it can't be isolated, and it immediately decomposes to liberate nitrogen gas and form an alcohol.

Primary aromatic amines form diazonium salts upon treatment with nitrous acid at 0°C. These crystalline salts are explosive when dry, but are somewhat more stable in solution than in the case of the aliphatic amines. The aromatic diazonium salts are usually kept in solution and reacted with various reagents to produce a large variety of products.

$$\text{aniline} + \text{HONO} \xrightarrow{0°C} \text{benzenediazonium chloride} + H_2O \qquad (16\text{-}46)$$

(from $NaNO_2$ + HCl)

On standing these diazonium salts will gradually evolve nitrogen gas as in the case of primary aliphatic amines. The use of diazonium salts in organic syntheses will be discussed in detail later in this chapter.

All secondary amines, whether aliphatic or aromatic, react with nitrous acid to produce *N*–nitroso compounds, which are usually yellow oils.

$$\underset{\text{secondary amine}}{R-\underset{\underset{R}{|}}{N}-\boxed{H + HO}\ NO} \longrightarrow H_2O + \underset{\textit{N}\text{–nitrosoamine}}{R-\underset{\underset{R}{|}}{N}-N=O} \qquad (16\text{-}47)$$

Tertiary amines react with nitrous acid to give complex products; aliphatic amines usually give a trialkyl ammonium nitrite salt that is soluble in water, whereas aromatic amines generally form a *para*–nitroso derivative, which is green in color.

$$\underset{\substack{\text{tertiary} \\ \text{aliphatic} \\ \text{amine}}}{R_3N} + HONO \longrightarrow \underset{\substack{\text{trialkylammonium} \\ \text{nitrite salt}}}{R_3\overset{\oplus}{N}HONO^{\ominus}} \qquad (16\text{-}48)$$

$$\underset{\textit{N,N}\text{–dimethylaniline}}{\text{[C}_6\text{H}_5\text{N(CH}_3)_2\text{]}} + HONO \xrightarrow{0°C} \underset{\substack{p\text{–nitroso–}\textit{N,N}\text{–} \\ \text{dimethylaniline} \\ \text{(green)}}}{\text{[p-ON-C}_6\text{H}_4\text{-N(CH}_3)_2\text{]}} \qquad (16\text{-}49)$$

16.8 Electrophilic Substitution Reactions of Aromatic Amines

The amino group (NH_2) strongly activates the aromatic ring towards electrophilic substitution reactions. The resonance effect of the amino group, specifically the donation of the unshared pair of electrons on the amino nitrogen atom into the aromatic ring, greatly increases the activity of the aromatic ring towards electrophilic reagents. The amino group is a very powerful *ortho–para*-directing group, and this is shown in the electrophilic substitution reactions of aniline.

The reactions of aniline, which is the simplest aromatic amine, are in general characteristic of other aromatic amines. Halogenation of aniline occurs quite readily, and as in the case of phenol forms a 2,4,6-trihalogenated derivative.

$$\text{aniline} + Br_2(H_2O) \longrightarrow \text{2,4,6-tribromoaniline} \qquad (16\text{-}50)$$

The increased activation of the benzene ring by the amino group causes the immediate formation of the trihalogenated compound. A problem thus arises if one desires to synthesize a monohalogenated derivative of aniline. In order to accomplish this it is necessary in some way to cut down the activity of the amino group. This can be done by acylating the amino group to an acetamido group

$$(-\overset{H}{\underset{|}{N}}-\overset{O}{\underset{\|}{C}}-CH_3)$$

which is *ortho–para*-directing, but much less of an activating group towards the benzene ring. (Why?) The acetamido group can eventually be hydrolyzed back to the original amino group. The overall synthetic sequence would be:

$$\text{aniline} \xrightarrow{(CH_3CO)_2O} \text{acetanilide} \xrightarrow[CH_3COOH]{Br_2} \text{p-bromoacetanilide} \xrightarrow[H^\oplus, \Delta]{H_2O} \text{p-bromoaniline} \qquad (16\text{-}51)$$

In effect what has been done is to temporarily protect the amino group by acetylation, at the same time decreasing the activity, and then, by hydrolysis of the acetamido group, to regenerate the original amino group back again in the final product. This same technique of protecting the amino group is used in other syntheses where it is feasible and necessary. For example, if one attempted to nitrate aniline directly in hopes of producing *p*-nitroaniline, the reaction would mainly yield black tarry oxidation products. This is due to the increased activity of the aromatic ring of aniline, which makes the molecule very susceptible to the oxidizing power of nitric acid. The nitration of aniline to form p-nitroaniline in good yield can best be attained by first protecting the amino group.

$$\text{aniline} \xrightarrow{(CH_3CO)_2O} \text{acetanilide} \xrightarrow[H_2SO_4,\text{ low temp.}]{HNO_3} \text{p-nitroacetanilide} \xrightarrow[H^\oplus, \Delta]{H_2O} \text{p-nitroaniline} \quad (16\text{-}52)$$

16.8-1 Sulfonation of Aromatic Amines

The sulfonation of aniline is interesting in that the salt that is formed initially rearranges on heating to the compound sulfanilic acid (p-aminobenzenesulfonic acid).

$$\text{aniline} + HOSO_3H \longrightarrow \text{anilinium hydrogen sulfate} \xrightarrow{180\text{-}200°C} \text{sulfanilic acid} + H_2O \quad (16\text{-}53)$$

Sulfanilic acid has the formula with the dipolar ion structure

[structure with $NH_3^\oplus$ and $SO_3^\ominus$] rather than [structure with NH_2 and SO_3H]

The dipolar ion structure has been verified by experimental evidence. This type of structure is sometimes called a **zwitterion,** and is found to occur in substances that have both an acidic and a basic functional group present in the molecule, as in the amino acids (containing carboxyl and amino groups) and in the proteins. Much of the chemistry of these substances can be explained on the basis of a dipolar ion or zwitterion structure.

Sulfanilic acid can be converted by a series of reactions to the compound

sulfanilamide, SO_2NH_2, which is the basic fundamental unit of the sulfa drugs.

Sec. 16.10 DIAZONIUM SALTS

The geometry and electron density of the sulfanilamide molecule is very similar to the compound *para*-aminobenzoic acid (PABA), $\overset{+}{NH_3}-C_6H_4-COO^{\ominus}$, which is part of an essential enzyme system in the metabolism of many types of bacteria. Can you suggest a reason why sulfa drugs are effective in controlling certain bacterial infections? (Hint: Note the similarity between the sulfanilamide and PABA molecules.)

16.9 Heterocyclic Amines

It should be briefly mentioned at this point that in addition to the aliphatic and aromatic amines, a large class of compounds that are often more complex in structure occurs in many materials in nature, namely, the **heterocyclic amines**. The chemistry of these amines is in most respects analogous to that of the other amines. A few examples of typical heterocyclic amines have been cited earlier, in Section 16.2. (The reader is also referred to Chapter 22.)

16.10 Preparation and Reactions of Diazonium Salts

As has already been mentioned, aromatic diazonium salts can be prepared by the reaction of a primary aromatic amine with nitrous acid at 0° C.

$$C_6H_5NH_2 \xrightarrow[HCl \text{ (or } H_2SO_4\text{), } 0°C]{NaNO_2} C_6H_5\overset{+}{N_2}Cl^{\ominus} \text{ (or } HSO_4^{\ominus}) + H_2O \quad (16\text{-}54)$$

aniline → benzene diazonium chloride

The diazonium ion (French, azote = nitrogen) is a resonance hybrid of the structures

$$C_6H_5-\overset{+}{N}\equiv N: \longleftrightarrow C_6H_5-\overset{+}{N}=\overset{..}{N}:$$

The great importance of the diazonium salts stems from their use in the syntheses of many organic compounds. They are extremely versatile in their reactions and often provide the only feasible synthetic route for a given compound.

16.10-1 Replacement by Halogen Diazonium salts lose nitrogen easily, and the nitrogen can be replaced by a variety of substituents. For example, potassium bromide or iodide will react with a diazonium salt to form bromobenzene or iodobenzene, respectively.

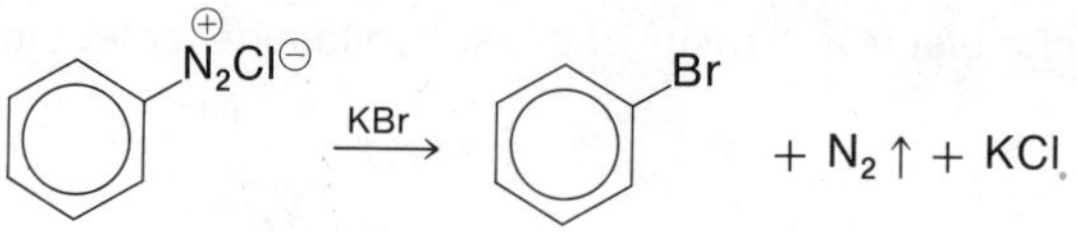

$C_6H_5N_2^+Cl^- \xrightarrow{KBr} C_6H_5Br + N_2\uparrow + KCl$ (16-55)

bromobenzene

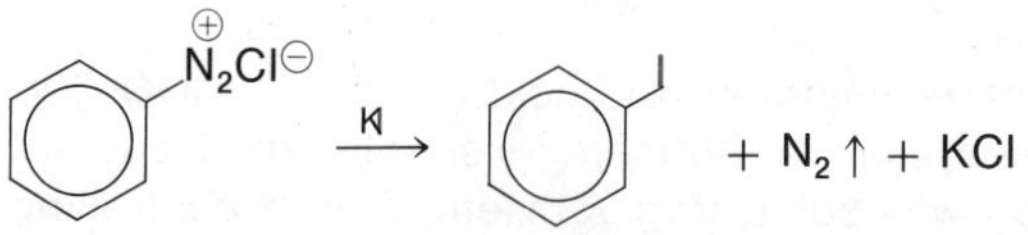

$C_6H_5N_2^+Cl^- \xrightarrow{KI} C_6H_5I + N_2\uparrow + KCl$ (16-56)

iodobenzene

16.10-2 The Sandmeyer Reaction The **Sandmeyer reaction** uses **cuprous salts** in the form of the halide or cyanide to synthesize halobenzenes and benzonitrile.

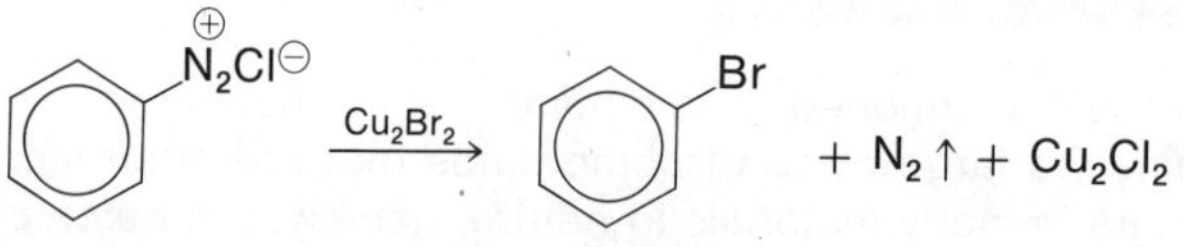

$C_6H_5N_2^+Cl^- \xrightarrow{Cu_2Br_2} C_6H_5Br + N_2\uparrow + Cu_2Cl_2$ (16-57)

bromobenzene

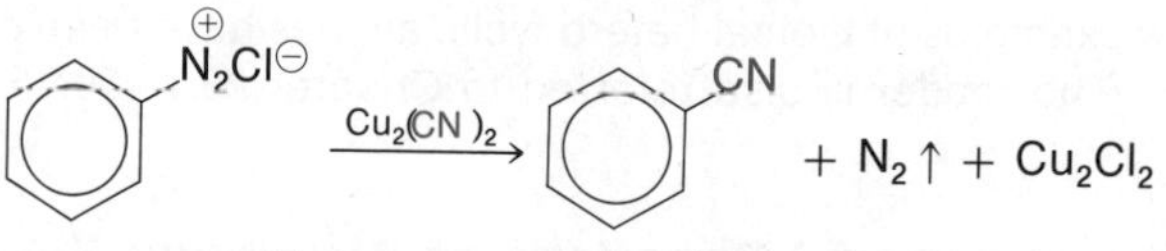

$C_6H_5N_2^+Cl^- \xrightarrow{Cu_2(CN)_2} C_6H_5CN + N_2\uparrow + Cu_2Cl_2$ (16-58)

benzonitrile

16.10-3 Replacement by Fluorine Fluorobenzene can be made by the reaction of a diazonium salt with sodium tetrafluoroborate, $NaBF_4$. This is the only known way of introducing a fluorine into a benzene ring.

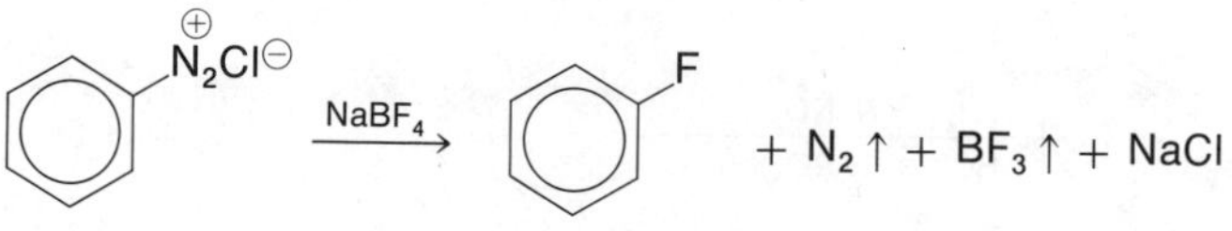

$C_6H_5N_2^+Cl^- \xrightarrow{NaBF_4} C_6H_5F + N_2\uparrow + BF_3\uparrow + NaCl$ (16-59)

fluorobenzene

16.10-4 Replacement by OH to Form Phenols Reaction of a diazonium salt with water is an excellent way to make phenol.

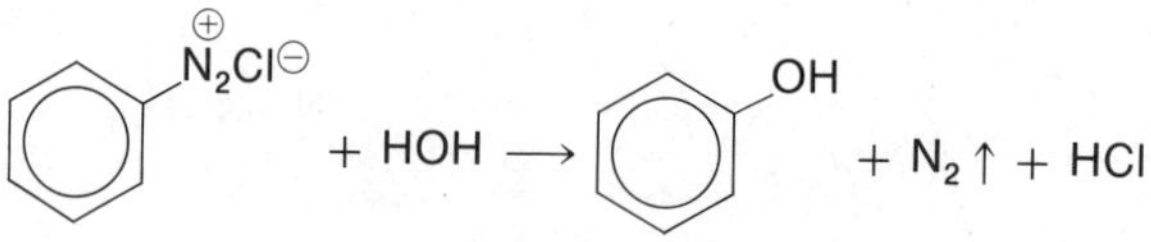

$C_6H_5N_2^+Cl^- + HOH \longrightarrow C_6H_5OH + N_2\uparrow + HCl$ (16-60)

phenol

This reaction often occurs as a side reaction in all diazonium salt replacement reactions since the reactions are run in an aqueous solution.

16.10-5 Replacement by H The compound H_3PO_2 or hypophosphorus acid is a good reducing agent and can replace the nitrogen of the diazonium salt with

Sec. 16.12 EXISTENCE OF BENZYNE

hydrogen. Hypophosphorous acid is often referred to as **"Kornblum's magic reagent"** after Professor Nathan Kornblum of Purdue University.

$$\text{C}_6\text{H}_5\text{N}_2^{+}\text{Cl}^{-} \xrightarrow[\text{H}_2\text{O}]{\text{H}_3\text{PO}_2} \text{C}_6\text{H}_6 + \text{N}_2\uparrow + \text{HCl} + \text{H}_3\text{PO}_3 \quad (16\text{-}61)$$

benzene

This reaction serves to remove an amino group by conversion into a diazonium salt followed by treatment with hypophosphorus acid.

16.11 Syntheses with Diazonium Salts

By utilizing these reactions via a diazonium salt a primary aromatic amine can be converted into many different types of compounds. Often a compound cannot be synthesized by any other simple reaction sequence than through a diazonium salt. As an illustration, consider the synthesis of *m*–bromotoluene from benzene. (Note that both the methyl group and the bromine are *ortho–para*-directing groups, and we desire these substituents to be *meta* in the product.)

$$\text{benzene} \xrightarrow[\text{AlCl}_3]{\text{CH}_3\text{Cl}} \text{toluene} \xrightarrow[\text{H}_2\text{SO}_4]{\text{HNO}_3} p\text{-nitrotoluene} \xrightarrow[\text{HCl}]{\text{Sn}} p\text{-aminotoluene} \xrightarrow{(\text{CH}_3\text{CO})_2\text{O}}$$

$$p\text{-CH}_3\text{C}_6\text{H}_4\text{NHCOCH}_3 \xrightarrow{\text{Br}_2} \text{2-Br-4-CH}_3\text{-C}_6\text{H}_3\text{NHCOCH}_3 \xrightarrow[\text{H}_2\text{O}]{\text{H}^{+}} \text{2-Br-4-CH}_3\text{-C}_6\text{H}_3\text{NH}_2$$

$$\xrightarrow[\text{HCl, 0°C}]{\text{NaNO}_2} \text{ArN}_2^{+}\text{Cl}^{-} \xrightarrow[\text{H}_2\text{O}]{\text{H}_3\text{PO}_2} m\text{-bromotoluene} \quad (16\text{-}62)$$

16.12 Evidence for the Existence of Benzyne

It is significant to note that a diazonium salt provided one of the most essential and conclusive experiments for the existence of benzyne. The work was done at the University of Michigan by Professors R. S. Berry and M. Stiles in the early

1960's. They reacted anthranilic acid (o-aminobenzoic acid) with nitrous acid, diazotizing the amino group to form a diazonium salt.

$$\underset{\text{anthranilic acid}}{\text{C}_6\text{H}_4(\text{NH}_2)(\text{COOH})} \xrightarrow[\text{HCl, 0° C}]{\text{NaNO}_2} \text{C}_6\text{H}_4(\text{N}_2^{\oplus})(\text{COO}^{\ominus})$$ (16-63)

The diazonium salt readily lost nitrogen and carbon dioxide when heated, and a product was isolated and identified as biphenylene.

$$\text{C}_6\text{H}_4(\text{N}_2^{\oplus})(\text{COO}^{\ominus}) \xrightarrow{\Delta} \text{N}_2\uparrow + \text{CO}_2\uparrow + [\underset{\text{benzyne}}{\text{C}_6\text{H}_4}] \longrightarrow \underset{\text{biphenylene}}{\text{C}_{12}\text{H}_8}$$ (16-64)

The biphenylene could only be formed by two very reactive benzyne fragments dimerizing together. No other species in the reaction mixture could account for the formation of biphenylene as a reaction product. An interpretation of experimental data such as this, and the use of mass spectroscopy, have clearly established the transitory existence of benzyne in many previously not understood reactions.

16.13 Coupling Reaction of Diazonium Salts

Diazonium salts (the diazonium ion) are weakly electrophilic species, but will react with strongly activated aromatic compounds such as phenols or amines. If a phenol or amine is added to a diazonium salt a coupling reaction occurs between the molecules to form an azo compound, containing a —N=N— bond in the molecule.

$$\underset{\substack{\text{benzene diazonium} \\ \text{chloride}}}{\text{C}_6\text{H}_5\text{–}\ddot{\text{N}}\text{=}\overset{\oplus}{\text{N}}\text{Cl}^{\ominus}} + \underset{\text{phenol}}{\text{C}_6\text{H}_5\text{–OH}} \longrightarrow \underset{p\text{-hydroxyazobenzene}}{\text{C}_6\text{H}_5\text{–N=N–C}_6\text{H}_4\text{–OH}}$$ (16-65)

The success of the reaction depends on the pH of the reaction medium to a large extent. Coupling occurs *para* to the hydroxy or amino group. If the *para* position is blocked by another substituent (than H) then coupling may occur in the *ortho* position. Azo compounds formed by the coupling reaction are colored and many are used to prepare azo dyes.

16.14 Color and Dyes

The process whereby a molecule absorbs light energy involves the promotion of an electron from a molecular orbital of lower energy to an orbital of higher energy

Sec. 16.14 COLOR AND DYES

content. It is easier to raise electrons that are not very tightly bound to a higher energy state. Thus, π electrons which are less firmly held than σ electrons should be elevated to a higher energy orbital more easily than σ electrons. Absorption of visible light therefore occurs readily by π electrons, especially in molecules that have many π electrons, such as those containing much unsaturation, or a substantial conjugated system. Compounds having such structural features are usually colored. However, not all colored substances are necessarily dyes. The substituents or structure in the molecule responsible for the production of color are called **chromophores**. A dye must have chromophores and also other functional groups that help fix the dye to the fabric, called **auxochromes**. Other groups may also be present that enhance or intensify the color, called **bathochromes** (e.g., —OH, —NR$_2$).

The actual application of a dye to a fabric may be accomplished in a number of ways. Dyes that color a fabric directly are called **direct dyes**. Some dyes are absorbed in a colorless form and upon the drying of the fabric in air are oxidized and turn color. These dyes are called **vat dyes**. Some dyes require heavy metal salts known as **mordants** to form insoluble compounds to help impregnate the fabric. Other types of dyeing are also used when required by the textile being dyed.

There are many different types of dyes, too numerous to mention in this text. The largest number of dyes belong to the so-called azo classification. A typical example is **para red,** made from the coupling reaction of p-nitrobenzenediazonium chloride and β-naphthol.

$$O_2N-\text{C}_6H_4-N{=}\overset{\oplus}{N}Cl^{\ominus} + \text{(β-naphthol)} \xrightarrow{\text{alk. sol.}} \text{(para red)} \quad (16\text{-}66)$$

p-nitrobenzene-
diazonium chloride

(para red)

Indigo, which is obtained from the indigo plant, has the structure

(indigo)

Other dyes that may be familiar to the reader are **malachite green,** a representative triphenylmethane dye, and the antiseptic **mercurochrome**.

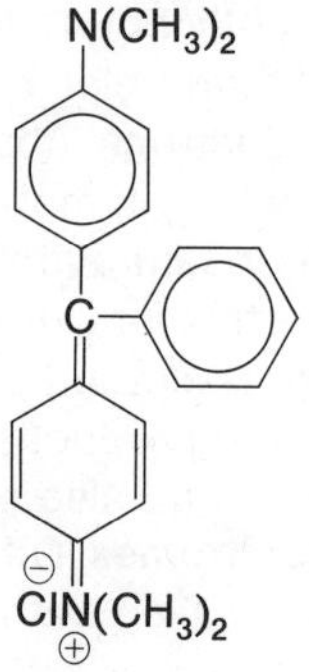

malachite green

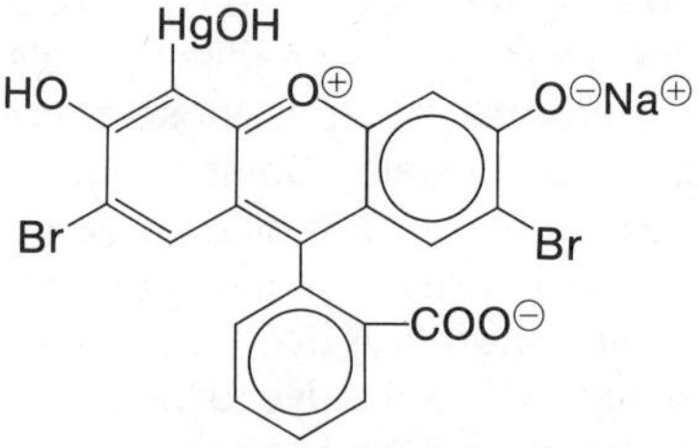

mercurochrome

Summary

- The amines are the most common organic compounds that behave as bases. They can be classified as primary, secondary, or tertiary amines on the basis of the number of groups attached to the nitrogen atom.

- The basicity of amines is due to their ability to form salts with acids. Basic strength has been discussed in Chapters 4 and 5 in relation to electronic and steric effects.

- Primary amines can be prepared by:

    (a) Reaction of alkyl halides with ammonia.

    (b) Reduction reactions of nitro compounds, amides, oximes, and nitriles.

    (c) Reductive amination of aldehyde and ketones.

    (d) Gabriel phthalamide synthesis.

    (e) Hofmann degradation of amides.

    (f) Reaction of aryl halides with sodamide in liquid ammonia (benzyne).

- The chemical reactivity of amines is due mainly to the tendency for the nitrogen to share its electron pair with various electron-seeking reagents:

    (a) Salt formation with acids.

    (b) Alkylation of amines can produce quaternary ammonium salts. Choline and lecithin are important biochemical substances that are quaternary ammonium hydroxides.

    (c) Acylation to amides.

    (d) Reaction with sulfonyl halides to produce sulfonamides (Hinsberg test for amines).

- Reaction with nitrous acid can be used as a test to distinguish amines:

    (a) 1° aliphatic ⟶ alcohol + N_2

    (b) 2° aliphatic ⟶ N-nitroso compound (yellow oil)

    (c) 3° aliphatic ⟶ salt

    (d) 1° aromatic ⟶ diazonium salt

- Aromatic amines undergo electrophilic substitution reactions. The NH_2 group greatly activates the aromatic ring and this often requires "protecting" the amino group in some syntheses.

PROBLEMS

- Sulfanilic acid has a zwitterion structure.
- Examples of heterocyclic amines include pyridine and pyrrole.
- Diazonium salts are extremely useful in synthesis;
  (a) Replacement by halogen.
  (b) Replacement by cyanide (Sandmeyer reaction).
  (c) Replacement by hydroxyl.
  (d) Replacement by H with H_3PO_2.
  (e) Coupling reactions.
- Color is due to the presence of chromophores in the molecule.
- Some representative dyes include para red, indigo, and mercurochrome.

PROBLEMS

1. Draw the correct structural formulas for each of the following compounds.
   (a) dimethylethylamine
   (b) 2-aminopentane
   (c) *N*-methyl-*N*-ethylaniline
   (d) 2,4-dibromopyridine
   (e) dimethylethylphenyl-ammonium iodide
   (f) *p*-aminoazobenzene
   (g) *o*-toluidine
   (h) 2,3-dinitroaniline
   (i) dimethyl *sec*-butylamine
   (j) triphenylamine
   (k) *tert*-butylamine
   (l) *p*-toluenediazonium fluoroborate

2. Draw the structural formulas for all the isomers having the molecular formula $C_4H_{11}N$. Name each compound by the IUPAC naming system and designate each amine as primary, secondary, or tertiary.

3. Complete each of the following equations by writing the correct structural formula(s) of the product(s) formed. If no reaction occurs, write N.R.

   (a) $CH_3NH_2 + HCl \longrightarrow$

   (b) *n*-butyronitrile $+ LiAlH_4 \xrightarrow{H_2O}$

   (c) *n*-heptaldehyde $\xrightarrow[H_2, \text{press.}]{NH_3}$

   (d) *o*-bromotoluene $+ NH_3 \longrightarrow$

   (e) *p*-nitrotoluene $+ Sn + HCl \longrightarrow$
   (f) $(CH_3)_3N + CH_3-CH_2I \longrightarrow$
   (g) propionamide $+ NaOH + Br_2 \longrightarrow$

   (h) *m*-toluidine $+ CH_3-\underset{\underset{O}{\|}}{C}-O-\underset{\underset{O}{\|}}{C}-CH_3 \longrightarrow$

(i) $CH_3-CH_2-NH_2 \xrightarrow{HONO}$

(j) Aniline + $Br_2(H_2O) \longrightarrow$

(k) 4-methyl-aniline (p-toluidine, CH_3 at top, NH_2 at bottom of benzene ring) $\xrightarrow[HCl,\ 0°C]{NaNO_2}$

(l) product from (k) + $Cu_2(CN)_2 \longrightarrow$
(m) product from (k) + $NaBF_4 \longrightarrow$
(n) product from (k) + $KI \longrightarrow$
(o) product from (k) + $H_2O \longrightarrow$
(p) product from (k) + $H_3PO_2 \xrightarrow{H_2O}$

(q) benzenediazonium chloride + phenol $\xrightarrow{NaOH}$

(r) m-toluidine + $C_6H_5-SO_2Cl \longrightarrow$

(s)* phthalimide (N–H) + NaOH + $Br_2 \longrightarrow$

(t) $CH_3-\underset{\underset{H}{|}}{\overset{\overset{O}{\|}}{C}}-N-CH_3$ + $LiAlH_4 \xrightarrow{H_2O}$

(u) $CH_3-\underset{\underset{H}{|}}{N}-CH_2-CH_2-CH_3 \xrightarrow{HONO}$

4 Perform all of the following syntheses from the indicated starting materials and any necessary inorganic reagents.

(a) 1-chloropropane to n-butylamine
(b) sec-butyl bromide to 2-aminobutane
(c) n-pentyl bromide to n-butylamine
(d) toluene to aniline
(e) benzene to sulfanilamide
(f) benzene to m-fluoroiodobenzene
(g) benzene to p-fluoroiodobenzene
(h) benzene to p-cresol
(i) *benzene to m-cyanotoluene
(j) benzene to p-hydroxyazobenzene
(k) benzene to N-phenylbenzamide
(l) ethylene to 1,4-diaminobutane
(m) acetaldehyde to 1-amino-3-hydroxybutane

PROBLEMS

5 Arrange the following compounds in order of increasing basic strength.

(a) n-propylamine
di-n-propylamine
aniline
cyclopentylamine
triphenylamine
propionamide

(b) aniline
p-toluidine
p-nitroaniline
p-chloroaniline
p-methoxyaniline
p-phenylenediamine.

6 *Explain why pyridine, 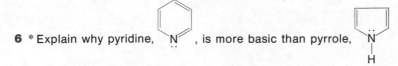, is more basic than pyrrole,

7 (a) Which is a stronger base, methylamine or p-toluidine? Explain.
(b) Why are aromatic amines generally less basic than aliphatic amines?

8 Name a chemical test or single chemical reagent which can be used to distinguish between the members of the following pairs of compounds.
(a) aniline and N-methylaniline
(b) methylamine and diethylamine
(c) pyridine and aniline
(d) tetraethylammonium iodide and aniline

9 *When bromobenzene is reacted with KNH_2 in liquid ammonia, aniline is formed; but 2,6-dimethoxybromobenzene does not react with KNH_2 in liquid ammonia. Explain.

10 (a) Suggest a detailed mechanism for the Hofmann degradation of benzamide in the presence of NaOH and Br_2.

*(b) Could the Hofmann rearrangement be run on N-methylbenzamide instead of benzamide? Explain.

*(c) What product would you expect to be formed in (a), if methyl alcohol were used as a solvent instead of water?

*(d) Arrange the following compounds in order of increasing rate of reaction with NaOH and Br_2:
benzamide
p-nitrobenzamide
p-methoxybenzamide.

11 Compound A, $C_9H_{12}NBr$, reacts with benzenesulfonyl chloride, in the presence of NaOH, to form a white precipitate, insoluble in both acid and base. When A is reacted with $KMnO_4$ under vigorous reaction conditions, an acid of neutralization equivalent = 83 is produced. The acid forms only one mononitro substitution product. Suggest a structural formula for compound A.

12 * Why is the Gabriel phthalimide synthesis limited only to the formation of primary aliphatic amines?

13 Would you expect p-nitroaniline to form a diazonium salt less readily than aniline? Explain.

14 Starting from ethylene synthesize the following amines:
(a) methylamine; (b) ethylamine; (c) n-propylamine.

17

Stereoisomerism— Optical Isomerism

Paper models of the asymmetric carbon atom, made by the young van't Hoff and sent to his friend G. J. W. Bremer in 1875.

Courtesy of the Rijks Museum in Leiden.

17.1 Introduction

We have already mentioned examples of the various types of isomerism (e.g., positional, branched-chain, functional group) in our earlier discussions. **Stereoisomerism** refers to a particular kind of isomerism dealing with the three-dimensional arrangement of the atoms within a molecule in space. The isomers differ only in the way the atoms are oriented in space. Two examples of stereoisomerism, conformational isomerism and geometrical or *cis–trans* isomerism have already been referred to in Sections 7.9 and 8.4. In this chapter we shall concentrate on still a third type of stereoisomerism, namely **optical isomerism.**

17.2 Optical Isomerism

Molecules exhibit optical isomerism when there is no element of symmetry in their structure. Many simple and some exceedingly complex molecules show optical activity, including some carbohydrates, amino acids and proteins, hormones, and natural products. In some cases it has been shown that optical isomerism will greatly affect or even prevent the biochemical action of certain materials.

Optical isomers show a characteristic behavior toward plane-polarized light, and can be detected because of this unusual feature.

17.3 Detection of Optical Activity— Plane-Polarized Light

The detection of optical isomers involves an understanding of the properties of plane-polarized light, some of which will be discussed at this time.

Sec. 17.3 DETECTION OF OPTICAL ACTIVITY

Ordinary light is made up of rays of many different wavelengths and is best understood by considering it to be a wave vibrating in all planes in space perpendicular to the direction in which the light waves are propagated. The term **monochromatic light** is used to describe light of a single, discrete wavelength that still vibrates in all possible planes. Thus ordinary light, whether normal white light, composed of many wavelengths, or monochromatic light, of a single wavelength, vibrates in an infinite number of directions. However, **plane-polarized light** is light whose waves vibrate in only one plane. Ordinary light can be converted into plane-polarized light by passing it through a lens made up of certain special materials, such as calcite (crystalline $CaCO_3$), arranged to form a Nicol prism, or through Polaroid.

An optically active substance rotates the plane of polarized light; that is, when plane-polarized (monochromatic) light is passed through an optically active material, the light is rotated and emerges from the material vibrating in a different plane than when it entered the optically active material. This phenomenon can be used experimentally to distinguish between certain substances by their behavior toward monochromatic plane-polarized light. Since optical properties are used to distinguish between the substances, the phrase optical isomers or optical activity has come into use.

The rotation of the plane-polarized light can be detected and measured by an instrument known as a **polarimeter.** The polarimeter is made up of several component parts: a light source, a lens system or polarizer to polarize the light, a sample tube, an analyzer to detect the amount and direction of rotation if any, and an eye piece through which to view the findings.

The light source is usually a sodium vapor lamp that produces monochromatic light of a single wavelength of 5893 Å; the wavelength is referred to as the D line of sodium. This light beam is passed through a Nicol prism or Polaroid lens to polarize the light, and the polarized light is then passed through a sample containing the substance to be examined dissolved in solution. The light that emerges is passed through an analyzer, which is really a prism similar to the polarizer. If the substance being examined is **optically inactive,** the light will not be rotated and will pass through the material emerging in the same plane as it entered. This is indicated when the polarizer and analyzer are parallel, with an angle $\alpha = 0°$ for the maximum light transmission. On the other hand, if the substance being examined is **optically active,** it will rotate the plane of polarized light a certain number of degrees, α. This will be indicated by the fact that the analyzer will have to be rotated an equal number of degrees (α) in order to achieve maximum light transmission, and the actual angle of rotation can be measured by means of a calibrated circular scale, viewed by the human eye through an eye piece. The angle of rotation of the analyzer is equal to the angle (α) that the plane of polarized light was rotated by the optically active substance. If the light is rotated clockwise to the right, the substance is said to be **dextrorotatory** (from the Latin, *dexter* = right) and the symbol small *d* or (+) is placed before the name of the substance to indicate that it is an optically active material and that the plane-polarized light is rotated a certain number of degrees (α) to the right. If the substance rotates the light counterclockwise to the left, it is said to be **levorotatory** (from the Latin, *laevus* = left) and is symbolized by a small *l* or (−) before the name of the substance.

17.3-1 Specific Rotation The amount of rotation depends upon several factors, all of which determine the number of particles the light comes in contact with while passing through the sample tube. The length of the tube, the concentration of material dissolved in solution, the temperature, and the wavelength of monochromatic light used, all affect the rotation to some extent. In order to eliminate these variables when comparing different substances and to be able to interpret and get meaningful results, it is necessary to standardize the influence of these variables so that the amount of rotation is only a characteristic dependent on each optically active substance being measured, and not a function of the other variables. For this purpose, the amount of rotation is measured as the **specific rotation** (α, the number of degrees), expressed as

$$[\alpha]_\lambda^t = \frac{100\alpha}{l \times c}$$

where t is the temperature in °C (usually 20° C), λ is the wavelength of light used (usually the so-called sodium D line), l is the length of the sample tube expressed in decimeters, and c is the concentration of the substance in solution expressed in the units of grams of solute per 100 ml of solvent. The specific rotation of an optically active substance is a distinctive physical property characteristic of the material, just as its density, melting and boiling points, and solubility.

17.4 Asymmetry and its Relation to the Tetrahedral Carbon Atom

The optical activity of the type described in the preceding section was discovered in certain materials by the French physicist Jean-Baptiste Biot in 1815. However, it was the renowned chemist Louis Pasteur who in 1847 postulated that optical isomers that are mirror images of each other occur in pairs. Based on his experimentation, Pasteur's statement formed the basis and foundation for stereoisomerism and stereochemistry as we understand them today.

What structural features in a molecule are responsible for optical activity? Why are some substances optically active whereas most substances are not?

The explanation to account for the phenomenon of optical activity is highly complex and mathematical. It is necessary to consider what happens when a beam of plane-polarized light is passed through a particular substance. The light beam interacts to some extent with the electrons in a molecule. For most substances, because of the random distribution of molecules within the sample, it can be expected that for every molecule the light encounters there is present in the sample another molecule, exactly identical, but oriented in space so as to be the mirror image of the first molecule, which completely cancels the effect of the light. The result is that the light passed through the molecules is *not* rotated, and the substance is said to be **optically inactive.** It must be understood that this effect is not due to the individual molecules themselves, but rather to the random distribution of molecules that can be mirror images of each other.

When molecules are asymmetric or dissymmetric, that is not superimposable on their mirror images, there will be an interaction of the electrons in the molecules with the light, resulting in rotation of the plane-polarized light when passed

Sec. 17.4 ASYMMETRY

Louis Pasteur (1822–95)—French chemist, professor at The Sorbonne, director of the Pasteur Institute, and one of the world's greatest scientists—is well-known for (among his vast accomplishments) his pioneer work in stereochemistry. (From the Dains Collection, courtesy of the Department of Chemistry, The University of Kansas.)

through such a substance. Since there is a rotation of the light the substance is said to be **optically active**.

In 1874, the Dutch chemist van't Hoff, and the French chemist Le Bel, each postulated independently that the carbon atom can form four bonds to substituents directed in space toward the corners of a tetrahedron. Upon further consideration of this geometrical arrangement of organic compounds the two chemists were able to explain the optical activity of some substances and the optical inactivity of others.

A knowledge of solid geometry shows that if there are *four different substituents* attached to a tetrahedral carbon atom there are but two possible arrangements in space.

$$\begin{array}{cc} a & a \\ | & | \\ c\overset{C}{\underset{d}{\diagdown}}b & b\overset{C}{\underset{d}{\diagdown}}c \end{array}$$

These two arrangements are actually mirror images and are non-superimposable on each other. (They are related to each other as our right and left hands are mirror images of each other.) An examination of molecular models, built so that four different substituents (represented by different colored balls or plastic tubing, etc.) are attached to a carbon atom, can easily verify this fact for the reader. If two or more of the substituents are identical, then the tetrahedral arrangement possesses a plane of symmetry and only one single arrangement of the molecule is possible. (Molecular models can substantiate this consequence of tetrahedral geometry also.)

It was suggested by van't Hoff and by Le Bel that the carbon atom was located in the center of a tetrahedron, with its four bonds directed towards the corners, occupied by the substituents.

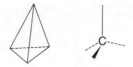

The normal tetrahedral bond angle present is the 109° 28', which has been verified by experimental evidence as occurring in many organic compounds. The importance of the geometrical arrangement of the tetrahedron is that a carbon atom that has four different substituents constitutes a center of asymmetry in the molecule. A carbon atom to which four different substituents are attached is called an **asymmetric carbon atom**.

$$d\text{---}^*C\text{---}b$$

(an asymmetric carbon represented by *C, with the substituents, a, b, c, and d)

17.5 Enantiomerism and the Tetrahedral Carbon Atom

According to our previous discussion, an asymmetric carbon atom has two possible arrangements in space, and these two spatial arrangements are non-superimposable, but are mirror images. They are related to each other in the same way as right and left hands, having similar properties, except in their rotation of plane-polarized light. One of the structures would rotate the light to the right a certain number of degrees, while the second structure would rotate the light in the opposite direction, to the left, exactly the same number of degrees. These structures, with their non-superimposable mirror images, are called **enantiomers** or **enantiomorphs**. Enantiomers (*enantio-* = opposite) are optical isomers that have identical physical properties (i.e., melting point, boiling point, density, etc.) except for the direction of their rotation of plane-polarized light. As a consequence of the work of van't Hoff and Le Bel, it can be asserted that any molecule containing at least one asymmetric carbon atom will have isomers that *may be* optically active.*

17.6 Criteria for Optical Activity

It should be mentioned at this point, however, that the presence of an asymmetric carbon atom in a molecule is not necessary for the existence of optical isomers or

---

*The terms **chiral** and **chirality** are becoming more prevalent in discussions on optical activity, replacing the older terms *asymmetric* and *asymmetry*.

Molecules that are not superimposable on their mirror images are said to be **chiral**. A compound whose molecules are chiral (having chirality) can exist as a pair of enantiomers. Thus, a carbon atom to which four different substituents are attached is called a **chiral center**, in place of the older term *asymmetric carbon atom*.

Sec. 17.7 FISCHER PROJECTION FORMULAS

optical activity. *A substance exhibits optical activity if the molecule is asymmetric or dissymmetric; that is, it contains some element of non-symmetry!* Dissymmetry here simply indicates that the substance is non-superimposable on its mirror image. Biphenyls substituted with bulky ortho substituents, substituted allenes, and spiranes are among the classes of compounds that exhibit optical activity although there is no asymmetric carbon atom present in these molecules.

2,2'-dinitro-6,6'-dicarboxy biphenyl   substituted allene   substituted spirane

Each of these molecules, having an isomer that is a non-superimposable mirror image, hence exhibits optical activity. Figure 17.1 shows an FMO model of the allene ($CH_2{=}C{=}CH_2$) molecule. Can you suggest an explanation for the optical activity of *substituted* allenes? (HINT: Look at Figure 17.1 and consider the bonding in the molecule.)

17.7 Fischer Projection Formulas

Although the presence or absence of an asymmetric carbon atom does not always determine whether or not a substance has optical activity, most optical isomers do

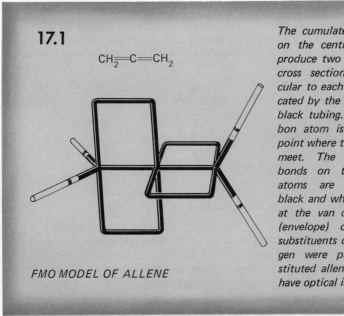

17.1

$CH_2{=}C{=}CH_2$

FMO MODEL OF ALLENE

The cumulated double bonds on the central carbon atom produce two π orbitals whose cross sections are perpendicular to each other, as is indicated by the two rectangles of black tubing. The central carbon atom is situated at the point where the two rectangles meet. The carbon-hydrogen bonds on the end carbon atoms are represented by black and white tubing ending at the van der Waals' radius (envelope) of hydrogen. If substituents other than hydrogen were present, the substituted allenes could possibly have optical isomers. Why?

contain one or more asymmetric carbon atoms in the molecule. For this reason we shall primarily concern ourselves with molecules that do contain one or more asymmetric carbon atoms.

In order to understand the concepts of stereochemistry and stereoisomerism it is essential that we be able to visualize in three dimensions the geometry and shape of molecules, as well as the spatial orientation of the atoms within the molecule. Some people can picture and visualize things in three dimensions more easily than others. Fortunately, the design and use of three dimensional molecular models serve as an aid in enabling us to build and visualize molecules in three dimensions, and thus to elucidate and solve many intricate stereochemical problems. Drawing a three-dimensional picture of a molecule is often extremely difficult, and so we face the problem of conveying in our two-dimensional drawings on paper what is actually present in three dimensions.

Let us consider the structural formula of the compound 2-bromobutane,

$$CH_3-\overset{H}{\underset{Br}{\overset{*}{C}}}-CH_2-CH_3.$$ The carbon atom is marked with an asterisk to indicate that

it is an asymmetric carbon atom with four different substituents attached to it (H, Br, CH_3, and CH_3-CH_2-). There are two possible arrangements of the substituents about the asymmetric carbon atom, resulting in two different molecules that are non-superimposable mirror images (enantiomers). These can be represented in three dimensions in the geometrical shape of a tetrahedron. By way of convention in drawing structural formulas, different symbols are used to indicate the three-dimensional character of the molecule; (—) lines indicate bonds in the plane of the molecule, (---) lines indicate bonds directed in a plane behind the molecule or away from the viewer as you view the molecule, and (▬) lines represent bonds directed forward from the plane of the molecule toward the viewer. The use of

mirror
2-bromobutane

so-called **Fischer projection formulas** enables us to represent molecular structures in two dimensions without losing sight of the three-dimensional geometry actually present in the molecule. Chemists have agreed on the following standard convention as a means of depicting Fischer projection formulas. A cross is drawn on paper, the asymmetric carbon atom being located where the lines cross and the four substituents present are attached to the four ends of the cross.

Sec. 17.8 ENANTIOMERS

In addition, if the asymmetric carbon atom is assumed to lie in the plane of the paper, the horizontal lines are to represent bonds coming out of the paper towards the viewer and vertical lines are to represent bonds going back behind the plane of the paper away from the viewer. Thus, in the diagram above, the bond lines for substituents *b* and *d* are coming towards us, whereas those for *a* and *c* are directed away from us in space. It can easily be seen by using molecular models that the interchanging of any two substituents in a Fischer projection formula gives an isomer that is a mirror image of the first structure.

The Fischer projection formulas for 2-bromobutane are

$$\begin{array}{ccc} & CH_3 & & CH_3 \\ H-\!\!\!\!-\!\!\!\!-\!\!\!\!-Br & & Br-\!\!\!\!-\!\!\!\!-\!\!\!\!-H \\ & CH_2CH_3 & & CH_2CH_3 \end{array}$$

Rotation of one of these Fischer projection formulas by 90° in the plane of the paper produces the other projection formula, which is the mirror image of the first; but a rotation of 180° leaves the structure unchanged.

17.8 Enantiomers

The mirror images of 2-bromobutane that we have just described are non-superimposable and hence are **enantiomers.** These enantiomers are actually optical isomers having the same physical properties except for their influence on a plane of polarized light. One of the enantiomers will rotate a plane of polarized light a certain number of degrees to the right. It is the dextrorotatory isomer and will be symbolized by the small letter *d* or (+) before the name of the compound (*d*-2-bromobutane or (+)-2-bromobutane). The other enantiomer will rotate the plane of polarized light the same number of degrees to the left, and is the levorotatory isomer symbolized by the small letter *l* or (−) before the name of the compound (*l*-2-bromobutane or (−)-2-bromobutane).

17.8-1 Racemic Modification If we have either one of these enantiomers in pure form then we have an optically active substance (*d* or *l*) that will rotate a plane of polarized light. However, if we have exactly equal amounts (50 %) of each of the enantiomers (50 % dextrorotatory + 50 % levorotatory isomer) the mixture will be optically inactive because the rotation due to the (+) isomer is exactly cancelled by the (−) isomer since there are equal numbers of molecules of both enantiomers present. A mixture of this composition (50 % *d* + 50 % *l*) is called a **racemic mixture,** or **racemic modification,** is represented by (*d, l*) or (±) and is *optically inactive.* The separation of a racemic mixture into its pure components, the *d* and *l* enantiomers, is known as **optical resolution,** and will be discussed in Section 17.12.

As another example, the compound lactic acid, CH_3—*CHOH—COOH, has an asymmetric carbon atom, and has two optical isomers that can be represented by the following Fischer projection formulas.

```
      COOH                COOH
       |                   |
  H ---+--- OH       HO ---+--- H
       |                   |
      CH₃                 CH₃
  (−)-lactic acid, or  (+)-lactic acid, or
     l-lactic acid        d-lactic acid
```

One of these isomers is dextrorotatory and the other is levorotatory. A mixture of equal amounts of the two isomers produces a racemic mixture or modification which is optically inactive, represented as (±)-lactic acid.

17.9 Absolute and Relative Configurations

The **configuration** of a molecule refers to the spatial arrangement of the atoms comprised by the molecule. Each molecular structure has its own configuration, as for example the d and l isomers that constitute a pair of enantiomers, in which the atoms are oriented differently in space.

In the previous section we drew two projection formulas representing the non-superimposable mirror images (enantiomers) of lactic acid. One of these structures represents the dextrorotatory isomer and the other the levorotatory isomer, but what is the spatial arrangement (configuration) of each isomer? That is, how do we know which structure (projection formula) represents the dextrorotatory and which the levorotatory isomer? The assignment of the structures is *not* arbitrary, but involves certain complex experimental techniques. It should be apparent that to assign a correct configuration to a molecule is an exceedingly difficult and complex task.

17.9–1 The D and L Configurations Because of the difficulties involved in assigning configurations, and also because of its influence on carbohydrate chemistry (Chapter 18) the chemists decided that the aldehyde, glyceraldehyde, be used as a standard reference substance. Glyceraldehyde,

$$H-\overset{\overset{O}{\|}}{C}-*CH(OH)-CH_2OH,$$

is an optically active compound, and in order to establish a systematic convention for compounds it was arbitrarily decided that the dextrorotatory isomer of glyceraldehyde be assigned a configuration of

```
        CHO
         |
    H ---C--- OH     D-(+)-glyceraldehyde
         |
        CH₂OH
```

to serve as a reference compound, and the levorotatory isomer be arbitrarily assigned the configuration

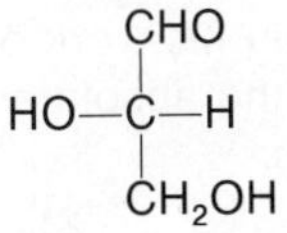

Sec. 17.9 ABSOLUTE AND RELATIVE CONFIGURATIONS

```
        CHO
         |
   HO—C—H     L-(−)-glyceraldehyde
         |
        CH₂OH
```

The configurations are as they are drawn in their Fischer projection formulas and are called the D and L configurations. The small-capital letters D and L refer to the arrangement of the atoms in space. The D configuration is represented by the isomer having the hydroxyl (OH) group on the asymmetric carbon next to the CH$_2$OH group on the right hand side of the molecule. The L configuration has the OH group on the asymmetric carbon atom next to the CH$_2$OH group drawn on the left hand side of the molecule. These assigned structures are done by convention in order to set up some standardized system of comparison between compounds. The D and L merely refer to the configuration of the atoms within the molecule; they indicate nothing at all about the direction of rotation of a plane of polarized light by an optical isomer represented by d or l or by (+) or (−).

Any compound found to be related to the D form of glyceraldehyde in its structure is assigned the D configuration, while any substance structurally similar to the L configuration of glyceraldehyde is assigned the L configuration. For example, the D isomer of glyceraldehyde can be converted by a series of steps to lactic acid. The reactions involved are conversion of the CHO group to a COOH group, and of the CH$_2$OH to a CH$_3$ group, while not in any way changing the configuration of the asymmetric carbon atom.

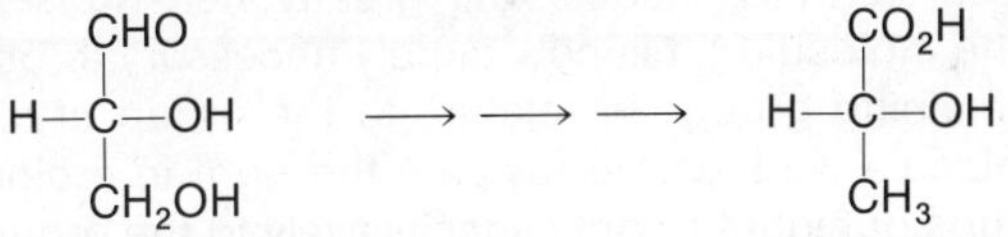

```
        CHO                              CO₂H
         |                                |
    H—C—OH      ⟶  ⟶  ⟶         H—C—OH
         |                                |
        CH₂OH                            CH₃
   D-(+)-glyceraldehyde              D-(−)-lactic acid
```

The lactic acid is said to be of the same *relative configuration* as the starting material in the reaction, and is therefore given the D configuration. However, the lactic acid formed in this reaction just happens by coincidence to be the levorotatory isomer, and this is represented by a minus (−) sign. The product is called D-(−)-lactic acid. Although we started with the dextrorotatory isomer of glyceraldehyde, we do *not* necessarily have to end up with a dextrorotatory product. The configuration of the product is related to the starting material and will be determined by what bonds are broken and formed in all the steps involved in the reaction sequence, but this does not in any way determine which enantiomer the product will be. The D and L configurations will be referred to again in Chapter 18 dealing with the carbohydrates.

Although the relative configuration of many compounds could be determined by relating the structures to a reference compound like D-(+)-glyceraldehyde, the problem of determining the **absolute configuration** of an optically active compound remained unanswered for many years. In 1949, J. M. Bijvoet determined the actual arrangement in space of the atoms in an optically active compound by using rather complex X-ray analysis techniques on crystal structure. It is interesting to mention that Professor Bijvoet was the Director of the van't Hoff Laboratory at the University of Utrecht, and that the first compound whose absolute configuration was determined was the same salt of dextrorotatory tartaric acid that Pasteur had investigated in his work on optical activity. Since 1949 when this technique was introduced the absolute configurations of many organic compounds have been determined.

17.10 The R and S Configurations

The use of D and L configurations does have its limitations. Although it is still used extensively for some classes of compounds today, the D and L designations for configuration do lead to some confusion in certain compounds, and furthermore it is necessary to draw a picture of the structural formula of a molecule before its configuration can be specified as being D or L.

To circumvent these disadvantages of the D,L notation for configuration, a more general and in many respects more satisfactory system was adopted, using the letters R and S rather than D and L. Recently, three prominent chemists, R. S. Cahn, Sir Christopher Ingold, and V. Prelog, suggested the more general R and S specification for configuration.

According to the system devised for the R and S configuration about an asymmetric carbon atom, each of the four atoms attached to the asymmetric carbon is assigned a sequence of priority. The priority depends on the atomic number of the atom, the higher the atomic number the higher the priority (e.g., I has a higher priority than Cl, which has a higher priority than H). Once the sequence of priority has been established, the molecule must be visualized in such a way that the group of lowest priority is directed in space away from the viewer. (The use of FMO models will greatly help in selecting the proper orientation of the molecule.) It is now merely necessary to observe the arrangement of the remaining groups in relation to the group of lowest priority. If, in viewing the molecule we must go towards the right in a clockwise direction in going from groups of higher priority successively to the group of lowest priority, then the molecule is said to be of the R configuration (Latin, rectus = right). On the other hand if we must go toward the left in a counterclockwise direction from the group of highest priority successively to the group of lowest priority, then the molecule is assigned the S configuration (Latin, sinister = left).

Let us illustrate the use of the R and S system of configuration with reference to a specific compound, chlorofluoroiodomethane, *CHFClI. According to the sequence rules, the iodine atom has the highest atomic number and therefore the highest priority. The order of priority for the substituents on the asymmetric carbon would be I > Cl > F > H. Now, we visualize or build a three-dimensional model of the compound so oriented that the group of lowest priority (the H atom) is directed away from us. The R and S configurations are assigned according to which direction we must go in viewing the I, Cl, and F substituents in that order.

R configuration

S configuration

The R and S notations are used before the name of the compound in place of D and L. The sequence rules are more involved when substituents containing more

Sec. 17.11 MORE THAN ONE ASYMMETRIC CARBON 295

than one atom, such as a CH_3, COOH, NO_2, SO_3H, etc., group, are attached to an asymmetric carbon. These are beyond the scope of this text, but suffice it to say that a priority can be established for any group of substituents. Molecules containing more than one asymmetric carbon are assigned an R or S configuration for every asymmetric carbon atom present in the molecule. Overall, the R and S system is much more general and more useful than the D and L designations for configuration, and is coming into more worldwide use from day to day.

17.11 Molecules with More than One Asymmetric Carbon Atom

The **van't Hoff rule** states that if a molecule has n asymmetric carbon atoms, there exists a total of 2^n possible isomers, where n is the number of asymmetric carbon atoms present in the molecule. The rule does not in any way claim that *all* the possible isomers *must* exist, it merely states the maximum number that *may* exist. Since there are but two possible arrangements or configurations possible for an asymmetric carbon atom, the reader can readily see the mathematical basis for the so-called 2^n rule.

According to the van't Hoff rule, for a molecule containing two asymmetric carbon atoms there would be 2^n, or $2^2 = 4$, possible isomers. A typical example would be the compound 3–chloro–2–butanol, CH_3—*CH—*CH—CH_3, with the
$\phantom{CH_3\text{—*CH—*CH—}CH_3,\text{ with the}}$ | |
$\phantom{CH_3\text{—*CH—*CH—}CH_3,\text{ with the }}$Cl   OH

two asymmetric carbon atoms indicated by asterisks. We will consider all the possible isomers of this compound and their relationship to one another. It will be necessary in our subsequent discussion to refer to a group of stereoisomers called diasteroisomers.

17.11-1 Diastereoisomers **Diastereoisomers** are stereoisomers that are *not* superimposable on one another and *not* mirror images. Since diastereoisomers are *not* mirror images, they differ from enantiomers in that they not only affect the rotation of plane-polarized light but they also have different physical properties, such as boiling point, melting point, density, solubility in various solvents, etc.

Let us draw Fischer projection formulas for the four possible isomers of 3–chloro–2–butanol.

```
      CH₃              CH₃              CH₃              CH₃
       |                |                |                |
  H——+——OH         HO——+——H          HO——+——H          H——+——OH
       |                |                |                |
  H——+——Cl         Cl——+——H           H——+——Cl         Cl——+——H
       |                |                |                |
      CH₃              CH₃              CH₃              CH₃
       I                II               III               IV
```

These formulas represent the only possible ways the substituents can be arranged about the asymmetric carbons. (This should be verified by the reader—no other formulas that are not equivalent to one of these structures are possible.)

Structures I and II represent a pair of enantiomers since they are non-superimposable mirror images. They only differ in the direction in which they rotate a plane of polarized light. Structures III and IV constitute a second pair of enantiomers, but differ from structures I and II as to the number of degrees they rotate the plane of polarized light. A mixture of equal amounts of I and II or of III and IV constitutes a racemic mixture.

Now let us discuss the relationship between isomers I and III, I and IV, II and III, and II and IV. All these pairs of isomers are related to each other as **diastereoisomers;** that is, they are non-superimposable and not mirror images.

As was mentioned previously, the van't Hoff rule merely indicates the maximum number of possible isomers. There are molecules that have fewer than 2^n isomers, due to certain special symmetry properties present in the molecule. The simplest case of this occurs in molecules containing two (or more) asymmetric carbon atoms that have similar substituents and are located in such a position that the molecule has a plane of symmetry. The compound tartaric acid, HOOC—*CH(OH)—*CH(OH)—COOH, is a classic example. Each of the asymmetric carbon atoms has the exact same substituents (H, OH, COOH, and the CH(OH)—COOH group). The Fischer projection formulas of the possible isomers of tartaric acid are

```
     COOH            COOH            COOH            COOH
  H──┼──OH       HO──┼──H         H──┼──OH       HO──┼──H      plane of
                                  ───┼───   ≡   ────┼────      symmetry
 HO──┼──H         H──┼──OH        H──┼──OH       HO──┼──H
     COOH            COOH            COOH            COOH
      V               VI              VII             VIII
                                         meso form
```

As can be seen, structures V and VI constitute a pair of enantiomers. Structures VII and VIII are actually identical and represent only one isomer. This can be seen if the Fischer projection formula of structure VIII is rotated by 180° in the plane of the paper. It will then be identical to structure VII, so that structures VII and VIII are actually one and the same.

17.11-2 Meso Structures Thus, in the case of tartaric acid there are only *three* isomers rather than the maximum possible number of four. Two of these isomers (structures V and VI) are a pair of enantiomers and are optically active. Of course, a mixture of 50 % isomer V and 50 % isomer VI constitutes a racemic mixture. The third isomer, represented by structure VII (or VIII), has a plane of symmetry through the center of the molecule. This plane of symmetry makes the top half of the molecule identical with the bottom half. The presence of a plane of symmetry means that each half of the projection formula is the mirror image of the other half. The two halves cancel each other so that this isomer will not rotate a plane of polarized light. This isomer is *optically inactive* and known as the *meso* form or a *meso* compound. (A *meso* compound is one whose molecules are superimposable on their mirror images even though they have asymmetric carbon atoms.) A *meso* form should not be confused with a racemic mixture; although each is optically inactive, a *meso* form is actually a single isomer, whereas a racemic mixture is made up of equal amounts of the dextrorotatory and levorotatory isomers constituting a pair of enantiomers.

The relationship between the *meso* form (VII) and structures V and VI is that of diastereoisomers, which can easily be verified upon inspection of the projection formulas.

Sec. 17.12 RESOLUTION OF RACEMIC MIXTURES

Compounds containing more than two asymmetric carbon atoms have many more possible isomers. The situation can become very complex and significant in problems involving the elucidation of molecular structures as will be seen in Chapter 18 on the carbohydrates.

17.12 Resolution of Racemic Mixtures

The separation of a racemic mixture into its components, the dextrorotatory (+) and levorotatory (−) isomers, is known as **optical resolution.** The actual process of resolving a racemic mixture is often quite complex and cumbersome. If one is exceedingly patient and has a crystalline racemic mixture he may separate the crystals mechanically by viewing the different crystalline forms through a magnifying glass and using forceps or tweezers. Pasteur used this painstakingly slow, but efficient process to separate a racemic mixture of sodium ammonium tartrate into its dextrorotatory and levorotatory components.

Today, the separation of a racemic mixture is usually accomplished by chemical and physical techniques rather than by mechanical methods. It will be recalled that enantiomers have similar physical properties with respect to boiling points, melting points, and solubility in various solvents. Thus, they cannot be separated into their dextrorotatory and levorotatory components by physical methods involving distillation or fractional crystallization, which depend on the physical properties of the material. However, it will also be recalled that diastereoisomers have different physical properties. The separation of a racemic mixture depends on the conversion of the enantiomers into diastereoisomers. This can only be accomplished by the reaction of the enantiomers with another optically active compound. The diastereoisomers, once formed, can then be separated by physical methods because they have different physical properties, such as different solubilities in various solvents. Most resolutions of racemic mixtures make use of this fact.

As an illustrative example let us discuss the process of resolving a racemic acid. The racemic acid is reacted with an optically active base, one of the enantiomers, either (+) or (−), to form a salt. Let us assume that the optically active base used is the pure (+) isomer; then the salts formed will be diastereoisomers as can be seen by the equation for the acid–base reaction.

$$\underbrace{\begin{matrix}(+)\text{-acid} \\ (-)\text{-acid}\end{matrix}}_{\text{racemic mixture}} + (+)\text{-base} \longrightarrow \underbrace{\begin{matrix}(+)\text{-acid} + (+)\text{-base} \\ (-)\text{-acid} + (+)\text{-base}\end{matrix}}_{\text{salts formed which are diastereoisomers}} \qquad (17\text{-}1)$$

The salts are not mirror images, nor are they superimposable, and thus they are diastereoisomers having different physical properties. They may be separated by fractional recrystallization from a solvent if their solubilities differ in that solvent. Each salt that has been separated by the fractional recrystallization technique can be converted into the optically pure dextrorotatory and levorotatory isomers of the original acid by treatment with HCl.

$$\underset{\text{salt}_1}{(+)\text{-acid} + (+)\text{-base}} + \text{HCl} \longrightarrow (+)\text{-acid} + (+)\text{-base} \cdot \text{HCl} \qquad (17\text{-}2)$$

$$\underset{\text{salt}_2}{(-)\text{-acid} + (+)\text{-base}} + \text{HCl} \longrightarrow (-)\text{-acid} + (+)\text{-base} \cdot \text{HCl} \qquad (17\text{-}3)$$

The racemic mixture has been resolved into its pure optically active, (+) and (−), components by this procedure. The optically active bases used in resolution are generally naturally occurring materials. The bases are amines obtained from plants and have rather complex structures. They are called alkaloids, and some representative bases used for resolution include brucine, cocaine, strychnine, quinine, and morphine. The reader will recognize the names of some of these bases (e.g., quinine, morphine) as having certain medicinal properties. Cocaine is used as a narcotic, and the compounds brucine and strychnine are extremely poisonous and must be handled very carefully. The bases can be recovered after the resolution has been completed by treating the base hydrochloride formed with a base, such as NaOH, to regenerate the free base as the amine. The amine can then be used to resolve another racemic mixture.

A basic racemic mixture can be resolved in an analogous manner by using an optically active acid. The resolution of some racemic organic compounds is limited, however, since the resolution process requires the presence of a functional group, such as an acidic or basic function, which can serve as a *"handle"* to react with the optically active reagent and form a pair of diastereoisomers, which can then be separated.

17.13 Resolution and Biochemistry

The resolution process is also of special significance in certain biochemical reactions. It is known for example that certain bacteria will interact with only one optical isomer, (+) or (−), of an enantiomer without regard to the other isomer. This specificity of reaction with only one optical isomer, or **stereoselectivity** as it is sometimes called, is especially apparent in many of the reactions involved in our metabolic processes. The enzymes involved in these metabolic reactions are actually specific catalysts, each enzyme catalyzing only one particular reaction in the chain of metabolic reactions, and the enzymes are usually optically active substances. The theory of the action of an enzyme on a substrate material is referred to as the **"template theory"** or **lock and key theory,** meaning that the enzyme and substrate must fit very precisely as a key fits into a lock in order for the substances to react. The molecular geometry of both the substrate and enzyme must be so exact that the active sites on the enzyme be properly oriented with respect to the substrate when the substrate is attached to the enzyme surface, so that reaction occurs. In view of this, it should not be too surprising that an enzyme may react with only one isomer, (+) or (−), of an enantiomer in preference to the other isomer just because of the precise geometrical features required in enzymatic reactions. Even an extremely minute difference in molecular structure, such as in the configurational differences between enantiomers, is significant enough to determine whether a reaction will occur.

Stereospecific reactions occur whenever an enzyme is reacted with a racemic mixture, so that only one of the components, (+) or (−), of the racemic mixture is acted upon by the enzyme, leaving the other pure optical isomer behind. This means of biochemical resolution is not feasible for very many reactions; but when it can be used, one of the enantiomers can be recovered, with the other used up by the action of the enzyme. This process is typical of many enzymatic reactions that occur in living cells.

17.14 A Brief Consideration of Some Reactions Involving Stereoisomers

Let us cite a few examples to illustrate some of the principles involved in stereochemical considerations with respect to certain reactions.

The stereochemistry of the free-radical chlorination of *n*-butane to form 2-chlorobutane, whose mechanism (the free-radical halogenation of alkanes) has been discussed in detail in Sections 6.8 and 7.7-1, will be considered first. The important point to remember is that the reaction involves free radicals that are flat, coplanar species with bond angles of 120° since they have sp^2 hybridization. In the course of the reaction sequence, the secondary butyl free radical, CH_3—$\dot{C}H$—CH_2—CH_3, is formed by abstraction of a hydrogen atom. The next propagation step involves the reaction of this free radical with a chlorine molecule to produce *sec*-butyl chloride or 2-chlorobutane.

$$CH_3-\dot{C}H-CH_2-CH_3 + Cl_2 \longrightarrow CH_3-\overset{H}{\underset{Cl}{\overset{*}{C}}}-CH_2-CH_3 \left(\text{or } CH_3-\overset{Cl}{\underset{H}{\overset{*}{C}}}-CH_2-CH_3 \right) + Cl\cdot$$

sec-butyl free radical   2-chlorobutane

(17-4)

The 2-chlorobutane formed is a racemic mixture containing 50% of the dextrorotatory and 50% of the levorotatory isomer. The *n*-butane (our starting material) has no asymmetric carbon atom, whereas the product, 2-chlorobutane, has one such asymmetric center that was created during the course of the reaction. The reason for the formation of a racemic mixture, rather than one particular enantiomer, as a reaction product, can be seen by an examination of the reaction of the *sec*-butyl free radical with chlorine. Since the free radical is flat and coplanar, there is an equal probability that the chlorine can attack the free radical from above or below the plane of the free radical. This results in the formation of the exactly 50 % *d* and 50 % *l* isomer that constitutes a racemic mixture.

Indeed, most reactions starting with optically inactive materials (no asymmetric carbon atom present) and generating a new asymmetric center in the product, will usually form a racemic mixture or result in partial racemization rather than a pure enantiomer. Many reactions of stereoisomers involving carbonium ions will form a racemic mixture or result in a partially racemized product. Why? (Consider the hybridization and structure of carbonium ions.)

Now let us devote our attention to a slightly different problem, that of seeing how the configuration of the products can be related to the reactants in a reaction.

The reaction of 2-methyl-1-butanol with phosphorus tribromide is a typical reaction of an alcohol to form an alkyl halide. Let us assume that we have available the pure dextrorotatory isomer of the alcohol and that it is of the *R* configuration. If this optically pure isomer is reacted with PBr_3, the resulting alkyl halide, 1-bromo-2-methylbutane, also has the *R* configuration.

The product has the configuration of the reactant, since *no* bond attached to the asymmetric carbon atom has been broken during the course of the reaction. Only the C—O bond was broken and a new C—Br bond formed at this

$$\underset{(R)\text{-}(+)\text{-}2\text{-methyl-1-butanol}}{CH_3-CH_2-\overset{*}{\underset{H}{\overset{CH_3}{C}}}-\underset{H}{\overset{H}{C}}-OH} \xrightarrow{PBr_3} \underset{(R)\text{-}(+)\text{-}1\text{-bromo-2-methylbutane}}{CH_3-CH_2-\overset{*}{\underset{H}{\overset{CH_3}{C}}}-\underset{H}{\overset{H}{C}}-Br} \qquad (17\text{-}5)$$

location in the molecule. Since we have not in any way destroyed the configuration of the groups attached to the asymmetric carbon atom, the configuration related to that carbon atom should be the same in the product as in the reactant. *Reactions that do not involve the breaking of bonds directly attached to an asymmetric carbon atom will result in the formation of a product having the same configuration about that asymmetric carbon atom as the reactant.* The relation between the direction and the amount of rotation of the plane of polarized light in the reactant and product is independent of any consideration with regard to configuration. Compounds of the same configuration do not necessarily rotate light in the same direction.

17.14–1 The Walden Inversion Let us now consider a reaction in which a bond attached to the asymmetric carbon is broken in the course of the reaction. Such a reaction is exemplified by the conversion of 2-bromooctane to 2-octanol by NaOH under S_N2-reaction conditions. (Indeed, this was one of the classical experiments that provided much information and evidence regarding the S_N2 mechanism.)

$$\underset{(S)\text{-}2\text{-bromooctane}}{CH_3-\overset{*}{\underset{Br}{\overset{H}{C}}}-C_6H_{13}} \xrightarrow[S_N2]{NaOH} \underset{(R)\text{-}2\text{-octanol}}{CH_3-\overset{*}{\underset{H}{\overset{OH}{C}}}-C_6H_{13}} \qquad (17\text{-}6)$$

If this reaction is started with the pure *S* isomer of 2-bromooctane, the product is the pure (100 %) *R* isomer of 2-octanol. The reaction is said to have gone with *complete inversion of configuration*. The product has a configuration opposite to that of the reactant. This complete inversion process, which is characteristic of all S_N2 reactions, is referred to as a *Walden inversion,* in honor of Paul Walden who discovered this phenomenon in 1896. The reason for the inversion of configuration can be understood if we recall the mechanism for the S_N2 reaction (Section 6.7). The reaction involves attack of a nucleophilic reagent (in this case $OH^{\ominus}$) at the back side of the carbon from where the leaving group ($Br^{\ominus}$) is being displaced. A C—Br bond is being broken at the same time that a HO—C bond is being formed at the asymmetric carbon atom. This can be represented by a diagram of the transition state for the reaction.

$$\begin{bmatrix} & H_3C_{\delta\oplus}H \\ HO & \cdots C \cdots Br \\ \delta^{\ominus} & | & \delta^{\ominus} \\ & C_6H_{13} & \end{bmatrix}$$

transition state for
S_N2 reaction

SUMMARY

The product will have the hydroxyl group attached to the asymmetric carbon atom on the opposite side of the molecule to where the bromine was present in the reactant, thus resulting in the change (inversion) of configuration in the product.

The breaking of a bond attached to an asymmetric carbon atom does not necessarily mean that the configuration of the product will differ from that of the reactant. *If the attacking reagent comes in from the back side of the asymmetric carbon to where the bond is being broken, then the product will have a configuration opposite to the reactant, but if the reagent attacks the asymmetric carbon on the same side of the molecule as where the bond is broken, then the product will have the same configuration as the reactant and the reaction is said to proceed with retention of configuration.*

Most reactions involving the breaking of a bond attached to an asymmetric carbon atom proceed with some inversion of configuration, but not always 100 % complete inversion, as in S_N2 reactions. Reactions that proceed with retention of configuration are much less common.

Summary

- Stereoisomers are compounds that differ only in the spatial arrangement of the atoms.
- Molecules exhibit optical isomerism when there is no element of symmetry in their structures.
- Optical activity can be detected with a polarimeter by measuring the angle of rotation of plane-polarized light passed through a sample. Dextrorotatory isomers rotate the light to the right (clockwise). Levorotatory isomers rotate the light to the left (counterclockwise).
- The amount of rotation is measured by the specific rotation.
- Asymmetry is related to a tetrahedral structure for the carbon atom.
- An asymmetric carbon atom (chiral center) has four different substituents attached to it.
- Enantiomers are non-superimposable mirror images.
- Fischer projection formulas are useful in understanding the concepts of stereoisomerism through drawing planar representations of three-dimensional structures.
- A racemic mixture or racemic modification contains equal amounts of dextrorotatory and levorotatory isomers, and is optically inactive.
- The spatial arrangement of the atoms in a molecule is called the configuration and can be expressed in either relative or absolute terms.
    (a) The D and L configurations used glyceraldehyde as a standard reference, and devised a technique of assigning relative configurations to compounds.
    (b) Recently, the R and S absolute configurations have come into prominence.
- The van't Hoff rule enables one to predict the maximum number of possible optical isomers.
- Compounds containing more than one asymmetric carbon atom can have diastereoisomers.
    The *meso* form is optically inactive.

- Racemic mixture can sometimes be resolved by conversion to diastereoisomers or by interaction of enantiomers with enzymes, a stereoselective process.
- Many organic reactions leading to the formation of asymmetric carbon atoms produce racemic mixtures.

In any reaction, the configuration of the products in relation to the reactants depends on the mechanism of the reaction, and whether any bonds attached to an asymmetric carbon atom are broken during the course of the reaction (Walden inversion, etc.).

PROBLEMS

1 Illustrate each of the following with a specific example.
   (a) asymmetric carbon atom
   (b) enantiomorphs
   (c) racemic mixture
   (d) diastereoisomers
   (e) *meso* form
   (f) resolution
   (g) Walden inversion
   (h) compound of S configuration

2 Which of the following compounds have geometrical and (or) optical isomers? For compounds having stereoisomers, draw the structural formulas for the *cis* and *trans* isomers, the enantiomorphs, diastereoisomers, *meso* forms, etc.

   (a) 2-butene
   (b) β-aminopropionic acid
   (c) 3-chloro-3-methylhexane
   (d) 3,3-dibromohexane
   (e) 3,4-dibromo-3-methylhexane

   (f) [biphenyl structure with HOOC, COOH, F, F substituents]

   (g) *trans*-1,2-cyclohexanediol

   (h) CH$_3$—CH=C—C—Br with H, H, CH$_3$ substituents

3 (a) A compound is named D-(+)-2-butanol. What does the D and the (+) indicate? Draw a Fischer projection formula of the compound.
   (b) Draw the Fischer projection formula for (R)-1-bromo-1-chloroethane.

4 Draw projection formulas for all the optical isomers of each of the following compounds.
   (a) HOOC—CH(CH$_3$)—CH(Br)—COOH
   (b) HO—CH$_2$—(CHOH)$_4$—CHO
   (c) 2,3-butanediol

Indicate which isomers constitute a pair of enantiomorphs, diastereoisomers, or *meso* forms. Which isomers are optically active, and which are optically inactive? Which isomers can form a racemic mixture?

5 Specify whether the following stereoisomers are of the R or of the S configuration.

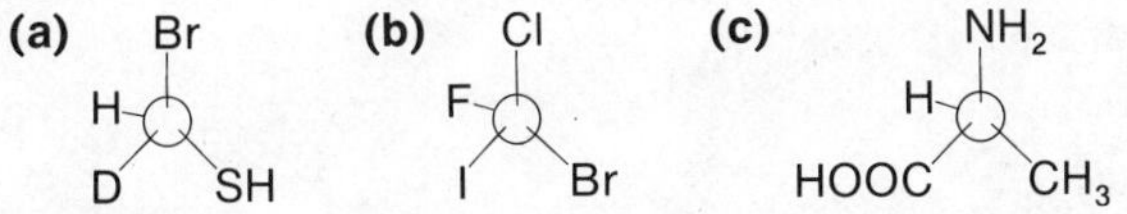

PROBLEMS

(a) Br, H, D, SH (b) Cl, F, I, Br (c) NH₂, H, HOOC, CH₃

6 Indicate the configuration of the major reaction product in each of the following reactions. Explain your answer.
(a) (S)-2-methyl-1-butanol + HCl ⟶ 1-chloro-2-methylbutane
(b) (R)-sec-butyl bromide + alc. KOH ⟶ ?
(c) (R)-2-methyl-2-chlorobutanoic acid + NH₃ $\xrightarrow{S_N 2 \text{ conditions}}$ (?)

7 Explain why each of the following reactions results in the formation of a racemic mixture.
(a) 2-butene + HI ⟶ 2-iodobutane
(b) (S)-(+)-1-chloro-2-methylbutane + Cl₂ $\xrightarrow{h\nu}$ (±)-1,2-dichloro-2-methylbutane
(c) acetaldehyde + HCN ⟶ acetaldehyde cyanohydrin

8 *Whenever an optically pure alkyl halide (such as 100% pure (+)-2-bromooctane) is reacted with a nucleophilic reagent under $S_N 1$ reaction conditions, the product formed is partially racemized. Give an explanation as to why all $S_N 1$ reactions proceed with partial racemization.

9 *Consider the reaction of (a) *cis*-2-butene, and (b) *trans*-2-butene with bromine in carbon tetrachloride. Recall that the reaction proceeds by electrophilic addition across the carbon–carbon double bond. The reaction proceeds by *trans* addition of the bromine to the alkene. Suggest a detailed mechanism for the reaction of (a) *cis*-2-butene, and (b) *trans*-2-butene with bromine, indicating the stereochemistry of the reaction products. Show clearly the course of the reaction. Which products are enantiomorphs, diastereoisomers, meso forms, racemic mixtures, etc.? Which products are optically active and which are optically inactive?

18 Carbohydrates

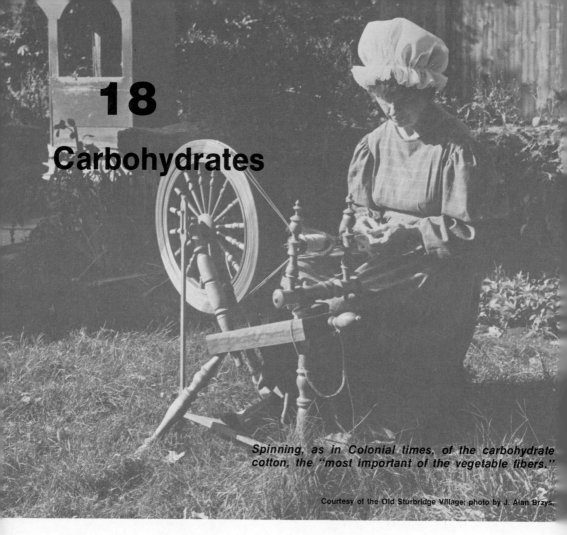

Spinning, as in Colonial times, of the carbohydrate cotton, the "most important of the vegetable fibers."

Courtesy of the Old Sturbridge Village; photo by J. Alan Brzys

18.1 Introduction

The carbohydrates are found throughout the plant and animal kingdom in the form of sugars, starches, cellulose, and many other compounds. The formulas of many carbohydrates are quite complex. For example, many of the components of paper are carbohydrates whose structures are not definitely known. Some carbohydrates have exceedingly simple structures such as the sugars ribose and deoxyribose, and yet they form an essential portion of two very complex and important molecules, RNA (ribosenucleic acid) and DNA (deoxyribosenucleic acid). Carbohydrates are essential in our metabolism as a source of energy and are important entities in many industrial processes.

18.2 Nomenclature and Classification of Carbohydrates

The name **carbohydrate** is used to designate a large class of substances, most of which have the empirical formula $C_n(H_2O)_m$. The compounds representing the carbohydrates all contain hydrogen and oxygen in the same ratio as is found in

Sec. 18.3 THE FISCHER–KILIANI SYNTHESIS

water, and the name results from the compounds' being considered as hydrates of carbon. The names of carbohydrates are usually the common names for the compounds, and many have the ending -*ose* (e.g., glucose, fructose, sucrose).

The carbohydrates are all polyhydroxyaldehydes or polyhydroxyketones, or substances that produce these compounds upon hydrolysis. In order to understand the chemistry of the carbohydrates it is therefore essential to know the chemistry of the aldehydes, ketones, and alcohols as well as to have a thorough knowledge of the formation and reactions of acetals and hemiacetals.

The simplest class of carbohydrates, in terms of molecular structure, are the **monosaccharides,** which can be subdivided into **aldoses** and **ketoses,** depending whether the carbonyl group is present in the form of an aldehyde or ketone. Further, the aldoses and ketoses are named according to the number of carbon atoms present in the molecule. For example, a ketotetrose would represent a four-carbon monosaccharide containing a carbonyl group in the form of a ketone. Most monosaccharides are pentoses and hexoses.

The term **disaccharide** is used to describe a molecule containing two monosaccharide units joined together. Hydrolysis of a disaccharide will yield the monosaccharides of which it is composed. The disaccharides may be considered to belong to a group of carbohydrates called **oligosaccharides.** These are carbohydrates that can be hydrolyzed to produce two to nine monosaccharide units each. **Polysaccharides** are actually gigantic molecules or polymers that contain many monosaccharide units joined together, sometimes in rather complex structures. Polysaccharides can be hydrolyzed to their monosaccharide components.

18.3 The Fischer–Kiliani Synthesis

The renowned German chemist, Emil Fischer (1852–1919) is the man primarily responsible for the synthesis and elucidation of many carbohydrate structures. He not only synthesized many carbohydrates, but he brilliantly determined the stereochemistry present in the molecules, and came to be known as the "Father of carbohydrate chemistry." For his outstanding work he received the Nobel Prize in 1902.

The Fischer–Kiliani synthesis is a classical procedure for the conversion of a sugar or carbohydrate into another carbohydrate containing one more carbon atom. The reaction sequence can be repeated over and over again so that a rather simple molecule can be converted into one containing more carbon atoms with a particular stereochemistry. Many of the steps in the Fischer–Kiliani synthesis are reactions we have discussed in earlier chapters. The starting material in the synthesis is either D-(+)-glyceraldehyde or L-(−)-glyceraldehyde. As will be recalled from Section 17.9-1, the D-(+)-glyceraldehyde is used as a reference compound in relating configurations of various substances similar in structure. Thus, sugars made from D-(+)-glyceraldehyde will be of the D series and are called D-sugars, while L-(−)-glyceraldehyde will produce L-sugars.

As an illustration of the Fischer–Kiliani synthesis, let us show the steps in the conversion of D-(+)-glyceraldehyde into two four carbon sugars, the aldotetroses, D-(−)-erythrose and D-(−)-threose.

$$\underset{\text{D-(+)-glyceraldehyde}}{\overset{\overset{\text{O}}{\underset{|}{\text{C-H}}}}{\underset{\text{CH}_2\text{OH}}{\overset{|}{\underset{|}{\text{H---OH}}}}}} \xrightarrow{\text{HCN}} \underset{\text{CH}_2\text{OH}}{\overset{\text{CN}}{\underset{|}{\overset{|}{\text{H---OH}}}\overset{|}{\underset{|}{\text{H---OH}}}}} + \underset{\text{CH}_2\text{OH}}{\overset{\text{CN}}{\underset{|}{\overset{|}{\text{HO---H}}}\overset{|}{\underset{|}{\text{H---OH}}}}} \xrightarrow{\text{H}^{\oplus},\text{H}_2\text{O}} \quad (18\text{-}1)$$

$$\xrightarrow{-\text{H}_2\text{O}} \underset{\text{CH}_2\text{OH}}{\overset{\text{COOH}}{\underset{|}{\overset{|}{\text{H---OH}}}\overset{|}{\underset{|}{\text{H---OH}}}}} + \underset{\text{CH}_2\text{OH}}{\overset{\text{COOH}}{\underset{|}{\overset{|}{\text{HO---H}}}\overset{|}{\underset{|}{\text{H---OH}}}}} \quad (18\text{-}2)$$

$$\longrightarrow \text{(lactones)} + \text{(lactones)} \xrightarrow{\text{Na(Hg)}} \quad (18\text{-}3)$$

$$\underset{\text{D-erythrose}}{\overset{\overset{\text{O}}{\underset{|}{\text{C-H}}}}{\underset{\text{CH}_2\text{OH}}{\overset{|}{\underset{|}{\text{H---OH}}}\overset{|}{\underset{|}{\text{H---OH}}}}}} + \underset{\text{D-threose}}{\overset{\overset{\text{O}}{\underset{|}{\text{C-H}}}}{\underset{\text{CH}_2\text{OH}}{\overset{|}{\underset{|}{\text{HO---H}}}\overset{|}{\underset{|}{\text{H---OH}}}}}} \quad (18\text{-}4)$$

The first step in the reaction sequence involves the formation of a cyanohydrin by HCN attacking the carbonyl group of D-(+)-glyceraldehyde. Since the carbonyl group is flat and coplanar (sp^2 hybridization) the HCN can attack from either side of the molecule, resulting in the formation of two cyanohydrins and the creation of a new asymmetric carbon atom. The two cyanohydrins are diastereoisomers of each other and the configurational differences present in them are preserved throughout the subsequent steps in the synthesis, so that two new isomeric sugars are obtained with one more carbon atom in the chain than was present in the D-(+)-glyceraldehyde. Since the sugars are derived from D-(+)-glyceraldehyde, the products are D-sugars; the D refers only to the configuration of the carbon atom next to the CH_2OH group derived from D-(+)-glyceraldehyde. An L-sugar would differ from a D-sugar in the configuration of the asymmetric carbon atom next to the CH_2OH group derived from L-(−)-glyceraldehyde. Compare D- and L-threose.

$$\underset{\text{D-(+)-glyceraldehyde}}{\overset{\text{CHO}}{\underset{\text{CH}_2\text{OH}}{\overset{|}{\underset{|}{\text{H---C---OH}}}}}} \longrightarrow \underset{\text{D-threose}}{\overset{\text{CHO}}{\underset{\text{CH}_2\text{OH}}{\overset{|}{\underset{|}{\text{HO---H}}}\overset{|}{\underset{|}{\text{H---OH}}}}}} \quad (18\text{-}5)$$

```
        CHO                    CHO
   HO—C—H      ⟶       H——OH                        (18-6)
       CH₂OH             HO——H
                              CH₂OH
    L-(−)-glyceraldehyde    L-threose
```

18.4 Monosaccharides

Most monosaccharides are optically active compounds, and many contain at least several asymmetric carbon atoms in the molecule. Thus, a particular monosaccharide is only one individual isomer of a group of possible isomers. The configurational relationship of one monosaccharide molecule to another is extremely important; and in particular, the configuration of each asymmetric center in a molecule is of prime significance in differentiating among various isomers. This can be seen by comparing representative formulas of some monosaccharides.

The D-ribose and the 2-deoxy-D-ribose are aldopentoses that are important constituents of RNA and DNA.

```
         O                      O
         ‖                      ‖
         C—H                    C—H
    H—C—OH                 H—C—H
    H—C—OH                 H—C—OH
    H—C—OH                 H—C—OH
        CH₂OH                  CH₂OH
       D-ribose           2-deoxy-D-ribose
```

RNA and DNA are polymers involved in the synthesis of proteins in the cell and the transfer of genetic information.

A typical ketohexose sugar is represented by D-(−)-fructose.

```
              CH₂OH
              C=O
         HO—C—H
          H—C—OH
          H—C—OH
              CH₂OH
          D-(−)-fructose
```

Three common aldohexoses are the sugars glucose, mannose, and galactose.

```
      O                    O                    O
      ‖                    ‖                    ‖
    ₁C—H                 ₁C—H                 ₁C—H
  H—₂C—OH            HO—₂C—H              H—₂C—OH
  HO—₃C—H            HO—₃C—H              HO—₃C—H
  H—₄C—OH             H—₄C—OH             HO—₄C—H
  H—₅C—OH             H—₅C—OH              H—₅C—OH
    ₆CH₂OH               ₆CH₂OH               ₆CH₂OH
   D-(+)-glucose       D-(+)-mannose        D-(+)-galactose
```

These three sugars are stereoisomers differing only in the configuration about one carbon atom. The D-(+)-glucose differs from D-(+)-mannose in the configuration about carbon 2, while D-(+)-galactose differs from D-(+)-glucose only in the configuration at carbon 4.

Perhaps the most important monosaccharide is glucose, which is found in many fruits and honey, and is an essential constituent of the blood. It is commonly called *dextrose* and occurs in nature principally as the optically active dextrorotatory isomer. A great deal of experimentation and research has been devoted to elucidating the structure of glucose, and the methods used are similar to those used in the elucidation of other carbohydrate structures. For that reason, it might be worthwhile to discuss the procedure involved in the structure determination of the glucose molecule.

18.5 Structure Determination of Glucose

Glucose can be shown by experiment to have an empirical formula of CH_2O and a molecular weight of 180, corresponding to a true molecular formula of $C_6H_{12}O_6$. The question to be answered at this point is, what is the actual arrangement of the atoms in the glucose molecule? When glucose is treated with hydrogen iodide, a reduction reaction occurs leading to the formation of 1-iodohexane and some 2-iodohexane as products.

$$C_6H_{12}O_6 \xrightarrow{HI} CH_3-(CH_2)_4-CH_2I + CH_3-\overset{I}{\underset{|}{CH}}-(CH_2)_3-CH_3 \quad (18\text{-}7)$$

glucose　　　1-iodohexane　　　　　2-iodohexane

This experiment leads to the conclusion that the six carbon atoms of glucose are arranged in a straight chain rather than a branched chain. The presence of an aldehyde group at the end of the carbon chain is confirmed by glucose forming a silver mirror when subjected to Tollens' reagent. The presence of the aldehyde function in the molecule is further substantiated by the formation of gluconic acid, $C_6H_{12}O_7$, when glucose is oxidized with bromine water. *Only an aldehyde functional group* can be oxidized to a carboxylic acid containing one more oxygen atom and the same number of hydrogen atoms. Gluconic acid can be oxidized by

Sec. 18.5 STRUCTURE DETERMINATION OF GLUCOSE

nitric acid to glucaric acid, which is a dicarboxylic acid with the formula $C_6H_{10}O_8$. This formula indicates the presence of a primary alcohol group ($-CH_2OH$) in glucose. (Why?)

$$C_6H_{12}O_6 \xrightarrow[H_2O]{Br_2} C_6H_{12}O_7 \xrightarrow{HNO_3} C_6H_{10}O_8 \qquad (18\text{-}8)$$

glucose   gluconic   glucaric
                 acid   acid

Reduction of glucose with sodium amalgam produces the compound sorbitol, $C_6H_{14}O_6$, containing six hydroxyl groups. Acetic anhydride forms a pentaacetate, $C_6H_7O(COOCH_3)_5$, with glucose, indicating the presence of five hydroxyl groups in the glucose molecule.

The above experiments and many similar additional experiments gave evidence that the structural formula of the glucose molecule must be a chain of six carbon atoms in a row, with an aldehyde group at the end of the chain and five hydroxyl groups attached to the other five carbon atoms of the chain.

$$\begin{array}{c}
\overset{O}{\underset{1}{C}}-H \\
_2CH(OH) \\
_3CH(OH) \\
_4CH(OH) \\
_5CH(OH) \\
_6CH_2OH
\end{array}$$

glucose

On the basis of this structure for glucose the previously discussed reactions can easily be understood. (See Formulas 18–9 to 18–12, page 310.)

Although the basic structural formula of glucose has been resolved, nothing has been said about the configurations of the individual carbon atoms, or in other words, about the stereochemistry of the glucose molecule. The structural formula of glucose indicates the presence of four asymmetric carbon atoms at positions 2, 3, 4, and 5. Thus, there are sixteen ($2^n = 2^4 = 16$) possible isomers according to van't Hoff's rule, only *one* of which is glucose. The remaining isomers are other carbohydrates having the same molecular formula as glucose, $C_6H_{12}O_6$, with a different arrangement of the atoms within the molecules.

By an application of the principles of stereochemistry it is possible to determine the configuration of the substituents about each asymmetric carbon atom in the glucose molecule as well as in other carbohydrates. The actual procedures involved are rather complex and beyond the scope of this discussion. The data obtained give conclusive evidence that the correct structural formula for the glucose molecule can be represented by the Fischer projection formula for D-(+)-glucose (on page 310) which clearly shows the configuration about the asymmetric carbon atoms at positions 2, 3, 4, and 5.

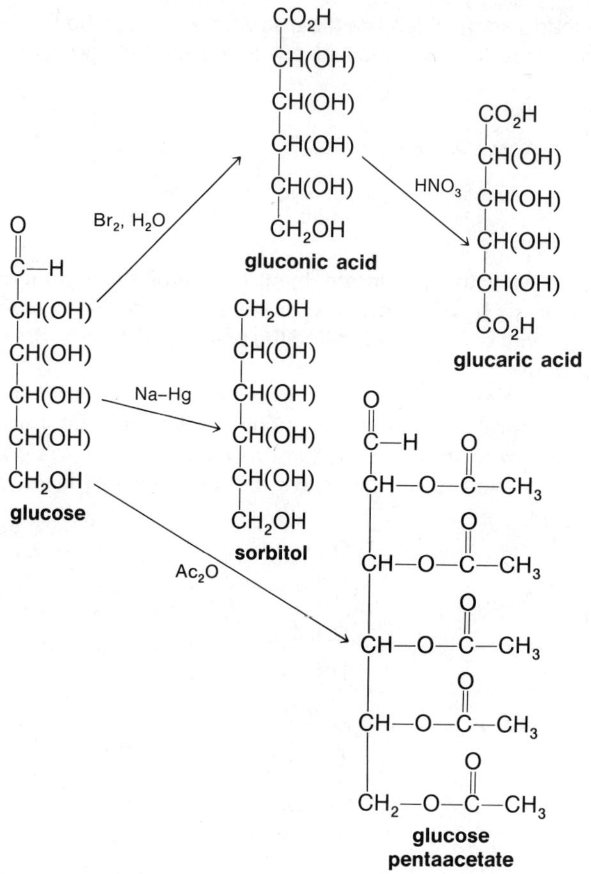

310   CARBOHYDRATES Ch. 18

$$\begin{array}{c}
CO_2H \\
CH(OH) \\
CH(OH) \\
CH(OH) \\
CH(OH) \\
CH_2OH \\
\text{gluconic acid}
\end{array}
\xrightarrow{HNO_3}
\begin{array}{c}
CO_2H \\
CH(OH) \\
CH(OH) \\
CH(OH) \\
CH(OH) \\
CO_2H \\
\text{glucaric acid}
\end{array}
\quad (18\text{-}9)$$

(18-10)

$$\begin{array}{c}
O \\
\parallel \\
C-H \\
CH(OH) \\
CH(OH) \\
CH(OH) \\
CH(OH) \\
CH_2OH \\
\text{glucose}
\end{array}
\xrightarrow{Br_2, H_2O}
\xrightarrow{Na-Hg}
\begin{array}{c}
CH_2OH \\
CH(OH) \\
CH(OH) \\
CH(OH) \\
CH(OH) \\
CH_2OH \\
\text{sorbitol}
\end{array}
\xrightarrow{Ac_2O}$$

(18-11)

$$\begin{array}{c}
O \\
\parallel \\
C-H \\
CH-O-\overset{O}{\underset{\parallel}{C}}-CH_3 \\
CH-O-\overset{O}{\underset{\parallel}{C}}-CH_3 \\
CH-O-\overset{O}{\underset{\parallel}{C}}-CH_3 \\
CH-O-\overset{O}{\underset{\parallel}{C}}-CH_3 \\
CH_2-O-\overset{O}{\underset{\parallel}{C}}-CH_3 \\
\text{glucose} \\
\text{pentaacetate}
\end{array}$$

(18-12)

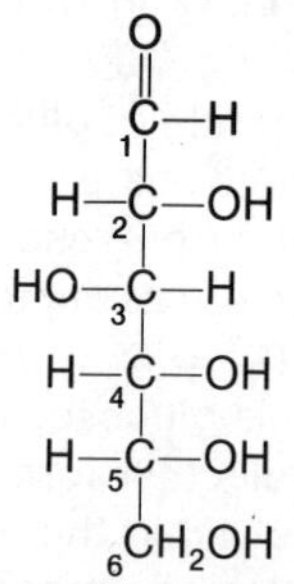

$$\begin{array}{c}
O \\
\parallel \\
\underset{1}{C}-H \\
H-\underset{2}{C}-OH \\
HO-\underset{3}{C}-H \\
H-\underset{4}{C}-OH \\
H-\underset{5}{C}-OH \\
\underset{6}{C}H_2OH
\end{array}$$
D-(+)-glucose

18.6 Evidence for a Cyclic Formula for Glucose

The open-chain formula for glucose indicates the presence of an aldehyde functional group in the molecule. One would therefore expect glucose to undergo the reactions typical of an aldehyde. Glucose does undergo many of the reactions

Sec. 18.6 CYCLIC FORMULA FOR GLUCOSE

of aldehyde, such as forming a cyanohydrin when treated with hydrogen cyanide, and the formation of an oxime when reacted with hydroxylamine. But, some of the other typical aldehyde reactions give rather unexpected reaction products. For example, when Emil Fischer attempted to convert glucose into its methyl acetal by treating glucose with two moles of methyl alcohol and dry HCl, he obtained *two* different reaction products, each containing only one methoxy group, but with characteristic properties typical of an acetal. It will be recalled that the reaction for acetal formation can be represented by the general equation

$$\underset{\text{aldehyde}}{R-\overset{O}{\overset{\|}{C}}-H} + \underset{\text{alcohol}}{R'OH} \underset{}{\overset{H^{\oplus}}{\rightleftharpoons}} \underset{\text{hemiacetal}}{R-\overset{OH}{\underset{OR'}{\overset{|}{\underset{|}{C}}}}-H} \overset{R'OH}{\rightleftharpoons} H_2O + \underset{\text{acetal}}{R-\overset{OR'}{\underset{OR'}{\overset{|}{\underset{|}{C}}}}-H} \quad (18\text{-}13)$$

The simplest interpretation of the data is that since glucose reacted with only one mole of alcohol to yield two acetals, the glucose molecule must already exist in the form of a hemiacetal. (As can be seen from the previous equation, a hemiacetal requires only one mole of alcohol to be converted into an acetal.) The question to be answered is, how can one depict a hemiacetal structure for glucose that is consistent with all the properties of the glucose molecule?

The answer to this question can be provided by an examination of accurate molecular models of the glucose molecule, and even more conclusively, by appropriate experimental evidence (see Figures 18.1 and 18.2).

Molecular models (such as the FMO models) of the glucose molecule show that the six-carbon chain is not straight, but rather zig-zag in shape. It can also be seen that a substituent on the 4 or 5 position of the chain is actually close to the number 1 position because of the zig-zag nature of a series of carbon atoms joined together by carbon–carbon single bonds. Because of this geometrical arrangement, the hydroxyl groups on carbons 4 or 5 are in close proximity of the carbonyl group at carbon number 1. The hydroxyl groups, which are alcohols, can react with the carbonyl group to form a hemiacetal since the distance between the OH and C=O functional groups is suitable for bonding to occur. Further, the formation of a bond between carbon number 1 and the oxygen atom of the hydroxyl group at position 4 or 5 results in the formation of a five- or six-membered ring that has little or no ring strain. The formation of a stable ring can be considered a driving force for the reaction to produce a cyclic hemiacetal structure.

Glucose and other carbohydrates exist in both five-membered and six-membered cyclic hemiacetal structures. It so happens that in the case of glucose the six-membered ring form of the hemiacetal is the predominant form. This ring is formed by the reaction between the hydroxyl group on carbon 5 and the carbonyl group at carbon 1. The cyclic hemiacetal structure of glucose creates a new asymmetric center at carbon 1, which accounts for the experimental observation by Emil Fischer that two acetals are formed upon treatment of glucose with alcohol. These acetals are optical isomers of each other, differing only with respect to the configuration about carbon 1 in the molecule, and are diastereoisomers.

18.1

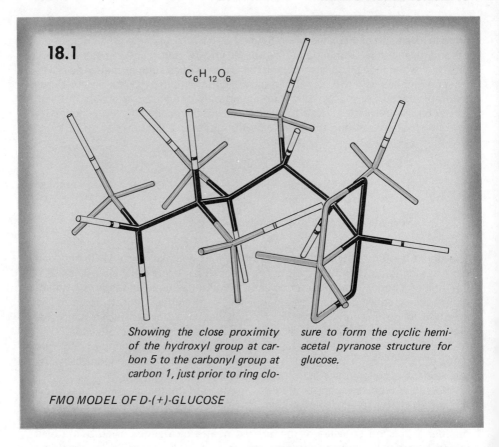

$C_6H_{12}O_6$

Showing the close proximity of the hydroxyl group at carbon 5 to the carbonyl group at carbon 1, just prior to ring closure to form the cyclic hemiacetal pyranose structure for glucose.

FMO MODEL OF D-(+)-GLUCOSE

One isomer, known as the α form, has a specific rotation of +113°, whereas the other isomer, the β form, has a specific rotation of +19°. However, when an aqueous solution of either the α or the β isomer is observed through a polarimeter the rotation of light gradually changes until a value of +52° is obtained. The change in the rotation of light of each isomer until an equilibrium value (or +52°) is reached is known as **mutarotation.** Some molecules of glucose are in the open-chain form, others being in either the α or the β cyclic hemiacetal form. The mutarotation process can be interpreted as the conversion of either isomer when dissolved in solution into an equilibrium mixture of all the forms of the glucose molecule, and the equilibrium mixture has a specific rotation of +52°. (The actual composition of the equilibrium mixture is approximately 37 % α form and 63 % β form; less than 0.1 % of the open-chain form is present at equilibrium. A very small number of the glucose molecules are in the form of five-membered rings also.)

Since the α and β cyclic hemiacetal forms are *diastereoisomers differing only in the configuration at carbon 1,* they are referred to as **anomers.** The α form is drawn, by convention, with the hydroxyl group on carbon 1 on the same side of the molecule as the oxygen bridge formed between carbons 1 and 5; in the β form, the hydroxyl group on carbon 1 is on the opposite side of the molecule from the oxygen bridge.

Sec. 18.6 CYCLIC FORMULA FOR GLUCOSE

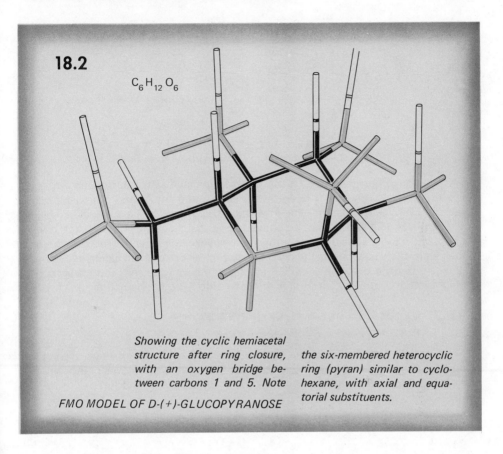

18.2

$C_6H_{12}O_6$

FMO MODEL OF D-(+)-GLUCOPYRANOSE

Showing the cyclic hemiacetal structure after ring closure, with an oxygen bridge between carbons 1 and 5. Note the six-membered heterocyclic ring (pyran) similar to cyclohexane, with axial and equatorial substituents.

(18-14)

β-D-(+)-glucose   open chain form of D-(+)-glucose   α-D-(+)-glucose

The two acetals that Emil Fischer isolated from the reaction of glucose and methyl alcohol can be clearly seen as arising from the reaction of the cyclic α and β hemiacetal forms with one mole of methyl alcohol. The reaction occurred at carbon 1 (which has the hemiacetal structure) and involved the introduction of a methoxyl group in place of the hydroxyl group on carbon 1. The acetals are called **glucosides**.

$$\begin{array}{c} \text{H—C—OCH}_3 \\ \text{H—C—OH} \\ \text{HO—C—H} \\ \text{H—C—OH} \\ \text{H—C} \\ \text{CH}_2\text{OH} \end{array} \quad \begin{array}{c} \text{CH}_3\text{O—C—H} \\ \text{H—C—OH} \\ \text{HO—C—H} \\ \text{H—C—OH} \\ \text{H—C} \\ \text{CH}_2\text{OH} \end{array}$$

methyl α-D-glucoside   methyl β-D-glucoside

The hydroxyl group at carbon 1 is the only true hemiacetal hydroxyl group present in the molecule, and so it is easily converted into the acetal (glucoside) by reaction with methyl alcohol. The other hydroxyl groups in the molecule are like ordinary alcohols and cannot be converted to acetals easily. Mutarotation does not occur in the glucosides, in contrast to the cyclic hemiacetals. This is because the hemiacetals can exist in the equilibrium with the open-chain aldehyde form of glucose, but the glucosides, which are really ethers (acetals), cannot be converted into the open-chain form with a free aldehydic functional group.

18.7 Cyclic Structure of Carbohydrates—Pyranose and Furanose Rings

Most carbohydrates exist in the form of five- or six-membered rings, having a hemiacetal grouping. The nomenclature for these cyclic structures of the carbohydrates comes from the name of the cyclic ethers **pyran** and **furan**.

pyran   furan

The six-membered cyclic ether pyran is used to describe carbohydrates having a six-membered cyclic hemiacetal structure, which is called **pyranose.** Furan, which has a five-membered ring, is used as a basis for naming carbohydrates having a five-membered cyclic hemiacetal structure, which is called **furanose.** The six-membered rings of the carbohydrates are very similar to cyclohexane. One should keep in mind that the chair form is the favored conformation and that the bulkier substituents would prefer the equatorial position rather than the axial. Whenever a choice of isomers exists, the one with the equatorial substituents will probably be the more stable. The conformations of the α and β cyclic hemiacetal forms of D-(+)-glucose can be drawn as

α-D(+)-glucopyranose   β-D(+)-glucopyranose   (18-A)

Sec. 18.8 FRUCTOSE

For convenience some chemists prefer to use so-called **Haworth projection formulas** to represent carbohydrate structures rather than the chair form of cyclohexane. The Haworth formulas show the H and OH groups above or below the plane of the ring.

For example, the Haworth structures of α-D-(+)-glucopyranose and β-D-(+)-glycopyranose would be

α-D(+)-glucopyranose   β-D(+)-glucopyranose

In the α-form the hydroxyl group at carbon 1 is drawn down, whereas in the β-form it is drawn up according to convention. Some typical furanose structures would be

α-D-deoxyribofuranose   α-D-fructofuranose

18.8 Fructose

The sugar D-(−)-fructose is an example of a ketohexose. It is sometimes called levulose or fruit sugar. Fructose is levorotatory and is related structurally to D-glyceraldehyde. The molecular formula of fructose is $C_6H_{12}O_6$. The open-chain structural formula of fructose has a carbonyl group at position 2, and has three asymmetric carbon atoms at positions 3, 4, and 5 that are similar to glucose with respect to their configuration.

β-D-fructose ⇌ D-(−)-fructose ⇌ α-D-fructose   (18-15)

Like glucose, fructose also exhibits the phenomenon of mutarotation. Fructose tends to form an oxygen bridge between carbons 2 and 6 to form a fructopyranose, and an oxygen bridge between carbons 2 and 5, forming a five-membered fructofuranose ring.

α-D-fructopyranose β-D-fructofuranose

18.9 Reactions of Monosaccharides

The monosaccharides undergo many of the characteristic reactions of the hydroxyl and carbonyl functional groups present in the molecule. Some of these reactions have already been mentioned in our discussion on the determination and proof of structure of the glucose molecule. Here we will discuss some other typical reactions of monosaccharides.

18.9-1 Monosaccharides as Reducing Agents Since the carbohydrates contain a carbonyl group in the form of an aldehyde or ketone one might expect that most carbohydrates should be reasonably good reducing agents. This is not completely true since only those sugars having a hydroxyl group on the carbon atom adjacent to the carbonyl group (that is α-hydroxyaldehydes or α-hydroxyketones) are **reducing sugars.** Sugars that do not have this type of structure (e.g., sucrose) are said to be **non-reducing sugars** and will not reduce the typical reagents used, such as Fehling's or Benedict's solution. The distinction whether a particular carbohydrate has reducing properties is often a useful means of determining certain structural features in the molecule.

It will be recalled that Fehling's and Benedict's solutions contain complex ions of $Cu^{\oplus\oplus}$, which are deep blue. A reducing agent, such as a carbonyl group (in an aldehyde) will reduce the $Cu^{\oplus\oplus}$ ion to cuprous oxide, Cu_2O, a red precipitate.

In considering which of the carbohydrates will reduce Fehling's or Benedict's solution, one must look at the structural features in the carbohydrate molecule in order to make an accurate prediction. Carbohydrates exist predominantly in a cyclic hemiacetal structure that is in equilibrium with the open-chain aldehyde or ketone form. It is the free carbonyl group (in the form of an aldehyde or ketone) that is responsible for the reducing ability of the carbohydrate. Thus, if the equilibrium exists so that the cyclic hemiacetal structure can be converted into the open-chain aldehyde or ketone, the carbohydrate will behave as a reducing agent. The presence of an α-hydroxy group adjacent to the carbonyl group in the molecule is essential for this ring-opening process to occur, to form a free carbonyl group that has reducing properties. This means that only carbohydrates having this structure have reducing ability. An inspection of the open-chain or cyclic hemiacetal formulas of glucose and fructose, for example, will reveal that both of these compounds are reducing sugars. Carbohydrates not having this

Sec. 18.9 REACTIONS OF MONOSACCHARIDES

structure, such as glucosides, can not be converted from their cyclic structures into an open-chain form containing a free carbonyl group, and are non-reducing sugars.

18.9-2 Glycosides We have already discussed the reaction of glucose with methyl alcohol to form two acetals, the α- and β-methyl glucosides. The glucosides belong to a larger class of compounds, the **glycosides,** which consist of compounds formed by the reaction of an alcohol or an amine with the hemiacetal oxygen atom. The reaction of a sugar with an amine produces an N-glycoside. The N-glycosides are important in understanding the chemistry of the nucleic acids, RNA and DNA.

$$\text{an } \alpha\text{-sugar} \xrightarrow{RNH_2} \text{an } \alpha\text{-N-glycoside} \qquad (18\text{-}16)$$

18.9-3 Formation of Osazones The reaction of monosaccharides with phenylhydrazine was extremely important in elucidating the structure of certain carbohydrates. The carbonyl group in the carbohydrate reacts with phenylhydrazine to form the expected phenylhydrazone derivative. However, if an excess of reagent is used, a second mole of phenylhydrazine reacts with the adjacent hydroxyl group, oxidizing it to a carbonyl group, which then reacts with a third mole of phenylhydrazine to form a second phenylhydrazone residue in the molecule. The product containing two phenylhydrazone residues in the molecule is a yellow crystalline compound called an **osazone**.

$$\begin{array}{c}\text{O}\\\|\\\text{C}-\text{H}\\|\\\text{CHOH}\\|\end{array} \xrightarrow{3C_6H_5-NHNH_2} \begin{array}{c}\text{CH}=\text{N}-\text{NH}-\text{C}_6\text{H}_5\\|\\\text{C}=\text{N}-\text{NHC}_6\text{H}_5\\|\end{array} + C_6H_5NH_2 + NH_3 \qquad (18\text{-}17)$$

sugar osazone aniline
(aldose or ketose)

In 1884, Emil Fischer reported that this reaction could be used as a tool in the determination of the structure of carbohydrates. Recently it has been shown that osazones have a chelate ring structure, which might offer an explanation as to why the reaction involves only the first two carbon atoms and doesn't proceed any further down the chain.

chelate structure of osazones

Fischer found that since the reaction involved only carbons 1 and 2 in the chain, osazone formation could be used not only to help identify carbohydrates, but also in determining the configurations at other carbon atoms in the molecule. Sugars with the same configurations at carbon atoms 3, 4, and 5 in the chain will give the same osazone when reacted with phenylhydrazine (regardless of any differences in configuration at carbons 1 and 2, which are converted into phenylhydrazone residues in the osazone).

For example, the aldohexoses, D-glucose and D-mannose, as well as the ketohexose sugar D-fructose all give the same osazone when treated with phenylhydrazine. This meant that these sugars differed in their structure only at carbon atoms 1 (and) or 2, and that the configurations of the remaining carbon atoms were identical in the three sugars.

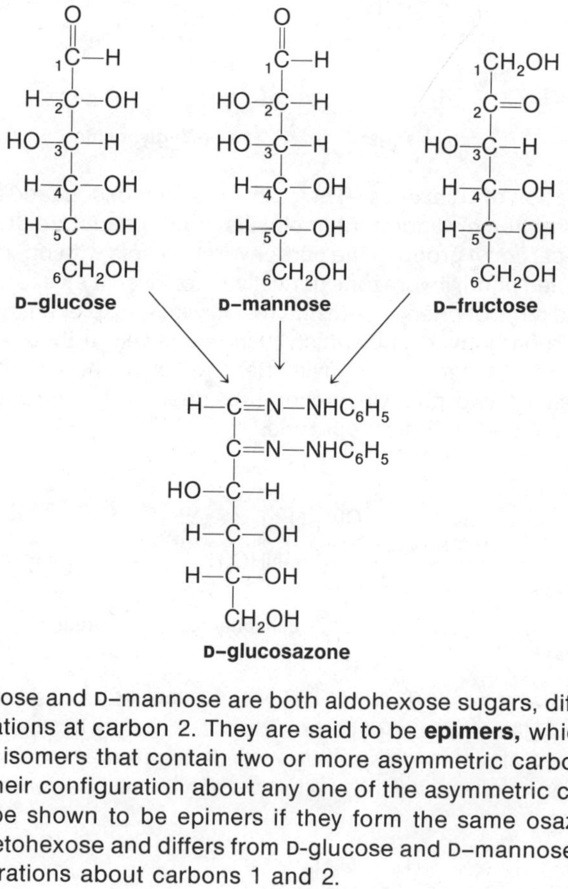

$$
\begin{array}{ccc}
\text{D-glucose} & \text{D-mannose} & \text{D-fructose} \\
\end{array}
$$

$$\downarrow$$

D-glucosazone

(18–18)

The D-glucose and D-mannose are both aldohexose sugars, differing only in their configurations at carbon 2. They are said to be **epimers,** which refers to a pair of optical isomers that contain two or more asymmetric carbon atoms that differ only in their configuration about any one of the asymmetric carbon atoms. Aldoses can be shown to be epimers if they form the same osazone. The D-fructose is a ketohexose and differs from D-glucose and D-mannose with respect to the configurations about carbons 1 and 2.

18.10 Disaccharides

The disaccharides are glycosides, which can be represented by the general formula

Sec. 18.11 LACTOSE

```
        OR
        |
    H—C———————
        |        |
       CHOH      |
        |        |
       CHOH      O
        |        |
       CHOH      |
        |        |
       CH————————
        |
       CH₂OH
```

where R is a monosaccharide that either has a potential free carbonyl group (as in maltose and lactose, reducing sugars) or has no such group (as in sucrose, a non-reducing sugar).

The disaccharides are composed of two monosaccharide units, one of which is a cyclic hemiacetal while the second monosaccharide provides an alcohol (OH) functional group to form the glycoside or acetal (hemiacetal + alcohol → acetal). The disaccharides can be considered to be acetals formed by two monosaccharide units condensing with the loss of water. The hydrolysis of a disaccharide yields the monosaccharide units of which it is composed. The structure of a disaccharide molecule must show which monosaccharide units are present, which groups are involved in the glycosidic linkage, the configuration and ring size of each monosaccharide unit, etc. We shall briefly discuss four important disaccharides: lactose, maltose, cellobiose, and sucrose.

18.11 Lactose

The sugar lactose is found in milk. When hydrolyzed by dilute acid, lactose forms equal amounts of glucose and galactose. This indicates that lactose is composed of one glucose unit and one galactose unit. The determination of the ring size in each monosaccharide unit is accomplished with the aid of a fundamental reaction characteristic of alcohols, the conversion of a hydroxyl group into a methyl ether when treated with dimethyl sulfate.

$$\text{ROH} \xrightarrow[\text{NaOH}]{(CH_3)_2SO_4} \text{R—O—CH}_3 \qquad (18\text{–}19)$$

alcohol   ether

When lactose is methylated with dimethyl sulfate in this manner, all the hydroxyl groups in the molecule are converted into methoxy (OCH_3) groups. Upon hydrolysis of the completely methylated lactose molecule, the monosaccharides obtained were 2,3,6–trimethylglucose and 2,3,4,6–tetramethylgalactose. The glucose fragment had methyl groups missing at positions 1, 4, and 5 while the galactose unit had methyl groups missing at positions 1 and 5. This meant that only two of these carbon atoms could be involved in the formation of a bond between the glucose and galactose units. Since lactose is a reducing sugar, it must have a free or potentially free carbonyl group in its structure. Consideration of this fact and the other properties of lactose leads to a formula for lactose with a 1,4–β linkage between the galactose and glucose units. In other words, carbon 1 of galactose is linked to carbon 4 of glucose (with the splitting out of a water molecule) to form the disaccharide lactose.

320   **CARBOHYDRATES Ch. 18**

[Structure: β-lactose showing D-galactose unit linked to D-glucose unit via 1,4-β linkage; β (can be α)]

β-lactose, 4-O-(α-D-galactopyranosyl)-β-D-glucopyranose

The reducing action of lactose is due to the free hydroxyl group at carbon 1 of the glucose unit that is in equilibrium with the open-chain aldehyde structure.

[Structure: open chain aldehyde form]

open chain aldehyde form

Lactose also forms an osazone typical of an α–hydroxyaldehyde structure.

18.12 Maltose

The disaccharide maltose yields two molecules of glucose upon hydrolysis. When maltose is methylated with dimethyl sulfate followed by hydrolysis, 2,3,4,6-tetramethylglucose and 2,3,6-trimethylglucose are formed. The structural formula, which is in accord with the properties of maltose, consists of two glucose units joined by a 1,4-α linkage between carbon 1 of one glucose unit and carbon 4 of the other glucose unit.

[Structure: α-maltose showing two D-glucose units linked by 1,4-α linkage; β (can be α), (potential free aldehyde group)]

α-maltose, 4-O-(α-D-glucopyranosyl)-β-glucopyranose

Maltose is a reducing sugar, because of the presence of a potential free aldehyde group at carbon 1 of the glucose unit. Maltose can be obtained by the partial hydrolysis of starch, either by dilute acid or by the enzyme diastase. It is fermented by yeast to glucose, and eventually to ethyl alcohol and carbon dioxide.

Sec. 18.16 STARCH

linkages to form a long chain. The chain also has branches composed of glucose units joined by a 1,6 linkage.

The starch in plants is composed of two polysaccharides, called **amylose** and **amylopectin**. When starch is heated with hot water it can be separated into these two components. The part that is soluble in water is about 15–25 % of the starch and is known as **amylose**. The remaining fraction, which is the major component of starch, 75–85 %, is called **amylopectin**.

18.16-1 Amylose

Amylose is the portion of starch responsible for the blue color obtained when starch is reacted with iodine. Amylose is the linear non-branched portion of the starch molecule. It is actually a polymer of many α–D-glucose units linked together by 1,4-α-glycosidic linkages. Each starch molecule consists of 200–1,000 glucose residues. Amylose comprises anywhere between 60 and 300 of these glucose units and has a molecular weight range of 10,000 to 50,000. The Haworth representation of a partial sequence of the amylose section of a starch molecule would look like

amylose section of starch molecule

18.16-2 Amylopectin

Amylopectin is the part of the starch molecule that is a branched polymer of α–D-glucose units. Amylopectin contains 1,6-branching linkages in addition to the 1,4 linkages in the straight-chain portion. The 1,6 linkages occur at intervals of about every 25 glucose units in the chain. Amylopectin comprises about 300 to 6,000 glucose units in a starch molecule and has a molecular weight range of 50,000 to 1,000,000. An amylopectin chain, with the amylose chains linked by a 1,6 linkage is represented by the Haworth formula.

an amylopectin chain showing 1,6 linkage

18.17 Glycogen

The polysaccharide glycogen is found in animal tissues, mainly in the liver and muscles. It is the storage form of carbohydrates, resembles starch, and usually has a higher average molecular weight. Glycogen is similar in structure to amylopectin in that it has long chains of glucose units bound by 1,4 linkages, with some 1,6 branching occurring at intervals in the chain. Glycogen has shorter linear chains, with 12 to 18 glucose units, than amylopectin, which has 24 to 30 glucose units, and has more frequent branching.

Glycogen is involved in maintaining the proper amount of glucose in the blood. The body regulates the amount of blood glucose by removing excess glucose and storing it in the form of glycogen, and supplying glucose to the blood, by breakdown of glycogen, when it is needed for conversion by the body into energy.

18.18 Cellulose

The polysaccharide cellulose is the structural material of plants. It is the major constituent of cell walls, cotton, wood pulp, and many other substances. Partial hydrolysis of cellulose yields cellobiose, which it will be recalled is a disaccharide composed of two glucose units joined by a 1,4-β linkage. Complete hydrolysis of cellulose produces glucose as the reaction product. The structure of the cellulose molecule is very similar to starch, in that the molecule has long chains of glucose units (as in cellobiose) joined together by an 1,4-β linkage, rather than an α linkage as in starch. In contrast to starch, there are no branched chains in cellulose. The molecular weight of cellulose varies between 300,000 and 500,000 and contains between 1,800 and 3,000 glucose units per molecule. Cellulose fibers are formed by hydrogen bonds between hydroxyl groups on adjacent chains, in bundles of chains held close together.

The Haworth projection formula for cellulose is:

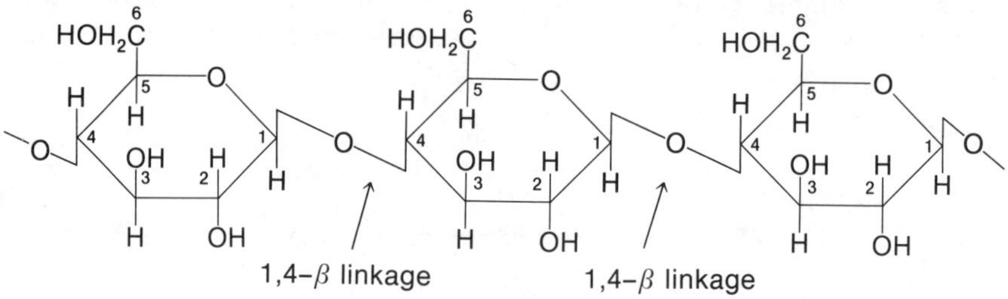

cellulose

As mentioned in our earlier discussion on cellobiose, man can not digest wood pulp or cellulose. The reason is that our digestive system does not have the enzymes capable of hydrolyzing β-glycosidic linkages. Humans can digest starch, since we do have enzymes that can hydrolyze the α linkage in the starch molecule. This is an example of the stereospecificity of enzymatic reactions and biochemical phenomena in general.

18.19 Photosynthesis in Plants

The process by which green plants synthesize carbohydrates, such as sugars, starch and cellulose, from carbon dioxide, water and inorganic salts (from the soil), is known as **photosynthesis**. As the name suggests, light is essential for the process, as well as a pigment that can absorb the light energy and transform it into some form usable for chemical reactions. In green plants, the pigment **chlorophyll** absorbs the sunlight, or other light of a suitable wavelength, enabling the synthesis of the organic compounds to take place. The overall photosynthetic process can be summarized as

$$n\text{CO}_2 + m\text{H}_2\text{O} \xrightarrow[\text{chlorophyll}]{\text{sunlight}} n\text{O}_2 + \underset{\text{carbohydrate}}{\text{C}_n(\text{H}_2\text{O})_m} \qquad (18\text{–}20)$$

It should be understood that this equation merely indicates the starting materials and final reaction products. In the course of photosynthesis many different organic compounds are formed, starting with a one-carbon *inorganic* reagent (CO_2) and terminating in the formation of a complex carbohydrate. The process actually involves numerous reactions catalyzed by many enzymes, each step in the sequence being a building block towards the end products. The determination of the actual intermediate compounds formed, and the sequence of steps in photosynthesis, is a complex research problem. Use of radioactive tracers in experiments, starting by exposing plants to carbon dioxide enriched in carbon[14] isotope ($C^{14}O_2$) results in the formation of labeled compounds containing C^{14}, which has led to a partial solution of the problem. The entire photosynthetic process is not completely understood. A great deal of progress towards the elucidation of the mechanism of photosynthesis has come from the work of Professor Melvin Calvin at the University of California at Berkeley. For his significant contributions in this area of research, Professor Calvin received the Nobel Prize in 1961.

18.20 The Metabolism of Carbohydrates in the Body

Whereas the photosynthetic process is endothermic and requires light energy to convert carbon dioxide and water to carbohydrates and oxygen, the metabolism of carbohydrate foods in the body is an exothermic process, where oxygen from the air we breathe helps convert the carbohydrates to carbon dioxide and water. The metabolism of the carbohydrates is actually the reverse of photosynthesis when one compares the starting materials and products of the two processes. However, the reactions are *not* the same, nor are the intermediate compounds formed during the various steps in the reaction sequences identical. The metabolism of carbohydrates in the body serves as a source of energy and is much more completely understood than is the process of photosynthesis in plants.

Polysaccharides and disaccharides are hydrolyzed, mainly by the enzymes of the saliva (such as ptyalin and maltase) and of the small intestine, to monosaccharides, such as glucose, fructose, and galactose. The monosaccharides react with certain phosphate-containing enzymes, such as ATP (adenosine triphosphate), and are converted to phosphate esters, which are transformed into pyruvic and lactic acids by muscular activity. The pyruvic and lactic acids are oxidized to carbon dioxide and water, which is an energy-yielding process. The overall metabolic process is exothermic and supplies energy for the body's needs.

18.20-1 The Krebs Cycle The entire sequence of reactions is rather complex, involving many steps, each catalyzed by certain enzymes. Some of the intermediate compounds formed are familiar to us as unsaturated acids or hydroxyacids. The entire sequence of reactions has been worked out by Krebs, and is known as the **Krebs cycle** or **citric acid cycle**. It is summarized in Figure 18.3, where we have used simple abbreviations for the enzymes* (e.g., CoA·SH for acetyl-coenzyme A; ATP for adenosine triphosphate; TPN for triphosphopyridine nucle-

---

*Strictly speaking, these substances are called **coenzymes** rather than enzymes. Each step in the Krebs cycle requires a specific enzyme (which we have not listed for the sake of simplicity). Many of these enzymes require the presence of certain complex organic **cofactors** or **coenzymes** in order to function. The cofactors or coenzymes constitute the non-protein part of the enzyme, and it is these substances that we have included in our Krebs cycle diagram.

18.3

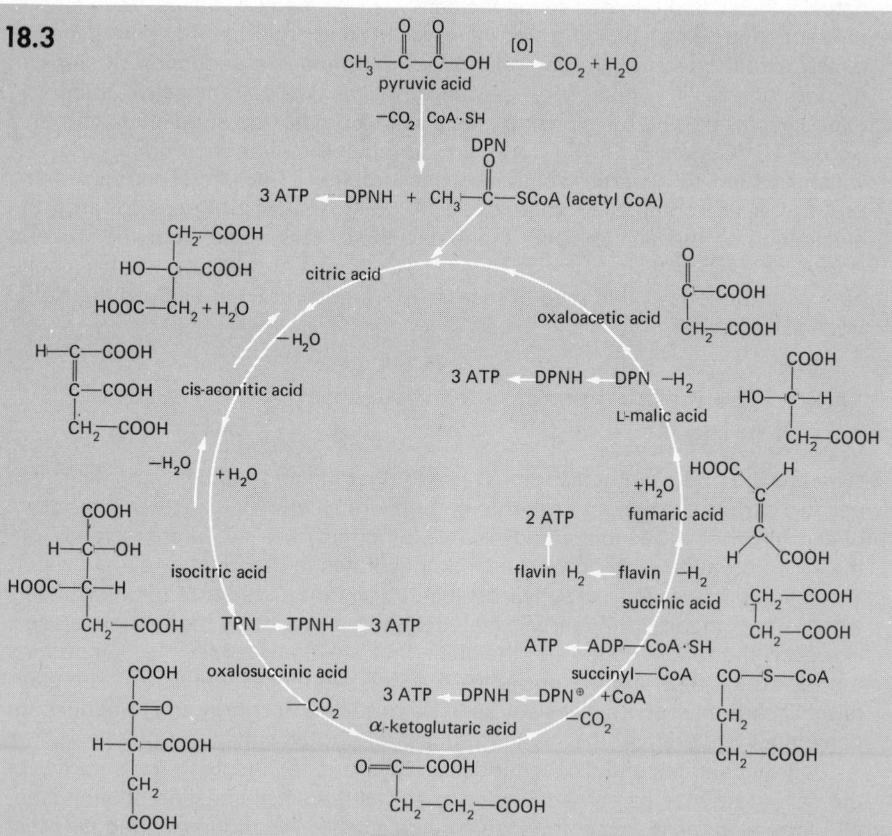

THE KREBS CYCLE

otide, etc.; note that Krebs' "DPN" has been corrected to NAD, and "TPN" to NADP) involved in the metabolic reactions rather than indicate their structural formulas. Most of the reactions involve addition or removal of water, carbon dioxide, or hydrogen.

The degradation of a molecule of pyruvic acid by the Krebs cycle yields 15 molecules of ATP (adenosine triphosphate) in the process. The ATP molecule contains so-called "high-energy" phosphate bonds, which provide the energy required by the human body to perform its functions.

Summary

- Carbohydrates are polyhydroxyaldehydes or polyhydroxyketones.

    (a) Carbohydrates can be classified as monosaccharides, disaccharides, or polysaccharides.
    (b) They can be further subdivided into aldoses, ketoses, pentoses, hexoses, etc.

- The Fischer–Kiliani Synthesis can be used to synthesize and elucidate the structure of many carbohydrates.

- Glucose, mannose, and galactose are all monosaccharides, and are examples of aldohexose sugars.

- Carbohydrates are found to exist as five- or six-membered cyclic hemiacetals, in equilibrium with the open-chain compound. The cyclic formulas are called furanose and pyranose, from the corresponding cyclic ethers. The six-membered pyranose structures exist primarily in the chair conformation.

- Fructose is a typical ketohexose.

- The reactions of the monosaccharides are due to the presence of —OH and
$$-\overset{O}{\underset{\|}{C}}-$$
 functional groups.

    (a) Reducing agents (Fehling's and Benedict's solutions).
    (b) Glycosides.
    (c) Osazone formation.

- Disaccharides include sucrose (a non-reducing sugar), lactose, maltose, cellobiose, etc. The disaccharides are made up of two monosaccharides linked together (by loss of H_2O) through a glycosidic linkage.

- Polysaccharides include starch, glycogen, and cellulose.

- The Krebs cycle describes the metabolic processes of carbohydrates in the body.

PROBLEMS

1 Illustrate each of the following with a specific example.
   (a) monosaccharide
   (b) disaccharide
   (c) polysaccharide
   (d) ketotetrose
   (e) aldopentose
   (f) *L*-pentose
   (g) a reducing sugar
   (h) a non-reducing sugar
   (i) an ethyl glucoside
   (j) two epimeric pentoses
   (k) two anomeric forms of galactose
   (l) an osazone

2 Draw Haworth projection formulas for each of the following structures.
  (a) β-D-galactopyranose
  (b) α-D-glucofuranose
  (c) methyl-α-D-mannopyranoside
  (d) methyl glycoside of a ketohexose sugar
  (e) 2,3,4,6-tetra-O-methyl-D-glucose

3 Write the correct structural formula for the product formed in each of the following equations by mannose reacting with the indicated reagent.
  (a) mannose + CH$_3$CH$_2$OH $\xrightarrow{HCl}$
  (b) mannose + HCN $\longrightarrow$
  (c) mannose + excess Ac$_2$O $\longrightarrow$
  (d) mannose + H$_2$/Ni $\longrightarrow$
  (e) mannose + HIO$_4$ $\longrightarrow$
  (f) mannose + phenylhydrazine $\longrightarrow$
  (g) mannose + Fehling's solution $\longrightarrow$
  (h) mannose + NH$_2$OH $\longrightarrow$

4 (a) Draw the structural formulas for all the isomeric aldohexoses, and indicate which of the sugars form identical osazones when reacted with phenylhydrazine.
  (b) What is the significance of sugars forming the same osazone?

5 *Why can glucose form a cyanohydrin, but no sodium bisulfite addition product?

6 Write a series of equations showing how you could convert the aldopentose, D-lyxose,

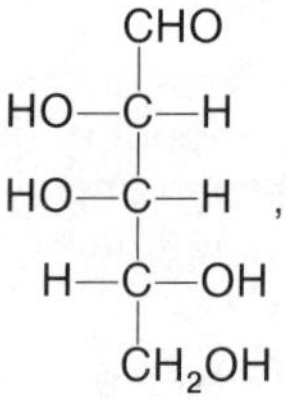

$$\begin{array}{c} CHO \\ HO-C-H \\ HO-C-H \\ H-C-OH \\ CH_2OH \end{array}$$

, into D-galactose.

7 *The sugar D-arabinose can be converted by the Fischer–Killani synthesis to D-glucose and D-mannose. Suggest a logical structural formula for D-arabinose.

8 Why is α-D-glucose a reducing sugar, whereas methyl-α-D-glucoside is a non-reducing sugar?

9 *The disaccharide trehalose, found in young mushrooms, is a non-reducing sugar. Trehalose yields only α-D-glucopyranose, on hydrolysis, as the sole product. Draw the Haworth projection formula for trehalose.

10 *The trisaccharide raffinose is found in sugar beets. Incomplete hydrolysis of raffinose yields sucrose and α-D-galactose as products. The complete hydrolysis of raffinose yields α-D-galactose, α-D-glucose, and β-D-fructose as products. It has been shown that the α-D-galactose is bonded to the α-D-glucose by a 1,6-linkage. Draw the Haworth projection formula for raffinose. Would you expect raffinose to reduce Fehling's solution? Explain.

19
Lipids—Fats and Oils

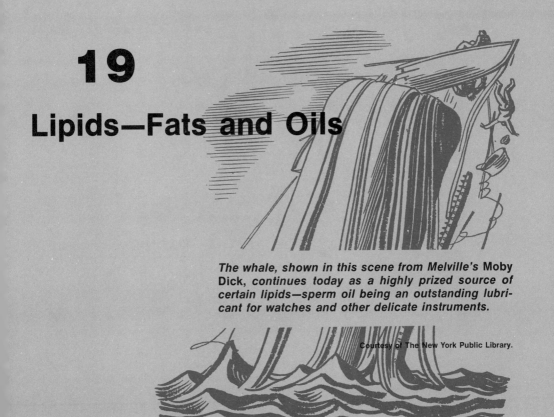

The whale, shown in this scene from Melville's Moby Dick, continues today as a highly prized source of certain lipids—sperm oil being an outstanding lubricant for watches and other delicate instruments.

Courtesy of The New York Public Library.

19.1 Introduction

The term **lipid** is used to refer to fats and fat-like substances that are naturally occurring materials found in plants or animals. Fats, oils, waxes, and many compounds having quite different molecular structures are grouped together as lipids because of their similar solubility characteristics. Lipids are generally soluble in non-polar organic solvents, and insoluble in water. Their solubility in various solvents differentiates lipids from carbohydrates or proteins. Among the various classes of lipids are the natural fats, phospholipids, cerebrosides, steroids, and lipoproteins. Other compounds that can be considered lipids are the vitamins A, D, E, and K. Some of these lipids will be discussed in this chapter; others will be referred to in Chapter 22, dealing with natural products.

The lipids are extremely important compounds. A number of lipids play important roles in the structure and function of cell membranes because of their solubility characteristics. The lipids serve as a barrier to the exchange of substances from within the cell to the extracellular fluid and also form lipoprotein complexes with proteins and various enzymes. The chemistry and function of cell membranes is very complex and is rapidly developing into a new area of research interest where any conclusive results will be most significant to mankind.

Michel Eugène Chevreul (1786–1889)—French chemist and professor, whose work with animal fats helped in the soap and candle industry—discovered and named olein and stearin. (Courtesy of the Division of Physical Sciences of the Smithsonian Institution.)

The lipids may be subdivided into two groups. One group, represented by the fats, oils, and waxes, are substances that can be hydrolyzed (saponified) by aqueous alkali. The second group of lipids are not saponified by base, as for example such steroids as cholesterol.

Fats are found in such materials as milk, butter, and vegetable fats and oils. Waxes and soaps are related in their structures to the fats and oils.

19.2 Neutral Fats and Oils

Neutral fats and oils are esters of glycerol and long chain fatty acids. Fatty acids are saturated and unsaturated carboxylic acids, usually between twelve to twenty carbon atoms in length. Since there are many fatty acids, there are numerous examples of animal fats and vegetable oils in nature. The three hydroxyl groups of glycerol, $\underset{\underset{OH}{|}}{CH_2}-\underset{\underset{OH}{|}}{CH}-\underset{\underset{OH}{|}}{CH_2}$, can be esterified by many different combinations of fatty acids to form either mono-, di-, or tri-glycerides. The esters of glycerol (fats and oils) are referred to as **glycerides.**

$$\begin{array}{l} CH_2-O-\overset{\overset{O}{\|}}{C}-R \\ \ \ | \\ CH-O-\overset{\overset{O}{\|}}{C}-R' \\ \ \ | \\ CH_2-O-\overset{\overset{O}{\|}}{C}-R'' \end{array}$$

a glyceride

Sec. 19.2 NEUTRAL FATS AND OILS

The term *fat* is used when the ester of glycerol is a solid, and oils are liquid esters of glycerol. The distinction between an oil and a fat is not really sharp, depending only on the melting point of the material, which in turn depends on the structure of the R groups of the fatty acid section of the molecule. Usually, when R is saturated the glycerides are solids, whereas when R is unsaturated the glycerides are liquids. In oils the R groups are usually highly unsaturated, containing many carbon–carbon double bonds; in fats relatively few double bonds are present in the R groups. Table 19.1 lists some common fatty acids that occur in fats and oils. Note that the fatty acids found in nature usually have an even number of carbon atoms.

A larger variety of fatty acids are found in animal fats than in vegetable fats. Animal fats usually contain palmitic or stearic acid, and some saturated and unsaturated fatty acids of twenty or more carbon atoms. Vegetable fats contain a large proportion of linoleic acid, which is found only in small amounts in the animal fats. The unsaturated fatty acids can exist in the form of geometrical isomers, due to the presence of carbon–carbon double bonds. Most of the

Table 19.1 Common Fatty Acids

Name	Formula	Some sources
Lauric acid	$CH_3(CH_2)_{10}COOH$	Spermaceti, coconut oil
Myristic acid	$CH_3(CH_2)_{12}COOH$	Coconut oil, nutmeg, butter
Palmitic acid	$CH_3(CH_2)_{14}COOH$	Animal and vegetable fats and oils
Stearic acid	$CH_3(CH_2)_{16}COOH$	Animal and vegetable fats and oils
Arachidic acid	$CH_3(CH_2)_{18}COOH$	Peanut oil
Cerotic acid	$CH_3(CH_2)_{24}COOH$	Beeswax, wool fat
Oleic acid (*cis* isomer)	$CH_3(CH_2)_7-\underset{10}{CH}=\underset{9}{CH}-(CH_2)_7COOH$	Animal and vegetable fats and oils
Linoleic acid	$CH_3(CH_2)_4-\underset{13}{CH}=\underset{12}{CH}-CH_2-\underset{10}{CH}=\underset{9}{CH}-(CH_2)_7COOH$	Linseed oil, cottonseed oil
Linolenic acid	$CH_3-CH_2-\underset{16}{CH}=\underset{15}{CH}-CH_2-\underset{13}{CH}=\underset{12}{CH}-CH_2-\underset{10}{CH}=\underset{9}{CH}-(CH_2)_7COOH$	Linseed oil
Ricinoleic acid (*cis* isomer)	$CH_3(CH_2)_5-\underset{OH}{CH}-CH_2-\underset{10}{CH}=\underset{9}{CH}-(CH_2)_7-COOH$	Castor oil
Arachidonic acid	$CH_3(CH_2)_4(CH=CH-CH_2)_4-(CH_2)_2COOH$	Lecithin, cephalin

common unsaturated fatty acids are the *cis* isomers. For example, ricinoleic acid, which is the *cis* isomer, is a constituent in castor oil.

19.3 Glycerides

The triesters of glycerol are known as **glycerides.** All three hydroxyl groups in glycerol can be esterified with the same acid forming a **simple glyceride,** or different acids can be used to esterify individual hydroxyl groups resulting in the formation of a **mixed glyceride.**

$$\begin{array}{l}CH_2-O-\overset{O}{\underset{\|}{C}}-(CH_2)_{14}-CH_3\\ CH-O-\overset{O}{\underset{\|}{C}}-(CH_2)_{14}-CH_3\\ CH_2-O-\overset{O}{\underset{\|}{C}}-(CH_2)_{14}-CH_3\end{array}$$

glyceryl tripalmitate (palmitin)

$$\begin{array}{l}CH_2-O-\overset{O}{\underset{\|}{C}}-(CH_2)_{16}-CH_3\\ CH-O-\overset{O}{\underset{\|}{C}}-(CH_2)_{10}-CH_3\\ CH_2-O-\overset{O}{\underset{\|}{C}}-(CH_2)_{7}-CH=CH-(CH_2)_{7}-CH_3\end{array}$$

glyceryl stearolauroleate

$$\begin{array}{ll}(\alpha) & CH_2-O-\overset{O}{\underset{\|}{C}}-(CH_2)_{14}-CH_3\\ (\beta) & CH-O-\overset{O}{\underset{\|}{C}}-(CH_2)_{14}-CH_3\\ (\alpha_1) & CH_2-O-\overset{O}{\underset{\|}{C}}-(CH_2)_{18}-CH_3\end{array}$$

α_1-arachido-α,β-dipalmitin

Both simple and mixed glycerides are constituents of naturally occurring fats and oils. Most natural fats and oils are a mixture of glycerides rather than a single pure glyceride. The composition of a fat is stated as to the percentages of the fatty acids it contains. Olive oil, for example, contains about 83 % oleic acid, 7 % linoleic acid, 6 % palmitic acid, and 4 % stearic acid. Butter contains many different fatty acids, including about 3 % of the low molecular weight fatty acid, butyric acid.

19.4 Some Reactions of Fats and Oils

19.4-1 Hydrolysis The triglycerides (fats and oils) are readily hydrolyzed by acid, base, superheated steam, and enzymes (lipases). Heating with an aqueous or alcoholic solution of potassium or sodium hydroxide hydrolyzes fats and oils to glycerol and salts of the fatty acids. The salts of the fatty acids are actually soaps, and so the alkaline hydrolysis of esters is called **saponification.** This reaction is used in commercial soap manufacture.

Sec. 19.4 SOME REACTIONS OF FATS AND OILS

$$\begin{array}{c}\text{CH}_2-\text{O}-\overset{\text{O}}{\overset{\|}{\text{C}}}-\text{R}\\|\\\text{CH}-\text{O}-\overset{\text{O}}{\overset{\|}{\text{C}}}-\text{R} + 3\text{KOH} \xrightarrow{\Delta}\\|\\\text{CH}_2-\text{O}-\overset{\text{O}}{\overset{\|}{\text{C}}}-\text{R}\\\text{triglyceride}\end{array} \quad \begin{array}{c}\text{CH}_2\text{OH}\\|\\\text{CHOH}\\|\\\text{CH}_2\text{OH}\\\text{glycerol}\end{array} + 3\text{R}-\overset{\text{O}}{\overset{\|}{\text{C}}}-\text{O}^{\ominus}\text{K}^{\oplus} \quad (19\text{-}1)$$

soap

$$\begin{array}{c}\text{CH}_2-\text{O}-\overset{\text{O}}{\overset{\|}{\text{C}}}-(\text{CH}_2)_{16}-\text{CH}_3\\|\\\text{CH}-\text{O}-\overset{\text{O}}{\overset{\|}{\text{C}}}-(\text{CH}_2)_{16}-\text{CH}_3 + 3\text{KOH} \xrightarrow{\Delta}\\|\\\text{CH}_2-\text{O}-\overset{\text{O}}{\overset{\|}{\text{C}}}-(\text{CH}_2)_{16}-\text{CH}_3\\\text{glyceryl tristearate}\\\text{(tristearin)}\end{array} \quad \begin{array}{c}\text{CH}_2\text{OH}\\|\\\text{CHOH}\\|\\\text{CH}_2\text{OH}\\\text{glycerol}\end{array} + 3\text{CH}_3-(\text{CH}_2)_{16}-\overset{\text{O}}{\overset{\|}{\text{C}}}-\text{O}^{\ominus}\text{K}^{\oplus}$$

potassium stearate
(a soap)

$$(19\text{-}2)$$

19.4-2 Hydrogenation of Fats and Oils Glycerides containing unsaturated R groups in the fatty acid portion of the molecule can add hydrogen across the carbon–carbon double bonds, resulting in the formation of saturated glycerides.

$$\begin{array}{c}\text{CH}_2-\text{O}-\overset{\text{O}}{\overset{\|}{\text{C}}}-(\text{CH}_2)_7-\text{CH}=\text{CH}-(\text{CH}_2)_7-\text{CH}_3\\|\\\text{CH}-\text{O}-\overset{\text{O}}{\overset{\|}{\text{C}}}-(\text{CH}_2)_7-\text{CH}=\text{CH}-(\text{CH}_2)_7-\text{CH}_3 + 3\text{H}_2 \xrightarrow[\Delta]{\text{Ni}}\\|\\\text{CH}_2-\text{O}-\overset{\text{O}}{\overset{\|}{\text{C}}}-(\text{CH}_2)_7-\text{CH}=\text{CH}-(\text{CH}_2)_7-\text{CH}_3\\\text{glyceryl trioleate}\\\text{(olein)}\end{array} \quad \begin{array}{c}\text{CH}_2-\text{O}-\overset{\text{O}}{\overset{\|}{\text{C}}}-(\text{CH}_2)_{16}-\text{CH}_3\\|\\\text{CH}-\text{O}-\overset{\text{O}}{\overset{\|}{\text{C}}}-(\text{CH}_2)_{16}-\text{CH}_3\\|\\\text{CH}_2-\text{O}-\overset{\text{O}}{\overset{\|}{\text{C}}}-(\text{CH}_2)_{16}-\text{CH}_3\\\text{glyceryl tristearate}\\\text{(stearin)}\end{array}$$

$$(19\text{-}3)$$

This process is used industrially in the hydrogenation of vegetable oils. The conversion of unsaturated to saturated bonds in the oil raises the melting point, so that eventually the oil is transformed into a solid fat. This *hardening* process is used to change liquid oils to solid fats. Crisco is made in this manner, and oleomargarine is an example of a partially hydrogenated mixture of oils.

The catalytic hydrogenation of vegetable oils produces solid fats that have the advantage of not turning rancid. The process of rancidity is much more significant in unsaturated than in saturated fats.

19.5 Waxes

Waxes differ from fats and oils in that they are esters, but *not* esters of glycerol. Waxes are esters of long-chain fatty acids and long-chain monohydric alcohols. Both the acid and alcohol portion of the ester can vary from sixteen to thirty-six carbon atoms in length, and both contain an even number of carbon atoms.

The general formula for a wax is the same as for an ester, $R-\overset{\overset{O}{\|}}{C}-OR'$. Waxes differ from fats and oils in that when they are saponified by alkali the products are often insoluble in water (especially the long-chain alcohols formed). Waxes may contain small amounts of other materials (e.g., fatty acids and steroids) along with the ester. The waxes are usually more brittle and harder than fats. They have numerous uses, such as in making polishes, cosmetics, and phonograph records.

Two typical waxes are beeswax and spermaceti. Beeswax has the formula $C_{15}H_{31}CO_2C_{30}H_{61}$, and is mainly the ester myricyl palmitate. The wax spermaceti is obtained from the head of the sperm whale and is largely composed of cetyl palmitate, $C_{15}H_{31}CO_2C_{16}H_{33}$.

19.6 Drying Oils

Some oils undergo certain changes when exposed to air. Rancidity is believed to result from these changes, and therefore must be most predominant in molecules that are highly unsaturated. Linseed, soy bean, and tung oils are among the oils that form a strong, waterproof, solid surface upon exposure to air. Apparently, the oils undergo oxidation and are polymerized. These so-called *drying oils* contain relatively large amounts of unsaturated fatty acids in the molecule. For example, linseed oil is approximately 50 % linoleic acid and 30 % linolenic acid. Oil paints are suspensions of finely divided pigments in linseed oil. Oilcloth and linoleum are other familiar products made from the drying oils.

19.7 Soaps and Detergents

The preparation of soaps by the saponification of fats and oils with alkali has already been discussed. The sodium salts of the fatty (long-chain) acids are used most extensively as soaps. The potassium salts are usually softer and more soluble than the sodium salts; therefore they are used mainly in shaving creams and liquid soaps. After saponification of a fat, salt is usually added to the reaction mixture to precipitate the soap from solution. The crude soap is then purified by a series of procedures, processed, and sold to the consumer.

The mechanism of the cleansing action of soaps is complex and will not be discussed here. What is important, is that one end of the soap molecule be highly polar or ionic, and the remainder of the molecule be non-polar (as in a hydrocarbon), for example sodium stearate.

$$\left(\underset{\text{non-polar}}{CH_3-(CH_2)_{16}} - \underset{\text{polar}}{\overset{\overset{O}{\|}}{C}-O^{\ominus}Na^{\oplus}} \right)$$

Sec. 19.8 SAPONIFICATION AND IODINE NUMBER

The polar end of the water molecule is essential to make the soap water soluble, while the non-polar end of the molecule makes the soap oil soluble, or attracted to fats. Since oils and water are immiscible, this type of molecular structure is necessary for a soap to remove dirt, fat, oil, and grease particles from a surface.

Synthetic detergents are being used more and more frequently. Many are as good or better cleansing agents than ordinary soaps, and have certain advantages (though if they contain phosphates, etc., they may also have serious disadvantages). Synthetic detergents are composed of several different types of compounds. Like the soaps, one end of the detergent must be polar, and the other end of the molecule non-polar. Detergents differ from ordinary soaps in that they do not form insoluble precipitates with $Ca^{\oplus\oplus}$, $Mg^{\oplus\oplus}$, or $Fe^{\oplus\oplus\oplus}$ ions. Hence, detergents can be used readily in either soft or hard water.

One type of detergent uses the sodium salts of alkyl sulfuric acids, made from long-chain alcohols and sulfuric acid. As was mentioned in section 11.11, the detergent sodium lauryl sulfate is made from lauryl alcohol and is sold under the trade name of Dreft.

$$CH_3(CH_2)_{10}CH_2O\boxed{H + HO}SO_3H \longrightarrow H_2O + CH_3(CH_2)_{10}CH_2OSO_2OH$$

lauryl alcohol   **lauryl hydrogen sulfate**

$$\downarrow NaOH$$

$$CH_3(CH_2)_{10}CH_2OSO_2\overset{\ominus}{O}Na^{\oplus} \quad (19\text{--}4)$$

sodium lauryl sulfate (Dreft)

Another type of detergent is the sodium alkylaryl sulfonate

$$R\text{—}\phenyl\text{—}SO_2\overset{\ominus}{O}Na^{\oplus} \quad (R = 12 \text{ carbon atoms})$$

19.8 The Saponification Number and the Iodine Number of Fats and Oils

The composition of fats and oils can vary as they are obtained from numerous natural sources. It is essential therefore that some analytical methods be devised that will give us accurate, quantitative information about the structural features in the lipid molecule. Two of the most useful analytical procedures in this regard determine the saponification number and the iodine number of a fat or oil.

Fats or oils may be treated with base and saponified. The saponification reaction is quantitative, that is to say, all of the fat or oil will be converted into glycerol and salts of fatty acids. From this experiment the so-called saponification number can be determined. The **saponification number** is the number of milligrams of potassium hydroxide required to hydrolyze one gram of fat or oil. Its numerical value gives us a good approximation of the average molecular weight of the fat or oil. This also gives some indication as to the average length of the carbon chains of the fatty acid portion of the molecule. The higher the numerical value of the saponification number, the greater the percentage of low-molecular-weight glycerides (containing lower fatty acids) present in the fat or oil.

The degree of unsaturation of a fat or oil is indicated by its iodine number. The **iodine number** is the number of grams of iodine that will combine with 100 grams of fat or oil. The iodine, in an alcoholic solution containing mercuric chloride, adds across the carbon–carbon double bonds present in the fatty acid portion of the molecule. The greater the unsaturation in the fat or oil, the higher will be the numerical value of the iodine number. A high iodine number indicates the presence of many carbon–carbon double bonds in the molecule.

19.9 The Metabolism of Lipids

Most of our body's energy comes from the oxidation of carbohydrates or fats, and the metabolism of these two foodstuffs is interrelated. The body not only oxidizes fats for energy, but continues at the same time to synthesize new body fats. Carbohydrates, fats, and proteins are continually being used up and synthesized in the body. In this section we will discuss the essential features of lipid metabolism.

Fats are not changed or digested to any great extent in the mouth or stomach. The triglycerides are absorbed primarily in the duodenum and jejunum (small intestine). The bile acids present in the small intestine emulsify the triglycerides, so that they can be hydrolyzed by various enzymes. For example, the enzyme pancreatic lipase hydrolyzes many triglycerides to free fatty acids and glycerol, other triglycerides are only partially hydrolyzed to monoglycerides. The fatty acids and glycerol may recombine and synthesize new body fats, or the glycerol may enter the normal carbohydrate metabolic cycles. The fatty acids may be broken down and converted to serve the body's energy requirements. We shall devote our subsequent discussion to the degradation of fatty acids in the body.

Much of our understanding of the oxidation of fatty acids in the body is due to the experimentation of Knoop in 1904. From his results, Knoop developed his **beta-oxidation theory,** which concluded that the oxidation of fatty acids occurs in such a manner that at each stage of the degradation process a loss of a two carbon atom fragment occurs, because oxidation occurs at the β carbon atom with respect to the carboxyl group.

That the common fatty acids contain an even number of carbon atoms, and that acetate (a two-carbon atom fragment) can be converted into larger fatty acid molecules in the body (as was determined by isotopic tracer experiments with carbon-14) gave much support to the β-oxidation theory.

The enzymes involved in the oxidation of fatty acids have been isolated and purified. However, instead of acetic acid being removed as the two-carbon fragment in the oxidation steps, it is the coenzyme, acetyl CoA, that is actually formed. Acetyl CoA is the vital constituent to the entire series of steps in the oxidation of fatty acids, and it is formed in the very first step where the fatty acid molecule is activated by conversion into a CoA derivative. Acetyl CoA is also involved in the Krebs cycle of carbohydrate metabolism. In 1947, Fritz Lipmann discovered the actual involvement of a substance, coenzyme A, as the acetyl group donor and acceptor that was vital to numerous biochemical processes in the body. Coenzyme A is a complex molecule of formula $C_{21}H_{36}O_{16}N_7P_3S$, containing a sulfhydryl (–SH) group. Coenzyme A is often abbreviated as CoA–SH. For his significant contribution Fritz Lipmann received the Nobel prize in 1953.

Sec. 19.9 THE METABOLISM OF LIPIDS

Let us illustrate a typical fatty acid degradation (palmitic acid to myristic acid) by the following set of equations.

$$CH_3(CH_2)_{12}\underset{\beta}{-}CH_2\underset{\alpha}{-}CH_2COOH + CoA-SH \longrightarrow H_2O + CH_3(CH_2)_{12}CH_2CH_2-\overset{O}{\underset{\|}{C}}-S-CoA$$

palmitic acid   **palmitic acid thioester**

$\downarrow -H_2$

$$CH_3(CH_2)_{12}-\overset{OH}{\underset{|}{CH}}-CH_2-\overset{O}{\underset{\|}{C}}-S-CoA \xleftarrow{+H_2O} CH_3(CH_2)_{12}-CH=CH-\overset{O}{\underset{\|}{C}}-S-CoA$$

$\downarrow -H_2$

$$CH_3(CH_2)_{12}-\underset{\beta}{\overset{O}{\underset{\|}{C}}}-\underset{\alpha}{CH_2}-\overset{O}{\underset{\|}{C}}-S-CoA \xrightarrow{CoA-SH} CH_3-\overset{O}{\underset{\|}{C}}-S-CoA + CH_3(CH_2)_{12}-\overset{O}{\underset{\|}{C}}-S-CoA$$

β-keto ester   **acetyl CoA**   **myristic acid thioester**

(etc.)

(19-5)

Coenzyme A activates the fatty acids by forming thioesters. The thioester of CoA-SH undergoes a series of enzymatic reactions involving other enzymes (not listed in our simplified schematic representation) and is oxidized (by loss of hydrogen), hydrated, oxidized to a β-keto ester, which is degraded to acetyl CoA and a thioester with two fewer carbon atoms in its structure. The reaction sequence is continued over and over again until the entire fatty acid chain has been degraded. Since the oxidation steps are exothermic, they provide the body with a source of energy.

The biosynthesis of fatty acids in the body is not a direct reversal of the fatty acid oxidation process, although acetyl CoA is the starting material. The acetyl CoA is carboxylated to form malonyl CoA in the presence of the enzyme acetyl CoA carboxylase, ATP, and the vitamin biotin. The subsequent steps include acetylation, decarboxylation, reduction, dehydration, and finally reduction to butyryl CoA, which can start the sequence over again with malonyl CoA. The process can continue over and over, resulting in the synthesis of fatty acids.

$$CH_3-\overset{O}{\underset{\|}{C}}-S-CoA \xrightarrow[\text{ATP, biotin}]{CO_2} HO-\overset{O}{\underset{\|}{C}}-CH_2-\overset{O}{\underset{\|}{C}}-S-CoA \xrightarrow{CH_3-\overset{O}{\underset{\|}{C}}-S-CoA} HO-\overset{O}{\underset{\|}{C}}-\underset{\underset{CH_3}{\overset{|}{\underset{|}{C=O}}}}{\overset{|}{CH}}-\overset{O}{\underset{\|}{C}}-S-CoA$$

acetyl CoA   **malonyl CoA**

$\downarrow -CO_2$

$$CH_3-\overset{O}{\underset{\|}{C}}-CH_2-\overset{O}{\underset{\|}{C}}-S-CoA \xrightarrow[\text{2. } -H_2O]{\text{1. } H_2} CH_3-CH_2-CH_2-\overset{O}{\underset{\|}{C}}-S-CoA \longrightarrow \text{(etc.)}$$
         acetoacetyl CoA   3. H_2 **butyryl CoA**

(19-6)

338   LIPIDS—FATS AND OILS   Ch. 19

Since the degradation and synthesis of fatty acids occur by different reaction paths, the two processes can be regulated by various cells according to the body's needs. This is of prime importance in the maintenance of good health in the human body.

Summary

- The fats and oils known as lipids are esters of long-chain carboxylic acids (fatty acids) and glycerol.
- The triesters of glycerol are known as glycerides. Both simple and mixed glycerides are known.
- The reactions of fats include saponification to soaps and hydrogenation of oils to produce solid fats.
- Waxes differ from fats and oils in that they are esters, but not esters of glycerol.
- The interaction of the polar and non-polar ends of soaps with oil and water explains their cleansing action.

Synthetic detergents.

- Fats and oils can be analyzed.

    (a) Saponification number gives a good approximation of average molecular weight.

    (b) Iodine number indicates degree of unsaturation.

- The Knoop β-oxidation theory and the role of coenzyme A are important in the metabolism of lipids and how they supply the body with energy.

Lipids can be synthesized by the body when necessary to maintain a proper balance.

PROBLEMS

1 Write correct structural formulas for the following substances. Indicate which are fats or oils, waxes, soaps, and detergents.

  (a) triolein   **(e)** sodium *p*–lauryl benzene sulfonate
  (b) lauryl stearate   **(f)** glyceryl laurostearopalmitate
  (c) cetyl palmitate   **(g)** glyceryl cerotoarachidonolaurate
  (d) potassium stearate   **(h)** β-stearo,α,α_1–dipalmitin

2 Name each of the following compounds.

(a)
$$\begin{array}{l} CH_2-O-\overset{\overset{\displaystyle O}{\|}}{C}-(CH_2)_{10}-CH_3 \\ CH-O-\overset{\overset{\displaystyle O}{\|}}{C}-(CH_2)_{10}-CH_3 \\ CH_2-O-\overset{\overset{\displaystyle O}{\|}}{C}-(CH_2)_{10}-CH_3 \end{array}$$

(b)

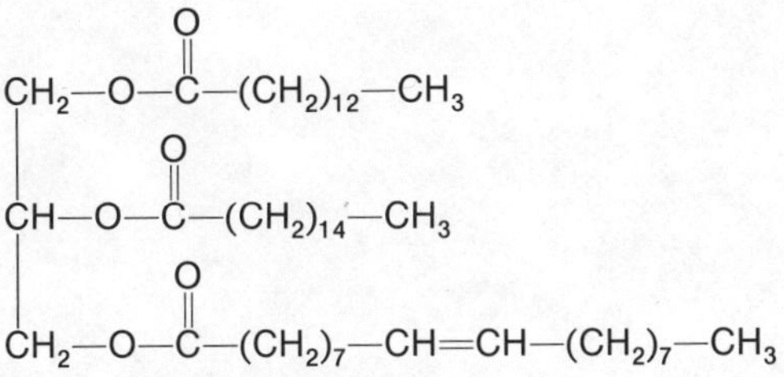

$$CH_2-O-\underset{\underset{O}{\|}}{C}-(CH_2)_{12}-CH_3$$
$$CH-O-\underset{\underset{O}{\|}}{C}-(CH_2)_{14}-CH_3$$
$$CH_2-O-\underset{\underset{O}{\|}}{C}-(CH_2)_7-CH=CH-(CH_2)_7-CH_3$$

(c)

$$CH_2-O-\underset{\underset{O}{\|}}{C}-(CH_2)_{12}-CH_3$$
$$CH-O-\underset{\underset{O}{\|}}{C}-(CH_2)_{12}CH_3$$
$$CH_2-O-\underset{\underset{O}{\|}}{C}-(CH_2)_{16}-CH_3$$

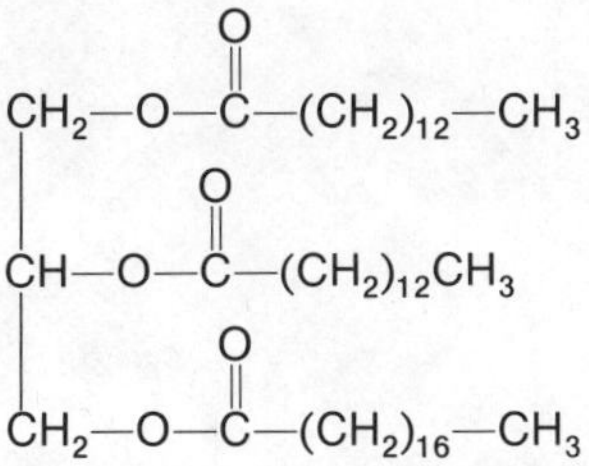

3. Write equations to illustrate the (a) saponification, and (b) hydrogenation, of glyceryl trilinolenate.

4. Calculate the saponification number of (a) stearin, (b) glyceryl tricerotate, and (c) glycerylmyristyllauropalmitate

5. Calculate the iodine number of (a) olein, (b) glyceryl trilinolenate, and (c) α-arachido-β,α_1-diolein

6. What is the significance of a high numerical value for an iodine number of a glyceride?

7. Calculate the molecular weight of a glyceride that has a saponification number of 240.

8. * Five grams of a fat react with 4.76 g of iodine. If the fat has a molecular weight of 1600, calculate the average number of double bonds per molecule of fat. Why is the value an *average* number?

9. What is the role of thioesters in lipid metabolism?

10. Illustrate by means of equations the fatty acid degradation of stearic acid according to the β-oxidation theory.

20

Amino Acids and Proteins

A Remington painting showing the buffalo of the early West and settlers for whom they were a prime source of protein—"the substance of life."

Courtesy of The New York Public Library

20.1 Introduction

Proteins are perhaps the most characteristic compounds found in a living cell. They are complex molecules and are the building blocks of animal tissue. Many common substances, such as hair, skin, nails, and feathers, are mainly proteins. The hemoglobin of the blood, enzymes, and viruses are protein in nature. The proteins have high molecular weights and actually are macromolecules or polymers, consisting of numerous amino acids linked together in a particular sequence. The hydrolysis of a protein (by acid or enzymatically) produces a mixture of the amino acids that make up the protein molecule. The α amino acids are the basic building blocks of all proteins, and despite the large number of proteins in nature, only about twenty-six different amino acids are involved in the synthesis of the many types of proteins.

20.2 Amino Acids

The hydrolysis of proteins yields a mixture of α–amino acids, which can be represented by the general structural formula R—$\underset{\underset{NH_2}{|}}{\overset{\alpha}{C}H}$—$\overset{\overset{O}{\|}}{C}$—OH. All of the naturally occurring amino acids except glycine (R=H) have at least one asymmetric carbon atom and are optically active compounds. All of the amino acids have the same absolute configuration, and are L–amino acids

Sec. 20.3 NOMENCLATURE AND CLASSIFICATION

$$\begin{array}{c} CO_2H \\ H_2N-\overset{|}{\underset{R}{C}}-H \end{array}$$

L-amino acid

and are related to the amino acid L-(−)-serine

$$\begin{array}{c} COOH \\ H_2N-\overset{|}{\underset{CH_2OH}{C}}-H \end{array}$$

L-(−)-serine

which serves as the standard compound for the amino acids, just as D-(+)-glyceraldehyde is the reference compound for the carbohydrates.

20.3 Nomenclature and Classification of the Amino Acids

Amino acids can be named as amino derivatives of the carboxylic acids, according to the usual rules of nomenclature for carboxylic acid derivatives. In addition, the amino acids have common names that are unrelated to their structure. Often the common names for amino acids are used in preference to the other nomenclature.

Most of the naturally occurring amino acids are so-called **neutral acids** (although they are not necessarily exactly neutral) containing one amino (basic) and one carboxyl (acidic) group in the molecule. Amino acids that contain two carboxyl groups and only one amino group are referred to as **acidic amino acids.** Some amino acids have more amino groups than carboxyl groups and are **basic amino acids.**

There are some amino acids that are required by human beings for proper metabolism and development but the human body cannot synthesize. These amino acids are referred to as **essential amino acids,** and must be incorporated into the body in the foods we eat. The essential amino acids include the following: phenylalanine, tryptophan, threonine, valine, leucine, isoleucine, lysine, and methionine. The reader may recognize the acid phenylalanine as being involved in the disease **phenylketonuria,** which can cause mental retardation in infants if not detected.

We will now list the structures of some of the important amino acids, classifying them into certain categories by structural similarity. Some of the names appear in italics, to indicate they are essential amino acids.

Aliphatic Monoamino-monocarboxylic Acids

Glycine (α-aminoacetic acid) H_2N-CH_2-COOH

Alanine (α-aminopropionic acid) $CH_3-\underset{\underset{NH_2}{|}}{CH}-COOH$

Valine (α-aminoisovaleric acid)

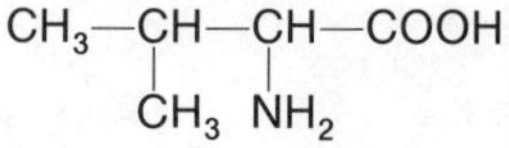

$$CH_3-CH-CH-COOH$$
$$||$$
$$CH_3NH_2$$

Leucine (α-aminoisocaproic acid)

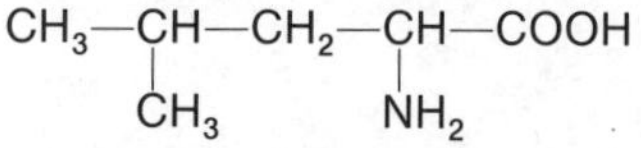

$$CH_3-CH-CH_2-CH-COOH$$
$$||$$
$$CH_3NH_2$$

Isoleucine (α-amino-β-methyl valeric acid, or α-amino-β-methyl-β-ethylpropionic acid)

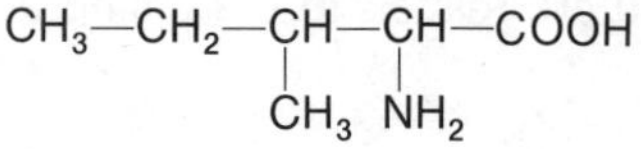

$$CH_3-CH_2-CH-CH-COOH$$
$$||$$
$$CH_3NH_2$$

Amino Acids Containing an Aromatic Ring

Phenylalanine (α-amino-β-phenylpropionic acid)

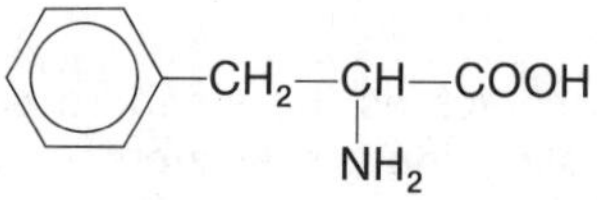

C$_6$H$_5$—CH$_2$—CH(NH$_2$)—COOH

Tyrosine (p-hydroxyphenylalanine)

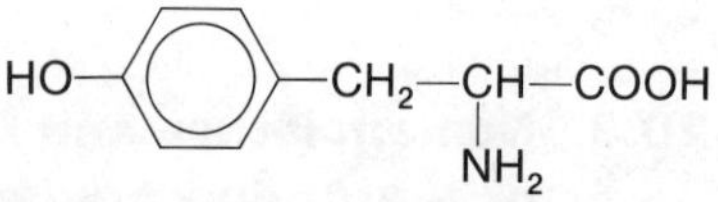

HO—C$_6$H$_4$—CH$_2$—CH(NH$_2$)—COOH

Tryptophan (α-amino-β-3-indole propionic acid)

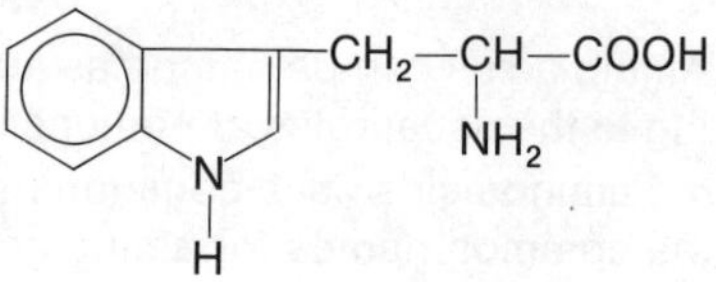

(indole)—CH$_2$—CH(NH$_2$)—COOH

Aliphatic Amino Acids Containing a Hydroxyl Group

Serine (α-amino-β-hydroxypropionic acid)

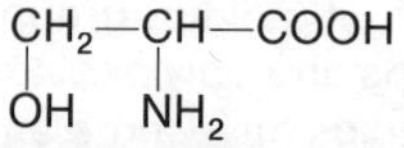

$$CH_2-CH-COOH$$
$$||$$
$$OHNH_2$$

Threonine (α-amino-β-hydroxybutyric acid)

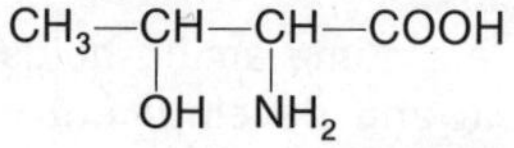

$$CH_3-CH-CH-COOH$$
$$||$$
$$OHNH_2$$

Sulfur-containing Amino Acids

Cysteine (α-amino-β-thiopropionic acid)

$$HS-CH_2-CH-COOH$$
$$|$$
$$NH_2$$

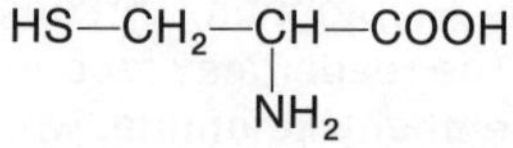

Cystine (di-α-amino-β-thiopropionic acid)

$$S-CH_2-CH-COOH$$
$$||$$
$$|NH_2$$
$$S-CH_2-CH-COOH$$
$$|$$
$$NH_2$$

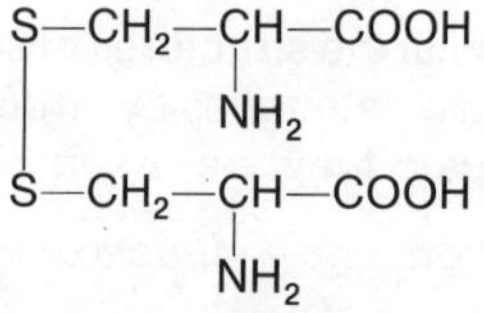

Methionine (α-amino-γ-methylthio butyric acid)

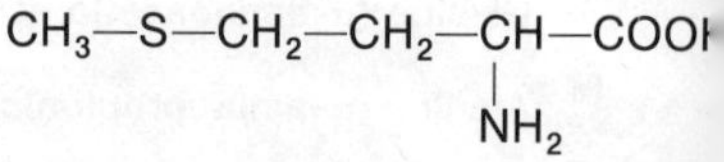

$$CH_3-S-CH_2-CH_2-CH-COOH$$
$$|$$
$$NH_2$$

Acidic Amino Acids (1 NH₂ group, 2 COOH groups)

Aspartic acid (α-aminosuccinic acid)

$$\text{HOOC—CH}_2\text{—CH(NH}_2\text{)—COOH}$$

Glutamic acid (α-aminoglutaric acid)

$$\text{HOOC—CH}_2\text{—CH}_2\text{—CH(NH}_2\text{)—COOH}$$

Basic Amino Acids (2 NH₂ groups, 1 COOH group)

Lysine (α,ε-diaminocaproic acid)

$$\text{CH}_2(\text{NH}_2)\text{—CH}_2\text{—CH}_2\text{—CH}_2\text{—CH(NH}_2)\text{—COOH}$$

Arginine (α-amino-δ-guanidylvaleric acid)

$$\text{HN=C(NH}_2\text{)—N(H)—CH}_2\text{—CH}_2\text{—CH}_2\text{—CH(NH}_2\text{)—COOH}$$

Histidine (β-4-imidazoylalanine)

(imidazole ring)—CH₂—CH(NH₂)—COOH

Secondary Amino Acids

Proline (pyrrolidine-α-carboxylic acid)

(pyrrolidine ring with COOH at α-carbon)

Hydroxyproline (4-hydroxypyrrolidine-α-carboxylic acid)

(4-hydroxypyrrolidine ring with COOH at α-carbon)

20.4 Synthesis of Amino Acids

The synthesis of α-amino acids has been briefly referred to in earlier chapters. The principal methods used to synthesize these compounds include amination of α-halogenated acids, the Strecker amino acid synthesis, and the malonic ester method.

20.4-1 Amination of α-Halogenated Acids It will be recalled that α-halogenated acids can be prepared from carboxylic acids by the Hell–Volhard–Zelinsky reaction. Treatment of the α-halogenated acid with an excess of ammonia produces the α amino acid in good yield.

$$R-CH_2COOH \xrightarrow[Cl_2]{P} R-\underset{Cl}{\underset{|}{CH}}-COOH \xrightarrow{\text{excess } NH_3} R-\underset{NH_2}{\underset{|}{CH}}-COOH \quad (20-1)$$

α–amino acid

$$CH_3COOH \xrightarrow[Cl_2]{P} ClCH_2-COOH \xrightarrow{\text{excess } NH_3} H_2N-CH_2-COOH \quad (20-2)$$

acetic acid **chloroacetic acid** **glycine**

20.4-2 The Strecker Amino Acid Synthesis The Strecker amino acid synthesis has been mentioned in conjunction with cyanohydrin formation of aldehydes and ketones (Section 14.8-1). An aldehyde or ketone can be converted to a cyanohydrin by addition of hydrogen cyanide across the carbonyl group. The cyanohydrin is reacted with ammonia and converted to an aminonitrile, and then the nitrile group is hydrolyzed to form the α–amino acid. The overall reaction sequence of the addition of hydrogen cyanide in the presence of ammonia (a solution of ammonium cyanide can be used), followed by hydrolysis, is known as the Strecker synthesis.

$$CH_3-\overset{O}{\underset{}{\overset{\|}{C}}}-H \xrightarrow{HCN} CH_3-\underset{CN}{\underset{|}{\overset{OH}{\overset{|}{C}}}}-H \xrightarrow{NH_3} CH_3-\underset{CN}{\underset{|}{\overset{NH_2}{\overset{|}{C}}}}-H \xrightarrow[H^+]{H_2O} CH_3-\underset{H}{\underset{|}{\overset{NH_2}{\overset{|}{C}}}}-COOH \quad (20-3)$$

acetaldehyde **alanine**

20.4-3 Malonic Ester Methods A third method for the synthesis of α–amino acids involves the reaction of an alkyl halide with malonic ester $\underset{COOC_2H_5}{\underset{|}{\overset{COOC_2H_5}{\overset{|}{CH_2}}}}$

The hydrogens of the methylene (CH_2) group of malonic ester are weakly acidic (Why?) and can be removed in the presence of a strong base such as sodium ethoxide to form a carbanion, $\ominus:\underset{COOC_2H_5}{\underset{|}{\overset{COOC_2H_5}{\overset{|}{CH}}}}$, which can then participate in nucleophilic substitution reactions with alkyl halides, displacing the halide ion to form an alkyl substituted malonic ester which can be converted to an α–amino acid.

$$\underset{COOC_2H_5}{\underset{|}{\overset{COOC_2H_5}{\overset{|}{CH_2}}}} \xrightarrow{NaOC_2H_5} \ominus:\underset{COOC_2H_5}{\underset{|}{\overset{COOC_2H_5}{\overset{|}{CH}}}} \xrightarrow{RX} R-\underset{COOC_2H_5}{\underset{|}{\overset{COOC_2H_5}{\overset{|}{CH}}}} + X^\ominus \quad (20-4)$$

malonic ester (diethyl malonate) **alkyl substituted malonic ester**

Sec. 20.5 PROPERTIES OF AMINO ACIDS

The entire reaction sequence of the malonic ester method is illustrated for the preparation of the amino acid, phenylalanine.

$$\underset{\text{malonic ester}}{\underset{|}{\overset{|}{\text{CH}_2}}\overset{\text{COOC}_2\text{H}_5}{\underset{\text{COOC}_2\text{H}_5}{}}} + \underset{\text{benzyl chloride}}{\text{C}_6\text{H}_5\text{-CH}_2\text{Cl}} \xrightarrow{\text{NaOC}_2\text{H}_5} \text{H}-\underset{\text{COOC}_2\text{H}_5}{\underset{|}{\overset{|}{\text{C}}}\overset{\text{COOC}_2\text{H}_5}{}}\text{-CH}_2\text{-C}_6\text{H}_5 \xrightarrow{\text{Br}_2}$$

$$\xleftarrow{\Delta, -\text{CO}_2} \text{Br}-\underset{\boxed{\text{COOH}}}{\overset{\text{COOH}}{\underset{|}{\overset{|}{\text{C}}}}}\text{-CH}_2\text{-C}_6\text{H}_5 \xleftarrow{\text{H}_2\text{O, H}^{\oplus}} \text{Br}-\underset{\text{COOC}_2\text{H}_5}{\overset{\text{COOC}_2\text{H}_5}{\underset{|}{\overset{|}{\text{C}}}}}\text{-CH}_2\text{-C}_6\text{H}_5$$

$$\text{Br}-\underset{\text{H}}{\overset{\text{COOH}}{\underset{|}{\overset{|}{\text{C}}}}}\text{-CH}_2\text{-C}_6\text{H}_5 \xrightarrow{\text{NH}_3} \underset{\text{phenylalanine}}{\text{HOOC}-\underset{\text{NH}_2}{\overset{|}{\text{CH}}}-\text{CH}_2\text{-C}_6\text{H}_5} \qquad (20\text{-}5)$$

20.5 Properties of Amino Acids

The amino acids have many properties that are *not* typical of most organic compounds. For example, most amino acids are soluble in water and insoluble in the common organic solvents, such as ether, benzene, and alcohol. They have rather high melting points, and in general their physical properties are similar to ionic compounds rather than covalent molecules. The reason for this behavior can be explained in terms of a zwitterion or dipolar ion structure for the amino acids (recall sulfanilic acid, Section 16.8-1). The strongly basic amino group accepts a proton from the acidic carboxyl group in the amino acid, so that the amino acids actually exist as internal salts (zwitterions) rather than the molecular structure we previously assigned to them.

$$\text{R}-\underset{\overset{|}{\overset{\oplus}{\text{NH}_3}}}{\text{CH}}-\overset{\overset{\text{O}}{\|}}{\text{C}}-\text{O}^{\ominus}$$

zwitterion structure of
an amino acid

This type of structure explains the physical properties of the amino acids, such as their relatively high melting points and solubility characteristics which closely resemble those of ionic compounds.

20.6 Isoelectric Point

The amino acids are **amphoteric,** that is, they can behave as both an acid and a base. In an acid solution they will behave as if they were bases and accept a proton from the acid. In basic solution the amino acids behave as acids, and donate a proton to the base. In acid solution, the zwitterion will be converted to a positively charged species, R—CH(⊕NH₃)—COOH; in basic solution it will be converted to a negatively charged species, R—CH(NH₂)—COO⁻.

IN ACID SOLUTION

$$\text{R—CH(}^{\oplus}\text{NH}_3\text{)—COO}^{\ominus} + \text{H}^{\oplus} \rightleftharpoons \text{R—CH(}^{\oplus}\text{NH}_3\text{)—COOH} \qquad (20\text{-}6)$$

IN BASIC SOLUTION

$$\text{R—CH(}^{\oplus}\text{NH}_3\text{)—COO}^{\ominus} + \text{OH}^{\ominus} \rightleftharpoons \text{H}_2\text{O} + \text{R—CH(NH}_2\text{)—COO}^{\ominus} \qquad (20\text{-}7)$$

If an amino acid that is solely in the form of its zwitterion, that is, (+) charge equals (−) charge, is subjected to an electrical field, it will *not* migrate toward any electrode. But, if an acidic solution of an amino acid is electrolyzed, the positively charged ion will migrate towards the negative electrode or cathode. Conversely, if a basic solution of an amino acid is placed in an electrical field, the negatively charged ion will migrate toward the positive electrode or anode. Between these two extremes, it is possible to have a solution of a particular pH at which the amino acid will *not* migrate toward either the positive or the negative electrode in an electrical field. This can only occur at a pH where the amino acid exists entirely in the form of its zwitterion, since the net charge on the molecule is zero. This pH, at which no migration will occur under the influence of an electrical field, is called the **isoelectric point** (IEP) and is characteristic of each amino acid (and protein). The IEP does not necessarily have to be at pH = 7.0 (a neutral solution). Basic amino acids will have isoelectric points in basic solutions, at a pH higher than 7.0. Acidic amino acids will have their isoelectric points in acidic solutions at a pH lower than 7.0. Neutral amino acids have isoelectric points near neutrality, but slightly acidic in the pH range 4.8–6.3.

The isoelectric point is a specific characteristic of every amino acid and protein molecule. Thus, it is possible to separate a complex mixture of amino acids (or proteins) by carefully regulating the pH of the solution and placing the solution in an electrical field. By varying the pH of the solution ever so slightly over intervals of time, different amino acids (or proteins) will migrate to electrodes (in the form of charged ions) at different pH's and be separated from the other components of the mixture. This technique, the migration of a protein molecule in an electrical field, is called **electrophoresis,** and is an extremely useful analytical tool for the biochemist.

20.7 Reactions of Amino Acids

The amino acids will undergo the characteristic reactions of both the amino and the carboxyl functional groups, present in the molecule. A few of these fundamental reactions will be presented as illustrative examples.

20.7-1 Reaction with Nitrous Acid

The reaction of a primary amino group with nitrous acid results in the formation of nitrogen gas.

Sec. 20.8 THE PEPTIDE BOND

$$CH_3-\underset{\underset{NH_2}{|}}{CH}-\overset{\overset{O}{\|}}{C}-OH + HONO \longrightarrow N_2\uparrow + H_2O + CH_3-\underset{\underset{OH}{|}}{CH}-\overset{\overset{O}{\|}}{C}-OH \quad (20\text{-}8)$$

alanine   lactic acid

The volume of nitrogen gas evolved can be measured, and from the data one can calculate the quantity of amino acid present. This technique was developed by D. D. Van Slyke and can be used to determine the number of free primary amino groups present in a protein molecule.

20.7-2 Reaction with Alcohols The carboxyl group of an amino acid will react with alcohols to form esters.

$$H_2N-CH_2-COOH + CH_3CH_2OH \xrightarrow{H^\oplus} H_2O + H_2N-CH_2-\overset{\overset{O}{\|}}{C}-O-CH_2-CH_3$$

glycine   ethyl ester

(20-9)

20.7-3 Reaction with Acyl Halides The reaction of the amino group of an amino acid with an acyl halide produces an amide.

$$H_2N-CH_2-COOH + C_6H_5-\overset{\overset{O}{\|}}{C}-Cl \longrightarrow HCl + C_6H_5-\overset{\overset{O}{\|}}{C}-\underset{\underset{}{|}}{\overset{\overset{H}{|}}{N}}-CH_2-COOH$$

glycine   benzoyl chloride   hippuric acid
                                                                     (benzoylglycine)

(20-10)

Hippuric acid is an *N*-substituted amide, found in the urine of horses. It is also the end product in the metabolism of benzoates by human beings, the benzoates being eliminated from the body in the form of hippuric acid (by combination of benzoic acid and glycine).

20.8 The Peptide Bond

The amino acids that constitute protein molecules are present in long chains in which the amino acids are joined together by amide linkages between the amino group of one acid and the carboxyl group of another acid. These linkages, $-\underset{\underset{}{|}}{\overset{\overset{H}{|}}{N}}-\overset{\overset{O}{\|}}{C}-$, are called **peptide bonds**. A protein contains many of these peptide bonds and is actually a **polypeptide**.

The geometry of the peptide bond is well known and is important in understanding the properties and chemistry of protein molecules. The amide group has sp^2 hybridization, is flat, and is coplanar. The carbon–nitrogen bond distance is 1.32 Å, which is shorter than the usual carbon–nitrogen single bond distance of 1.47 Å. This suggests that the amide structure is actually a resonance hybrid of several contributing structures, one of which has some carbon–nitrogen double bond character to account for the shorter than normal carbon–nitrogen single bond distance present in the amides.

$$\left[\begin{array}{c} \overset{\overset{\displaystyle :\overset{\ominus}{O}:}{\|}}{-C-\overset{|}{N}-} \end{array} \longleftrightarrow \begin{array}{c} \overset{\overset{\displaystyle :\overset{\ominus}{O}:}{|}}{-C=\overset{\oplus}{\underset{|}{N}}-} \end{array} \right]$$

If two amino acids are joined together in a chain, the structure contains one peptide linkage and is called a **dipeptide**.

$$R-\underset{NH_2}{\overset{H}{\underset{|}{C}}}-\boxed{\overset{O}{\underset{}{\overset{\|}{C}}}\xrightarrow{1.32\ \text{Å}}\overset{H}{\underset{H}{\underset{|}{N}}}}-\overset{R'}{\underset{|}{\underset{|}{C}}}-COOH$$

peptide bond → a dipeptide

A chain of three amino acids is called a **tripeptide**, etc. Proteins are molecules containing over 50 amino acid units, with molecular weights greater than 10,000, and are called **polypeptides**. Occasionally, however, the term polypeptide is used to describe a molecule consisting of a long chain containing a relatively large number of, but less than 50, amino acid units.

The amino acid that furnishes the free amino group in a peptide is called the N–*terminal acid,* and the amino acid that provides the free carboxyl group is called the C–*terminal acid*. The free amino group is written at the left end of the protein chain and the free carboxyl group at the right end, as agreed to by convention.

20.9 Determination of the Structure of Polypeptides

The structure determination of a polypeptide or protein is essential to understand the chemistry of the molecule in question, and provides the information as to which amino acids are present, and in what quantity, and perhaps most significant, the sequence of the amino acids in the chain.

The analysis as to which amino acids are present, and in what amount they are present, is a relatively simple task. However, the determination of the sequence of the amino acids in the peptide chain is often very difficult. For this reason, the complete structures of many proteins are virtually unknown to us.

20.9-1 The Ninhydrin Test The presence of an amino acid can be detected in a number of ways. One of the most sensitive tests for the detection of amino acids involves the compound, ninhydrin, which produces a blue color in the presence of amino acids. The **Ninhydrin test** is so sensitive that one can determine the presence of 10^{-6} moles of an amino acid quantitatively.

Sec. 20.9 THE STRUCTURE OF POLYPEPTIDES

$$2 \text{ ninhydrin} + R-CH(NH_2)-COOH \xrightarrow{\text{base}}$$

$$\text{blue ion} + H_3O^{\oplus} + CO_2\uparrow + R-\overset{O}{\underset{\|}{C}}-H \quad (20\text{-}11)$$

In order to determine which amino acids are present and in what quantity, the polypeptide or protein is completely hydrolyzed into its constituent amino acids. The various amino acids can then be separated by a technique known as paper chromatography (or by column chromatography), or by automated ion-exchange chromatography in an amino acid analyzer, and the relative amounts and composition of each acid present can be determined. This information and a knowledge of the molecular weight of the original polypeptide or protein which can be determined, enables us to calculate the true molecular formula of the molecule. Now the problem remains to determine the sequence in which the amino acids are joined together to form the peptide chain. This is a far more difficult and complex problem.

20.9-2 Sequence of Amino Acids in Polypeptides The general procedure is to partially hydrolyze the polypeptide or protein to a series of smaller peptides ranging in size from two to five amino acid units. These di- to penta-peptides can be separated by chromatographic techniques, and hopefully can be identified. Then, the problem is to fit the pieces of the puzzle together and establish the complete sequence of amino acids by examining the overlapping sequences of the peptides formed in the partial hydrolysis. This involves combining the techniques of partial hydrolysis with terminal residue analysis. Terminal residue analysis is the determination of which amino acid residue is present at one end of a polypeptide chain (N-terminal acid and C-terminal acid), which because of its structure has a free amino group at one end and a free carboxyl group at the other end of the chain.

20.9-3 The Sanger Method The N-terminal amino acid in a polypeptide is usually determined by the Sanger method (developed in 1945), which involves reacting the peptide with the reagent 2,4-dinitrofluorobenzene. This reagent has the ability to undergo nucleophilic aromatic substitution reactions with primary amino groups. Hydrolysis of the product results in the formation of a mixture of amino acids, only *one* of which has a 2,4-dinitrophenyl group attached to its amino nitrogen atom. This "tagged" amino acid can be easily identified as being the N-terminal acid which is present at one end of the peptide chain.

$$O_2N-\underset{NO_2}{\bigcirc}-F + H_2N-\underset{R}{CH}-\underset{\parallel}{C}-\underset{\parallel}{N}-\underset{R'}{CH}-\underset{\parallel}{C}-\underset{\parallel}{N}-(etc.)$$
$$\underbrace{}_{\text{peptide chain}} \quad \Big\downarrow -HF$$

$$O_2N-\underset{NO_2}{\bigcirc}-\underset{H}{N}-\underset{R}{CH}-\underset{\parallel}{C}-\underset{H}{N}-\underset{R'}{CH}-\underset{\parallel}{C}-\underset{H}{N}-$$

$$\Big\downarrow \overset{H^\oplus}{HCl} \text{ (complete hydrolysis)}$$

$$O_2N-\underset{NO_2}{\bigcirc}-\underset{H}{N}-\underset{R'}{CH}-COOH + \text{(other amino acids)}$$
$$\underbrace{}_{\text{N-terminal acid}} \qquad (20\text{-}12)$$

Sanger used this technique to determine the complete amino acid sequence in the hormone, insulin, for which he received the Nobel prize in 1958. To appreciate this feat, let us state that insulin has a molecular weight of 5,734 and contains 48 amino acid units made up of 16 different amino acids (cysteine being considered to be the same as cystine).

20.9–4 The Edman Procedure Another method for identifying the N–terminal residue in a polypeptide was developed by Edman in 1950. The Edman procedure is more helpful in the analysis of small peptides than the Sanger method because amino acid residues can be removed from the N–terminal position one at a time, so that one can identify what residue was next to the N–terminal residue, and establish a partial sequence of amino acids by repeating the reaction on the remaining peptide chain until all the units in the chain have been determined. Unfortunately, experimental procedures limit the Edman procedure in that at most a sequence of five or six amino acids in succession can be established.

In the Edman procedure the peptide is reacted with the compound phenyl-isothiocyanate and then treated with dilute acid, liberating the N–terminal residue as the phenylthiohydantoin, while the remaining peptide chain has acquired a new N–terminal residue, so the reaction sequence can be repeated over again.

$$R-\underset{NH_2}{CH}-\underset{\parallel}{C}-\underset{H}{N}-(etc.) + \bigcirc-N=C=S \longrightarrow \bigcirc-\underset{H}{N}-\underset{\parallel}{C}-\underset{H}{N}-\underset{R}{CH}-\underset{\parallel}{C}-\underset{H}{N}-(etc.)$$

peptide   phenyliso-
               thiocyanate

Sec. 20.9 THE STRUCTURE OF POLYPEPTIDES

H_2N—(etc.) +

phenylthiohydantoin
(can be identified and N-terminal acid determined)

(20-13)

20.9-5 Selective Enzymatic Hydrolysis The best procedure for determining the C-terminal amino acid in a polypeptide is by selective enzymatic hydrolysis. The enzyme carboxypeptidase (found in the pancreas) has the ability to selectively hydrolyze amide linkages adjacent to free or terminal carboxyl groups in the polypeptide chain. This procedure permits us to degrade the peptide chain one amino acid unit at a time from the carboxyl end, and establish a partial sequence of amino acids.

$$(etc.)-\overset{O}{\underset{}{C}}-\overset{H}{\underset{}{N}}-\overset{}{\underset{R'}{CH}}-\overset{O}{\underset{}{C}}-OH \xrightarrow{\text{carboxypeptidase}} (etc.)-\overset{O}{\underset{}{C}}-OH + H_2N-\overset{}{\underset{R}{CH}}-\overset{O}{\underset{}{C}}-OH$$

peptide　　　　　　　　　　　　　　　　　**C-terminal acid**
　　　　　　　　　　　　　　　　　　　　　(can be identified)

(20-14)

Other enzymes that are highly specific in their action can also be used to selectively break down protein molecules and help to determine the structure of the polypeptide.

The complete structure including the sequence of amino acids has been determined for a number of peptides by using the methods and principles just discussed. Among the more important substances whose complete structure has been elucidated are oxytocin (a hormone of the pituitary gland), insulin, certain cytochrome enzymes, and the enzyme chymotrypsinogen-A. Nevertheless, the complete structure of most polypeptides and proteins is unknown.

As a simple illustration of the difficulties involved in elucidation of structure let us consider the possible sequences of amino acids that can occur in a very small molecule, such as a tripeptide composed of three different amino acids, A, B, and C. There are *six* possible structures for tripeptides composed of these three amino acids. The structures for the six possible tripeptides are: A-B-C, A-C-B, B-A-C, B-C-A, C-A-B, and C-B-A. One can now appreciate how complex the problem is, when considering a much more complex molecule, such as a polypeptide or protein, composed of many different amino acids and many amino acid units. The sequence determination of amino acids is quite a task, when one considers all the possible amino acid sequences in terms of the mathematical probability, permutations, and combinations. It is no small wonder that the synthesis of polypeptides and proteins poses quite a challenge to the research chemist!

20.10 Synthesis of Polypeptides

The formation of a peptide bond between two amino acids *appears* to be a relatively simple process, involving a reaction between the amino group of one amino acid molecule and the carboxyl group of another amino acid, resulting in the loss of water and the formation of an amide linkage or peptide bond between the amino acid units. Such is *not* the case as no peptide bond can be formed by reacting two amino acids directly; the formation of a peptide bond is a much more involved process. The synthesis of peptides is important for many reasons (e.g., investigations on structure proof and biological activity), and can be accomplished in a number of ways. Only one method for the synthesis of peptides will be considered here.

Let us consider the synthesis of the dipeptide glycylalanine

$$H_2N-CH_2-\overset{O}{\underset{}{C}}-\overset{H}{\underset{}{N}}-\underset{CH_3}{CH}-\overset{O}{\underset{}{C}}-OH$$

$$\underbrace{}_{\text{glycyl unit}} \underbrace{}_{\text{alanyl unit}}$$

from glycine and alanine. The problem is of course to make sure that the amide linkage (peptide bond) is formed between the *carboxyl group of glycine* and the *amino group of alanine*. In order to make sure that we synthesize the desired dipeptide rather than some other dipeptide or side-product, it is essential that we protect the amino group of glycine in some way to prevent it from reacting (instead of the amino group of alanine), while being able to remove the protecting group at the end of the synthesis and regenerate the free amino group of glycine, still keeping intact the peptide bond that was formed.

20.10-1 The Bergmann Synthesis

The **Bergmann synthesis** or carbobenzoxy method is an excellent way to form peptide bonds. The key to the synthesis is the selection of the carbobenzoxy (sometimes called carbobenzyloxy) group as a protecting group to block the amino group (of glycine). Benzyl alcohol is reacted with phosgene to form the reagent benzyl chloroformate, which is the carbobenzoxy derivative required for the peptide synthesis.

$$\text{C}_6\text{H}_5\text{-CH}_2\text{O}\boxed{\text{H}} + \boxed{\text{Cl}}\text{-}\overset{O}{\underset{}{C}}\text{-Cl} \xrightarrow{-20°\text{C}} \text{HCl} + \text{C}_6\text{H}_5\text{-CH}_2\text{-O-}\overset{O}{\underset{}{C}}\text{-Cl} \quad (20\text{-}15)$$

benzyl alcohol   phosgene   benzyl chloroformate (carbobenzoxy chloride)

The benzyl chloroformate reacts with amino acids, as illustrated, for the reaction with glycine by the following equation.

Sec. 20.10 SYNTHESIS OF POLYPEPTIDES

$$\text{C}_6\text{H}_5\text{-CH}_2\text{-O-C(=O)-}\boxed{\text{Cl}} + \boxed{\text{H}}_2\text{N-CH}_2\text{-COOH} \xrightarrow{\text{dil. NaOH}}$$

$$\text{C}_6\text{H}_5\text{-CH}_2\text{-O-C(=O)-N(H)-CH}_2\text{-COOH} \quad (20\text{-}16)$$

carbobenzyloxyglycine

The amino group of glycine has been protected, while the carboxyl group can be converted into a more reactive form (such as an acyl halide) necessary for the subsequent step in the reaction sequence. The treatment of carbobenzyloxyglycine with phosphorus pentachloride converts it to the more reactive acid chloride

$$\text{C}_6\text{H}_5\text{-CH}_2\text{-O-C(=O)-N(H)-CH}_2\text{-COOH} \xrightarrow{\text{PCl}_5} \text{C}_6\text{H}_5\text{-CH}_2\text{-O-C(=O)-N(H)-CH}_2\text{-C(=O)-Cl}$$

$$(20\text{-}17)$$

which reacts with the next amino acid to follow in the peptide sequence, in this case, alanine.

$$\text{C}_6\text{H}_5\text{-CH}_2\text{-O-C(=O)-N(H)-CH}_2\text{-C(=O)-}\boxed{\text{Cl}} + \boxed{\text{H}}_2\text{N-CH(CH}_3\text{)-COOH}$$

$$\text{C}_6\text{H}_5\text{-CH}_2\text{-O-C(=O)-N(H)-CH}_2\text{-}\boxed{\text{C(=O)-N(H)}}\text{-CH(CH}_3\text{)-COOH} \quad (20\text{-}18)$$

↑
peptide bond
carbobenzyloxyglycylalanine

The peptide bond is formed in this step, and the resulting dipeptide contains the protected amino group of glycine, which must be removed without destroying the peptide linkage.

The carbobenzoxy protecting group can be removed by reduction with hydrogen and palladium black, to toluene and carbon dioxide, along with the formation of the dipeptide, glycylalanine (see Formula 20–19).

Hydrolysis is *not* needed to remove the carbobenzoxy group, which is a distinct advantage to the procedure as hydrolysis would also hydrolyze and destroy the peptide bond.

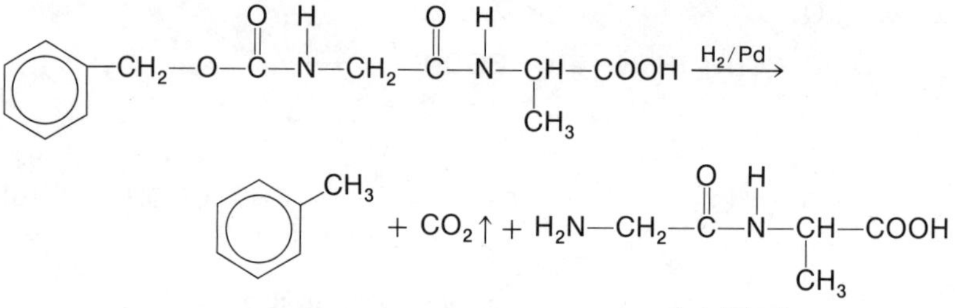

$$\underset{}{\bigcirc}-CH_2-O-\overset{O}{\underset{}{C}}-\overset{H}{\underset{}{N}}-CH_2-\overset{O}{\underset{}{C}}-\overset{H}{\underset{}{N}}-\underset{CH_3}{\underset{|}{CH}}-COOH \xrightarrow{H_2/Pd}$$

$$\underset{}{\bigcirc}-CH_3 + CO_2\uparrow + H_2N-CH_2-\overset{O}{\underset{}{C}}-\overset{H}{\underset{}{N}}-\underset{CH_3}{\underset{|}{CH}}-COOH \quad (20\text{-}19)$$

glycylalanine

In naming peptides, the amino acids are named in sequence, each unit ending in *-yl* rather than *-ine,* except the last amino acid unit, which is given its full common name. For example, the tripeptide, phenylalanylmethionyltyrosine has the following structure.

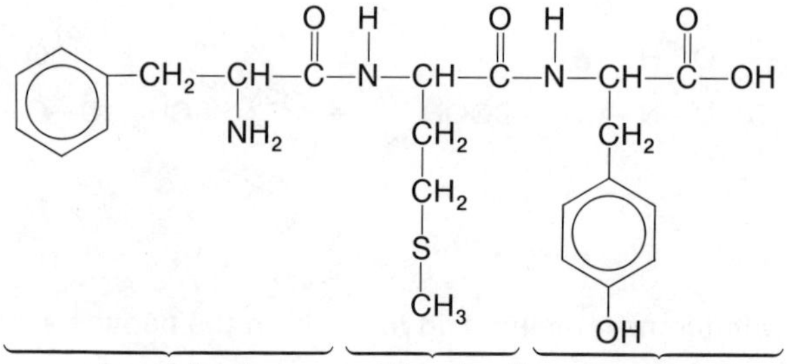

| phenylalanyl unit | methionyl unit | tyrosyl unit |

If one desires to synthesize a higher peptide, the chain is built up gradually according to the reaction sequence just outlined, but the protecting group is not removed until the very last step in the synthesis.

R. B. Merrifield and his colleagues have developed the **solid-phase technique** for chemical synthesis of polypeptides. In this procedure, the synthesis of the polypeptide chain is built, residue by residue, starting from the C–terminal amino acid that is attached to an insoluble solid resin particle.

20.11 Classification of Proteins

Proteins can be subdivided into two large groups: **simple proteins,** and **conjugated proteins.** The simple proteins are composed of chains of amino acids joined by peptide bonds, whereas the conjugated proteins have some non-protein entity as a part of the overall structure. These non-protein entities are bonded at certain sites to the peptide chain, and are called **prosthetic groups.** Examples of conjugated proteins include enzymes, nucleoproteins, and hemoglobin, which transports oxygen in the blood. Hemoglobin consists of a protein called **globin** attached to a prosthetic group, **hemin,** which has a complex structure called a **porphyrin.**

Sec. 20.12 STRUCTURE OF PROTEINS

hemin

Proteins may also be classified as being **fibrous** or **globular**. Fibrous proteins are the structural proteins present in animal tissue. They have a fibrous structure and are insoluble in all solvents except strong acids and bases. Some typical examples of fibrous proteins are **keratin** (from hair, nails, wool, feathers), **collagen** (from tendons, gelatin), **fibroin** (from silk), and **myosin** (from muscles).

Globular proteins are more complex in structure than the fibrous proteins. They are soluble in water and dilute salt solutions, and are sensitive to changes in temperature. The globular proteins can be further subdivided into several subclasses, depending on their solubility in certain solvents. Examples of globular proteins are all enzymes, all antibodies, hemoglobin, and albumin.

20.12 Structure of Proteins

The structure of proteins is far more complex than we have indicated thus far. It is extremely important to understand the structure of proteins in order to elucidate many complex biochemical problems. We will merely indicate some of the fundamental concepts and structural relationships that prevail in the protein molecule.

Let us begin by saying that there are several levels of organization present in the protein molecule that contribute to its composite three-dimensional configuration.

Only one of these, the primary organization, involving the sequence of amino acids in the molecule, has been considered up to this point. The amino acid sequence in a protein is referred to as the **primary structure.**

The **secondary structure** concerns itself with the shapes assumed by the peptide chains, or the preferred conformations of the chain. This is perhaps more important than the primary structure in explaining many properties of protein molecules. X-ray diffraction studies of crystals provide us with much of our information concerning the secondary structure of proteins. Experimental evidence indicates that hydrogen bonds are formed in the protein molecule between the NH groups and carbonyl groups present in the chain. These hydrogen bonds are believed to be responsible for the secondary structure of a protein. The most common arrangement of the chain in the secondary structure of proteins is the so-called α-*helix*. In the α-*helix* arrangement the *hydrogen bonds are formed between NH groups and carbonyl groups on the same peptide chain*. These

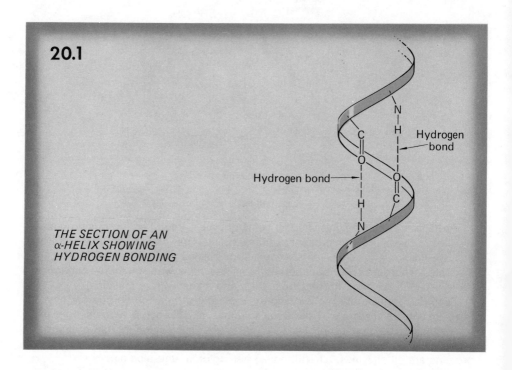

20.1

THE SECTION OF AN
α-HELIX SHOWING
HYDROGEN BONDING

hydrogen bonds are believed to be responsible for holding the peptide chain in the helical coil arrangement (see Figure 20.1). The α-helix is a right-handed helix that contains 3.6 amino acid residues per turn, with hydrogen bonds formed by the NH group and carbonyl oxygen located four peptide bonds apart. It has been determined that the spacing between turns of the helix is about 5.4 Å. The α-helix arrangement is found in many proteins such as α-keratin, and myoglobin. However, not all proteins have the helical structure. Some proteins have a so-called *sheet* or *β-structure* in which the helices become uncoiled and the chains stretch out side by side to give a sheet-like appearance. In the *sheet structure* of a protein there are *hydrogen bonds between the NH groups and carbonyl groups on different peptide chains,* resulting in an extended sheet-like structure, which is kept somewhat rigid by the hydrogen bonds formed between adjacent chains. The sheet structure is not as common as the α-helix structure in proteins. This is probably due in part to repulsions between the R groups on adjacent chains, which are in close proximity of one another in the sheet structure. The sheet structure is common in proteins when the R groups are small (e.g., R=H as in glycine), that is to say, when a large percentage of the protein is made up of small amino acids such as glycine (see Figure 20.2).

There are still higher levels of organization concerned with the structure of proteins. The **tertiary structure** deals with the ways in which coiled chains are folded and solvated in the protein molecule, while the **quaternary structure** is still more complex. Not much is known about these levels of organization in the protein molecule in comparison to the information available on the primary and secondary structures of proteins.

20.2

SHEET OR β STRUCTURE OF PROTEINS

20.13 Chemical Behavior of Proteins

In many respects the properties of proteins are very similar to those of the amino acids. For example, proteins are amphoteric, can be acidic or basic depending on their amino acid content, and have characteristic isoelectric points just like the amino acids. Proteins can also exhibit optical activity due to the presence of many asymmetric centers in the molecule.

When proteins are subjected to heat, or ultraviolet light, or are treated with concentrated acids or bases, certain organic solvents or other chemicals, they undergo an irreversible change known as **denaturation.** Perhaps the most familiar example of denaturation is the coagulation of egg white (albumin) when an egg is heated or cooked.

Denaturation involves certain complex changes in the structure of the protein molecule. No change occurs in the primary structure of the protein, but the protein shows profound changes in its physical properties once it is denatured. For example, the denatured protein is less soluble in such solvents as water, than the protein was before denaturation. The crystalline structure of a protein is often lost because of denaturation. Perhaps most significant, denaturation always leads to a complete loss in the physiological activity of the protein molecule (as when the protein is part of an enzyme).

The nature and mechanism of the denaturation process are not completely understood. It is known that the secondary structure of the protein becomes more random. Probably the helical structure in the protein becomes uncoiled and the sheet structure becomes more random and less rigid in appearance. It is also believed that denaturation affects the tertiary structure of proteins in some way related to the degree of hydration of the protein molecule.

20.14 Metabolism of Proteins

We will discuss only a few of the more important processes of the metabolism of proteins. The proteins are hydrolyzed in the body by various digestive enzymes (such as pepsin and rennin) to smaller peptides and amino acids. These products are absorbed and then transported by the blood throughout the body. Proteins in the body are undergoing a continuous process of degradation and at the same time are constantly being resynthesized. These processes involve the elimination of amino acids from proteins during degradation, as well as the need for amino acids when new proteins are being synthesized in the body. Certain amino acids required for growth and protein synthesis cannot be synthesized in the body. These acids must be incorporated in the diet, and are the essential amino acids referred to earlier in the chapter. Non-essential amino acids can be synthesized in the body.

20.14-1 Transamination

One extremely important reaction which occurs in protein metabolism is *transamination*. As the name suggests, an amino group can be transferred from one molecule to another. The transamination process converts α-amino acids to α-keto acids and vice versa. It is particularly significant because it shows that the metabolism of carbohydrates and proteins are clearly related in the body. As an illustration, consider the reaction between pyruvic acid and glutamic acid to form alanine and α-ketoglutaric acid.

$$\underset{\substack{\text{pyruvic}\\\text{acid}}}{CH_3-\overset{O}{\overset{\|}{C}}-\overset{O}{\overset{\|}{C}}-OH} + \underset{\substack{\text{glutamic}\\\text{acid}}}{\begin{array}{c}COOH\\|\\CHNH_2\\|\\CH_2\\|\\CH_2\\|\\COOH\end{array}} \overset{\text{transaminase}}{\rightleftharpoons} \underset{\text{alanine}}{CH_3-\underset{\underset{NH_2}{|}}{CH}-\overset{O}{\overset{\|}{C}}-OH} + \underset{\substack{\alpha\text{-ketoglutaric}\\\text{acid}}}{\begin{array}{c}COOH\\|\\C=O\\|\\CH_2\\|\\CH_2\\|\\COOH\end{array}}$$

(20-20)

It will be recalled also that many α-keto acids are formed during the Krebs cycle in the metabolism of carbohydrates (Section 18.22-1). Thus, the transamination reaction provides a link between carbohydrate and protein metabolism in the body.

The final end product of normal protein metabolism in the body is urea, which is excreted in the urine. The formation of urea involves the deamination of amino acids (from the hydrolysis of proteins) by enzymes in the liver to form keto acids and ammonia, the latter of which reacts with carbon dioxide to form urea.

$$\underset{\alpha\text{-amino acid}}{R-\underset{\underset{NH_2}{|}}{CH}-COOH} \overset{\text{liver enzyme}}{\rightleftharpoons} R-\underset{\underset{NH}{\|}}{C}-COOH \overset{H_2O}{\longrightarrow} \underset{\text{keto acid}}{R-\overset{O}{\overset{\|}{C}}-CO_2H} + NH_3$$

(20-21)

PROBLEMS

$$2NH_3 + CO_2 \longrightarrow H_2N-\overset{\overset{O}{\|}}{C}-NH_2 + H_2O \qquad (20\text{-}22)$$
urea

Summary

- Proteins are polymers of various α-amino acids. All the naturally occurring amino acids have the L-configuration.
- The nomenclature of amino acids is discussed in Section 20.3. Many are known by common names. The amino acids can be classified as being acidic, basic, or neutral.
- Some essential amino acids cannot be synthesized by the human body and must be incorporated in the diet.
- The α-amino acids may be synthesized by:
   (a) Amination of α-halogenated acids.
   (b) Strecker amino acid synthesis.
   (c) Malonic ester method.
- Amino acids exist as dipolar ions, are amphoteric and have characteristic isoelectric points.
- Amino acids undergo the characteristic reactions of the amino and carboxyl functional groups.
- The formation of amide or peptide bonds between α-amino acids forms polypeptides or proteins.
- The structure determination of a protein is a complex and difficult problem.
   The sequence of amino acids in polypeptides is determined by the Sanger method (N-terminal amino acids) or the Edman Procedure, and selective enzymatic hydrolysis (C-terminal amino acids).
- Partial synthesis of polypeptides can be accomplished by the Bergmann synthesis (use of carbobenzoxy protecting group) and by the Merrifield procedure.
- Proteins can be classified as simple or conjugated proteins, fibrous or globular.
- The primary structure in a protein is the amino acid sequence.
   The secondary structure refers to the α-helix shape assumed by the peptide chains.
- Denaturation involves certain complex changes in the structure of the protein molecule, affecting its chemical behavior.
- Transamination is an important reaction in the metabolism of proteins.

PROBLEMS

1 Draw the correct structural formulas for the following peptides.
   (a) *l*-leucylleucine   (c) tyrosyltryptophan
   (b) serylmethionine   (d) alanylglycylleucine

2 Synthesize the amino acid phenylalanine from benzene and any necessary inorganic reagents. Use a Strecker synthesis as part of your reaction sequence.

3 Perform the following syntheses from the indicated starting materials and any necessary inorganic reagents.
(a) ethylene to glycine
(b) ethylene to aspartic acid
(c) *i*–butyl bromide to valine
(d) *i*–butyl bromide to leucine

4 Predict the pH corresponding to the approximate isoelectric points of the following amino acids. Explain.
(a) aspartic acid
(b) lysine
(c) alanine
(d) arginine

5 Write the structural formula(s) of the product(s) formed in each of the following reactions, between alanine and the indicated reagents.
(a) alanine + HCl $\longrightarrow$
(b) alanine + NaOH $\longrightarrow$
(c) alanine + CH_3CH_2OH $\xrightarrow{H^{\oplus}}$
(d) alanine + HONO $\longrightarrow$
(e) alanine + Ac_2O $\longrightarrow$
(f) alanine + $SOCl_2$ $\longrightarrow$
(g) alanine + NH_3 $\xrightarrow{\Delta}$

6 Starting from coal tar and inorganic reagents, synthesize the dipeptide phenylalanylglycine.

7 *A certain peptide is hydrolyzed completely and found to contain the following amino acids: aspartic acid, histidine, methionine, phenylalanine, and valine. The application of the techniques of terminal residue analysis and partial hydrolysis produced the following series of dipeptides: (valine + aspartic acid), (methionine + histidine), (aspartic acid + methionine), and (phenylalanine + valine). Work out the sequence of amino acids in the peptide; draw its correct structural formula, and name it.

8 Explain how some proteins can be quite acidic or quite basic.

9 Illustrate the transamination reaction by writing an equation between oxalacetic acid and glutamic acid in the presence of the enzyme glutamic-oxalacetic aminotransferase.

10 Of What biochemical significance is the reaction referred to in Problem 9?

21
Nucleic Acids

The double-helix structure of DNA proposed by Watson and Crick—the polynucleotide chains are fastened by hydrogen bonds along cross-chain bridges between the pyrimidine and purine bases.

From Goodman & Morehouse, *Organic Molecules in Action*, Gordon & Breach, 1973.

21.1 Introduction

Perhaps the most important developments in all of science during the past twenty years have been in the elucidation of the chemistry of the nucleic acids. The nucleic acids are the naturally occurring polymers found in all living cells and are perhaps the fundamental substances controlling life and death. These amazing molecules have many functions, among which are the transmission of genetic information from generation to generation (as genes) and the regulation of protein synthesis in the cell. In this chapter, we will endeavor to make clear some of the basic structural features and chemistry of these macromolecules.

Historically, the nucleic acids were first reported in 1871 by Miescher, who extracted them from cell nuclei as an acidic substance he called nuclein. Further investigation showed that the nucleic acids were present in all living cells, as macromolecules in the free state, or combined with proteins in the form of so-called *nucleoproteins*.

We will refer to two types of nucleic acids, deoxyribonucleic acids (DNA) and ribonucleic acids (RNA). A combination of analytical and histochemical techniques has led to the positive identification of DNA in the nucleus of the cell as part of the chromatin material or chromosomes. (Hence, the involvement of DNA in the genetic code.) RNA has been found to be mainly present in the cell cytoplasm, although small quantities of RNA have recently been discovered in the nucleus also. DNA is also known to be present in the mitochondria and chloroplasts of the cell, but in relatively small amounts.

21.2 Constituents of the Nucleic Acids

The fundamental structure of the nucleic acids includes a polymer, containing repeating sugar units in the form of phosphate esters, with each sugar residue being attached to one of a group of heterocyclic bases as an N-glycoside.

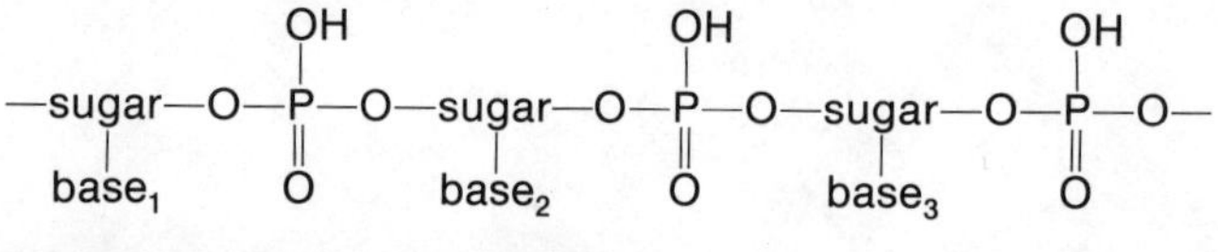

$$-\text{sugar}-O-\overset{OH}{\underset{\underset{O}{\|}}{P}}-O-\text{sugar}-O-\overset{OH}{\underset{\underset{O}{\|}}{P}}-O-\text{sugar}-O-\overset{OH}{\underset{\underset{O}{\|}}{P}}-O-$$
$$\text{base}_1 \qquad\qquad \text{base}_2 \qquad\qquad \text{base}_3$$

The sugar present in the nucleic acids may be either of two pentoses, D-*ribose* or D-*2-deoxyribose*.

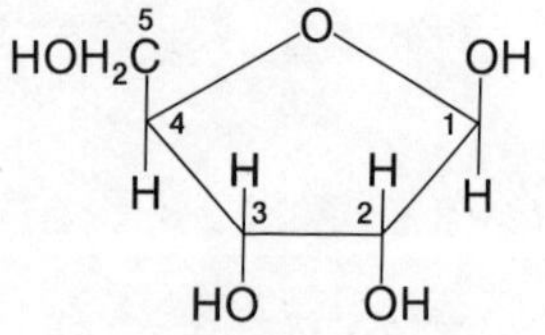

β-D-ribose
(β-D-ribofuranose)

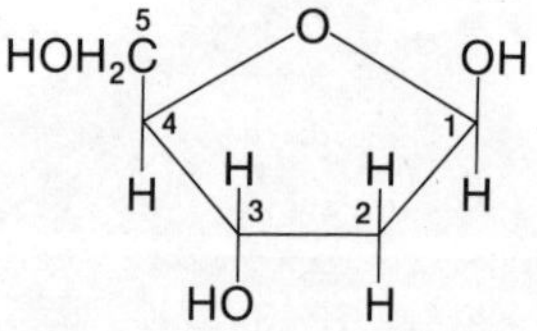

β-D-2-deoxyribose
(β-D-2-deoxyribofuranose)

The group of nucleic acids referred to as deoxyribonucleic acids contain the sugar D-2-deoxyribose, where the ribonucleic acids contain the sugar D-ribose. The most important property of these pentose sugars (in relation to the nucleic acid structure) is that the hydroxyl groups can form phosphate esters with phosphoric acid, especially the hydroxyl groups at carbons 3 and 5.

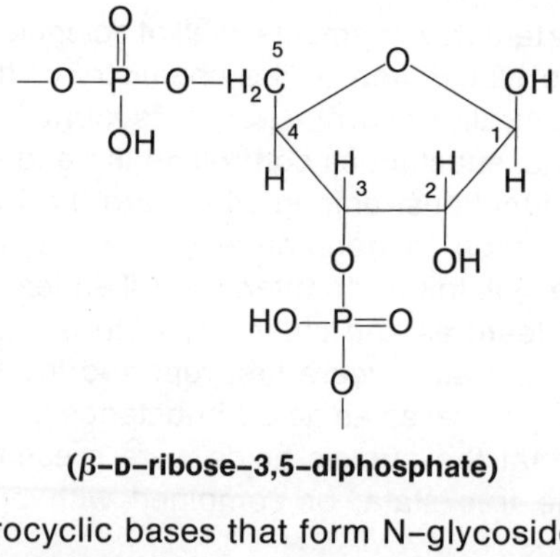

(β-D-ribose-3,5-diphosphate)

The common heterocyclic bases that form N-glycosides with the sugars are related to the parent compounds purine or pyrimidine. The five most common bases found in the nucleic acids are *adenine, guanine, cytosine, uracil,* and *thymine.* Adenine and guanine are derivatives of the parent compound purine.

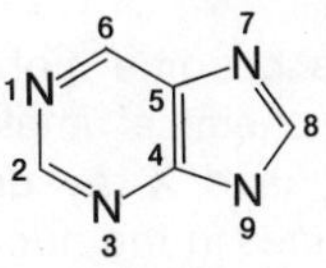

Sec. 21.2 CONSTITUENTS OF NUCLEIC ACIDS

purine

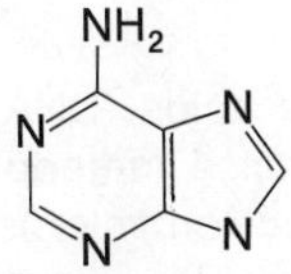

adenine
(6-aminopurine)

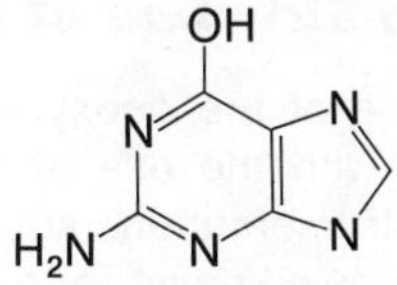

guanine
(2-amino-6-hydroxypurine)

The three other nitrogenous bases, cytosine, uracil, and thymine are derivatives of the parent compound pyrimidine.

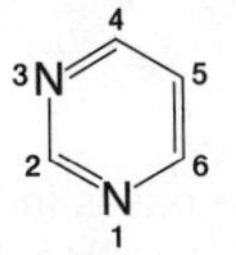

pyrimidine

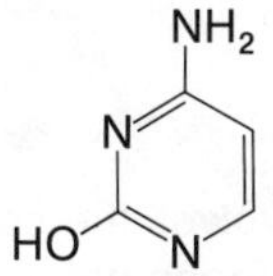

cytosine
(2-hydroxy-4-aminopyrimidine)

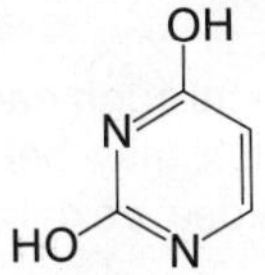

uracil
(2,4-dihydroxypyrimidine)

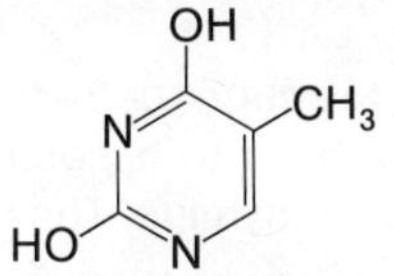

thymine
(2,4-dihydroxy-5-methylpyrimidine)

It will be noted that four of the five bases contain hydroxyl group(s) in the molecule. These hydroxyl groups are significant in that when these bases form N-glycosides with the sugars in the nucleic acids, it is essential that the base undergo tautomerism from the enol to the keto or amide form before combining with the sugar molecule.

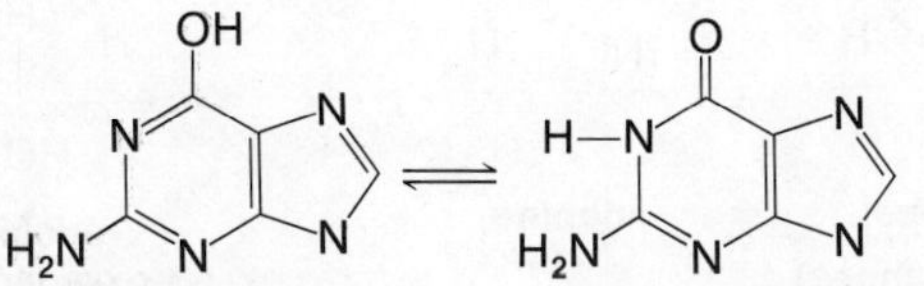

guanine, enol form ⇌ guanine, keto form (21-1)

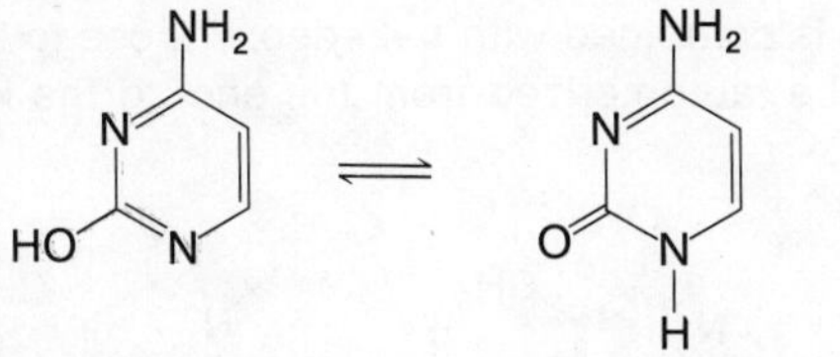

cytosine, enol form ⇌ cytosine, keto form (21-2)

In the deoxyribonucleic acids (DNA) the bases adenine, guanine, cytosine, and thymine are present, while in the ribonucleic acids (RNA), the thymine is replaced by uracil. These bases form an N-glycoside with carbon 1 of the sugar molecule, called a **nucleoside**.

21.3 The Structure of DNA

The structure of the deoxyribonucleic acids (DNA) has become more clearly understood with the use of more precise methods of chemical analysis and modern instrumentation, such as electron microscopy and X-ray diffraction studies. It is now known for example, that the chromosomes in the nuclei of the cell are actually macromolecules of DNA attached to proteins. The molecular weight of DNA molecules is in the order of several million, and they have been seen under the electron microscope. As in the case of the proteins, the nucleic acids can be considered as being composed of a primary, a secondary, and a tertiary structure.

21.4 The Primary Structure

The *primary structure* involves the composition of the bases in the molecule, as well as a nucleotide sequence determination. The DNA molecules are polymers consisting of repeating sugar units in the form of phosphate esters (at positions 3 or 5), the sugar being bonded to heterocyclic bases at carbon 1 to form N-glycosides.

DNA contains the sugar D-2-deoxyribose bonded through carbon 1 (which is the hemiacetal carbon) to a heterocyclic base, which may be either adenine, guanine, cytosine, or thymine. The resulting N-glycosides are called **nucleosides.**

β-D-2-deoxyribose + adenine $\xrightarrow{-H_2O}$ adenine deoxyribose (21-3)
(β-D-2-deoxyribofuranose)   (deoxyadenosine, a nucleoside)

If the base thymine is combined with D-2-deoxyribose to form a nucleoside, the thymine must first be tautomerized from the enol to the keto form

thymine, enol form $\rightleftharpoons$ thymine, keto form   (21-4)

Sec. 21.4 THE PRIMARY STRUCTURE

β-D-2-deoxyribose + thymine, keto form $\xrightarrow{-H_2O}$ thymine deoxyribose (thymidine, a nucleoside) (21-5)

The hydroxyl group at carbon 5' of the sugar in the nucleoside can be esterified with orthophosphoric acid (H_3PO_4) to form a phosphate ester, which is called a **nucleotide**. A nucleotide is actually a sugar in the form of a phosphate ester, the sugar also being bonded to a heterocyclic amine as an N-glycoside.

adenine deoxyriboside (a nucleoside) $\xrightarrow[-H_2O]{H_3PO_4}$ adenine deoxyribonucleotide (adenine-5'-monophosphate, a nucleotide) (21-6)

We have previously referred to several molecules that are *nucleotides,* although this term was not used to describe them. Among the particularly important nucleotides are adenosine-5'-triphosphate (ATP), which plays a vital role in the conservation and utilization of energy liberated in the cell during carbohydrate metabolism, and coenzyme A, so important in the metabolism of lipids and fatty acids. The structures for these two molecules are as follows.

ATP (adenosine-5'-triphosphate)

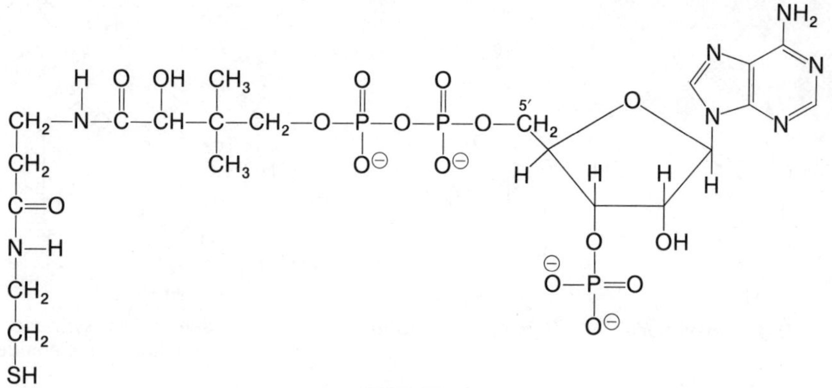

coenzyme A

The important part of the CoA molecule is the terminal sulfhydryl group (SH) which forms thioesters with fatty acids, as was discussed in Section 19.9 on lipid metabolism.

The nucleotides are the fundamental monomer units from which the giant macromolecules (polymers) of nucleic acids are built. The nucleic acids are actually polynucleotides. The nucleotide units are linked to each other by a phosphate ester formed between the hydroxyl group at carbon 3′ of one nucleotide and the phosphate group at carbon 5′ of another nucleotide. This can be shown experimentally by selective hydrolysis experiments on nucleic acids, which confirm that the nucleotide residues are held together by phosphate linkages between carbon 3′ of a ribose residue and carbon 5′ of an adjacent ribose residue

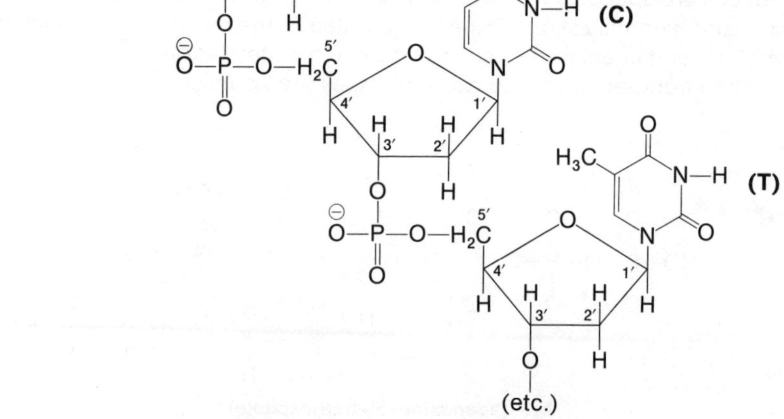

A = adenine
C = cytosine
T = thymine
G = guanine

structure of a section of a DNA chain

Sec. 21.5 SECONDARY AND TERTIARY STRUCTURES

in the primary structure. The macromolecule of DNA (or RNA) may contain as many as several million nucleotide units. The sequence of the bases in the chain of nucleotide units will vary somewhat depending from what source the DNA was obtained. A section of a typical DNA chain is shown on the preceding page.

It is believed that the sequence of bases along the chain determines how and in what way the nucleic acids function. The sequence of bases is extremely important in that it appears to control the characteristics of the cell, such as genetic information and cell metabolism, by some complex mechanism. It seems rather astonishing that the sequence of bases in the polynucleotide chain can transmit certain metabolic and genetic information, and control the vital functions of the cell itself. How this actually occurs is not completely understood. However, the secondary and tertiary structures of DNA seem to provide a partial explanation as to how the nucleic acids transmit information.

21.5 The Secondary and Tertiary Structures

Although the *exact sequence* of bases in the chain of DNA is not completely known, a certain very definite relationship exists among certain pairs of bases. For example, in all DNA molecules the number of guanine (G) groups is exactly equal to the number of cytosine (C) groups, and the number of adenine (A) groups is the same as the number of thymine (T) groups present in the macromolecule. This relationship is referred to as uniformity of base composition or as base-pairing, that is to say, $G = C$ and $A = T$ residues in DNA. The explanation for these rather unusual features and for many properties of the DNA molecules can be found in the secondary and tertiary structures of DNA.

21.5-1 The Double Helix X-ray diffraction studies on DNA have revealed that it has a helical configuration. Watson and Crick proposed a theory to account for the uniformity of base composition in DNA based on a helical configuration for the molecule. They proposed that the structure of DNA exists as two identical polynucleotide chains, helically entwined around each other to form a double strand, or **double helix,** with an average molecular diameter of about 20 Å. According to the Watson–Crick hypothesis, the double helix has the sugar-phosphate chain on the outside of the helix and the bases directed in space on the inside of the double helix, the two strands or chains being held together by *hydrogen bonding between appropriate base-pairs* (see Figure 21.1). It is essen-

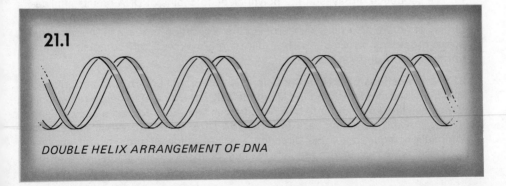

21.1

DOUBLE HELIX ARRANGEMENT OF DNA

tial that specific bases be exactly opposite each other in the two strands of the double helix in order for hydrogen bonding to occur. Thus, guanine in one strand can only form hydrogen bonds with a cytosine in the other strand, and adenine can only form hydrogen bonds with thymine in the other strand. Any other combination of bases cannot form hydrogen bonds since their molecular geometry is unsuitable for the proper double helix structure. The sequence of atoms in the individual strands (of the double helix) run in opposite directions. This permits a base on one strand to form hydrogen bonds to a corresponding base on the other strand, and in this way maintain the integrity of the double helix structure. Each turn of the helix contains ten nucleotide units. Studies by Pauling and Corey indicate that adenine and thymine are held together by two hydrogen bonds, whereas guanine and cytosine are attached by three hydrogen bonds.

thymine–adenine base-pair showing two hydrogen bonds

cytosine–guanine base-pair showing three hydrogen bonds

As a further consequence of their theory, Watson and Crick suggested that there is room between the two strands of the double helix for a polypeptide or protein to wind around the axis of the helix. This may account for the formation of nucleoproteins.

The sequence of bases in the polynucleotide chains of DNA has its implications in the genetic code and the field of chemical genetics. The secondary and tertiary structure also play a vital role in the replication process in the cell. The actual processes involved in DNA replication are exceedingly complex and are not completely understood.

21.5–2 Functions of DNA in the Cell It is known that DNA has to perform two vital functions in maintaining the life of the cell. First, it must undergo replication during cell division in order to be able to transmit an exact duplicate copy of itself to each of the daughter cells formed. The DNA must also provide essential information required for the direction and regulation of protein and enzyme synthesis, which will in turn regulate the metabolic processes in the new growing cell.

Sec. 21.6 THE STRUCTURE OF RNA

In the replication process the two strands of the double helix are believed to untwist and form two separate, individual strands, each of which can serve as a template for the formation of its counterpart, which will have a structure identical to that of the original strand to which it was bonded through hydrogen bonding. Other possible mechanisms have been suggested for DNA replication also.

21.6 The Structure of RNA

We have mentioned the function of DNA in storing genetic information and referred to its role in protein and enzyme synthesis. However, DNA can't be directly involved in any of these processes, since they all occur in the cellular cytoplasm which does not contain any DNA. The role of DNA must involve the transferring of information from DNA to other substances that actually perform the specific function at a particular site in the cell. Since the ribonucleic acids (RNA) are largely present in the cellular cytoplasm, this would suggest that they are the substances directly involved in the aforementioned vital processes.

The structure of RNA is similar to that discussed for DNA, except for several major differences. In RNA the sugar present is D-ribose rather than D-2-deoxyribose as in DNA. The bases in RNA include adenine, cytosine, and guanine, but uracil in place of thymine. Furthermore, the base-pairing observed in DNA is not nearly as predominant in RNA. This suggests that RNA has a partial helical structure and is composed of a single strand. Some base-pairing can occur, and in the large viral RNA molecules (viruses) double strands are known to exist. A typical RNA nucleotide unit would look very similar to a DNA nucleotide, except for the presence of the D-ribose sugar instead of D-2-deoxyribose.

cytosine riboside
(cytidine-5′-monophosphate, a nucleotide)

The synthesis of proteins directly involves the RNA molecules. Actually, the DNA present in the nucleus of the cell controls the structure of the RNA molecules produced in the cell, which carry information and are involved in protein synthesis at sites known as **ribosomes**.[*]

21.6-1 Varieties of RNA At least three distinct varieties of single-strand RNA molecules are known to be involved in protein synthesis. These RNA are *ribosomal RNA, messenger RNA,* and *soluble* or *transfer RNA*.

---

[*] Dr. Gobind Khorana, of the University of Wisconsin and MIT, who won a Nobel Prize in 1968 for work in this field, has reported the complete synthesis of a gene containing 77 nucleotide units.

370 NUCLEIC ACIDS Ch. 21

The **ribosomal RNA** is combined with proteins to form the ribosomes, which are actually ribonucleoproteins, and are the sites where protein synthesis occurs in the cell. The function of the **messenger RNA** is to relay information regarding what protein is to be synthesized, while the **soluble** or **transfer RNA** controls the amino acid sequence in protein synthesis by steering each amino acid to its proper position in the protein. This last process is believed to occur on the surface of the ribosome.

Although a great deal of the chemistry of DNA and RNA is still a puzzle to us, using modern research techniques devoted research investigators are gradually solving this vast mystery, which determines our very existence and destiny.

Summary

- Two types of nucleic acids, DNA and RNA, are found in all living cells. These polymers embody the genetic code of heredity.
- The fundamental structure of nucleic acids consists of a polymer, containing repeating sugar units (D-ribose or D-2-deoxyribose) in the form of phosphate esters, with each sugar residue being attached to one of a group of nitrogen heterocyclic bases (adenine, guanine, cytosine, uracil, thymine) as an N-glycoside.
- The primary structure of DNA involves the composition of the bases in the molecule, as well as a nucleotide sequence determination.
- The secondary and tertiary structure of DNA refers to the so-called double helix, the two strands of chains being held together by hydrogen bonding between appropriate base-pairs.
- RNA is somewhat similar in structure to DNA except that the sugar present is D-ribose rather than D-2-deoxyribose.

   (a) RNA has a single-strand helix.

   (b) Varieties of RNA include ribosomal RNA, messenger RNA, and soluble or transfer RNA.

PROBLEMS

1. Draw the correct structural formula of the nucleoside between the members of the following pairs.

   (a) D-ribose and guanine   **(c)** D-2-deoxyribose and adenine
   (b) D-ribose and uracil   **(d)** D-ribose and thymine

2. Draw the correct structural formula for the following nucleotides.

   (a) adenosine-5'-phosphate   **(c)** *adenosine-2',3'-cyclic monophosphate
   (b) guanosine-5'-phosphate   **(d)** thymosine-3'-phosphate

3. * Draw a segment of a DNA chain using 5-methylcytosine (which occurs in some DNA) instead of cytosine.

4. Draw a segment of an RNA chain.

5. The hydrolysis of some nucleic acids has led to the isolation of many purines as products. Among the purines that have been isolated are xanthine or 2,6-dihydroxypurine, and hypoxanthine or 6-hydroxypurine. Draw the structural formulas showing the keto form of these two compounds.

22
Natural Products

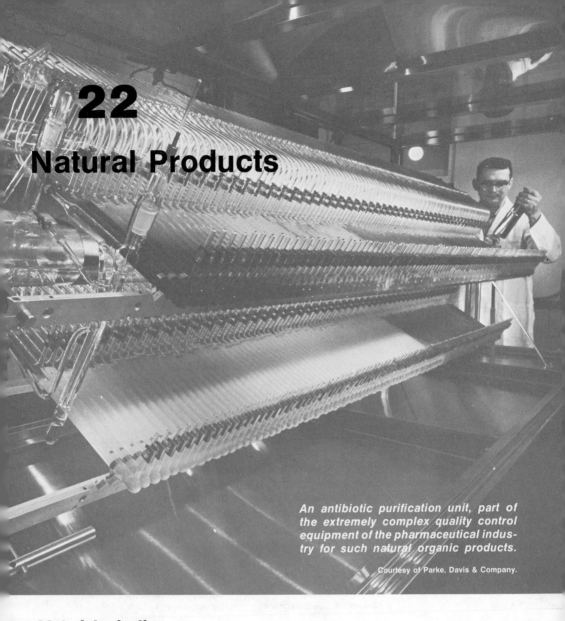

An antibiotic purification unit, part of the extremely complex quality control equipment of the pharmaceutical industry for such natural organic products.

Courtesy of Parke, Davis & Company.

22.1 Introduction

The term *natural products* refers to substances occurring in nature. We have already devoted an entire chapter to a group of natural products, the nucleic acids, because of the intense interest and importance of these compounds. Most natural products have complex structures and contain several functional groups in the molecule. Many natural products exhibit some type of biological or physiological activity. It is difficult to classify many natural products as belonging to a particular class of compounds because of the complex structural features and numerous functional groups often present in the molecule. However, it is possible to divide the natural products into several classes based on the composition of the main skeletal structure of the molecule. Thus, in this chapter we will discuss two general classes of natural products: (1) molecules whose main skeleton

Robert Burns Woodward (1917—)—professor of chemistry at Harvard University—is widely known for the chemical synthesis of such natural products as quinine, cholesterol, chlorophyll, and tetracycline. (Courtesy of Professor Woodward.)

contains only carbon atoms, in the form of either a chain or a ring, known as the *terpenes* and *steroids;* and (2) heterocyclic natural products, in which rings are present containing an atom other than carbon, such as oxygen, nitrogen, or sulfur.

22.2 Terpenes

The **terpenes** are a very large group of molecules which are characterized by their relation to isoprene, a constituent of natural rubber (Section 8.18). All terpenes have the common structural feature of containing one or more isoprene units in their carbon skeleton. For this reason they are sometimes referred to as isoprenoids.

$$\underset{1}{CH_2}=\underset{2}{\overset{\overset{\displaystyle CH_3}{|}}{C}}-\underset{3}{CH}=\underset{4}{CH_2}$$

isoprene
(2-methyl-l,3-butadiene)

abbreviated formula
for an isoprene unit

Isoprene itself is not found free in nature and is not the biological precursor of the terpenes, despite being an integral part of their molecular structure.

The terpenes are the constituents responsible for the fragrances of many plants. Many oils familiar to us are terpenes. Some perfumes, medicines, and solvents are terpenes. The word *terpene* comes from *terpentin,* which is an older form of the word *turpentine.*

22.3 Monoterpenes

The simplest terpenes, the monoterpenes, are actually dimers of isoprene. They are unsaturated compounds with the basic molecular formula $C_{10}H_{16}$, composed of two five-carbon isoprene units usually linked together at carbons 1 and 4.

Sec. 22.3 MONOTERPENES

Most monoterpenes are volatile compounds and are constituents of the essential oils in plants. These oils are responsible for the odors or flavors of the plants. The individual isoprene units within the molecule are sometimes indicated by a dashed line showing the bonds linking the isoprene units together. It is apparent that a great variety of molecules can be derived from the basic isoprene structure.

Myrcene, a typical monoterpene, is obtained from bayberry wax or bay oil.

Limonene is found in the oil of citrus fruits such as the lemon and orange.

Menthol is the chief constituent of peppermint oil.

Piperitone is a constituent of eucalyptus oil.

α-Pinene is a bicyclic monoterpene and the chief constituent of turpentine.

Camphor, obtained from the camphor tree in nature, can be synthesized in the laboratory from pinene.

Neptalactone, a monoterpene, is interesting in that its odor is attractive to cats, and it is the active component in catnip.

22.4 Sesquiterpenes

The sesquiterpenes are composed of three isoprene units, and are C_{15} compounds of quite varied structure.

Farnesol is a sesquiterpene acyclic alcohol isolated from citronella oil. It has an odor similar to *lily of the valley*, and is an important reaction intermediate in the biosynthesis of triterpenes and steroids.

Humulene has a rather unusual and complex structure, including an eleven-membered ring. It is a constituent of oil of hops.

Caryophyllene, obtained from oil of cloves, has a still more interesting structure, as does the following—

Longifolene, a sesquiterpene, from pine oil.

22.5 Diterpenes and Triterpenes

Diterpenes are compounds consisting of four isoprene units (tetramers of isoprene) and contain twenty carbon atoms, and the triterpenes are thirty carbon atom compounds composed of six isoprene units.

Representative diterpenes include **vitamin A** which is involved in the synthesis of certain pigments essential to vision and proper eyesight, and **abietic acid** (pine resin) obtained as a residue from turpentine.

vitamin A   abietic acid

Two important triterpenes are **squalene** and **lanosterol.** Squalene is obtained from shark liver oil, while lanosterol is obtained from sheep wool fat. Both compounds are important in the biosynthesis of steroids.

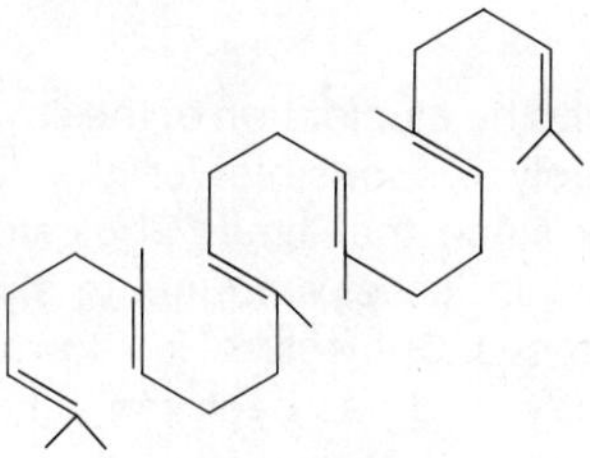

squalene

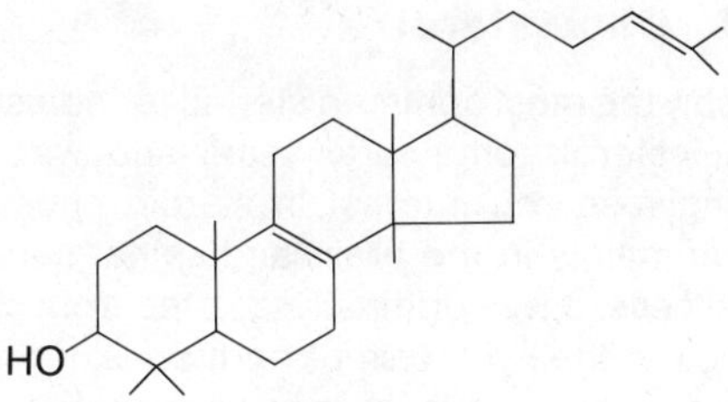

lanosterol

22.6 Steroids

The steroids are the group of lipids that are not saponified by base to fatty acids and glycerol. They are constituents of animal and plant fats, mainly in the form of solid alcohols called sterols. The steroids are a group of molecules that have numerous biological and physiological functions. The basic structure of the steroids is a fused-ring system of three six-membered rings and one five-membered ring, containing a total of seventeen carbon atoms in the skeleton. The tetracyclic steroid ring system is referred to as the perhydrocyclopentanophenanthrene nucleus.

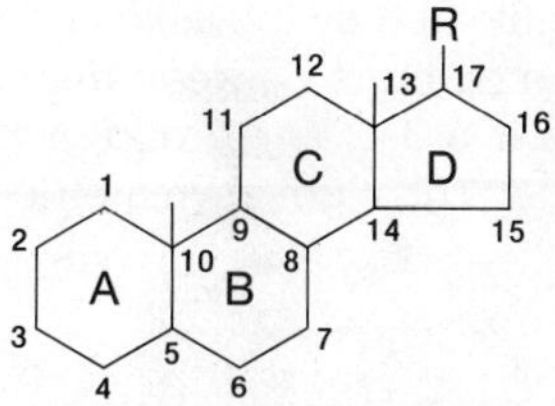

steroid ring system

The rings are referred to by the letters A, B, C, and D. Most steroids have two methyl groups attached to carbons 10 and 13, called **angular methyl groups.** A side chain or other functional group is usually present at carbon 17.

The structural formula for a steroid contains three six-membered cyclohexane rings. These rings, A, B, and C, are all in the chair conformation and are joined together by a so-called *trans* (or *cis*) ring fusion. The angular methyl groups are axial and are on the upper or "top" side of the molecule, called the β side of the molecule. A β substituent is up with respect to the plane of the molecule. Substituents that are on the lower side or α side of the molecule and extend down from the plane of the molecule are said to be alpha (α).

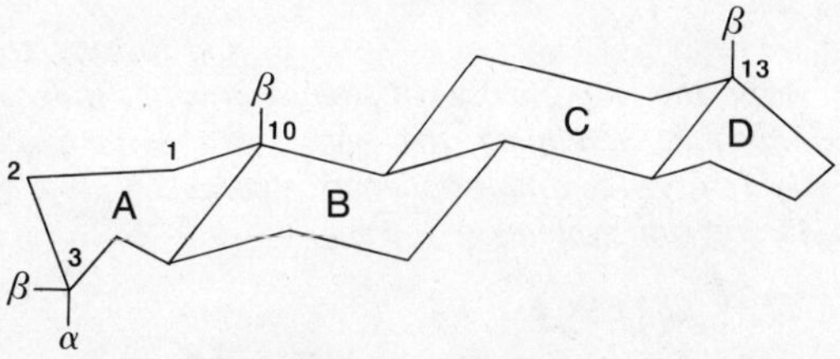

steroid ring system

22.7 Cholesterol

Probably the most common steroid is **cholesterol.** The elucidation of the structure of cholesterol some forty years ago was largely responsible for the rapidly growing research interest in steroid chemistry since that time. Cholesterol is present mainly in the brain and spinal cord, and in lesser amounts in all living animal cells. It was originally isolated from gallstones. Cholesterol is known to be involved in the synthesis of hormones in the body, and has been the subject of much debate as to its role in arteriosclerosis.

Cholesterol has the same basic steroid structure discussed in the preceding section. The cholesterol molecule has eight asymmetric carbon atoms, and therefore is but one of a possible 2^8 or 256 stereoisomers. The structural formula of cholesterol is as follows.

Note the *trans*-ring fusion indicated by the solid and dashed lines of substituents attached to the ring-junction carbons between rings B and C, and C and D. The hydroxyl group at carbon 3 is a 3–β–OH group, and an eight carbon alkyl side chain is present at carbon 17. The eight asymmetric carbon atoms are indicated

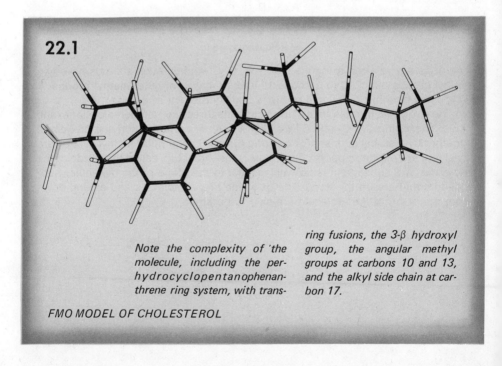

22.1

Note the complexity of the molecule, including the perhydrocyclopentanophenanthrene ring system, with trans-ring fusions, the 3-β hydroxyl group, the angular methyl groups at carbons 10 and 13, and the alkyl side chain at carbon 17.

FMO MODEL OF CHOLESTEROL

Sec. 22.8 SOME OTHER IMPORTANT STEROIDS

by asterisks. Figure 22.1 shows an FMO model of the cholesterol molecule, with the structural details of the cholesterol molecule in three dimensions. It is no small wonder that cholesterol with its complex molecular structure (i.e., *trans*-ring fusions, α and β substituents, one out of 256 possible stereoisomers, etc.) posed quite a problem for the synthetic organic chemist. Although cholesterol occurs in nature and can be synthesized by the living cell in a few minutes, chemists were unable to synthesize cholesterol in the laboratory until the 1950's.

22.8 Some Other Important Steroids

Many other steroids have important biological functions in the body. For example, the steroid **ergosterol** is a precursor of vitamin D, which controls the amount and ratio of calcium and phosphorus in the blood.

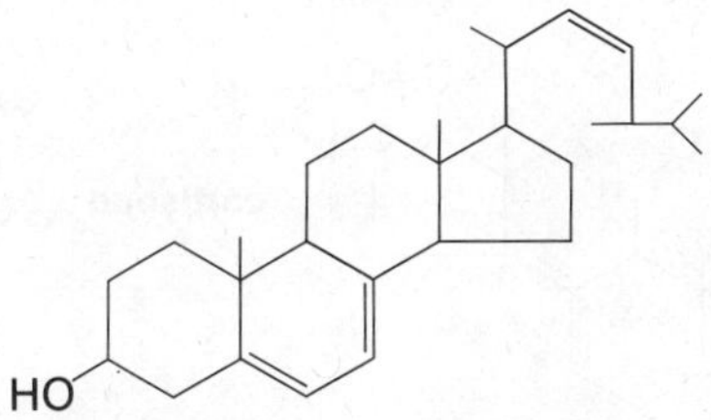

ergosterol

Many hormones contain the steroid structure, in particular the sex hormones and adrenal cortex hormones.

The sex hormones refer to substances produced in the gonads (ovaries and testes) that control reproduction and the secondary sex characteristics in human beings. The sex hormones are also important in the maintenance of good health and normal growth. The female sex hormones are called estrogens, of which **estrone** is a typical example.

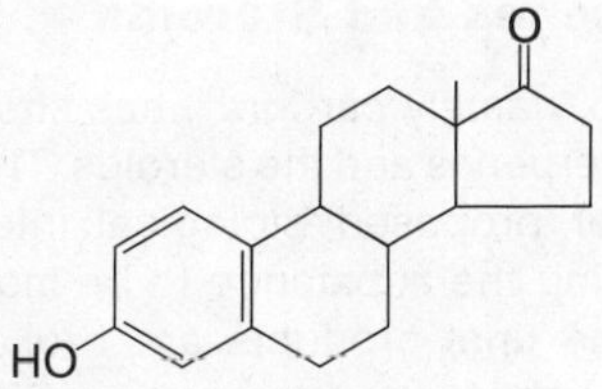

estrone

Note that ring A in estrone is aromatic and that a keto group is present at carbon 17. In addition to the estrogens, another female sex hormone is **progesterone**, which is required for pregnancy.

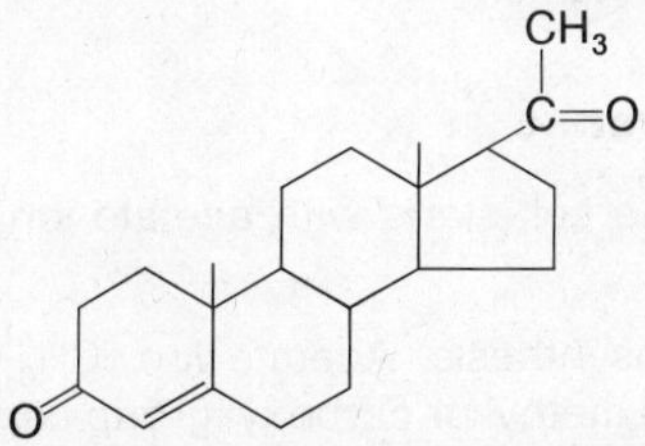

progesterone

The male sex hormones include **testosterone** which regulates the development of the male reproductive organs and secondary sex characteristics.

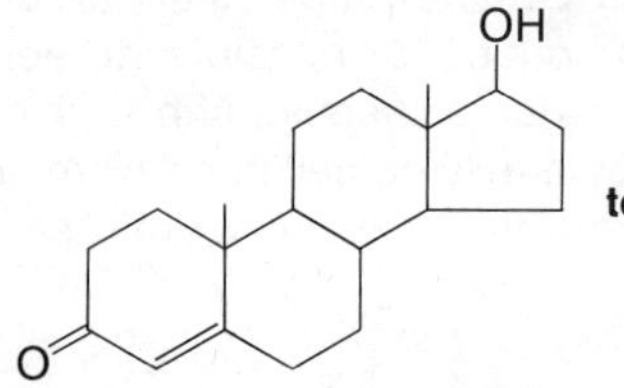

testosterone

Over thirty different hormones of the adrenal cortex gland have been isolated. Perhaps foremost among all these hormones is **cortisone,** which has been used in the treatment of rheumatoid arthritis and rheumatic fever.

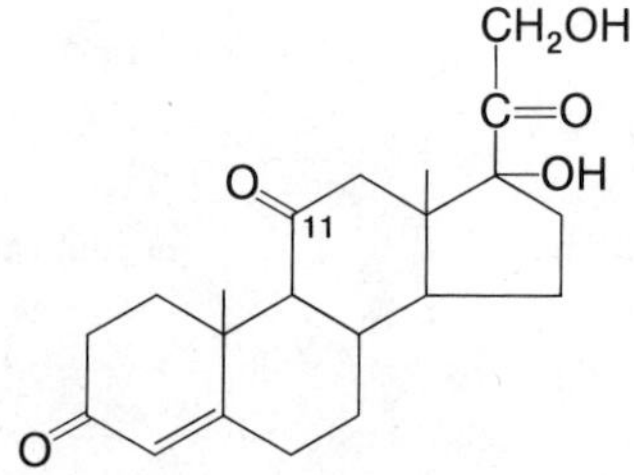

cortisone

A rather unusual feature is the presence of a keto group at carbon 11 in cortisone.

Cortisone and its various derivatives are used to treat many ailments, such as helping to bring relief to allergy sufferers. The pharmacological use of corticosteroids to treat these disorders is due to their anti-inflammatory activity and their ability to reduce swelling. Some steroids have been used in the treatment of birth control problems.

22.9 Biogenesis of Terpenes and Steroids

The use of radioactive isotopes, namely carbon^{14}, has shown that a biogenetic relationship exists between the terpenes and the steroids. The principal technique involved is to add a known or proposed biological intermediate containing carbon^{14} to an organism, allowing the substance to be metabolized, and finally isolating the end products. The final products are broken down into simple molecules, and the presence or absence of the radioactive isotope carbon^{14} in various places in the molecules gives information concerning the metabolic pathways and biosyntheses that occurred. What has essentially been done is to trace what happens to the various carbon atoms of the starting material through the sequence of biochemical reactions.

22.10 Biogenesis of Terpenes

The biogenesis of terpenes in the cell starts with acetate ion (actually acetate salts) as a precursor for the biosynthesis. Acetate ion, $CH_3-\overset{O}{\underset{\|}{C}}-O^{\ominus}$, can be labeled with carbon^{14} either in the methyl or carboxyl group carbon atom, so that

Sec. 22.10 BIOGENESIS OF TERPENES

the complete reaction sequence and the ultimate "fate" of the radioactive carbon tracer can be followed. In the cell the enzymes, which catalyze all the biochemical reactions, want to form a bond with the acetate and start the biogenetic sequence. This can only occur if the enzymes have the proper "handle" to enable them to react with acetate ion. As we saw in the metabolism of lipids, the substance coenzyme A forms an acetyl derivative, acetyl CoA, to start the reaction sequence.

$$CH_3-\overset{O}{\underset{\|}{C}}-O^{\ominus} + H^{\oplus} + CoA-SH \longrightarrow CH_3-\overset{O}{\underset{\|}{C}}-S-CoA + H_2O \quad (22\text{-}1)$$

$$\text{acetate} \qquad \text{coenzyme A} \qquad \text{acetyl CoA}$$

The next step in the sequence involves a Claisen condensation between two moles of acetyl CoA to give acetoacetyl CoA.

$$2CH_3-\overset{O}{\underset{\|}{C}}-S-CoA \longrightarrow CH_3-\overset{O}{\underset{\|}{C}}-CH_2-\overset{O}{\underset{\|}{C}}-S-CoA \longrightarrow$$

$$\text{acetoacetyl CoA}$$

A third mole of acetyl CoA adds across the carbonyl group of the acetoacetyl CoA via an aldol condensation to form

$$\xrightarrow{CH_3-\overset{O}{\underset{\|}{C}}-S-(CoA)} CH_3-\underset{\underset{\underset{(CoA)}{S}}{\underset{\|}{C=O}}}{\overset{OH}{\underset{|}{C}}-CH_2-\overset{O}{\underset{\|}{C}}-OH$$

which is reduced to mevalonic acid.

$$\xrightarrow{\text{reduction}} CH_3-\underset{\underset{OH}{\underset{|}{CH_2}}}{\underset{\underset{|}{CH_2}}{\overset{OH}{\underset{|}{C}}}}-CH_2-\overset{O}{\underset{\|}{C}}-OH \quad (22\text{-}2)$$

mevalonic acid

Mevalonic acid is the first important intermediate formed in the synthesis of terpenes. It has two alcoholic hydroxyl groups in the molecule, and these are converted to pyrophosphate esters by ATP, which is an excellent phosphorylating agent in biochemical reactions. The pyrophosphate ester loses carbon dioxide to form isopentenyl pyrophosphate, which is the so-called active isoprene unit in the biogenesis of terpenes.

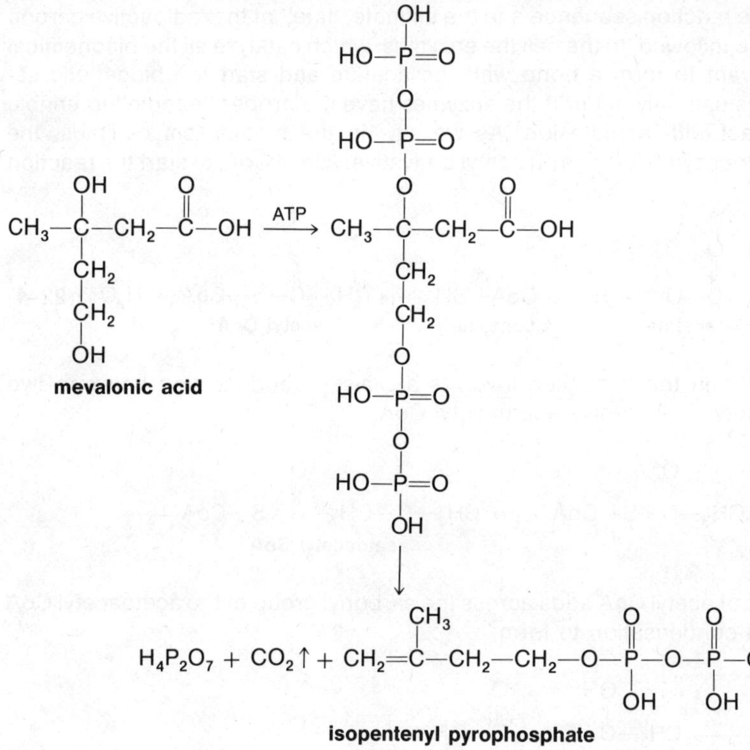

$$\text{mevalonic acid} \xrightarrow{\text{ATP}} \text{(phosphorylated mevalonate)} \rightarrow$$

$$H_4P_2O_7 + CO_2\uparrow + CH_2=\underset{CH_3}{C}-CH_2-CH_2-O-\underset{OH}{\overset{O}{P}}-O-\underset{OH}{\overset{O}{P}}-OH \quad (22\text{-}3)$$

isopentenyl pyrophosphate

It will be recalled that isoprene is not found uncombined in nature, and it is the compound isopentenyl pyrophosphate that is considered to be the biological precursor of the terpenes. The monoterpenes can be seen as being formed by a combination of two isopentenyl pyrophosphate units. For example, the monoterpene **geraniol** (found in oil of geranium) can be formed by one isopentenyl pyrophosphate unit undergoing an allylic rearrangement, prior to its combining with a second isopentenyl pyrophosphate unit. Subsequent hydrolysis of the pyrophosphate group yields geraniol as the product.

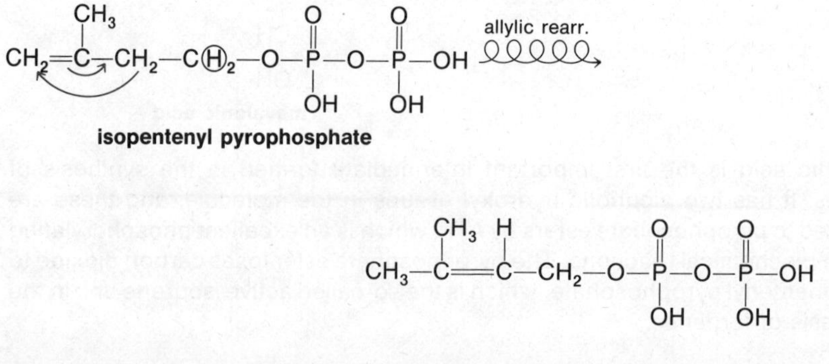

$$CH_2=\underset{CH_3}{C}-CH_2-CH_2-O-\underset{OH}{\overset{O}{P}}-O-\underset{OH}{\overset{O}{P}}-OH \xrightarrow{\text{allylic rearr.}}$$

isopentenyl pyrophosphate

$$CH_3-\underset{CH_3}{C}=\underset{H}{C}-CH_2-O-\underset{OH}{\overset{O}{P}}-O-\underset{OH}{\overset{O}{P}}-OH \quad (22\text{-}4)$$

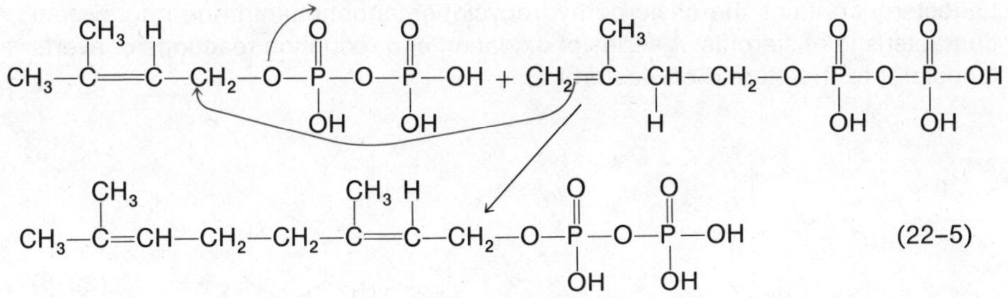

Sec. 22.11 BIOGENESIS OF STEROIDS

$$CH_3-\underset{CH_3}{\overset{}{C}}=\overset{H}{C}-CH_2-O-\overset{O}{\underset{OH}{P}}-O-\overset{O}{\underset{OH}{P}}-OH + CH_2=\underset{H}{\overset{CH_3}{C}}-CH-CH_2-O-\overset{O}{\underset{OH}{P}}-O-\overset{O}{\underset{OH}{P}}-OH$$

$$CH_3-\underset{CH_3}{\overset{}{C}}=CH-CH_2-CH_2-\underset{CH_3}{\overset{}{C}}=\overset{H}{C}-CH_2-O-\overset{O}{\underset{OH}{P}}-O-\overset{O}{\underset{OH}{P}}-OH \qquad (22\text{-}5)$$

geraniol pyrophosphate

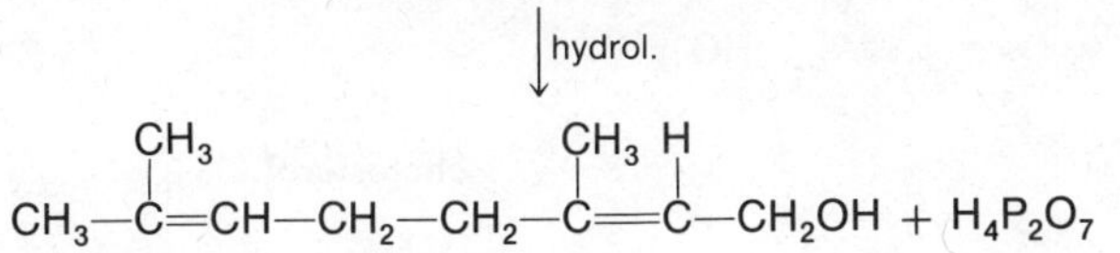

↓ hydrol.

$$CH_3-\underset{CH_3}{\overset{}{C}}=CH-CH_2-CH_2-\underset{CH_3}{\overset{}{C}}=\overset{H}{C}-CH_2OH + H_4P_2O_7 \qquad (22\text{-}6)$$

geraniol

Other monoterpenes are prepared by variations of the above process. The addition of a third isopentenyl pyrophosphate to geraniol pyrophosphate gives the compound **farnesyl pyrophosphate** (a sesquiterpene), which is a key intermediate in the biogenesis of steroids.

22.11 Biogenesis of Steroids

The biogenesis of steroids can be considered to be related to the synthesis of triterpenes. The key compound, **farnesyl pyrophosphate** undergoes a reductive coupling reaction to form the acyclic triterpene, **squalene.** By a two-step sequence which is rather complex, squalene is converted to the triterpene **lanosterol.**

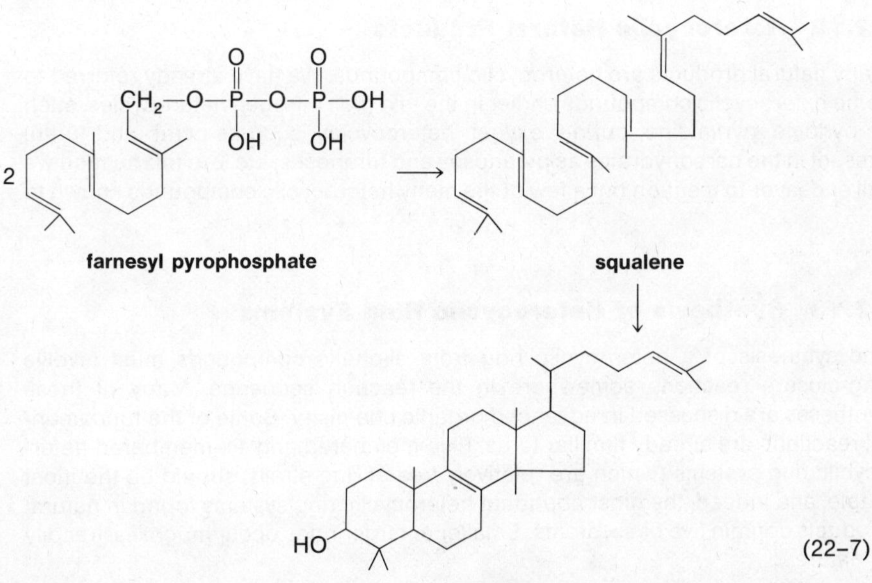

farnesyl pyrophosphate   squalene

↓

lanosterol   (22-7)

Lanosterol contains the basic perhydrocyclopentanophenanthrene ring system characteristic of steroids. A series of oxidation and reduction reactions converts lanosterol to the steroid **cholesterol**.

lanosterol → cholesterol   (22-8)

Use of radioactive isotopic labeling experiments have clearly established that cholesterol is synthesized from acetate, via the reaction sequence shown, going through farnesyl pyrophosphate, the precursor squalene and lanosterol. The conversion of lanosterol to cholesterol is interesting in that the three methyl groups shown (at carbon 4 and 14) in lanosterol are oxidized and lost as carbon dioxide. The side-chain carbon–carbon double bond is reduced, and the double bond between carbons 8 and 9 migrates between carbons 5 and 6, resulting in cholesterol as a final product. Other oxidation and reduction reactions will lead to the synthesis of other steroids. Our understanding of the process in the biogenesis of cholesterol is due largely to the research work of Bloch and Lynen, for which they received the Nobel Prize in 1964.

22.12 Heterocyclic Natural Products

Many natural products are heterocyclic compounds. We have already referred to some heterocyclic compounds earlier in the text (i.e., nitrogen heterocycles, such as pyridine, pyrimidine, purine; oxygen heterocycles, such as pyran and furan, present in the carbohydrates as pyranoses and furanoses; etc.). In this section we will endeavor to mention but a few of the many heterocyclic compounds known to man.

22.13 Synthesis of Heterocyclic Ring Systems

The synthesis of a heterocyclic ring from aliphatic compounds must involve ring-closure reactions somewhere in the reaction sequence. Many of these syntheses are discussed in advanced organic chemistry. Some of the fundamental reactions are already familiar to us. Five-membered and six-membered heterocyclic ring systems (which are relatively free of ring strain) should be the most stable, and indeed, the most abundant heterocyclic ring systems found in natural products contain five or six atoms. Smaller or larger rings occur much less readily in nature.

22.14 Oxygen Heterocycles

We have already referred to some oxygen heterocycles in the form of the pyranoses and furanoses of the carbohydrates. Two other natural products that also are oxygen heterocycles are **ascorbic acid, (vitamin C),** and **vitamin E,** which is composed of a group of compounds called **tocopherols.**

Ascorbic acid is actually a derivative of a carbohydrate (L–gluconic acid), and it is well known that a dietary deficiency in ascorbic acid leads to the disease scurvy. Ascorbic acid is found in citrus fruits and tomatoes. The ascorbic acid

ascorbic acid

molecule contains a five-membered ring containing a hetero oxygen atom. The ring is actually a cyclic ester or **lactone.** Ascorbic acid can be easily oxidized and is involved in various biochemical oxidation-reduction reactions in the body.

Vitamin E exists in several forms of tocopherols. The exact biological function of vitamin E is not known. It has been shown to be involved in reproduction (as an antisterility factor) in rats. Some chemists believe that vitamin E has some function in cellular oxidation reactions. The α–tocopherol form has the highest activity.

α–**tocopherol**
(vitamin E)

The compound **coumarin,** , is an oxygen heterocycle found in grass and clover. It has a pleasant odor and was used rather extensively in perfumes and food flavorings, but recently was shown to be carcinogenic, which has reduced its applicability. Coumarin can be synthesized from readily available materials such as from phenol. The reaction sequence involves two name reactions we have previously discussed: the Reimer–Tiemann reaction, and the Perkin reaction. Coumarin is actually a cyclic ester or *lactone,* which helps explain the ring closure in the last step to form the six-membered heterocyclic ring containing oxygen.

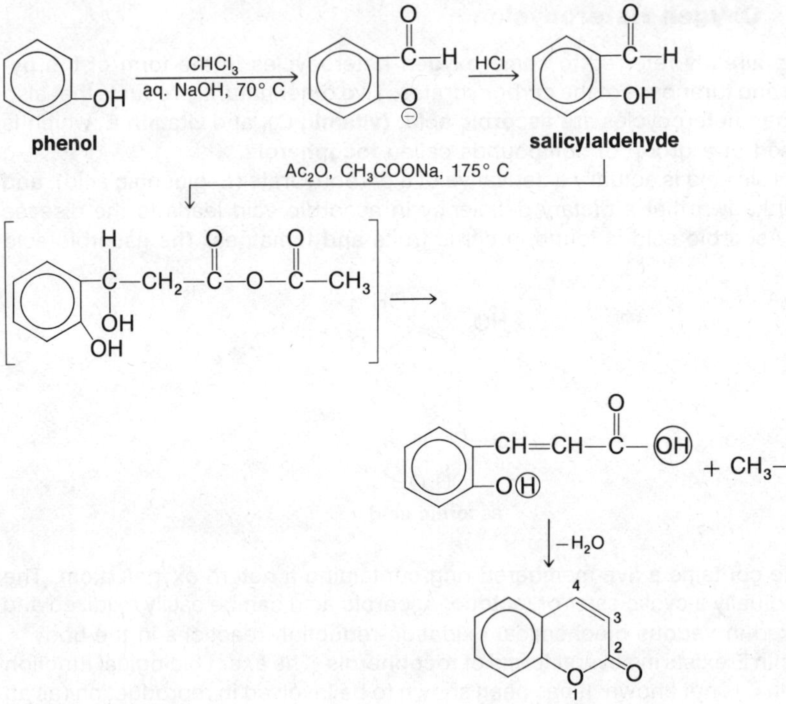

$$\text{phenol} \xrightarrow[\text{aq. NaOH, 70° C}]{\text{CHCl}_3} \text{[intermediate]} \xrightarrow{\text{HCl}} \text{salicylaldehyde} \quad (22\text{-}9)$$

$$\text{salicylaldehyde} \xrightarrow{\text{Ac}_2\text{O, CH}_3\text{COONa, 175° C}} \left[\text{o-HOC}_6\text{H}_4\text{-CH(OH)-CH}_2\text{-CO-O-CO-CH}_3 \right] \longrightarrow$$

$$\text{o-(HO)C}_6\text{H}_4\text{-CH=CH-C(O)-OH} + \text{CH}_3\text{-COOH} \xrightarrow{-\text{H}_2\text{O}} \text{coumarin} \quad (22\text{-}10)$$

Compounds similar to coumarin, but with the carbonyl group at carbon 4 are responsible for some of nature's beautiful colors. They are called **chromones.**

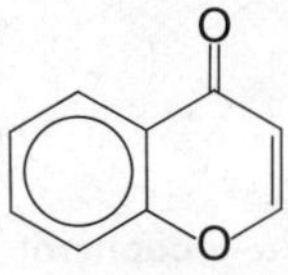

chromone

The flavones (which are derivatives of 2-phenyl chromone) are responsible for the colors of various flowers, fruits, and berries. The flavone **quercitin,** for example, is found in the bark of the beautiful Douglas fir.

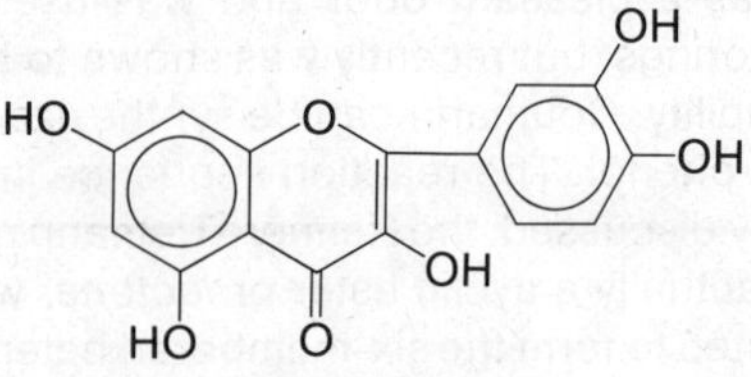

quercitin
(a flavone)

22.15 Nitrogen Heterocycles

The pyrrole ring system, [pyrrole structure], is very common among natural products. It is often present in the form of a rather complex structure known as a **porphyrin,** which is composed of four pyrrole rings. The parent compound, porphyrin, does not occur free in nature, but slight alterations in the porphyrin result in the formation of some extremely important natural products.

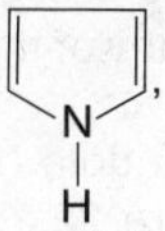

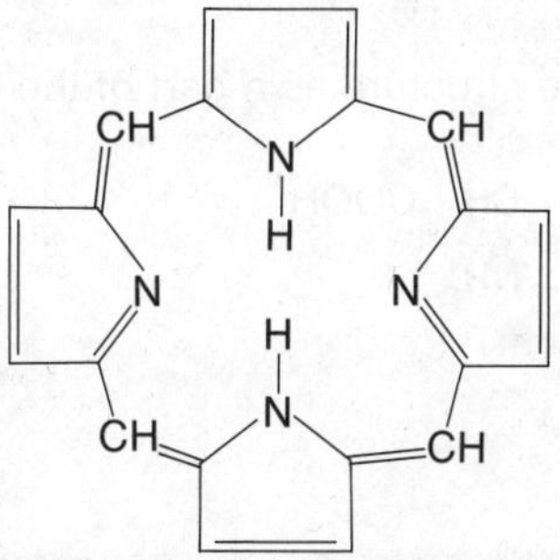

porphyrin

For example, the heme of the hemoglobin present in red blood cells contains the porphyrin structure (Section 20.11). **Chlorophyll,** the green coloring matter in plants, has a magnesium atom in the center of the porphyrin structure, and is essential to the process of photosynthesis.

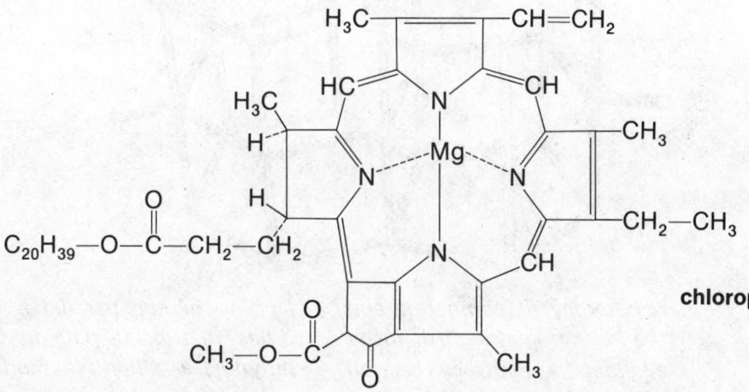

chlorophyll-a

The total synthesis of chlorophyll was finally performed in 1960 by Professor Robert B. Woodward (1917–) and co-workers at Harvard University. Professor Woodward has also synthesized numerous other natural products, including many steroids, strychnine, reserpine, and lysergic acid. For his outstanding contributions to organic chemistry, Professor Woodward received the Nobel Prize in 1965.

Another porphyrin derivative is **vitamin B_{12},** associated with pernicious anemia. It has a cobalt atom in place of the magnesium in chlorophyll, and a much more complex overall structure.

22.15-1 Alkaloids By far, the largest group of nitrogen heterocycles are the **alkaloids,** which are basic (alkaline) substances present in plants. Most of the alkaloids have rather complex structures, and many of them show some biological or physiological activity.

The so-called **indole alkaloids** all have the indole (benzopyrrole) ring as an essential part of their structures (see also Figure 22.2).

indole

We have already seen the indole structure as a part of the essential amino acid

tryptophan, —CH_2—CH—$COOH$.
NH_2

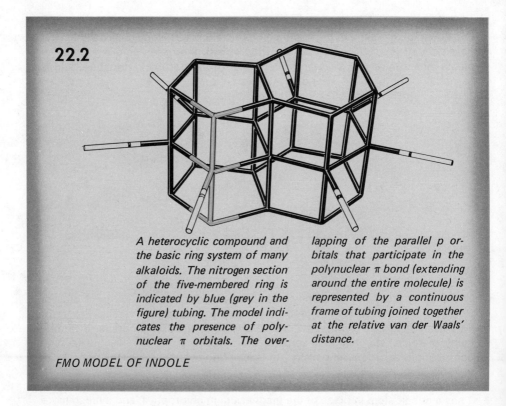

22.2

A heterocyclic compound and the basic ring system of many alkaloids. The nitrogen section of the five-membered ring is indicated by blue (grey in the figure) tubing. The model indicates the presence of polynuclear π orbitals. The overlapping of the parallel p orbitals that participate in the polynuclear π bond (extending around the entire molecule) is represented by a continuous frame of tubing joined together at the relative van der Waals' distance.

FMO MODEL OF INDOLE

Sec. 22.15 NITROGEN HETEROCYCLES

One of the most fundamental indole alkaloids is the compound **serotonin**,

HO—[indole]—CH$_2$—CH$_2$—NH$_2$,

which occurs in many plants and animals. Serotonin has been shown to have some relationship to mental diseases, in particular to schizophrenia, but its exact function in relation to mental health is not completely understood.

Perhaps the most famous indole alkaloid is **lysergic acid,** whose diethylamide (LSD) is one of the most potent hallucinogenics known.

lysergic acid

Strychnine and **reserpine** (or serpasil) are two other indole alkaloids with exceedingly complex structures.

strychnine   **reserpine**

Strychnine is familiar to us as an exceptionally powerful poison, apparently attacking the central nervous system. Reserpine is the active ingredient of the Indian snake root plant, which grows wild in the Himalayas. It has been used for centuries by the natives as a remedy to treat insanity, epilepsy, hysteria, snake bites, etc. Reserpine is used in the treatment of high blood pressure and as a tranquilizer in the treatment of mental diseases. However, its mode of action is still not completely understood.

22.15–2 Pyridine-related Compounds Six-membered rings containing nitrogen are very common among heterocyclic natural products. Among the simple compounds related to pyridine are **nicotine** and **nicotinic acid.**

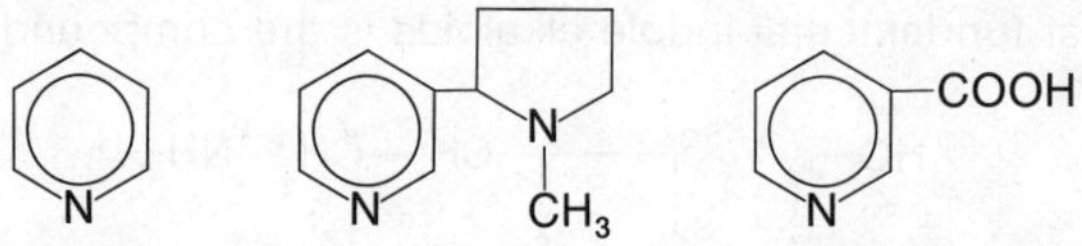

pyridine   **nicotine**   **nicotinic acid**

Nicotine is, of course, found in tobacco. Oxidation of nicotine produces nicotinic acid, which is better known as **niacin,** used in the treatment of pellagra.

Many alkaloids contain reduced ring systems, such as a reduced pyridine ring known as **piperidine.** An example is **cocaine,** used in medicine in small doses, and perhaps better known as a narcotic.

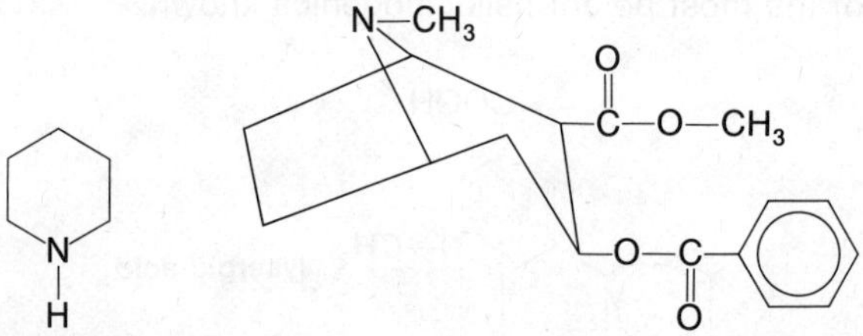

piperidine   **cocaine**

Atropine is used in dilute solution to dilate the pupil of the eye prior to its examination.

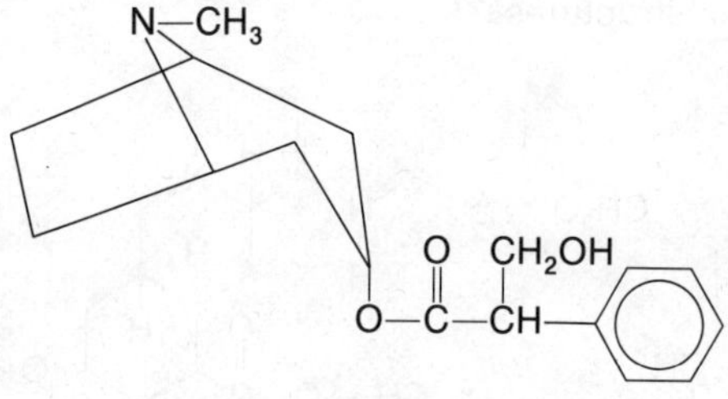

atropine

22.15-3 The Quinoline Ring System   The **quinoline** ring system occurs frequently in heterocyclic natural products.

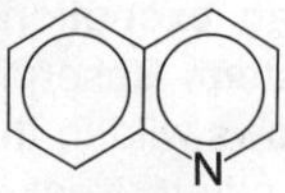

quinoline

Quinoline, found in coal tar, contains a benzene ring and a pyridine ring fused together. The most useful method for synthesizing quinoline or substituted quinolines is known as the **Skraup synthesis.** As a typical example, quinoline can be prepared by reacting aniline with glycerol, nitrobenzene, concentrated sulfuric acid, and ferrous sulfate.

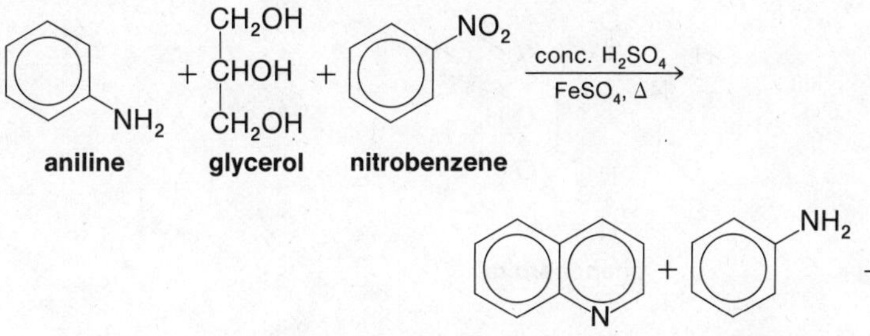

Sec. 22.16 MORE THAN ONE HETEROATOM

aniline + glycerol + nitrobenzene $\xrightarrow[\text{FeSO}_4, \Delta]{\text{conc. H}_2\text{SO}_4}$ quinoline + aniline + H_2O (22-11)

The best known of the quinoline alkaloids is **quinine,** found in the bark of the cinchona tree and used to treat malaria. The synthetic drug *chloroquine* is also used extensively in place of the natural product quinine to treat malaria.

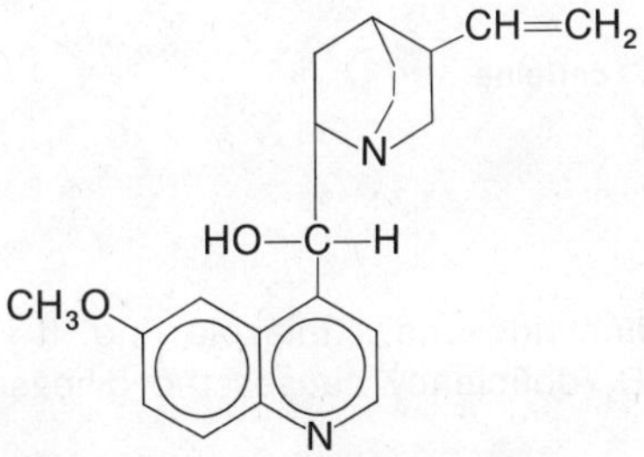

quinine

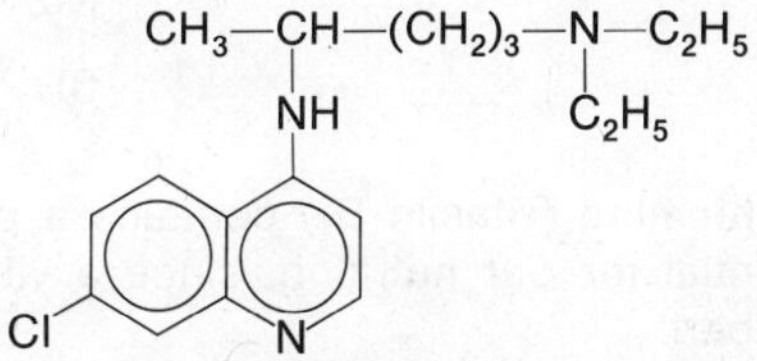

chloroquine

The reduced isoquinoline ring system 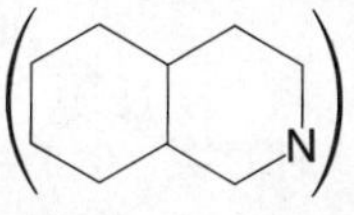 is present in **morphine** and **codeine,** isolated from the opium poppy.

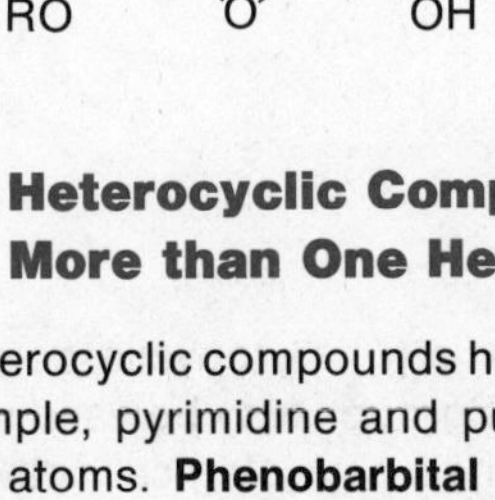

R = H, morphine
R = CH$_3$, codeine
OR and OH replaced by OCOCH$_3$, heroin

22.16 Heterocyclic Compounds with More than One Heteroatom

Many heterocyclic compounds have ring systems with more than one heteroatom. For example, pyrimidine and purine (Chapter 21) have ring systems with two nitrogen atoms. **Phenobarbital** is a typical heterocyclic compound related to pyrimidine.

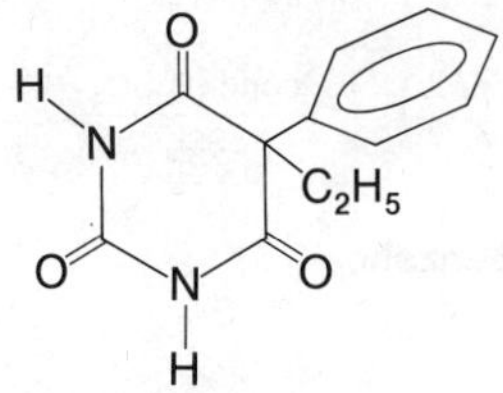

phenobarbital

Caffeine is an alkaloid with the purine ring skeleton. It is a constituent of the coffee bean and tea plant, and is responsible for the stimulating effect one gets from drinking these beverages.

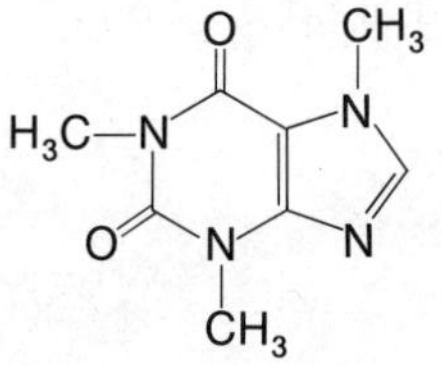

caffeine

Thiamine (vitamin B_1) contains a pyrimidine ring and a thiazole ring. It is essential for our nutrition, since a vitamin B_1 deficiency causes the disease beri-beri.

thiazole ring

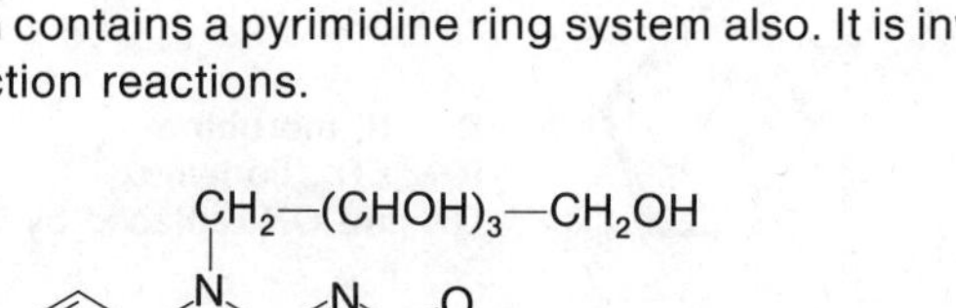

thiamine

Vitamin B_2 or riboflavin contains a pyrimidine ring system also. It is involved in biological oxidation-reduction reactions.

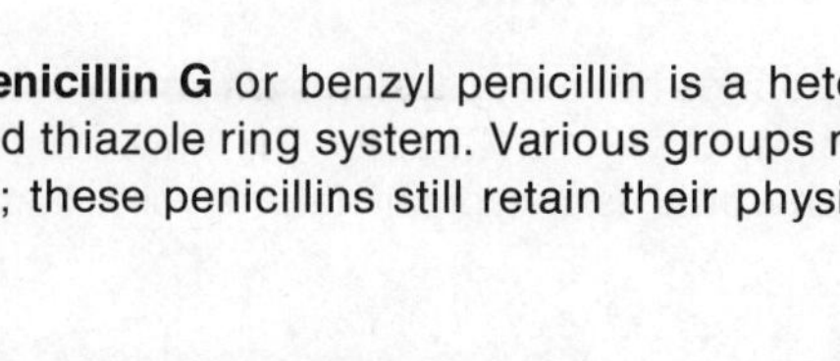

riboflavin

The antibiotic **penicillin G** or benzyl penicillin is a heterocyclic compound containing a reduced thiazole ring system. Various groups may be used in place of the benzyl group; these penicillins still retain their physiological activity.

PROBLEMS

penicillin G
(or benzyl penicillin)

The structures of such other antibiotics as the mycins, chloromycetin, etc., are not related to or similar to that of penicillin. All antibiotics function in the same way, however, by inhibiting essential metabolic processes in certain microorganisms, which stops their growth or kills them.

Summary

- The term *natural products* refers to substances that occur in nature. These include:
  (a) Terpenes and steroids.
  (b) Heterocyclic products.
- The terpenes are all related to isoprene and are built up by combinations of isoprene units:
  (a) Monoterpenes.
  (b) Sesquiterpenes.
  (c) Diterpenes and triterpenes.
- Steroids are the group of lipids that are not saponified by base to fatty acids and glycerol.
  Important steroids include cholesterol, ergosterol, the hormones estrone, progesterone, testosterone, and cortisone.
- The biogenesis of terpenes in the cell starts with the precursor acetate ion.
- The biosynthesis of steroids can be considered to be related to the synthesis of triterpenes.
- There are various heterocyclic ring systems containing oxygen, nitrogen, sulfur atoms, etc.
- Alkaloids are nitrogen heterocycles that are basic substances present in plants.

PROBLEMS

1. What products would you expect to be formed by reacting myrcene, $C_{10}H_{16}$, with ozone, followed by hydrolysis?

2. Write equations for the reaction of geraniol with each of the following reagents:
   (a) Ac_2O (d) Br_2/CCl_4
   (b) $SOCl_2$ (e) hot $KMnO_4$
   (c) H_2/Pt (f) $CH_3-CH_2-\overset{\overset{O}{\|}}{C}-OH + H^{\oplus}$

3   Write equations for the reaction of piperitone with each of the following reagents.
    (a) NH_2OH (e) H_2/Pt
    (b) HCN   (f) Br_2/CCl_4
    (c) $LiAlH_4$ (g) cold, dilute, neutral $KMnO_4$
    (d) Zn/Hg + HCl

4   Write equations for the reaction of cholesterol with each of the following reagents.
    (a) Ac_2O (c) $LiAlH_4$
    (b) H_2/Ni (d) NBS

5   *Starting with the carboxyl carbon of acetate labeled with C^{14}, trace the biogenesis of terpenes and steroids, and indicate where you would expect the labeling to occur in each of the following.
    (a) mevalonic acid   (c) geraniol
    (b) isopentenyl pyrophosphate   (d) farnesol

6   (a) *Would you expect the pyrrole and furan ring systems to be more or less reactive than benzene towards electrophilic substitution reactions? Explain.
    (b) At what position in these rings would you expect that electrophilic substitution would be most likely to occur? Explain.

7   *Chromones are somewhat basic and can be protonated by acids on the carbonyl oxygen atom. How do you account for this?

8   *Starting from benzene and any necessary inorganic reagents, synthesize the following compounds.
    (a) 8-nitrocoumarin   (c) 6-nitroquinoline
    (b) 6-methylquinoline

9   *Piperine, $C_{17}H_{19}O_3N$, is found in peppers. When piperine is treated with aqueous NaOH, piperidine ($C_5H_{11}N$) and piperic acid ($C_{12}H_{10}O_4$) are formed. Piperic acid reacts with two moles of bromine in CCl_4 to form a compound of formula $C_{12}H_{10}O_4Br_4$. When piperic acid is heated with fused KOH, the compounds acetic acid, oxalic acid, and 3,4-dihydroxybenzoic acid are formed. Piperic acid reacts with $KMnO_4$ to produce a compound, $C_8H_6O_4$, which undergoes hydrolysis in acid solution to formaldehyde and 3,4-dihydroxybenzoic acid. Suggest a possible structure for piperine based on this information.

23
Synthetic Polymers

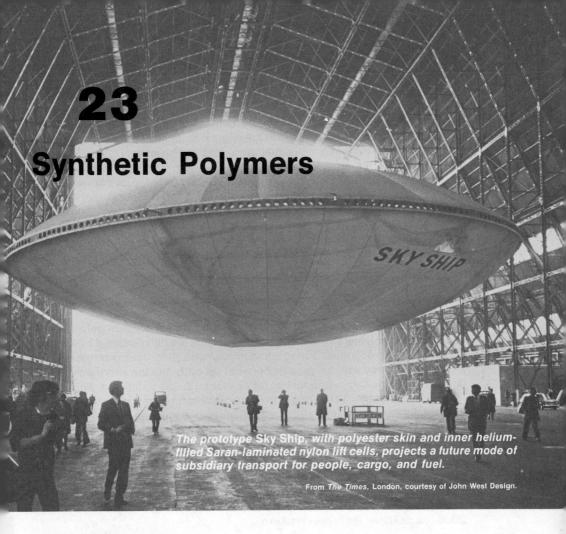

The prototype Sky Ship, with polyester skin and inner helium-filled Saran-laminated nylon lift cells, projects a future mode of subsidiary transport for people, cargo, and fuel.

From *The Times*, London, courtesy of John West Design.

23.1 Introduction

We have devoted much space in the text for discussions of naturally occurring polymers, such as the polysaccharides, proteins, and nucleic acids. We have referred only briefly, however, to synthetic polymers produced by chemists in the laboratory. In this chapter we will outline the general procedures used in preparing synthetic polymers, and cite a few specific examples.

23.2 General Characteristics of Polymers The word *polymer* means many units. Each individual unit that is formed into the polymer is called a *monomer* or simply a *mer*. *Poly* means many; hence the word polymer refers to very large molecules composed of many units, having gigantic dimensions. A **polymer** or **macromolecule** is a substance consisting of a long chain of repeating units (although for the proteins a series of different amino acids forms the polypeptide chain), and having a high molecular weight. Because of their size, these macromolecules have certain physical properties that clearly differentiate them from ordinary molecules.

The basic structure of polymers is their long chain of repeating units. The configuration of the chain may be in the form of a rigid rod, a helical coil, or a coil of some random arrangement. The physical properties of the polymers are determined to a large extent by the configuration of the molecule. Furthermore, the fundamental laws of physical chemistry that are applicable to molecules of ordinary dimensions, do not apply to these gigantic molecules, and new models must be used to explain the unusual properties of polymers.

Polymers are either very viscous liquids or amorphous or crystalline solids. They are non-volatile and have very limited solubilities in various solvents. The viscosity of a polymer (in solution) depends mainly on the chain length of the molecules, with a longer chain length resulting in a higher viscosity. Most polymers contain molecules of varying chain lengths within a given sample. For this reason, we refer to an *average chain length* and *average molecular weight* of a polymer, instead of just the chain length or the molecular weight.

23.3 General Classification of Synthetic Polymers

Synthetic polymers can be divided into two large classes, **addition polymers** and **condensation polymers.** The two types of polymers are differentiated by the ways in which they were formed. During addition polymerization, each discrete polymer molecule is formed within a period of several seconds, but the entire polymerization process may take much longer (many hours) to be completed. This means that during the course of the polymerization process the reaction mixture consists of some polymer molecules and some still unreacted monomer units. In condensation polymerization, the average size of the polymer is continually increasing as the polymerization proceeds. Thus, the final high molecular weight polymer is not present in the reaction mixture in any large amount until the polymerization process has almost arrived at completion.

23.4 Addition Polymerization

We have referred to addition polymerization in connection with the polymerization of the alkenes (Section 8.11). Polyethylene, for example, is formed by the polymerization of ethylene in this way. There are two general mechanisms by which addition polymerization is believed to occur. These are by free-radical polymerization, and by ionic polymerization.

23.4–1 Free-Radical Mechanism The mechanism for the free-radical polymerization process involves the initiation, propagation, and termination steps common to all free-radical reactions. The initiator, which is used to generate the free radical, is usually some such organic peroxide as benzoyl peroxide. The free radical then adds to the carbon–carbon double bond of an alkene, generating another free radical, which keeps the chain reaction going by attacking another alkene molecule, which in turn attacks another alkene, continually increasing the chain length until a polymer is formed. Thousands of monomer units (alkenes) combine to form a chain in several seconds. The propagation steps keep the reaction going until two free radicals combine, upon which the chain growth ceases and the reaction is terminated. The proposed steps for the mechanism of free-radical addition polymerization are:

Sec. 23.4 ADDITION POLYMERIZATION

INITIATION STEPS

$$R-O-O-R \xrightarrow{\Delta \text{ or } h\nu} 2RO\cdot \qquad (23\text{-}1)$$

organic peroxide → peroxy free radicals

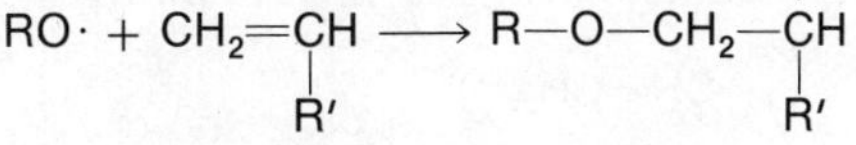

$$RO\cdot + CH_2=\underset{R'}{CH} \longrightarrow R-O-CH_2-\underset{R'}{\overset{\cdot}{C}H} \qquad (23\text{-}2)$$

(Rad. = RO·)

PROPAGATION

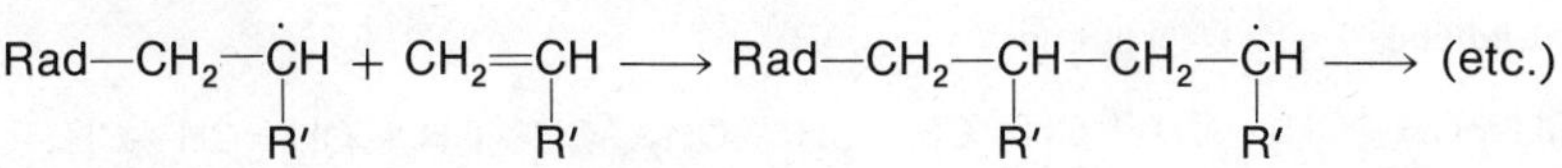

$$Rad-CH_2-\underset{R'}{\overset{\cdot}{C}H} + CH_2=\underset{R'}{CH} \longrightarrow Rad-CH_2-\underset{R'}{CH}-CH_2-\underset{R'}{\overset{\cdot}{C}H} \longrightarrow (\text{etc.}) \quad (23\text{-}3)$$

TERMINATION

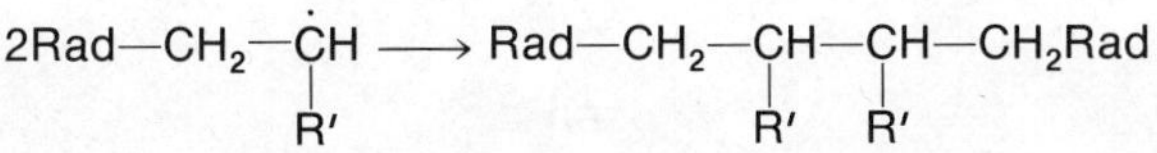

$$2Rad-CH_2-\underset{R'}{\overset{\cdot}{C}H} \longrightarrow Rad-CH_2-\underset{R'}{CH}-\underset{R'}{CH}-CH_2Rad \qquad (23\text{-}4)$$

Many familiar substances are prepared by the polymerization of various substituted alkenes in this way. Orlon is a polymer of acrylonitrile.

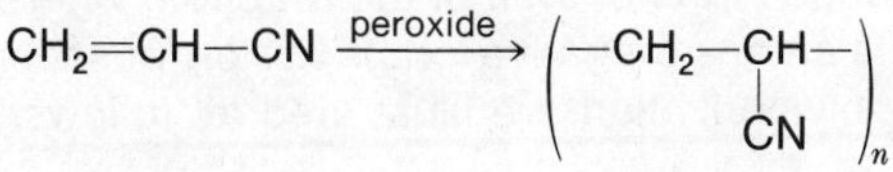

$$CH_2=CH-CN \xrightarrow{\text{peroxide}} \left(-CH_2-\underset{CN}{CH}-\right)_n \qquad (23\text{-}5)$$

acrylonitrile → Orlon

Lucite or plexiglas is actually a polymer of methyl methacrylate.

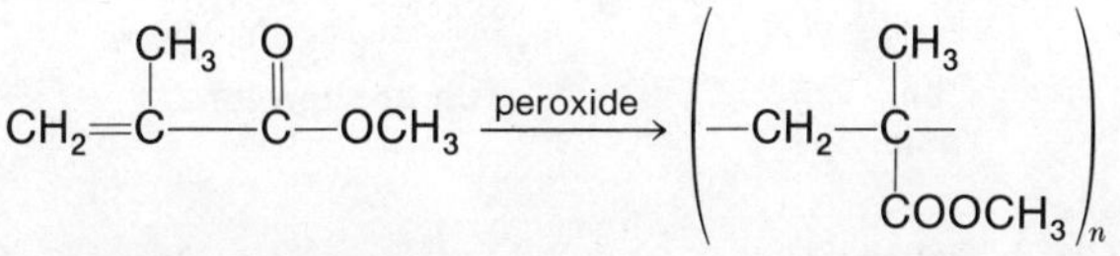

$$CH_2=\underset{CH_3}{\overset{CH_3}{C}}-\overset{O}{\overset{\|}{C}}-OCH_3 \xrightarrow{\text{peroxide}} \left(-CH_2-\underset{COOCH_3}{\overset{CH_3}{C}}-\right)_n \qquad (23\text{-}6)$$

methyl methacrylate → Lucite

Polystyrene is made by polymerizing styrene, and poly(vinyl acetate) is made from the monomer vinyl acetate.

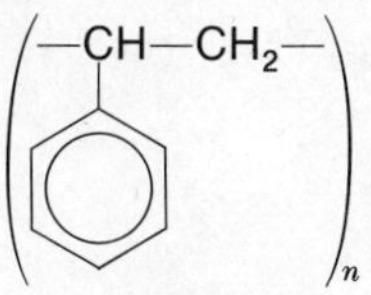

polystyrene 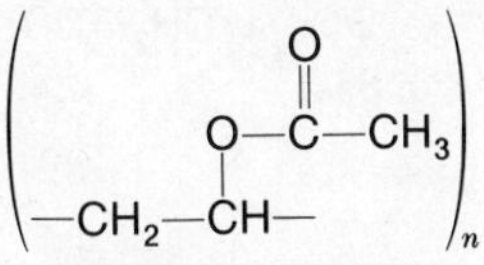   poly(vinyl acetate)

23.4-2 Copolymerization The properties of a polymer derived from one particular monomer can be modified somewhat by the process of copolymerization. **Copolymerization** is a process whereby two different monomers are allowed to react and form an addition polymer. The resulting *copolymer* usually possesses certain physical properties different from those of polymers derived from one of the monomer units undergoing self-polymerization.

During World War II, a shortage of natural rubber, which was essential to the war effort, occurred. Many methods were devised for the production of synthetic rubbers. One of the methods developed to replace the unavailable natural rubber was to copolymerize 1,3–butadiene and styrene to form a synthetic rubber, SBR.

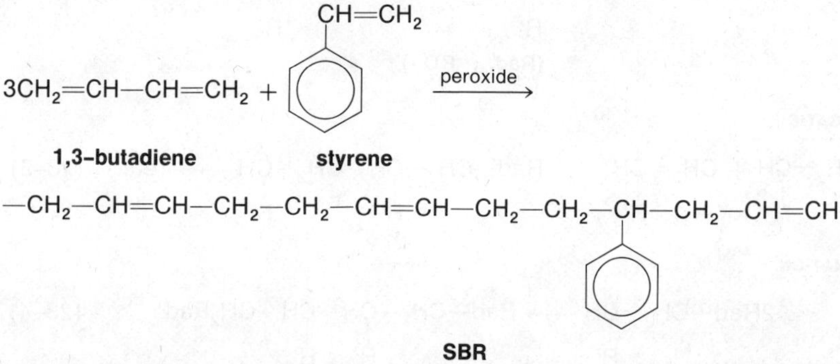

$$3CH_2=CH-CH=CH_2 + \underset{\text{styrene}}{C_6H_5-CH=CH_2} \xrightarrow{\text{peroxide}}$$

1,3-butadiene   **styrene**

$$-CH_2-CH=CH-CH_2-CH_2-CH=CH-CH_2-CH_2-CH(C_6H_5)-CH_2-CH=CH-CH_2- \quad (23\text{-}7)$$

SBR

23.4-3 Ionic Mechanism

It is also possible to have addition polymerization proceeding by an ionic mechanism. When an acid is used as a catalyst, cations (positive ions) are formed in the course of the reaction, whereas a basic catalyst produces intermediate anions (negative ions). The proposed mechanisms for the two modes of ionic polymerization are illustrated as follows.

CATIONIC POLYMERIZATION

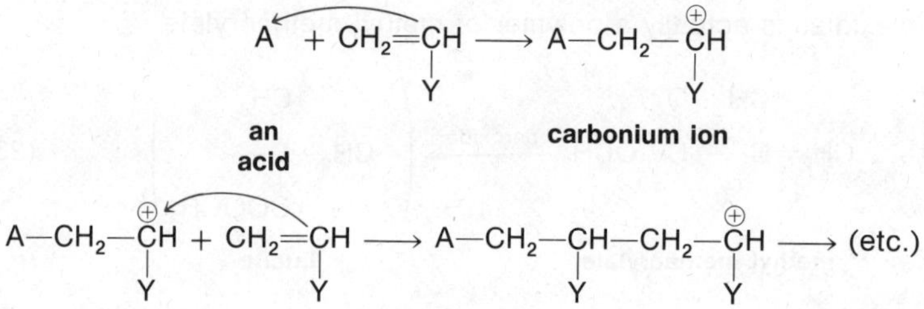

$$A^- + CH_2=CH(Y) \longrightarrow A-CH_2-\overset{\oplus}{C}H(Y) \quad (23\text{-}8)$$

an acid   carbonium ion

$$A-CH_2-\overset{\oplus}{C}H(Y) + CH_2=CH(Y) \longrightarrow A-CH_2-CH(Y)-CH_2-\overset{\oplus}{C}H(Y) \longrightarrow (\text{etc.}) \quad (23\text{-}9)$$

ANIONIC POLYMERIZATION

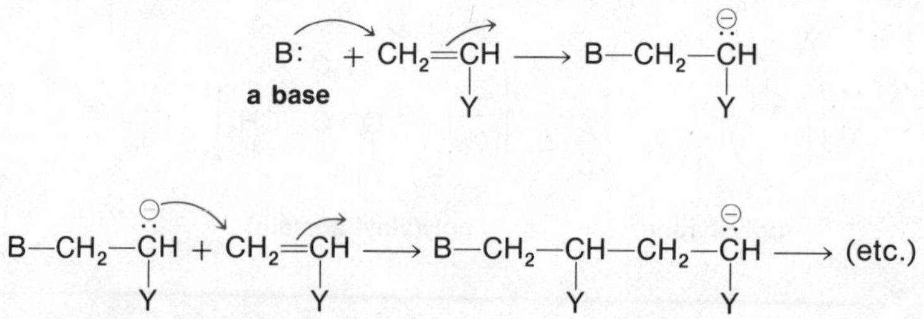

$$B: + CH_2=CH(Y) \longrightarrow B-CH_2-\overset{\ominus}{C}H(Y) \quad (23\text{-}10)$$

a base

$$B-CH_2-\overset{\ominus}{C}H(Y) + CH_2=CH(Y) \longrightarrow B-CH_2-CH(Y)-CH_2-\overset{\ominus}{C}H(Y) \longrightarrow (\text{etc.}) \quad (23\text{-}11)$$

The cationic polymerization of isobutylene was discussed in Section 8.11. An example of anionic polymerization is the conversion of methyl methacrylate to poly(methyl methacrylate) (Lucite) in the presence of a strong base such as sodamide.

$$CH_2=\overset{\underset{\displaystyle CH_3}{|}}{C}-\overset{\underset{\displaystyle }{\overset{\displaystyle O}{\|}}}{C}-O-CH_3 \xrightarrow{NaNH_2} \left(-CH_2-\overset{\underset{\displaystyle COOCH_3}{|}}{\overset{\displaystyle CH_3}{\underset{\displaystyle }{|}}}C-\right)_n \qquad (23\text{-}12)$$

methyl methacrylate	poly(methyl methacrylate) (Lucite)

23.5 Stereochemical Control of Addition Polymers

The field of polymer chemistry was revolutionized in 1955 by a series of discoveries by Karl Ziegler in Germany and Giulio Natta in Italy, both of whom received the Nobel Prize in 1963 for their work. These scientists developed a group of catalysts that permit the control of the ionic polymerization process to an extremely high degree, never possible previously. The catalysts included complexes of organoaluminum alkyls (R_3Al) with $TiCl_4$ or $VOCl_3$, such as triethylaluminum-titanium trichloride complex. (The catalysts are sometimes referred to as *Ziegler catalysts*.)

The influence of these catalysts on polymerization will become apparent with a few specific examples. Polyethylene, which is prepared by free-radical polymerization of ethylene, has highly branched chains. This is due to the fact that the free-radical process is run at high temperatures, which enables the free radicals not only to add to another molecule of ethylene, but also to abstract a hydrogen atom from a chain that was formed. The result is a product with a great deal of chain branching. The highly branched polyethylene molecules can not fit together too well, resulting in a somewhat random arrangement of molecules. The product therefore is a substance of little crystalline character. It has a low melting point, and is mechanically weak in structure.

However, by employing the unusual Ziegler catalysts, ethylene can undergo ionic polymerization to polyethylene. The reaction occurs at much lower temperatures, under milder reaction conditions, and since it is not a free-radical reaction, the polyethylene formed is unbranched. This polyethylene polymer is highly crystalline, has a higher melting point, and is much more mechanically sound.

Polypropylene or polystyrene prepared by ionic polymerization, using similar catalysts, were crystalline, high melting point polymers, in contrast to the usual viscous liquids obtained by the usual free-radical polymerization of propylene or styrene. The crystallinity of the polymers is attributed to the configuration of the polymer chains. Ordinary polypropylene and polystyrene are polymers whose chains are in a random coil arrangement, whereas the crystalline polymers have their chains in the form of a helical coil.

Perhaps the most significant development in ionic polymerization, as a result of the Ziegler–Natta technique, is in the field of stereochemistry. It is possible to prepare **stereospecific addition polymers** and have **stereochemical control** of the reaction.

Propylene, for example, can polymerize to form polypropylene of three different arrangements. The three possible arrangements of the polymer are called *isotactic, syndiotactic,* and *atactic.* The **isotactic polymer of polypropylene** has all the methyl groups (or comparable groups in other polymers) on the same side of the extended chain of the polymer. **Syndiotactic polypropylene** has the methyl groups alternating from side to side, while the **atactic polymer** has the methyl groups distributed at random along the chain. The isotactic polymer has an orderly or helical coil chain configuration, whereas an atactic polymer has a random orientation. By the use of appropriate catalysts mounted on crystalline solids, such isotactic polymers as highly crystalline polypropylene can be made. Atactic polymers can be made by using catalysts mounted on an amorphous solid surface. Atactic polypropylene produced in this way is a soft, elastic, rubber-like material.

It is possible to polymerize isoprene to a synthetic rubber, practically identical with naturally occurring rubber, by employing the Ziegler–Natta technique. This is an example of a stereospecific or stereoselective synthesis (since natural rubber contains only *cis* configurations linking the isoprene units), as is the formation of any isotactic polymer.

23.6 Condensation Polymerization

The process of condensation polymerization involves a condensation reaction between two monomer molecules, resulting in the formation of a polymer, accompanied by the loss of some small stable molecule (such as water or methyl alcohol), at the same time as bonds are being formed to produce the polymer molecules. The most common types of condensation polymers are polyamides and polyesters.

Nylon is a condensation polymer produced by the reaction of adipic acid and hexamethylenediamine at a high temperature.

In the formation of nylon, the adipic acid reacts with the hexamethylenediamine to form a salt, which when heated loses water to form amide bonds. Nylon is a polyamide.

$$\text{HOOC}-(CH_2)_4-\text{COOH} + \text{H}_2\text{N}(CH_2)_6-\text{NH}_2 \longrightarrow$$
adipic acid   hexamethylenediamine

$$^{\ominus}\text{OOC}-(CH_2)_4-\text{COOH}_3\overset{\oplus}{\text{N}}-(CH_2)_6-\overset{\oplus}{\text{NH}}_3 \xrightarrow[-H_2O]{\Delta}$$
nylon salt

$$\left(\begin{matrix} \text{O} & & \text{O} & \text{H} & & \text{H} \\ \| & & \| & | & & | \\ -\text{C}-(CH_2)_4-\text{C}-\text{N}-(CH_2)_6-\text{N}- \end{matrix} \right)_n \quad (23\text{-}13)$$

Nylon 66
(a polyamide)

Dacron is an example of a condensation polymer that is a polyester. It is formed by heating ethylene glycol with terephthalic acid or its dimethyl ester. Methyl alcohol is lost during the condensation process to form Dacron.

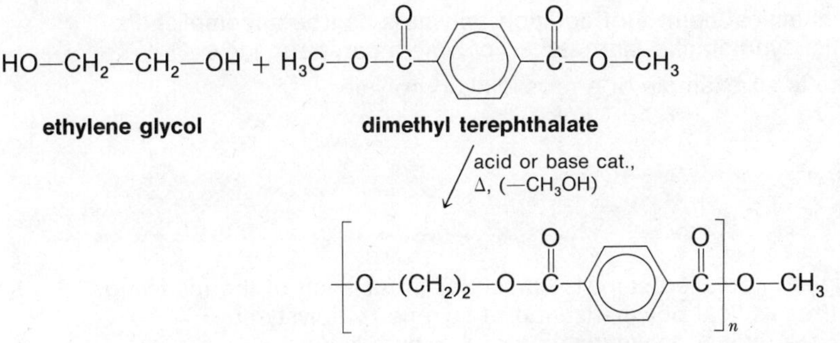

SUMMARY

HO—CH$_2$—CH$_2$—OH + H$_3$C—O—C(=O)—C$_6$H$_4$—C(=O)—O—CH$_3$

ethylene glycol   **dimethyl terephthalate**

↓ acid or base cat., Δ, (—CH$_3$OH)

[—O—(CH$_2$)$_2$—O—C(=O)—C$_6$H$_4$—C(=O)—]$_n$ O—CH$_3$ (23-14)

Dacron
(a polyester)

23.6–1 Cross-linked Polymers If one uses a monomer containing three functional groups (rather than two) in a condensation polymerization, the product will not only contain a long chain of linked monomer units, but will also have chains that will cross-link various polymer chains together. If the cross-linking is present to a large extent, the polymer will in effect be tied together, all the individual polymer chains being part of one gigantic cross-linked system. The result is a **cross-linked condensation polymer,** which usually is insoluble in all solvents and does not melt. These polymers are said to thermoset, whereas other polymers are thermoplastic. Examples of cross-linked condensation polymers are urea-formaldehyde resins and the synthetic plastic Bakelite, made by the cross-linked condensation of phenol and formaldehyde.

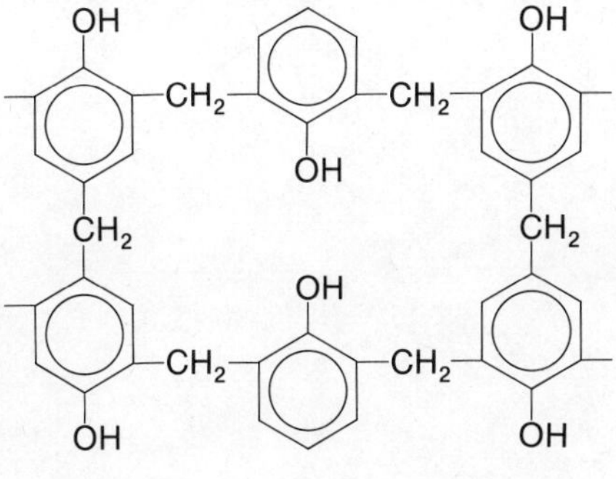

Bakelite

Summary

- The term polymer refers to very large molecules composed of many units, having gigantic dimensions.
- Synthetic polymers can be divided into two large classes, addition polymers and condensation polymers.
- Addition polymers can be formed by either a free-radical or an ionic mechanism.
- Copolymerization is a process whereby two different monomers are allowed to react and form an addition polymer.

- Stereochemical control of addition polymers can be accomplished. Isotactic, syndiotactic, and atactic polymers can be made.
- Bakelite is an example of a cross-linked polymer.

PROBLEMS

1. Suggest a detailed stepwise mechanism for each of the following.
   (a) Free radical polymerization of styrene to polystyrene.
   (b) Free radical polymerization of 1,3-butadiene.
   (c) Free radical polymerization of tetrafluoroethylene to Teflon.
   (d) Acid catalyzed cationic polymerization of isobutylene (2-methylpropene).

2. *Suggest a mechanism for the formation of Bakelite by base-catalyzed anionic polymerization.

3. The polymer *glyptal* is formed by the condensation polymerization of phthalic anhydride and glycerol. Suggest a mechanism for the formation of glyptal.

4. *Can you suggest why polymerization reactions should take place to produce polymers with regularly alternating groups?

5. Would you expect isobutylene (2-methylpropene) to give stereoisomeric polymers as are formed in the polymerization of propylene with Ziegler catalysts? Explain.

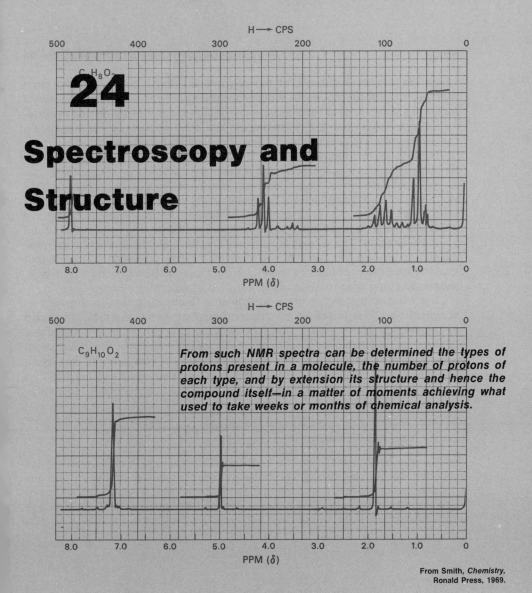

24
Spectroscopy and Structure

From such NMR spectra can be determined the types of protons present in a molecule, the number of protons of each type, and by extension its structure and hence the compound itself—in a matter of moments achieving what used to take weeks or months of chemical analysis.

From Smith, *Chemistry*, Ronald Press, 1969.

24.1 Introduction

In our survey of the various classes of organic compounds, we have discussed the properties and characteristics of molecules containing numerous functional groups. Some of these molecules have been quite simple and others exceedingly complex in their molecular structure and geometry.

Methods useful in the elucidation of molecular structure were rather limited in scope until about thirty years ago; typical of the techniques used were elemental analysis and molecular weight determination of the compound, solubility classifications, and qualitative chemical tests to detect the presence or absence of functional groups, such as the Tollens test for an aldehyde or the Hinsberg test for an amine. Degradative procedures such as ozonolysis were also employed, from which the smaller fragments of the degradation could be pieced and fitted together to give the complete molecular structure. In recent years, no single factor has done more to revolutionize the procedure and time involved in structure determination than the development of modern methods of instrumentation that investigate the spectral properties of organic substances.

Throughout this text we have often indicated the presence or absence of a functional group in a molecule by qualitative chemical tests that are characteristic of the functional group. Note, however, that these tests are *not* always 100 % accurate nor always conclusive. Other techniques, such as chemical degradation procedures or the total synthesis of a compound to prove the structure of a molecule, are time-consuming and may take months or even years to produce conclusive results. The availability of modern instrumentation, which gives reliable and accurate measurements in a few minutes, has provided information that enables the organic chemist to solve even the most complex structural problems, especially in the area of structure elucidation of natural products.

The most important of these instrumental techniques are in the field of spectroscopy. These methods have become such powerful tools that, when used in conjunction with the older classical methods of organic analysis, they have become the predominant techniques in the determination of the structure of organic compounds. We will present in this chapter a brief survey and discussion of some of the most widely used spectroscopic techniques as they apply to organic compounds: ultraviolet–visible, infrared, nuclear magnetic resonance, and mass spectroscopy.

24.2 The Electromagnetic Spectrum

Energy can be transmitted in the form of *electromagnetic radiation*. Molecules can interact with electromagnetic radiation by absorption of *quanta* or *photons* (small bundles of energy) according to the quantum theory, where the energy of a photon depends on the particular wavelength (τ) of the radiation, according to the equation

$$E = \frac{hc}{\tau} = h\nu$$

where E = energy of photon, h = Planck's constant (6.6×10^{-27} erg-sec), τ = wavelength in cm, ν = frequency of the wavelength expressed in Hz or cycles/sec, and c is the velocity of light (3×10^{10} cm/sec).

The electromagnetic spectrum includes cosmic rays, gamma rays, X-rays, ultraviolet rays, visible light, infrared radiation, and radio and radar waves. These are simply the components of a spectrum that ranges from cosmic or gamma rays whose wavelengths are often much smaller than 1 Angstrom unit (1 Å = 10^{-8} cm), to radio waves or radar whose wavelength may be several meters or kilometers in length. All of these forms of electromagnetic radiation travel in waves (p. 403) and all travel at the speed of light (3×10^{10} cm/sec). The reader should convince himself from the equation $E = hc/\tau = h\nu$ that *the shorter the wavelength of the radiation, the higher the frequency*. Figure 24.1 shows the various components of the electromagnetic spectrum.

Sec. 24.2 THE ELECTROMAGNETIC SPECTRUM

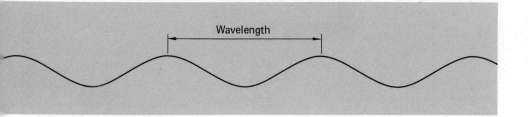

Instead of the Angstrom unit, the ultraviolet and visible regions of the spectrum are often measured in units of millimicrons (1 mμ = 10^{-7} cm), and the infrared region may be measured in microns (1 μ = 10^{-4} cm) or wave numbers (cm^{-1}), the number of waves per centimeter.

When electromagnetic radiation is passed through a substance, the radiation may or may not be absorbed, depending on the frequency of the radiation and the molecular structure of the species encountered. The mechanism of the absorption of energy is different in various regions of the electromagnetic spectrum. The fundamental process, however, whether in the ultraviolet, infrared, nuclear magnetic resonance, or any other region of the spectrum, is the absorption of a certain amount of energy, given by the equation $E = h\nu$, where E is the energy required for the transition from a lower energy state to a higher energy state, and is directly related to the frequency, ν, of electromagnetic radiation that causes the transition.

The absorption of electromagnetic radiation by organic molecules is not continuous over all wavelengths (or frequencies), nor is it a random process. For any excitation process, a molecule absorbs only one discrete amount of energy corresponding to radiation of a specific wavelength of one frequency in the electromagnetic spectrum.

For example, absorption of ultraviolet or visible light causes excitation of electrons, absorption in the infrared region of the spectrum is due to changes in the energy states of molecular vibrations (or rotations) of the atoms in the bonds, and the nuclear magnetic resonance spectrum is attributed to nuclear spin transitions of certain atomic nuclei within the molecule.

24.1

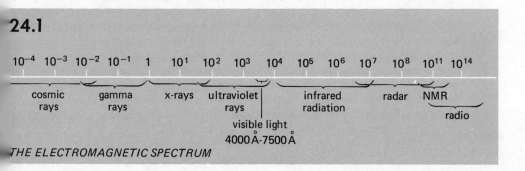

THE ELECTROMAGNETIC SPECTRUM

24.2

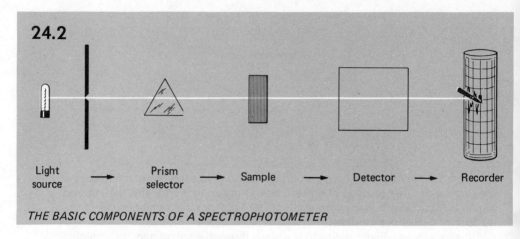

THE BASIC COMPONENTS OF A SPECTROPHOTOMETER

24.3 The Spectrophotometer

Spectrophotometers are instruments that have been developed for measuring the absorption of various types of radiation (ultraviolet, infrared, etc.) by molecules. A spectrophotometer consists of a light source of radiation, with a prism that can select the desired wavelengths, which are passed through a sample of the compound being investigated. The *radiation that is absorbed by the sample* can be detected, analyzed, and recorded and *is known as the spectrum of the compound* (see Fig. 24.2).

24.4 Ultraviolet–Visible Spectroscopy (UV–VIS)

The visible region of the electromagnetic spectrum comprises a very small portion of the electromagnetic spectrum, between 4000 Å and 7500 Å (400–750 mμ). Visible white light is actually composed of various colors and can be dispersed by a prism into components of red, orange, yellow, green, blue, and violet light. Beyond the red end of the visible spectrum (τ greater than 7500 Å) is the infrared region of the electromagnetic spectrum, which will be examined in Section 24.5. Below the violet end of the visible spectrum (τ less than 4000 Å) lies the ultraviolet region (2000–4000 Å), which will be discussed in this section.

The UV–VIS spectrophotometer (spectrometer) can irradiate a sample by varying radiation between 2000 Å and 7500 Å. The absorption of radiation in this region is sufficient to allow electrons to move to a higher energy orbital, further away from the nucleus. The electron is promoted from the normal ground state electronic configuration of the molecule to a higher energy level or excited state. Only quanta (or photons) of exactly the right wavelength or frequency of energy will be absorbed by a compound to promote specific electrons into an excited state. The energy required to promote an electron to an excited state (higher energy orbital) will depend on the structure of the molecule, and the type of orbital the electron occupies. The electrons that require the least amount of energy to be promoted to a higher energy level are those found in pi-(π)-bonds since they are less tightly held than the more stable sigma-(σ)-bond electrons. Pi-bond electrons that are part of conjugated systems where the electron cloud is dispersed over a

Sec. 24.4 ULTRAVIOLET-VISIBLE SPECTROSCOPY

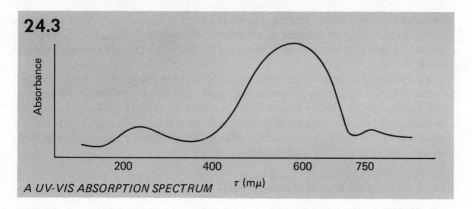

24.3
A UV-VIS ABSORPTION SPECTRUM τ (mμ)

greater region of the molecule require even less energy (lower frequencies or longer wavelengths) to raise an electron to an excited state. Conjugated systems as well as aromaticity help to stabilize the excited state by delocalizing the higher energy of the promoted electron.

The UV (or visible) spectrum is composed of only a few bands of absorption and usually shows only a few broad humps (see Fig. 24.3). The wavelength of maximum absorbance is referred to as τ_{max} and the intensity of its absorbance is called the molar absorptivity or molar extinction coefficient, ε_{max}. The UV spectrum is used to examine the electronic characteristics of a molecule; in other words it measures electronic transitions from lower to higher energy states.

Usually, compounds that contain pi-bonds absorb light in the ultraviolet region of the electromagnetic spectrum, whereas those compounds containing sigma bonds do not absorb ultraviolet radiation. Conjugated pi-bonds require less energy than compounds containing isolated double bonds to promote electrons to excited states, and absorb ultraviolet light of longer wavelength. (Compare (CH_2=CH_2, τ_{max} = 171 mμ, and CH_2=CH—CH=CH_2, τ_{max} = 217 mμ.) Absorption of ultraviolet light in the range 2150 Å (215 mμ) to 4000 Å (400 mμ) is usually good evidence for the presence of pi-bonds, or a conjugated, or aromatic system in the molecule. As the conjugated system in the molecule is extended (the electrons being held less tightly and becoming more delocalized), promotion of a pi-electron is much easier and requires less energy (longer wavelengths). Very highly conjugated molecules with extensive conjugated systems absorb wavelengths of light in the visible region of the electromagnetic spectrum (4000–7500 Å), and these compounds appear colored (see Section 16.14). The color of the compound corresponds to the complementary color remaining after the wavelength of the absorbed light has been subtracted out. For example, β-carotene is the yellow plant pigment found in carrots and is a precursor of vitamin A. It is made up of a

β–carotene

series of isoprene units, and contains eleven carbon–carbon double bonds in conjugation. β-Carotene absorbs radiation at 415 mμ (τ_{max}) at the violet end of the visible spectrum. It is yellow in color since yellow is the complementary color of violet.

Many other factors are involved in UV–VIS spectroscopy which we have not discussed. Although these spectra are often difficult to interpret, they provide information about the electronic environments in various molecules and the relationships between functional groups such as various conjugated systems and aromatic rings. The infrared spectrum shows the presence or absence of individual functional groups in a molecule.

24.5 Infrared Spectroscopy (IR)

The portion of the infrared region of the electromagnetic spectrum of most interest to organic chemists is in the range 2.5 to 15 microns (μ) (2.5×10^4 to 1.5×10^5 Å). The wavelength, expressed in microns, or the wave number expressed as cm^{-1} (in the range 4000–667 cm^{-1}) is used to measure the position of a peak of infrared absorption in a spectrum. In contrast to the relatively few peaks found in the UV–VIS spectra of most organic compounds, the infrared spectrum of a compound often shows a large number of peaks, characteristic of the various types of bonds in the molecule. An infrared spectrum is a highly characteristic property of a molecule and it can provide a wealth of structural information about a molecule if read and interpreted correctly.

The infrared region of the electromagnetic spectrum lies beyond the red part of the visible spectrum, at greater wavelengths. The energy of the infrared radiation at these greater wavelengths is too low to raise pi-electrons into excited states, but can affect the chemical bonds in the molecules, since infrared light is of the proper energy to cause covalent bonds to bend or stretch. Covalent bonds are not rigid and can be pictured as springs, connecting two particles. Both sigma (σ) and pi (π) bonds can absorb infrared radiation of appropriate wavelengths, that will cause the bonds to stretch (be lengthened or shortened), bend, vibrate, or rotate. Different types of bonds absorb infrared radiation of different wavelengths, and give rise to characteristic absorption bands (peaks). For example, any molecule containing an —OH group (such as the alcohols) will absorb infrared light somewhere in the region of 2.8–3.1 μ (3200–3600 cm^{-1}) causing the O—H bond to stretch. Any amine will show a characteristic N—H stretch band at 2.9–3.2 μ (3300–3500 cm^{-1}). Some other characteristic absorption frequencies observed in the IR region include the C—O bond stretch between 5.7 and 6.0 μ (1690–1760 cm^{-1}), characteristic of the carbonyl $\left(\mathrm{C{=}O}\right)$ group found in aldehydes, ketones, carboxylic acids, and esters, and a C—H stretching frequency (alkanes) at 3.3–3.5 μ (3030–2860 cm^{-1}), and two C—H bending frequencies at 6.85 μ (1460 cm^{-1}) and 7.28 μ (1374 cm^{-1}).

Quite often, the interpretation of an infrared spectrum is not simple, since some bands may be obscured by the overlapping of other bands having similar infrared absorption frequencies. Since the infrared spectrum shows many peaks, however, each characteristic of a particular type of bond or functional group, a

Sec. 24.5 INFRARED SPECTROSCOPY

Table 24.1 Some Characteristic IR Absorption Bands

Bond	Range in μ	Range in cm^{-1}
C—H (CH$_3$)	3.36–3.39	2972–2952
	3.47–3.49	2882–2865
	6.76–7.02	1478–1424
	7.14–7.44	1400–1345
—CH=CH$_2$	3.33–3.88	3004–2580
	6.92–7.23	1446–1384
	10.92–11.39	916–878
	14.28–16.00	700–624
C=C (alkene aromatic (phenyl) ring)	5.95–6.17	1680–1620
	6.2, 6.3	1610, 1590
	6.7, 6.9	1491, 1449
O—H stretch	2.74–2.90	3650–3450
O—H (1° alcohol)	9.30–10.00	1075–1000
O—H (2° alcohol)	8.60–9.90	1163–1010
O—H (3° alcohol)	8.20–8.90	1220–1123
O—H (phenol)	7.80–8.70	1282–1150
N—H (1° amine)	2.82–2.92	3550–3420
	2.90–3.03	3450–3300
	6.06–6.28	1650–1590
C=O (aldehyde)	5.75–5.90	1740–1695
C=O (ketone)	5.81–6.01	1720–1680
C=O (carboxylic acid)	5.75–5.95	1740–1680
C=O (amide)	5.92–6.06	1690–1650
	6.13–6.17	1630–1620

good spectroscopist can interpret the spectrum, with which, along with other experimental data, he can deduce and elucidate the structures of even very complex organic molecules. *No two compounds have exactly identical infrared spectra.* A comparison of the IR spectrum of an unknown compound with the IR spectrum of an authentic sample of a known compound can be used as evidence to help establish the identity of the unknown structure. The appearance of a discrete absorption band in an infrared spectrum is conclusive evidence that the particular bond or functional group is present in the molecule, whereas the absence of such a band clearly indicates that the functional group is not present in the molecule.

The region between 7 and 11 μ (1430–910 cm^{-1}) contains many absorption bands caused by C—C, C—O, and C—N stretching and bending vibrations. This region of an infrared spectrum is usually very complex, and two different compounds will usually show some differences in this region of the spectrum. For this reason, it is frequently called the *fingerprint region* and is unique for each molecule. Table 24.1 lists some of the characteristic infrared absorption bands for various types of bonds.

Figure 24.4 is an IR spectrum for the aromatic hydrocarbon toluene. This can be deduced by the absorption bands at 3.3, 6.3, and 6.7 μ which are indicative of a benzene ring, the band at 7.3 μ represents the methyl group, the assignments at

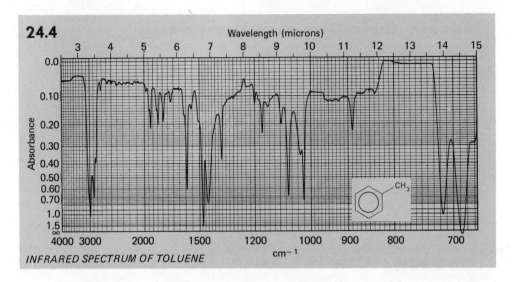

INFRARED SPECTRUM OF TOLUENE

9.3 and 9.8 μ are characteristic of a substituted benzene ring, and the bands at 13.8 and 14.4 μ represent an aromatic —CH group.

The reader should examine the IR spectra in Figures 24.5, 24.6, and 24.7, and by referring to Table 24.1 and other material in this section, he should convince himself why the spectra correspond to the compounds cyclohexene, *n*-butyl alcohol, and butanone, respectively.

24.6 Nuclear Magnetic Resonance Spectroscopy (NMR)

The nuclei of certain atoms possess a mechanical spin and behave as small magnets. One such nucleus that possesses these magnetic properties, and is of most interest to organic chemists, is the ordinary hydrogen nucleus, ^{1}H, the proton.

When a proton or a molecule containing hydrogen atoms is placed in a strong external magnetic field, the magnetic moment of the spinning hydrogen nuclei (according to quantum mechanics) can align itself in either of two ways, oriented with or against the external field. In other words, the spinning nucleus in a magnetic field produces a magnetic moment in a direction parallel to the magnetic field (called the low-energy state) and a magnetic moment opposed to the external field (called the high-energy state). The alignment of the nuclei with the external magnetic field is the more stable (lower energy state) and energy must be absorbed to "flip" the nuclei over to the less stable (higher energy state) orientation against the external field. The energy required to flip the nuclei over depends on the strength of the external magnetic field. If electromagnetic radiation of the proper energy interacts with the molecules, the hydrogen atoms (protons), behaving as small bar magnets, absorb the energy, and change their orientation with respect to the external field.

Let us picture the proton as a small, spinning bar magnet. In a magnetic field the hydrogen nuclei (protons) will tend to spin more in the direction parallel to the

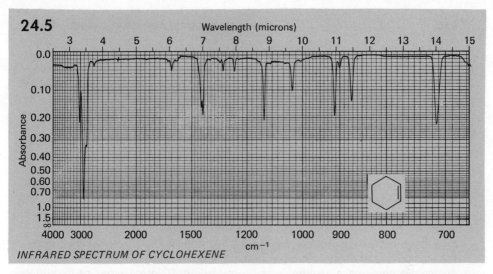
24.5 INFRARED SPECTRUM OF CYCLOHEXENE

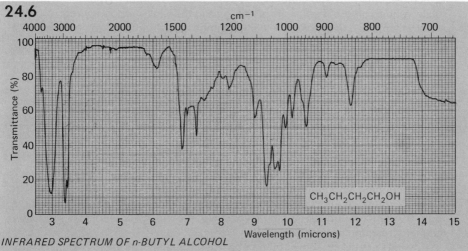
24.6 INFRARED SPECTRUM OF n-BUTYL ALCOHOL

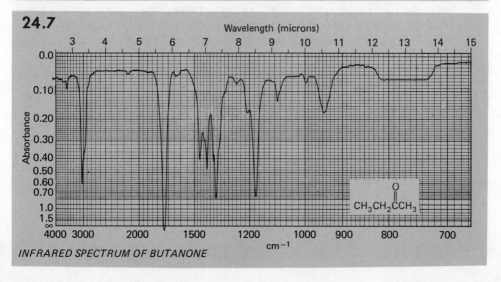
24.7 INFRARED SPECTRUM OF BUTANONE

external field, since this is the lower energy state. Now, if the protons are placed in a beam of radiation in the radio-frequency range of the electromagnetic spectrum, and the externally applied magnetic field is varied in strength, there will be some value of the field strength where the radiation will be of the proper energy required to cause the nuclei (protons) to "flip" over from their more-stable orientation (parallel to the field), to the less-stable orientation (opposed to the field). At this frequency the protons absorb energy and a signal (resonance) is observed that can be recorded. The spectral data obtained are referred to as a nuclear magnetic resonance (NMR) spectrum.

The simplified presentation here implies that all the protons in any organic molecule absorb energy at exactly the same field strength, and that a nuclear magnetic resonance spectrum consists of only a single signal or peak, which would tell us virtually nothing about the structure of the molecule. This is not the case, because the absorption of energy by hydrogen nuclei (protons) depends on the molecular and electronic environment (in their vicinity) as determined by the adjacent atoms and substituents in the molecule. Different types of hydrogen nuclei require different amounts of energies in the radio-frequency region of the electromagnetic spectrum to ("flip") change the orientation of protons with respect to the external magnetic field. For example, the hydrogen atoms of the

CH_3 group in toluene [structure of toluene with H's on ring and CH_3 group] absorb radiation of a slightly different

energy than the aromatic hydrogens directly attached to the benzene ring. The NMR technique is so sensitive and useful a tool in structure determination that it can differentiate among the different "kinds" of protons in a molecule. The NMR study of protons requires an external magnetic field of about 14,000 gauss and a radio-frequency of about 60 megacycles per second (60 Mc) to give a nuclear magnetic resonance spectrum.

An NMR spectrum shows many absorption peaks. The relative positions of these peaks can clearly differentiate the environments of the protons and can give very detailed information about the structure of a molecule. The types of information that can be obtained from an NMR spectrum include:

1. The number of signals, which indicates the number of different "kinds" of protons present in a molecule;
2. The positions of the signals in the spectrum, which tells us about the electronic environment of each proton;
3. The intensities of the signals (areas under the peaks), which indicates how many protons of each kind there are present in the molecule;
4. The splitting of a signal into more than one peak, which tells us about the environment of a proton with respect to other nearby protons.

24.6–1 NMR—Number of Signals In any molecule, protons that are in the same molecular environment absorb radiation at the same applied field strength. Protons in different environments absorb radiation at different applied field

Sec. 24.6 NUCLEAR MAGNETIC RESONANCE

strengths. Protons having the same molecular envionment are said to be *equivalent*, those in different molecular envionments are called *non-equivalent*. The number of signals in an NMR spectrum indicates the different "kinds" of protons, and tells us which are equivalent, and which are non-equivalent.

As an illustrative example, let us consider the following compounds: equivalent protons are indicated by the same letter, non-equivalent protons by different letters.

CH_3—Cl	CH_3—CH_2—Cl	CH_3—CH_2—CH_2—Cl	CH_3—$\overset{\overset{\displaystyle Cl}{\|}}{CH}$—$CH_3$
a	a b	a b c	a b a
1 signal	2 signals	3 signals	2 signals
methyl chloride	**ethyl chloride**	**n-propyl chloride**	**i-propyl chloride**

We see that *n*-propyl chloride can be readily distinguished from its isomer, *i*-propyl chloride, by merely examining the NMR spectra of the two compounds. (*n*-Propyl chloride gives three NMR signals and has three non-equivalent sets of protons, whereas *i*-propyl chloride gives two NMR signals and has two sets of equivalent protons and one non-equivalent proton.)

24.6-2 NMR—Positions of Signals The number of signals in an NMR spectrum indicates how many kinds of different (non-equivalent) protons a given molecule contains. The position of these signals helps us to identify what type of protons they are—aromatic, aliphatic, primary, secondary, or tertiary; whether they are adjacent to a halogen, an alkene, or to the carbonyl group of an aldehyde; and so on. All the different kinds of protons have different electronic environments. The magnetic field used to obtain an NMR spectrum will cause the electrons in a molecule to circulate, generating secondary magnetic fields. The circulation of the electrons around a naked proton generates a field that usually is in a direction opposed to the direction of the externally applied field. The field strength felt by this proton is diminished, and the proton is said to be *shielded*. The secondary fields generated by the circulation of electrons in nearby nuclei to a proton in a molecule can either enhance the effective field felt by the proton (*deshield*), or diminish the effective field felt by the proton (*shield*). These effects are characteristic and dependent upon the molecule being studied. When compared to a naked proton, shielded protons require a higher applied field strength (greater energy), and deshielded protons require a lower applied field strength (lower energy) to get NMR absorption. Thus, the electronic environment and electronic effects within a given molecule can influence and shift the position of NMR signals. Shielding effects shift absorption to higher energies (up-field), and deshielding shifts the position of NMR absorption to lower energies (down-field). These shifts are called *chemical shifts*.

The direction and magnitude of the chemical shift are measured from a reference point using a standard substance. The compound $(CH_3)_4Si$, tetramethylsilane (TMS), is the reference compound most frequently used (one reason for its preference being that it gives a single sharp NMR absorption signal at a higher frequency than most types of protons found in organic compounds). The chemical shift of a particular type of proton is measured as the difference between the

Table 24.2 Proton Chemical Shifts in NMR Spectroscopy

Type of Proton	Chemical Shift in ppm (δ) from TMS
RCH_3	0.9
R_3-C-H	1.5
$C=C-H$	4.6–5.9
$C\equiv C-H$	2–3
$C=C-CH_3$	1.7
$H-C-F$	4–4.5
$H-C-Cl$	3–4
$H-C-OH$	3.4–4
$H-C-COOH$	2–2.6
$R-CHO$	9–10
$R-OH$	1–5.5
$R-COOH$	10.5–12
$R-NH_2$	1–5

NMR resonance line of tetramethylsilane and that of the proton being investigated. The positions of most proton signals are found down-field (less shielded) from tetramethylsilane.

The most commonly used scale in NMR spectroscopy is the delta (δ) scale. The position of the TMS signal is taken as zero (0) δ or 0.0 ppm (parts per million). Most chemical shifts have δ values between 0 and 10. A small δ value indicates that the proton in question is highly shielded, and a larger δ value is indicative of a less-shielded proton, further down-field from TMS. (Occasionally the τ (tau scale) is used where the TMS signal is 10.0 ppm. Hence $\tau = 10 - \delta$).

Table 24.2 lists some characteristic chemical shifts of protons from TMS.

Figure 24.8 shows the NMR spectrum for *t*-butyl alcohol. Refer to Table 24.2 and predict the positions of the signals in the spectrum, then look at Figure 24.8, and see if you were correct in your interpretations.

24.6–3 NMR—Peak Area and the Number of Protons More information can be obtained if one examines the relative intensities, that is the areas under the various absorption peaks. The area is proportional to the number of that kind of proton present in the molecule. These areas are usually determined automatically by electronically integrating the areas under the peaks in the NMR spectrum. For example, the compound ethyl benzene, $C_6H_5-CH_2-CH_3$, gives three signals in the NMR spectrum: the aromatic hydrogens, the methylene hydrogens, and the methyl hydrogens. Integration of these three peaks (signals) is in the ratio 5:2:3, and tells us how many protons of each kind are present in the ethyl benzene

Sec. 24.6 NUCLEAR MAGNETIC RESONANCE

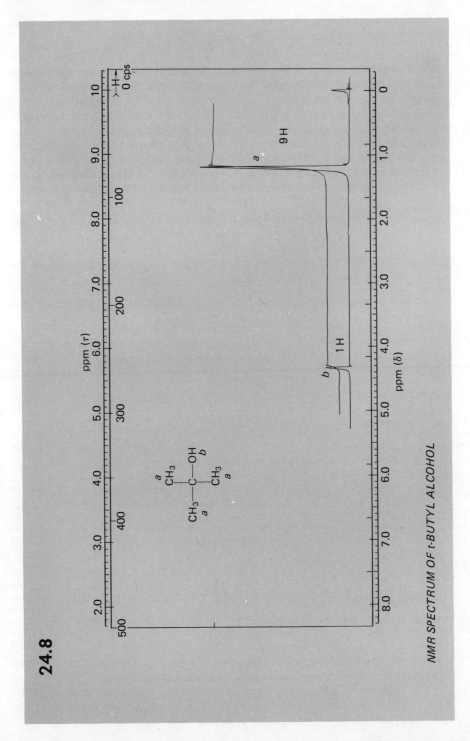

24.8 NMR SPECTRUM OF t-BUTYL ALCOHOL

molecule. Integration of the NMR spectrum of *t*-butyl alcohol (see Figure 24.8) gives the expected ($a = 9$, $b = 1$) ratio.

24.6–4 NMR—Splitting of Signals, Spin–Spin Coupling Under appropriate conditions (high resolution NMR), the absorption signal of a given type of hydrogen (proton) may be split into several other peaks. Instead of having only single peaks for each kind of proton, the spectrum appears to be much more complex because of this fine structure. The presence of these multiple peaks, such as doublets and triplets, is attributed to the interaction between hydrogen atoms on adjacent carbon atoms. In general, a set of *n* equivalent protons will split a signal of protons on adjacent carbon atoms into *n + 1 peaks*. This so-called multiplicity of the peaks reflects the number of neighboring protons on adjacent carbon atoms. For example, a molecule which has as part of its structure —C̲H—CH$_2$— will give an NMR spectrum containing a triplet and a doublet, as predicted by the *n* + 1 rule of neighboring protons. The —C̲H— proton should be split into a triplet

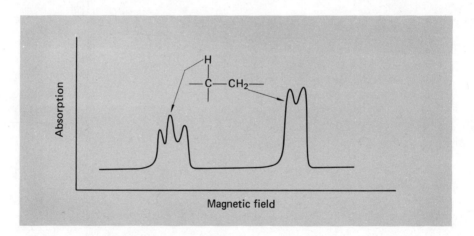

by the neighboring —CH$_2$— protons ($n = 2$), $2 + 1 = 3$, and the —CH$_2$— protons should be split into a doublet ($n = 1$), $1 + 1 = 2$, by the neighboring —C̲H— proton. We may expect to observe this so-called *spin–spin splitting* (or spin–spin coupling) only between non-equivalent neighboring protons.

Figure 24.9 is the NMR spectrum of an alkyl halide, having the formula C_3H_7Br. Using your understanding of the material in the previous sections, predict the structural formula of the compound in Figure 24.9.

Sec. 24.6 NUCLEAR MAGNETIC RESONANCE 415

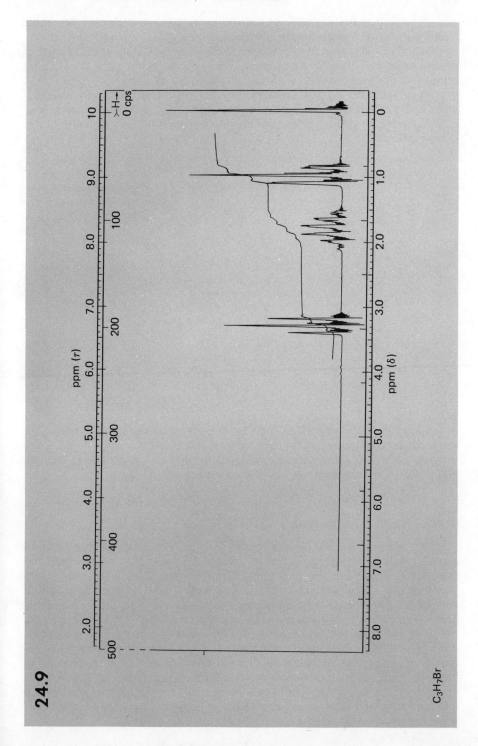

24.9   C₃H₇Br

24.6-5 Applications of NMR Spectroscopy to Solving Complex Structural Problems

The technique of NMR spectroscopy in solving even the most complex stereochemical and structural problems is perhaps most evident in the field of natural products. As a typical example, the detailed conformational structural formula of the compound 1,2-O-isopropylidene-3-O-benzoyl-5-deoxy-β-L-arabinose has been determined largely from NMR spectral data. The chemical shifts of the different kinds of protons, the multiplicity of the peaks, and other information obtained from NMR data provide evidence that the structure of this compound is

The presence of the substituted *isopropylidene* $\left(\begin{array}{c}HCH_3\\ C=C\\ RCH_3\end{array}\right)$ type moiety

in this carbohydrate was established by interpretation of the NMR data. The isopropylidene group occurs in many natural products.

Many other complex structural problems, involving such molecules as steroids and proteins, have been solved by NMR spectroscopy.

24.7 Mass Spectrometry

The field of mass spectrometry, as applied to organic compounds, has grown at a rapid rate in the past few years. It is based on different principles from the spectroscopic methods discussed in the previous sections. In the mass spectrometer, the technique used involves vaporizing organic molecules in a high vacuum, and bombarding the molecules with a beam of high-energy electrons. This results in some electrons being ejected from the organic molecule, with the molecule being ionized and broken up into many fragments, including some positive ions ($M^\oplus$). The positively charged ions are accelerated toward a negatively charged plate by passing them through a magnetic field, which deflects the particles. Particles of lighter mass are deflected more than heavier particles. Each kind of ion has a particular mass to charge (m/e) ratio. For most ions produced in the fragmentation of the molecule, the charge is $+1$, so that m/e usually represents the mass of the ion.

Sec. 24.7 MASS SPECTROMETRY

The set of ions formed from a molecule can be analyzed since each has its own m/e value, and produces a signal whose intensity is due to the relative abundance of that ion. The largest peak found in a *mass spectrum* (that of highest intensity) is called the *base peak,* and is given a numerical value of 100. The intensities of all other peaks are expressed relative to the height of the base peak. A mass spectrum is highly characteristic of a particular compound.

When one electron is removed from the parent molecule, a *molecular ion* or *parent ion* is produced. The m/e value of the molecular ion peak is the molecular weight of the compound, and is extremely accurate. The molecular ion peak may or may not be the peak of highest intensity (the base peak). In interpreting mass spectral data, it is essential that a knowledge of the relative natural abundance of the various isotopes of the elements be known. This is necessary in order to determine an accurate molecular formula for a compound.

The use of mass spectral data is not limited to the determination of the molecular weight of compounds. It also has great utility in the elucidation of the structure of a molecule. The fragmentation processes of organic molecules follow certain patterns (mechanisms) in that the fragments broken off are related to bond strengths and to the stability of the species (smaller molecules such as H_2O, NH_3, CO, CO_2 or carbonium ions) formed. Since certain structural features in molecules produce definite characteristic fragmentation patterns, the identification of these fragments and their relative intensities in a mass spectrum can be used to help determine the structure of organic molecules.

As an illustration, n-octane, $CH_3-CH_2-CH_2-CH_2-CH_2-CH_2-CH_2-CH_3$, can be easily distinguished from its structural isomer 4-methylheptane,
$CH_3-CH_2-CH_2-\underset{\underset{CH_3}{|}}{CH}-CH_2-CH_2-CH_3$, since the methyl group branched at
C_4 in the latter compound, is easily broken off as a $CH_3^\oplus$ fragment and will give a peak of large intensity of $m/e = 15$.

As another illustrative example, let us consider the mass spectrum shown below. The interpretation of the fragmentation pattern and the relative abundance of the peaks is as follows. A molecular weight of 32, which corresponds to O_2, is ruled out, since the only other possible fragment would be O with a

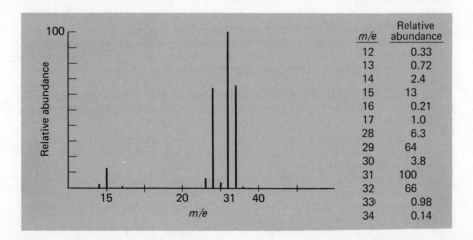

m/e	Relative abundance
12	0.33
13	0.72
14	2.4
15	13
16	0.21
17	1.0
28	6.3
29	64
30	3.8
31	100
32	66
33	0.98
34	0.14

peak of relatively high intensity at $m/e = 16$. (An m/e peak is present at a value of 16, but has a very low relative intensity, 0.21). The peak at $m/e = 15$ is almost certainly due to a $CH_3^{\oplus}$ fragment, and the overall spectrum indicates an elemental composition of CH_4O. Consideration of the other peaks present leads one to the conclusion that the only possible structural assignment for the molecule producing this spectrum is that of CH_3OH or methyl alcohol.

Although the compound neopentane, $CH_3-\overset{\overset{\displaystyle CH_3}{|}}{\underset{\underset{\displaystyle CH_3}{|}}{C}}-CH_3$, has a molecular weight of 72, its mass spectrum produces a base peak at $m/e = 57$. What would you predict is the species responsible for this base peak? A consideration of some of the possible fragmentation patterns will lead the reader to the interpretation that this peak at $m/e = 57$ is almost certainly due to the loss of a CH_3 fragment with a mass of 15 ($72 - 15 = 57$) producing the stable $(CH_3)_3-C^{\oplus}$, t-butyl carbonium ion. This peak is much more intense than the molecular ion peak of neopentane.

Summary

• The elucidation of molecular structure by spectroscopic techniques such as ultraviolet, infrared, nuclear magnetic resonance absorption and mass spectrometry, is prevalent in organic chemistry today.

• The electromagnetic spectrum ranges from exceedingly small cosmic and gamma rays to much longer wavelengths such as radar and radio waves.

   (a) All interactions of molecules with electromagnetic radiation in any region of the electromagnetic spectrum are based on the quantum theory, according to the equation, $E = h\nu$.
   (b) Radiation of shorter wavelength has a higher frequency or energy than radiation of a longer wavelength.

• Spectrophotometers are instruments for measuring the absorption of various types of radiation by a molecule.

• Ultraviolet–visible spectroscopy involves electronic transitions in molecules from lower-energy to higher-energy (excited) states.

   (a) Less tightly held electrons require less energy to be promoted into excited states than more tightly bound electrons.
   (b) The UV–VIS spectrum usually consists of only a few bands of absorption; the wavelength of maximum absorption is referred to as τ_{max}, and the intensity of that absorbance is called the molar absorptivity or molar extinction coefficient, ε_{max}.
   (c) Molecules that absorb radiation in the visible region of the electromagnetic spectrum will exhibit the complementary color to the wavelength of light being absorbed.

• The infrared spectrum consists of many peaks, characteristic of the various types of bonds (functional groups) present in the molecule, and involves vibrational and rotational transitions in the chemical bonds.

   (a) Every chemical bond has its own characteristic absorption frequency in the IR; different functional groups absorb IR radiation at different wavelengths, and their presence or absence in a molecule can be determined by examination of an IR spectrum.

(b) The region between 7–11 μ (1430–910 cm^{-1}) is known as the fingerprint region in the IR spectrum, and can be used to distinguish between many similar molecular structures.

- Nuclear magnetic resonance spectroscopy measures the properties of nuclei which have a mechanical spin in a magnetic field, the hydrogen nucleus or proton being of most interest in organic chemistry.

(a) An NMR spectrum provides information with regard to the different "kinds" of protons present in a molecule, the electronic environment of each proton, how many protons of each kind are present in the molecule, and spin–spin splitting in spectra tells us about the environment of a proton with respect to other non-equivalent neighboring protons.

(b) Tetramethylsilane (TMS) is used as a standard reference substance in NMR spectroscopy. Chemical shifts of protons are measured relative to TMS.

- The technique of mass spectrometry involves molecules being bombarded by a high-energy electron beam in a mass spectrometer.

(a) The peak of highest intensity in a mass spectrum is referred to as the base peak.

(b) Fragmentation processes can produce numerous fragments, from which the structures of even very complex molecules can be deduced. When one electron is removed from a molecule, a molecular ion is produced. The m/e value of the molecular ion peak is the molecular weight of the compound being investigated. The molecular weights obtained by mass spectrometry are extremely accurate.

PROBLEMS

1 Would you expect the compound 3-methyl-5-ethyldecane to give an UV absorption spectrum?

2 Predict the approximate value of τ_{max} absorption for the compound 1,3,5-hexatriene in the UV region of the electromagnetic spectrum.

3 Would you expect that the compounds 2-pentanone and 3-pentanone could be easily differentiated by their IR spectra? Explain. What type of spectroscopy would be best to differentiate between the two compounds? Explain.

4 How many signals would you expect the compound neopentane,

$$CH_3-\underset{\underset{CH_3}{|}}{\overset{\overset{CH_3}{|}}{C}}-CH_3,$$ to give in its NMR spectrum? What would the approximate

chemical shift of the signal(s) be relative to TMS?

5 Which protons would absorb more downfield to TMS in an NMR spectrum, those in CH_4 or those in CH_3F? Explain your answer.

6 *Predict what the complete NMR spectrum of the compound i-propyl bromide would be. Do the same for the compound ethyl bromide.

7 The compounds toluene, p-xylene, and mesitylene (1,3,5-trimethylbenzene) all give very similar NMR spectra with respect to the chemical shifts of the methyl protons. Explain how integration of the areas under all the NMR peaks and proton counting could be used to differentiate between these compounds.

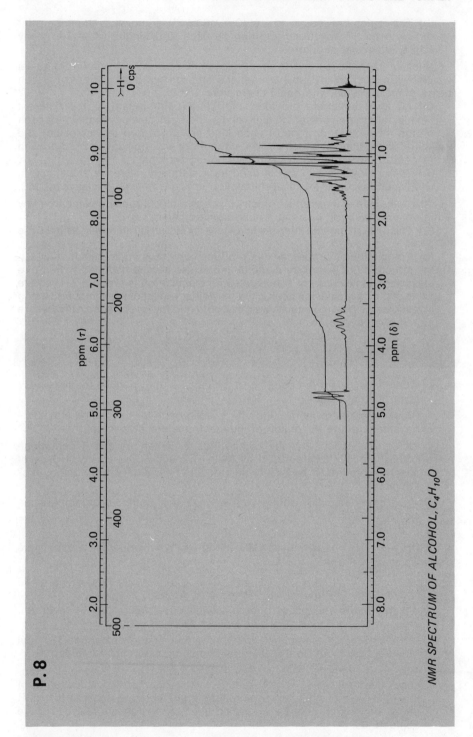

P.8 NMR SPECTRUM OF ALCOHOL, $C_4H_{10}O$

PROBLEMS

8 An alcohol, $C_4H_{10}O$, produces the NMR spectrum shown. Deduce and draw the structure of the alcohol.

9 *Consider the mass spectrum of ethylbenzene. What causes the peak of $m/e = 106$? The base peak in the mass spectrum is at $m/e = 91$. What species do you believe has a $m/e = 91$? Explain. The mass spectrum also contains m/e peaks at 78 and 77. Explain.

10 *An unknown compound of formula C_6H_{10} has the following IR and NMR spectra. It reacts with H_2O in the presence of $HgSO_4$ and H_2SO_4 to produce a ketone. What is a reasonable structure for the compound?

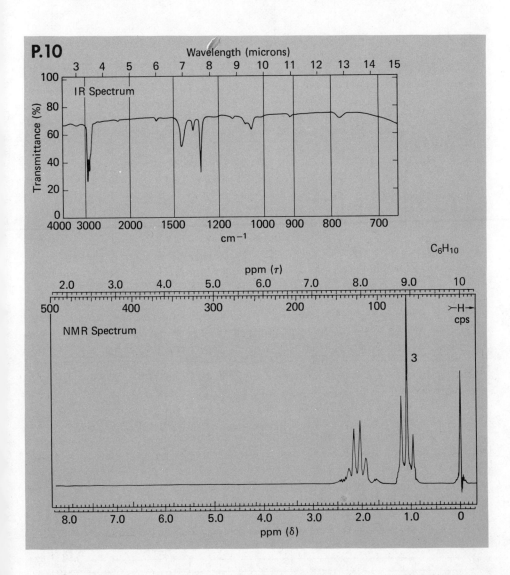

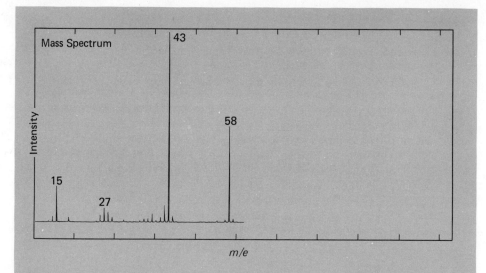
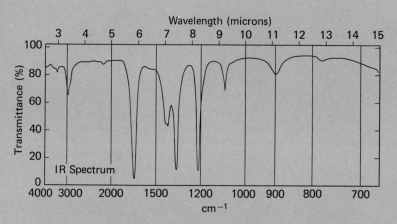
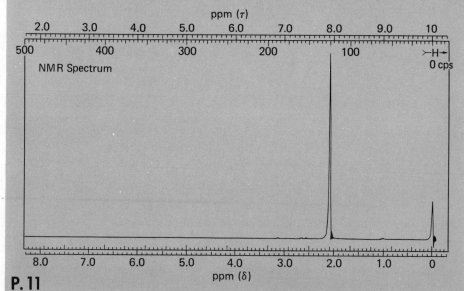

P.11

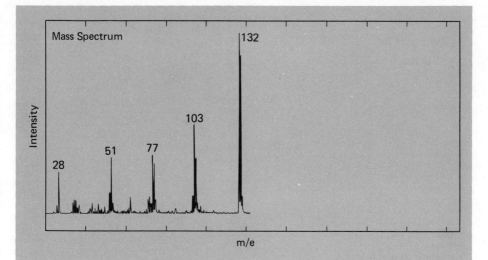
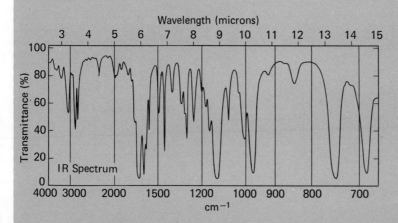
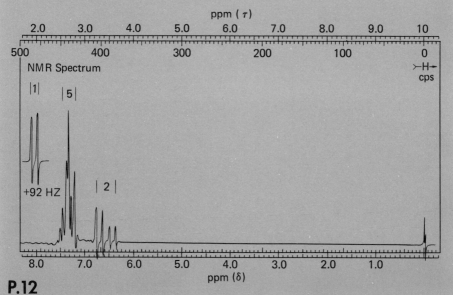

P.12

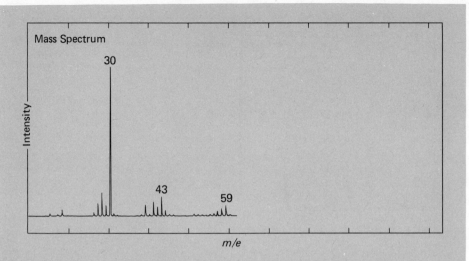

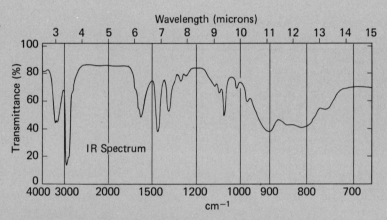

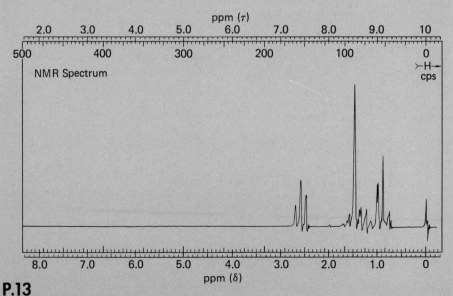

P.13

PROBLEMS

11 * An unknown, low-boiling liquid (b.p. 56°) and a common reagent and solvent found in all laboratories, has the following mass, infrared, and NMR spectral data. Identify the compound.

12 * The unknown compound is a pleasant-smelling liquid (b.p. 251°) and is a component of a common flavoring agent. It can be prepared by a crossed-aldol condensation of acetaldehyde with benzaldehyde. What is the structural formula of the compound from the following spectral data?

13 * The unknown compound is a pungent liquid that reacts with benzene sulfonyl chloride to yield a sulfonamide soluble in NaOH. Identify the unknown from the following spectral data.

Solutions to Selected Problems

Chapter 1

2. (a) $1s^2\ 2s^2\ 2p^6\ 3s^1$ (b) $1s^2\ 2s^2\ 2p^6$

4. (d) $Mg^{\oplus\oplus} + {:}\overset{..}{\underset{..}{O}}{:}^{\ominus\ominus}$ (e) $H{:}\overset{H}{\underset{H}{C}}{:}Cl{:}$

6. $\underbrace{Cl_2 \cong CF_4}_{\text{non-polar covalent}} < \underset{\delta\oplus\ \ \delta\ominus}{I{-}Cl} < \underset{\delta\oplus\ \ \delta\ominus}{IF_7} < \underset{\delta\oplus\ \ \delta\ominus}{H{-}Cl} < \underset{\delta\oplus\ \ \ \ \delta\ominus}{H{-}\overset{H}{\underset{H}{C}}{-}Cl} < \underbrace{Mg^{\oplus\oplus} + 2Cl^{\ominus},\ Na^{\oplus} + Cl^{\ominus}}_{\text{ionic}}$

7. Linear

9. SiF_4, $CHCl_3$, $NH_4^{\oplus}$, tetrahedron; BF_3, $H{-}\overset{\overset{O}{\|}}{C}{-}H$, $CH_3^{\oplus}$, flat, angular (120°), coplanar; HCN, linear; $CH_3^{\ominus}$, collapsed tetrahedron (pyramidal)

10. (a)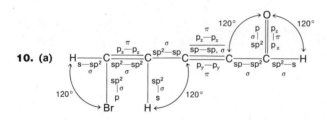

13. Empirical formula is CH_3O; true molecular formula is $C_2H_6O_2$
*14. $C_2H_4Cl_2$

Chapter 2

1. $CH_3{-}CH_2{-}\overset{\overset{O}{\|}}{C}{-}H$ aldehyde $CH_3{-}CH{=}CH{-}CH_3$ alkene

$CH_3{-}CH_2Br$ halide $CH_3{-}CH_2{-}NH_2$ amine

$CH_3{-}CH_2OH$ alcohol

$CH_3{-}C{\equiv}C{-}CH_3$ alkyne $CH_3{-}\overset{\overset{O}{\|}}{C}{-}NH_2$ amide

$CH_3{-}CH_2{-}CH_3$ alkane $CH_3{-}CH_2{-}O{-}CH_2{-}CH_3$ ether

CHAPTER 4

arene (aromatic) $CH_3-\overset{\overset{O}{\|}}{C}-CH_3$ ketone

$CH_3-\overset{\overset{O}{\|}}{C}-OH$ carboxylic acid

2. $CH_3-CH_2-CH_2-CH_2-CH_2-CH_3$

$CH_3-\overset{\overset{CH_3}{|}}{\underset{\underset{H}{|}}{C}}-CH_2-CH_2-CH_3$ $CH_3-CH_2-\overset{\overset{CH_3}{|}}{\underset{\underset{H}{|}}{C}}-CH_2-CH_3$ $CH_3-CH_2-\overset{\overset{CH_3}{|}}{\underset{\underset{CH_3}{|}}{C}}-CH_3$

$CH_3-\underset{\underset{CH_3}{|}}{CH}-\underset{\underset{CH_3}{|}}{CH}-CH_3$

4. (a) branched chain (b) positional (c) positional (d) functional group (e) positional

Chapter 3

1. (a) $CH_3-CH_2-\underset{\underset{Br}{|}}{CH}-CH_3$ (b) $CH_3-\overset{\overset{O}{\|}}{C}-H$ (e) $CH_3-CH_2-\underset{\underset{OH}{|}}{CH}-CH_2-CH_2-CH_3$

(g) $CH_3-\underset{\underset{Cl}{|}}{CH}-CH_2-\overset{\overset{O}{\|}}{C}-OH$ (j) [3-nitro-benzene with NO$_2$ groups] (n) $CH_3-\overset{\overset{O}{\|}}{C}-\underset{\underset{CH_3}{|}}{CH}-(CH_2)_3-CH_3$

(o) [cyclopentene with CH$_2$Cl substituent]

2. (a) 4-chloro-4-*i*-propyl-5,5-dimethyloctane (f) 1-nitro-4-chlorobenzene (i) 2-chloroethanoic acid (l) dimethylethylamine

3. (a) Too high number for substituent; correct name, 3-methylhexane (f) Longest chain is five carbon atoms; correct name, 3-methyl-3-ethylpentane (g) Double bond and substituent numbered incorrectly; correct name, 2-methyl-2-butene

5. (a) $CH_3-CH_2-\overset{\overset{CH_2-CH_3}{|}}{\underset{\underset{H}{|}}{C}}-\overset{\overset{CH_3}{|}}{\underset{\underset{CH_3}{|}}{C}}-CH_3$

Chapter 4

1. (a) $HS^{\ominus}$ (b) $NH_2^{\ominus}$ (d) $CH_3O^{\ominus}$
2. (a) H_2SO_4 (d) $C_6H_5NH_3^{\oplus}$ (e) C_6H_5OH
3. (a) $HPO_4^{\ominus} + HPO_4^{\ominus} \rightleftharpoons H_2PO_4^{\ominus} + PO_4^{\ominus}$

5. A. $HOOC-CH_2-COOH + H_2O \rightleftharpoons H_3O^{\oplus} + HOOC-CH_2-COO^{\ominus}$
B. $HOOC-CH_2-COO^{\ominus} + H_2O \rightleftharpoons H_3O^{\oplus} + {}^{\ominus}OOC-CH_2-COO^{\ominus}$ It is easier to remove a proton from a neutral molecule than from a negatively charged species.
6. No; the strength of an acid depends on how easily a proton can be removed, not

on the number of protons present in the substance. A polyprotic acid has more protons than a monoprotic acid and may be a weaker acid if these protons are not removed easily.

7. (a) $H_3SbO_4 < H_3AsO_4 < H_3PO_4 < HNO_3$

9. butanoic acid < ethanoic acid < 3-chloropropanoic acid < 2-hydroxypropanoic acid < 2-cyanoethanoic acid < 2,2,2-trifluoroethanoic acid.

Chapter 5

1. (a) H—O—N(⊕)(=O:)(—O:⊖) ⟷ H—O—N(⊕)(—O:⊖)(=O:) ⟷ etc.

(e) Br-phenyl ⟷ :Br⊕= cyclohexadienyl carbanion ⟷ :Br⊕= ⟷ :Br⊕= ⟷ etc.

3. Resonance stabilization of the acetate ion results in two equivalent carbon–oxygen bonds.

5. (a) 4-nitrobenzoic acid (COOH, NO₂) **(b)** 2,6-dinitrophenol (OH, O₂N, NO₂, NO₂) **(c)** 4-methylaniline (NH₂, CH₃) **(d)** CH₃NH₂ **(e)** N,N-dimethylaniline (N(CH₃)₂)

6. (a) 4-aminobenzoic acid (COOH, NH₂) < benzoic acid (COOH) < 4-chlorobenzoic acid (COOH, Cl) < 4-nitrobenzoic acid (COOH, NO₂)

(b) Butanoic acid < 4-bromobutanoic acid < 4,4-dichlorobutanoic acid < 3-bromobutanoic acid < 2-bromobutanoic acid **(c)** 6-chloroheptanoic acid < propanoic acid < ethanoic acid < 2-cyanoethanoic acid < 2,2-dichloropropanoic acid < 2,2,2-trichloroethanoic acid.

7. (b) aniline (NH₂) < N-methylaniline (NHCH₃) < $(CH_3)_3N < CH_3NH_2 < CH_3\text{—}\underset{H}{N}\text{—}CH_3$

8. Hydrogen bonding is present in the *ortho* isomer due to the close proximity of the OH to the COOH group. This facilitates proton removal. Hydrogen bonding is not possible in the *para* isomer since the OH group is far removed from the COOH.

salicylic acid with intramolecular O—H⋯O **(hydrogen bond)**

CHAPTER 7

Chapter 6

1. (a) electrophilic aromatic substitution (b) addition (c) nucleophilic substitution (d) elimination (f) nucleophilic addition (h) free radical substitution (j) nucleophilic substitution

2. (a) Nucleophilic substitution.

$$S_N2 \text{ Mechanism: } HO:^{\ominus} + H-\underset{\underset{H}{|}}{\overset{\overset{CH_3}{|}}{C}}-Br \xrightarrow{slow} \left[HO\overset{\delta\ominus}{\cdots}\underset{\underset{CH_3}{|}}{\overset{\overset{H\ \ \ H}{|}}{C}}\overset{\delta\ominus}{\cdots}Br \right] \xrightarrow{fast} HO-\underset{\underset{H}{|}}{\overset{\overset{CH_3}{|}}{C}}-H + Br^{\ominus}$$

(b) Free radical substitution.
Mechanism

Initiation: $Br_2 \xrightarrow{h\nu} 2Br\cdot$
Propagation: $CH_3Br + Br\cdot \longrightarrow HBr + \cdot CH_2Br$
$\cdot CH_2Br + Br_2 \longrightarrow CH_2Br_2 + Br\cdot$, etc.
Termination: $Br\cdot + Br\cdot \longrightarrow Br_2$
$\cdot CH_2Br + Br\cdot \longrightarrow CH_2Br_2$
$\cdot CH_2Br + \cdot CH_2Br \longrightarrow Br-CH_2-CH_2-Br$

(c) Nucleophilic substitution.

S_N1 Mechanism:

$$CH_3-\underset{\underset{CH_3}{|}}{\overset{\overset{CH_3}{|}}{C}}-I \underset{}{\overset{slow}{\rightleftharpoons}} CH_3-\underset{\underset{CH_3}{|}}{\overset{\overset{CH_3}{|}}{C}}{}^{\oplus} + I^{\ominus}$$

$$CH_3-\underset{\underset{CH_3}{|}}{\overset{\overset{CH_3}{|}}{C}}{}^{\oplus} + SH^{\ominus} \xrightarrow{fast} CH_3-\underset{\underset{CH_3}{|}}{\overset{\overset{CH_3}{|}}{C}}-SH$$

Chapter 7

1. (a) $CH_3-CH_2-\underset{\underset{}{|}}{\overset{\overset{CH_3}{|}}{CH}}-\underset{\underset{}{|}}{\overset{\overset{Br}{|}}{CH}}-CH_2-CH_2-CH_3$ (b) $CH_3-\underset{\underset{CH_3}{|}}{\overset{\overset{CH_3}{|}}{C}}-CH-\underset{\underset{CH_3}{|}}{\overset{\overset{}{|}}{CH}}-CH_3$
$\underset{}{|}\underset{}{|}$
CH_3I

(c) [cyclopentane with Cl, Cl on one carbon and CH_3, H on adjacent carbon]

2. (a) $CH_3-CH_2-\underset{\underset{CH_3}{|\ \ 1°}}{\overset{\overset{H}{|}}{\underset{3°}{C}}}-\underset{\underset{CH_3}{|\ \ 1°}}{\overset{\overset{Cl}{|}}{\underset{3°}{C}}}-\underset{\underset{H}{|}}{\overset{\overset{CH_3\ 1°}{|}}{\underset{3°}{C}}}-CH_3$
${}_{1°}{}_{2°}$

2,3,4-trimethyl-3-chlorohexane

4. ethane < pentane < 2-methylpentane < n-hexane < 2,2-dimethylpentane

5. (a) $CH_3-CH_2Br + HBr$ (i) $CH_3-CH_2-CH_3 + Mg(OH)Br$

7. CH₃—CH₂—CH₂—CH(Br)—CH₂—CH₃   3-bromohexane

*9. (a) CH₄ **(b)** △ **(c)** CH₃—CH₂—CH₂—CH₃ or CH₃—CH(CH₃)—CH₃ **(d)** CH₃—C(CH₃)₂—CH₃

(e) CH₃—CH₂—CH₂—CH₂—CH₃ **(f)** CH₃—CH(CH₃)—CH₂—CH₃ **(g)** CH₃—CH(CH₃)—CH(CH₃)—CH₃

11. (a) staggered (most stable)   eclipsed

(b) staggered or anti (most stable)   eclipsed

13. (a) △ + Br₂ —hv→ Br—CH₂—CH₂—CH₂Br

(b) ⬠ + Br₂ —hv→ ⬠-Br + HBr

In reaction (a) the cyclopropane ring breaks open due to Baeyer ring strain. The opening of the ring changes the bond angles from 60° to 109° 28′ and relieves the strain. Reaction (b) occurs as a free-radical substitution reaction, typical of alkanes, since the cyclopentane ring has bond angles of 108° and is relatively free from any ring strain.

Chapter 8

1. (b) CH₃—CH(Br)—CH=CH—CH₂—CH₃   **(e)** CH₂=CH—CH₂—CH₂—CH=CH₂

(h) cyclohexene with Cl   **(l)** CH₃—CH=C(H)—CH₂—CH₃

2. (a) 1–chloro–1–butene **(c)** 4,4–dimethyl–2–pentene **(g)** 3–bromo–1,2–pentadiene
(h) 2–methyl–1,3–cyclohexadiene

3. (a) Numbered incorrectly; should be numbered from other end so double bond

takes lower number; correct name is 1-pentene **(d)** No *cis* or *trans* isomers; correct name is 3-methyl-1-butene
5. (b), (f), (g), and (h) exhibit geometrical isomerism.

6. (b) [cyclohexane with CH₃ and Cl on one carbon, Cl and H on adjacent carbon]

(e) $CH_3-CH_2-CH_2-\underset{I}{CH}-CH_3$ (g) $CH_3-\underset{H}{\overset{CH_3}{C}}-CH_2Br$

(i) 2-bromopropane + alc. KOH ⟶ $CH_3CH=CH_2$

*(p) [1,4-naphthoquinone partially reduced structure]   (r) [cyclohexane with CH₃ and OH on adjacent carbons]

7. (a) $CH_3-CH_2-CH_2-CH_2Br \xrightarrow{alc., KOH} CH_3-CH_2-CH=CH_2 \xrightarrow{Br_2/CCl_4} CH_3-CH_2-\underset{Br}{CH}-CH_2Br$

(c) $CH_3-\underset{OH}{CH}-CH_3 \xrightarrow{conc. H_2SO_4} CH_3-CH=CH_2 \xrightarrow{NBS} BrCH_2-CH=CH_2 \xrightarrow{HOCl} Br-CH_2-\underset{OH}{CH}-CH_2Cl$

(e) $CH_3-\underset{CH_3}{CH}-CH_2OH \xrightarrow[\text{or } Al_2O_3, \Delta]{conc. H_2SO_4} CH_3-\underset{CH_3}{C}=CH_2 \xrightarrow[H_2O_2]{HBr} CH_3-\underset{H}{\overset{CH_3}{C}}-CH_2Br$

(f) [cyclohexanol] $\xrightarrow{conc. H_2SO_4}$ [cyclohexene] $\xrightarrow{Br_2, CCl_4}$ [1,2-dibromocyclohexane]

8. (a) Br_2/CCl_4; or cold, dilute $KMnO_4$; or conc. H_2SO_4 (b) Baeyer test; or conc. H_2SO_4 (c) Br_2/CCl_4; or cold, dilute $KMnO_4$; or conc. H_2SO_4
10. (a) propylene (b) $CH_2=CH_2$ (c) $CH_3-CH=CH_2$

11. A $CH_3-CH_2-\underset{H}{\overset{CH_3}{C}}=\underset{}{\overset{}{C}}-CH_3$ 2-methyl-2-pentene

  B $CH_3-\underset{H}{\overset{H}{C}}=\underset{}{\overset{H}{C}}-\underset{CH_3}{C}-CH_3$ 4-methyl-2-pentene

C $CH_2=CH-CH_2-CH=\overset{H}{C}-CH_3$ 1,4-hexadiene

D (cyclohexene structure)   cyclohexene

*12. (a) In *cis*-1,3-dimethylcyclohexane the two methyl groups are both equatorial, whereas in the *trans* isomer one methyl group is axial and one is equatorial. Therefore, the *cis* isomer is more stable.

cis isomer   *trans* isomer

(b) The *trans* isomer is more stable in the 1,2 and 1,4 derivatives since both methyl groups are equatorial, whereas in the *cis* isomers one methyl group is axial and one is equatorial.

14. $CH_2=CH-CH=CH-CH=CH_2 + HOCl$
$$\downarrow$$
$$HOCH_2-CH=CH-CH=CH-CH_2Cl$$
major product (1,6 addn.)
Adds to ends of conjugated system

*16. $R-O-OR \xrightarrow{h\nu} 2RO\cdot$
$RO\cdot + CHCl_3 \longrightarrow ROH + \cdot CCl_3$
$\cdot CCl_3 + CH_3-(CH_2)_5-CH=CH_2 \longrightarrow CH_3-(CH_2)_5-\dot{C}H-CH_2-CCl_3$
$CH_3-(CH_2)_5-\dot{C}H-CH_2-CCl_3 + CHCl_3 \longrightarrow CH_3-(CH_2)_5-CH_2-CH_2-CCl_3$
$+$
$\cdot CCl_3$, etc.

*17. (series of carbocation rearrangement structures)

$CH_3-\underset{CH_3}{\underset{|}{C}}-\underset{OH}{\underset{|}{C}}(H)(CH_3)-CH_3 \xrightleftharpoons{H^\oplus} CH_3-\underset{CH_3}{\underset{|}{C}}-\underset{OH_2^\oplus}{\underset{|}{C}}(H)(CH_3)-CH_3 \longrightarrow H_2O + CH_3-\underset{CH_3}{\underset{|}{C}}-\underset{\oplus(2°)}{\underset{|}{C}}(H)(CH_3)-CH_3$

rearr. ↘   ↓ ROH

$CH_3-\underset{\oplus(3°)}{\underset{|}{C}}(CH_3)-\underset{CH_3}{\underset{|}{C}}(H)-CH_3$ $ROH_2^\oplus + CH_3-\underset{CH_3}{\underset{|}{C}}(H)-C(CH_3)=CH_2$

                                                    minor product

ROH ↓

$ROH_2^\oplus + CH_3-\underset{CH_3}{\underset{|}{C}}=\underset{CH_3}{\underset{|}{C}}-CH_3 + CH_3-\underset{CH_3}{\underset{|}{C}}(H)-\underset{=CH_2}{C}-CH_3$

major product (since most   minor product
substituted alkene)

*18. A (cyclohexanol, OH)   B (cyclohexene)   C (3-chlorocyclohexene, Cl)   D (cyclohexene)   E (3,4-dibromocyclohexene, Br, Br)

Chapter 9

1. (a) CH$_3$—CH(Br)—C≡C—H

2. (a) 4-chloro-4-methyl-2-pentyne

3. (b) N.R.   (d) CH$_3$—C(=O)—CH$_2$—CH$_3$ (e) C$_2$H$_5$—C(H)=C(H)—C$_2$H$_5$ *trans*-3-hexene

   (f) CH$_3$—C≡C—Cu

4. (a) CH$_3$—CH(OH)—CH$_3$ $\xrightarrow{\text{conc. H}_2\text{SO}_4}$ CH$_3$—CH=CH$_2$ $\xrightarrow[\text{CCl}_4]{\text{Br}_2}$ CH$_3$—CH(Br)—CH$_2$Br $\xrightarrow[\text{2. NaNH}_2]{\text{1. alc. KOH}}$ CH$_3$—C≡C—H

   (b) (i) CaC$_2$ + H$_2$O ⟶ H—C≡C—H   (ii) CO + H$_2$ $\xrightarrow[\text{cat.}]{\Delta,\text{ press.}}$ CH$_3$OH $\xrightarrow{\text{PBr}_3}$ CH$_3$Br

   (iii) H—C≡C—H + CH$_3$Br + Na ⟶ CH$_3$—C≡C—H + HI ⟶ CH$_3$—C(I)=CH$_2$ $\xrightarrow{\text{HCl}}$ CH$_3$—C(I)(Cl)—CH$_3$

   (c) CaC$_2$ + H$_2$O ⟶ H—C≡C—H $\xrightarrow{\text{H}_2/\text{Pt}}$ CH$_2$=CH$_2$ $\xrightarrow{\text{HBr}}$ CH$_3$CH$_2$Br $\xrightarrow[\text{Na}]{\text{H—C≡C—H}}$

   CH$_3$—CH$_2$—C≡C—H $\xrightarrow{\text{H}_2/\text{Pt}}$ CH$_3$—CH$_2$—CH=CH$_2$ $\xrightarrow[\text{2. H}_2\text{O}_2,\text{ OH}^\ominus]{\text{1. B}_2\text{H}_6}$ CH$_3$—CH$_2$—CH$_2$—CH$_2$OH

*6. A CH$_3$—CH=CH—CH(CH$_3$)(H)—CH$_3$ B CH$_3$—CH(Cl)—CH(Cl)—CH(CH$_3$)—CH$_3$

   C CH$_3$—C≡C—CH(CH$_3$)—CH$_3$ D CH$_3$—C(CH$_3$)(H)—C(=O)—OH   E CH$_3$—C(=O)—OH

Chapter 10

1. (a) 3-chlorotoluene (methyl + Cl meta on benzene)

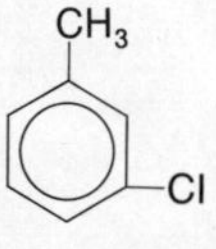

   (f) 4-iodobenzenesulfonic acid (SO$_3$H and I para on benzene)

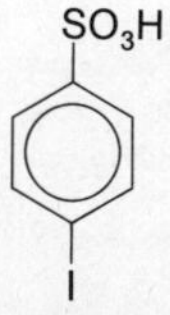

   (j) 3,5-dichlorobenzoic acid (COOH with Cl at 3,5 on benzene)

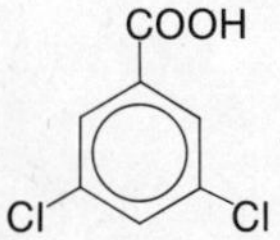

2. (a) o-nitroiodobenzene **(d)** m-xylene **(i)** 1,3-dichloro-2-nitro-5-bromobenzene
5. 3 × (+28.6) kcal/mole = +85.8 kcal/mole. 85.8 − 49.8 = 36 kcal/mole; resonance energy–resonance stabilization in benzene is 36 kcal/mole. Benzene contains 36 kcal/mole less energy than predicted due to resonance stabilization.

6. (b) C$_6$H$_5$CH$_2$Br

(c) C$_6$H$_5$CH$_2$–CH$_3$ **(d)** o-NO$_2$-C$_6$H$_4$-COOH   **(i)** p-CH$_3$-C$_6$H$_4$-Tl(OOCCF$_3$)$_2$ **(l)** 2-bromo-4-nitrotoluene

7. (a) benzene $\xrightarrow{CH_3Cl, AlCl_3}$ toluene $\xrightarrow{K_2CrO_7, H_2SO_4}$ benzoic acid $\xrightarrow{HNO_3, H_2SO_4}$ m-nitrobenzoic acid

(e) benzene $\xrightarrow{CH_3Cl, AlCl_3}$ toluene $\xrightarrow{HNO_3, H_2SO_4}$ p-nitrotoluene $\xrightarrow{Fe, Br_2}$ 2-bromo-4-nitrotoluene $\xrightarrow{K_2Cr_2O_7, H_2SO_4}$ 2-bromo-4-nitrobenzoic acid

(i) benzene $\xrightarrow{CH_3Cl, AlCl_3}$ mesitylene(1,3,5-trimethylbenzene) $\xrightarrow{3H_2, Pt}$ 1,3,5-trimethylcyclohexane

(j) toluene + Tl(OOCCF$_3$)$_3$ $\xrightarrow{CF_3COOH}$ p-CH$_3$-C$_6$H$_4$-Tl(OOCCF$_3$)$_2$ $\xrightarrow{KI}$ p-iodotoluene

8. (a) CHCl$_3$ and AlCl$_3$ or conc. H$_2$SO$_4$
(b) Br$_2$ in CCl$_4$ or cold, dil. KMnO$_4$
(c) Br$_2$ in CCl$_4$
(d) Br$_2$ in CCl$_4$ or Ag(NH$_3$)$_2$$^\oplus$
(e) NaOH or H$_2$O (litmus paper)

9. C$_6$H$_5$-N(CH$_3$)$_3$$^\oplus$ < C$_6$H$_5$-NO$_2$ < C$_6$H$_6$ < C$_6$H$_5$-CH$_3$ < C$_6$H$_5$-OH

CHAPTER 11

10. (a) [resonance structures of aniline showing NH$_2$ group with positive charge delocalized, with negative charges at ortho and para positions] ⟷ etc.

NH$_2$ group has electron-donating resonance effect. Attack by electrophilic reagent occurs at *ortho* and *para* positions since these are the positions of highest electron density.

12. $CH_3-CH-CH_2Cl + ZnCl_2 \longrightarrow ZnCl_3^{\ominus} + CH_3-CH-CH_2^{\oplus}$
     | |
     CH_3 CH_3

$\underset{1°}{CH_3-CH-CH_2^{\oplus}} \xrightarrow{\text{rearr.}} \underset{3°}{CH_3-\overset{CH_3}{\underset{\oplus}{C}}-CH_3}$
     |
     CH_3

[benzene] + $CH_3-\overset{CH_3}{\underset{\oplus}{C}}-CH_3 \longrightarrow \left[\text{arenium ion with H and C(CH}_3)_3\right]^{\oplus} \xrightarrow{ZnCl_3^{\ominus}} HZnCl_3 +$ [t-butylbenzene, $C(CH_3)_3$]

$HZnCl_3 \longrightarrow HCl + ZnCl_2$

*13. The *t*-butyl group is an *ortho–para* director in electrophilic aromatic substitution reactions. The steric factors due to the bulky *t*-butyl group prevent nitration at the *ortho* position.

14. [phenol with HO$_3$S group para to OH] $\xrightarrow[Br_2]{Fe}$ [phenol with HO$_3$S para to OH and Br ortho to OH]
(major product)

Br takes the position shown since OH is an *ortho–para* director and SO$_3$H group is a *meta* director.

18. (a) Nitro group is an electron-withdrawing group which deactivates the benzene ring so that it isn't reactive enough to be attacked by the electrophilic reagent in the Friedel-Crafts reaction. (b) and (c) The unshared electron pair on the N in aniline and O in phenol tie up the Lewis acid catalyst by forming a Lewis acid–base complex, and prevent the catalyst from functioning as it should in the Friedel-Crafts reaction.

Chapter 11

1. (a) $CH_3-\overset{CH_3}{\underset{OH}{C}}-CH_3$ (c) [2-chlorophenol: benzene with OH and Cl ortho] (h) [4-methylphenol: benzene with OH para to CH$_3$] (i) $HO-CH_2-CH_2-C\equiv C-CH_2-CH_3$

2. (a) 3-methyl-1-butanol (f) 2-methyl-4-pentene-2-ol
       primary alcohol   tertiary alcohol

3. (a) Use of *iso* prefix is for common names; here it is mixed with IUPAC.-ol ending. Should use numbers for substituents; correct name, 2-propanol. (d) Longest chain is

five carbon atoms, not four; correct name, 2-methyl-2-pentanol. **(e)** Ring numbered incorrectly; can use lower numbers; correct name, 2-nitrophenol.

6. phenol < ethanol < cyclohexanol < 3-hexanol < *t*-butyl alcohol < benzyl alcohol

7. (a) CH_3-CH_2-I **(b)** $CH_3-CH=CH_2$ **(f)** 2,4,6-tribromophenol (Br, OH, Br, Br on ring) **(g)** $CH_3-\overset{O}{\underset{\|}{C}}-CH_2-CH_3$

(j) $CH_3-CH_2-CH_2-CH_2-CH_2-\overset{O}{\underset{\|}{C}}-OH$ or $CH_3-(CH_2)_4-\overset{O}{\underset{\|}{C}}-OH$

8. (a) $CaC_2 + H_2O \longrightarrow H-C\equiv C-H \xrightarrow{H_2}{Pt} CH_2=CH_2 \xrightarrow{H_2O}{H^+} CH_3-CH_2OH$

(e) benzene $\xrightarrow{H_2SO_4, \Delta}$ benzene-1,3-disulfonic acid (SO_3H, SO_3H) $\xrightarrow{NaOH, \Delta}$ disodium salt (ONa, ONa) $\xrightarrow{H^{\oplus}}$ resorcinol (OH, OH)

(f) benzene $\xrightarrow{CH_3Cl/AlCl_3}$ toluene (CH_3) $\xrightarrow{Cl_2, h\nu}$ benzyl chloride (CH_2Cl) $\xrightarrow{aq.\ NaOH}$ benzyl alcohol (CH_2OH)

$\left(CO + H_2 \xrightarrow[cat.]{\Delta,\ press.} CH_3OH \xrightarrow{PCl_3} CH_3Cl \right)$

(j) cyclopentene $\xrightarrow{NBS}$ 3-bromocyclopentene (Br) $\xrightarrow{cold,\ dil.,\ neutral\ KMnO_4}$ bromo-diol cyclopentane (OH, OH, Br)

(k) $CH_3-\overset{OH}{\underset{|}{CH}}-CH_3 \xrightarrow{conc.\ H_2SO_4} CH_3-CH=CH_2 \xrightarrow[H_2O_2]{HBr} CH_3-CH_2-CH_2Br$

10. (a) Lucas test **(b)** $NaOH$ or $Br_2(H_2O)$

12. **A** $CH_3-\overset{CH_3}{\underset{OH}{C}}-\overset{CH_3}{\underset{H}{C}}-CH_3$ **B** $CH_3-\overset{CH_3}{\underset{Br}{C}}-\overset{CH_3}{\underset{H}{C}}-CH_3$ **C** $CH_3-\overset{CH_3}{C}=\overset{CH_3}{C}-CH_3$

D $CH_2=\overset{CH_3}{C}-\overset{CH_3}{\underset{H}{C}}-CH_3$

Chapter 12

1. **(b)** $CH_3-CH_2-O-CH_2-CH_3$ **(c)** $CH_3-CH_2-O-CH_2-CH_2-CH_2-CH_3$ **(f)** 4-bromo-1-methoxybenzene (p-BrC₆H₄OCH₃)

2. **(a)** 2-methoxypentane **(c)** 3,4-dinitroanisole **(e)** methyl phenyl sulfide

4. **(a)** $CH_3-CH_2-O-CH_3$ **(b)** N.R. **(d)** $CH_3-CH_2-S-CH_3$ **(e)** phenol (C₆H₅OH) + CH_3Br

(g) $CH_3-CH(OH)-CH(OCH_2CH_3)-CH_3$

5. **(c)** $CH_3-CH_2-CH=CH_2 \xrightarrow[H_2O_2]{HBr} CH_3-CH_2-CH_2-CH_2Br$

$(CH_3)_2C=CH_2 \xrightarrow[H^{\oplus}]{H_2O} (CH_3)_3C-OH \xrightarrow{Na} (CH_3)_3C-ONa$

$(CH_3)_3C-ONa + CH_3-CH_2-CH_2-CH_2Br \longrightarrow CH_3-CH_2-CH_2-CH_2-O-C(CH_3)_3$

(d) C₆H₆ $\xrightarrow[\Delta]{H_2SO_4}$ C₆H₅-SO₃H $\xrightarrow[\Delta]{NaOH}$ C₆H₅-ONa $\xrightarrow{CH_3Cl}$ C₆H₅-OCH₃

$(CO + H_2 \xrightarrow[cat.]{\Delta, press.} CH_3OH \xrightarrow{PCl_3} CH_3Cl)$

(f) $CaC_2 + H_2O \longrightarrow H-C\equiv C-H \xrightarrow{H_2}_{Pt} CH_2=CH_2 \xrightarrow[H^{\oplus}]{H_2O} CH_3-CH_2OH$

$CH_2=CH_2 \xrightarrow[\Delta, press., cat.]{Ag, O_2} \text{ethylene oxide} \xrightarrow[H^{\oplus}]{CH_3CH_2OH} CH_2(OH)-CH_2(OCH_2-CH_3)$

(h) $CH_2=CH_2 \xrightarrow{HOCl} Cl-CH_2-CH_2OH \xrightarrow[\Delta]{H_2SO_4} Cl-CH_2-CH_2-O-CH_2-CH_2Cl$
$\xrightarrow{\text{alc. KOH}} CH_2=CH-O-CH=CH_2$

*7. **(a)** Because the tertiary halide would be dehydrated to an alkene (as the major product) in the presence of the strong alkoxide base (get elimination rather than substitution). **(b)** No; the alcohol would undergo elimination (dehydration) to an alkene instead.

(c) $CH_3-\underset{\underset{CH_3}{|}}{C}=CH_2 + H^{\oplus} \longrightarrow CH_3-\underset{\underset{CH_3}{|}}{\overset{\oplus}{C}}-CH_3$

$\downarrow CH_3-\underset{\underset{CH_3}{|}}{\overset{\overset{CH_3}{|}}{C}}-OH$

$CH_3-\underset{\underset{CH_3}{|}}{\overset{\overset{CH_3}{|}}{C}}-\underset{\underset{H}{|}}{\overset{\oplus}{O}}-\underset{\underset{CH_3}{|}}{\overset{\overset{CH_3}{|}}{C}}-CH_3 \longrightarrow H^{\oplus} + (CH_3)_3C-O-C(CH_3)_3$

*9. The $Cl^{\ominus}$ ion is a weak nucleophilic reagent, whereas $Br^{\ominus}$ and $I^{\ominus}$ ions are good nucleophiles.

*11. The presence of electron-withdrawing nitro groups facilitates the reaction. They stabilize the intermediate carbanion formed during the reaction.

12. Ethylene oxide has a three-membered ring, hence ring strain. This ring strain can be relieved by reaction when the ring opens up to achieve bond angles near the normal tetrahedral angle for carbon. Since most cyclic ethers have little or no ring strain they would behave as ordinary ethers, and be fairly unreactive (since they are essentially hydrocarbon in nature with unshared electrons on the oxygen atom).

13. (a) S_N2 *Mechanism:*

$CH_3-CH_2-CH_2-O^{\ominus} + CH_3-CH_2-CH_2I \xrightarrow{slow}$

$\left[CH_3-CH_2-CH_2-\overset{\delta\oplus}{O}\cdots\underset{\underset{H\ \ H}{|}}{\overset{\overset{CH_2-CH_3}{|}}{\underset{\delta\ominus}{C}}}\cdots\overset{\delta\ominus}{I} \right] \xrightarrow{fast} I^{\ominus} + CH_3-CH_2-CH_2-O-CH_2-CH_2-CH_3$

14. A: CH$_2$Br, OCH$_3$ (on benzene ring, ortho)
B: COOH, OCH$_3$ (on benzene ring, ortho)
C: COOH, OH (on benzene ring, ortho)

Chapter 13

1. (a) $Br-CH_2-CH=CH_2$ **(c)** CH_2Cl on benzene ring with NO_2 para

2. (a) 3,4,4-trichloro-1-pentene **(b)** *m*-iodotoluene **(d)** 1,3-dibromo-1-methylcyclohexane

3. (c) $CH_3-CH_2-\underset{\underset{}{\overset{\overset{I}{|}}{}}}{CH}-CH_3$

(e) [2-chlorotoluene] + [4-chlorotoluene]

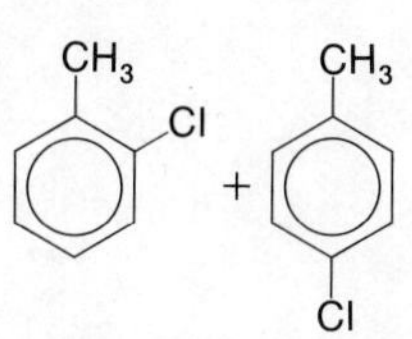

(f) CH_3-CH_2-I

(g) $F_3C-\underset{CF_3}{\underset{|}{C}}-\underset{CF_3}{\underset{|}{C}}-\underset{F}{\underset{|}{C}}-CF_3$ (with F on top of each C)

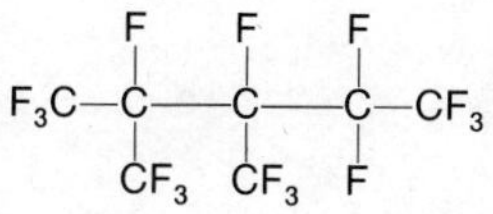

(o) $EtCOO-\underset{\underset{CH_3}{|}}{\underset{CH_2}{|}}{CH}-COOEt$

4. (a)

$CH_3-CH_2-CH_2-CH_2Br \xrightarrow{\text{alc. KOH}} CH_3-CH_2-CH=CH_2 \xrightarrow{Br_2} CH_3-CH_2-\underset{\underset{}{|}}{\overset{Br}{\underset{|}{CH}}}-CH_2Br \xrightarrow[\text{2. NaNH}_2]{\text{1. alc. KOH}} CH_3-CH_2-C\equiv C-H \xrightarrow[\text{2. HBr}]{\text{1. HCl}} CH_3-CH_2-\underset{\underset{Br}{|}}{\overset{Cl}{\underset{|}{C}}}-CH_3$

(d)

Benzene $\xrightarrow[AlCl_3]{CH_3Cl}$ toluene $\xrightarrow[Br_2]{Fe}$ p-bromotoluene $\xrightarrow[h\nu]{Br_2}$ p-Br-C$_6$H$_4$-CH$_2$Br

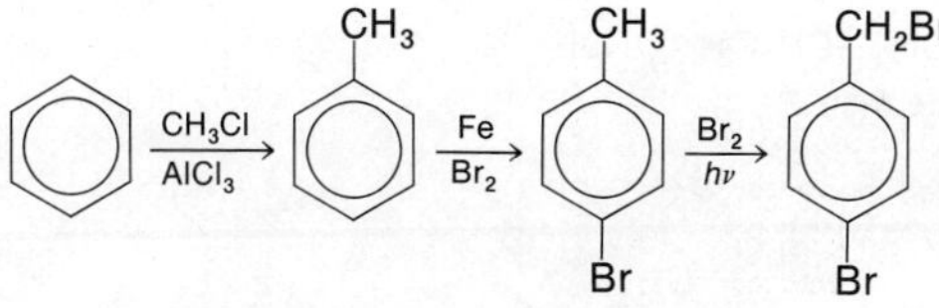

$CO + H_2 \xrightarrow[\text{cat.}]{\Delta, \text{press.}} CH_3OH \xrightarrow{PCl_3} CH_3Cl$

(f)

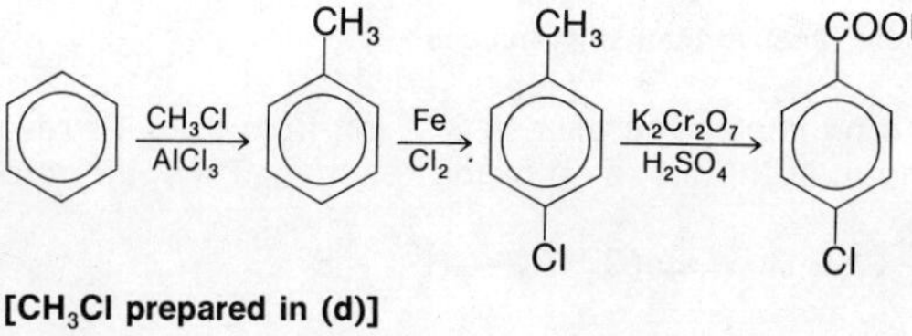

Benzene $\xrightarrow[AlCl_3]{CH_3Cl}$ toluene $\xrightarrow[Cl_2]{Fe}$ p-chlorotoluene $\xrightarrow[H_2SO_4]{K_2Cr_2O_7}$ p-Cl-C$_6$H$_4$-COOH

[CH$_3$Cl prepared in (d)]

(h) $CH_3-CH_2-CH_2Cl \xrightarrow{\text{alc. KOH}} CH_3-CH=CH_2 \xrightarrow{HI} CH_3-\underset{\underset{I}{|}}{CH}-CH_3$

(j) $CaC_2 + H_2O \longrightarrow H-C\equiv C-H \xrightarrow[Na]{CH_3Cl} CH_3-C\equiv C-H \xrightarrow{H_2/Pt}$

$ClCH_2-CH=CH_2 \xleftarrow[600°C]{Cl_2} CH_3-CH=CH_2$

[CH$_3$Cl prepared in (d)]

5. (a) $CH_3-CH_2-CH_2-CH_2Cl$ 1-chlorobutane

$CH_3-CH_2-\underset{\underset{}{|}}{\overset{Cl}{\underset{|}{CH}}}-CH_3$ 2-chlorobutane

$$CH_3-\underset{\underset{CH_3}{|}}{\overset{\overset{CH_3}{|}}{C}}-Cl \quad \text{2-chloro-2-methylpropane}$$

$$CH_3-\underset{\underset{H}{|}}{\overset{\overset{CH_3}{|}}{C}}-CH_2Cl \quad \text{1-chloro-2-methylpropane}$$

*(b) 2-*chloro*-2-methylpropane (*t*-butyl chloride) since it forms the most stable carbonium ion of all these isomers.

*(c) No; can't remove halogen from aromatic ring since this would destroy the resonance stabilization in the molecule, requiring a great deal of energy.

9. $CH_2=CHCl < CH_3Cl < CH_3CH_2Cl < [(CH_3)_3-C-CH_2]_2-\underset{\underset{}{|}}{\overset{\overset{CH_3}{|}}{C}}-Cl < Cl-CH_2-CH=CH_2$

10. $CH_3-\underset{\underset{CH_3}{|}}{\overset{\overset{CH_3}{|}}{C}}-Cl < CH_3-CH_2-\overset{\overset{Cl}{|}}{CH}-CH_3 < \triangle\!-Cl < CH_3-\underset{\underset{CH_3}{|}}{\overset{\overset{CH_3}{|}}{C}}-CH_2Cl <$

$CH_3-CH_2-CH_2Cl < CH_3Cl$

*11. (a) $NaOH + CH_3CH_2I \longrightarrow CH_3-CH_2OH + NaI$
$OH^\ominus$ is a better nucleophile than $F^\ominus$.

(b) $KI + CH_3CH_2Br \longrightarrow CH_3-CH_2I + KBr$
$Br^\ominus$ is a better leaving group than $Cl^\ominus$.

(c) $NaOCH_2CH_3 + (CH_3)_3CBr \longrightarrow CH_3-\overset{\overset{CH_3}{|}}{C}=CH_2 + NaBr + CH_3CH_2OH$
Elimination occurs readily when tertiary halides are in the presence of a strong base.

(d) $CH_2=CH-CH_2Cl + AgNO_3 \longrightarrow CH_2=CH-CH_2NO_3 + AgCl\downarrow$
Allyl halides are much more reactive than vinyl halides.

13. This is due to LeChatelier's principle. The excess $Br^\ominus$ ion from the LiBr reverses the formation of the carbonium ion in the first step of the reaction. Thus, the rate of the reaction is slowed down.

$$(C_6H_5)_2-CH-Br \rightleftharpoons (C_6H_5)_2-CH^\oplus + Br^\ominus$$

Chapter 14

1. (b) $CH_3-CH_2-\overset{\overset{O}{||}}{C}-CH_2-CH_2-CH_3$ (c) [p-nitrobenzaldehyde structure: benzene ring with H-C=O at top and NO₂ at bottom] (h) [benzophenone oxime: (C₆H₅)₂C=N-OH]

(k) $CH_3-(CH_2)_5-\underset{\underset{O-CH_2-CH_3}{|}}{\overset{\overset{O-CH_2-CH_3}{|}}{C}}-H$

2. (a) *m*-nitrobenzaldehyde (b) 2,4-dimethylpentanal (f) acetaldehyde,dimethyl acetal (1,1-dimethoxyethane) (g) acetophenoneoxime (l) 2-heptanone

CHAPTER 14

3. (a) $CH_3-CH_2-\underset{CN}{\underset{|}{C}}(OH)-H$

(e) $C_6H_5-CH=N-OH$

(f) $H-\underset{\underset{}{}}{\overset{O}{\overset{\|}{C}}}-O^{\ominus}Na^{\oplus} + CH_3OH$

(i) $C_6H_5-CH_2-C_6H_5$

(n) $(C_6H_5)_2C=CH-\overset{O}{\overset{\|}{C}}-H$

(o) catechol (1,2-dihydroxybenzene) $+ CH_3-\overset{O}{\overset{\|}{C}}-H$

(s) $CH_3-\underset{CH_3}{\underset{|}{C}}(OH)-CH_3$

4. (b) $CH_3-CH=CH_2 \xrightarrow[2.\ H_2O_2,\ OH^{\ominus}]{1.\ B_2H_6} CH_3-CH_2-CH_2OH \xrightarrow[\Delta]{Cu} CH_3-CH_2-\overset{O}{\overset{\|}{C}}-H \xrightarrow{OH^{\ominus}}$

$H_3C-\underset{H}{\underset{|}{\overset{H}{\overset{|}{C}}}}-\underset{H}{\underset{|}{\overset{OH}{\overset{|}{C}}}}-\underset{H}{\underset{|}{\overset{CH_3}{\overset{|}{C}}}}-\overset{O}{\overset{\|}{C}}-H$

(d) $C_6H_5-\overset{O}{\overset{\|}{C}}-H \xrightarrow{HCN} C_6H_5-\underset{CN}{\underset{|}{C}}(OH)-H \xrightarrow[H^{\oplus}]{H_2O} C_6H_5-\underset{H}{\underset{|}{C}}(OH)-\overset{O}{\overset{\|}{C}}-OH$

*(e) $CaC_2 + H_2O \longrightarrow H-C\equiv C-H \xrightarrow[Pt]{H_2} CH_2=CH_2 \xrightarrow{HBr} CH_3-CH_2Br$

$H-C\equiv C-H + CH_3-CH_2Br + Na \longrightarrow H-C\equiv C-CH_2-CH_3 \xrightarrow[Pt]{H_2}$

$HO-CH_2-CH_2-CH_2-CH_3 \xleftarrow[2.\ H_2O_2,\ OH^{\ominus}]{1.\ B_2H_6} H_2C=CH-CH_2-CH_3$

$\downarrow Cu, \Delta$

$H-\overset{O}{\overset{\|}{C}}-CH_2-CH_2-CH_3 \xrightarrow[\Delta]{OH^{\ominus}} CH_3-CH_2-CH_2-\underset{H}{\underset{|}{C}}=\underset{}{\overset{CH_2-CH_3}{\overset{|}{C}}}-C=O \xrightarrow{LiAlH_4}$

$CH_3-CH_2-CH_2-CH_2-\underset{}{\overset{CH_2-CH_3}{\overset{|}{CH}}}-CH_2OH \xleftarrow[Pt]{H_2} CH_3-CH_2-CH_2-CH=\overset{CH_2-CH_3}{\overset{|}{C}}-CH_2OH$

(k) $CaC_2 + H_2O \longrightarrow H-C\equiv C-H \xrightarrow[Pt]{H_2} CH_2=CH_2 \xrightarrow[H^{\oplus}]{H_2O} CH_3-CH_2OH$

$\downarrow H_2O,\ HgSO_4,\ H_2SO_4$

$CH_3-\overset{O}{\overset{\|}{C}}-H$

$CH_3-\overset{O}{\overset{\|}{C}}-H + 2CH_3CH_2OH \xrightarrow{H^{\oplus}} CH_3-\underset{OCH_2-CH_3}{\underset{|}{\overset{H}{\overset{|}{C}}}}-OCH_2-CH_3$

(l)

$$\underset{\text{Br}}{C_6H_5} \xrightarrow[\text{anhy. ether}]{Mg} \underset{\text{MgBr}}{C_6H_5} \xrightarrow[H^\oplus]{H_2O} C_6H_6 \xrightarrow[Pt]{2H_2} \text{cyclohexene}$$

$$\text{cyclohexanone oxime} \xleftarrow{NH_2OH} \text{cyclohexanone} \xrightarrow[H_2SO_4]{K_2Cr_2O_7} \text{cyclohexanol} \xrightarrow[H^\oplus]{H_2O}$$

5. (a) $CH_3-CH_2OH \xrightarrow{PBr_3} CH_3-CH_2Br \xrightarrow[\text{anhy. ether}]{Mg} CH_3-CH_2-MgBr$

$$CH_3-CH_2-CH_2OH \xrightarrow[\Delta]{Cu} CH_3-CH_2-\overset{O}{\underset{\|}{C}}-H$$

$$CH_3-CH_2-\overset{O}{\underset{\|}{C}}-H + CH_3CH_2MgBr \xrightarrow[H^\oplus]{H_2O} CH_3-CH_2-\underset{H}{\overset{OH}{\underset{|}{C}}}-CH_2CH_3$$

(c) $CH_3-CH_2-CH_2OH \xrightarrow{PBr_3} CH_3-CH_2-CH_2Br \xrightarrow[\text{anhy. ether}]{Mg} CH_3-CH_2-CH_2MgBr$

$$CH_3-CH_2-\overset{OH}{\underset{|}{CH}}-CH_3 \xrightarrow[H_2SO_4]{K_2Cr_2O_7} CH_3-CH_2-\overset{O}{\underset{\|}{C}}-CH_3$$

$$CH_3-CH_2-\overset{O}{\underset{\|}{C}}-CH_3 + CH_3-CH_2-CH_2MgBr \xrightarrow[H^\oplus]{H_2O} CH_3-CH_2-\underset{CH_3}{\overset{OH}{\underset{|}{C}}}-CH_2-CH_2-CH_3$$

6. (a) Fehlings or Tollens' test or iodoform test **(b)** Iodoform test **(c)** Lucas test, $K_2Cr_2O_7/H_2SO_4$ **(d)** Tollens' test, Fehlings test, $KMnO_4/H_2SO_4$ **(e)** iodoform test

8. In the compound trifluoroacetaldehyde, the influence of the three fluorine atoms which exert an electron-withdrawing inductive effect make the carbonyl carbon atom more positive than in acetaldehyde, and more susceptible to attack by nucleophilic reagents.

9. The aldehyde group might be oxidized to a carboxylic acid. To prepare 2,3–dihydroxybutanal from crotonaldehyde, first protect the aldehyde group by converting it to an acetal, then react the C=C with cold $KMnO_4$ to form the glycol, finally hydrolyzing the acetal back to the aldehyde.

11. The methylene ($-CH_2-$) hydrogens between the two carbonyl groups are somewhat acidic because of the electron-withdrawing effect of the neighboring carbonyl groups. Acetone has only one carbonyl group to exert an inductive effect on the two CH_3 groups and also, the enolate anion of acetylacetone formed by removal of a proton is resonance stabilized. Thus, acetylacetone is more acidic than acetone.

12. $CH_3-CH_2-\overset{O}{\underset{\|}{C}}-H + OH^\ominus \rightleftharpoons H_2O + CH_3-\underset{\ominus}{CH}-\overset{O}{\underset{\|}{C}}-H$

$$C_6H_5-\overset{O}{\underset{\|}{C}}-H + CH_3-\underset{\ominus}{CH}-\overset{O}{\underset{\|}{C}}-H \rightleftharpoons C_6H_5-\underset{H}{\overset{O^\ominus}{\underset{|}{C}}}-\underset{CH_3}{\overset{}{\underset{|}{CH}}}-\overset{O}{\underset{\|}{C}}-H$$

$$\underset{\underset{H}{|}\ \underset{CH_3}{|}}{C_6H_5-\overset{O^{\ominus}}{\underset{|}{C}}-CH-\overset{O}{\underset{|}{C}}-H} + H_2O \rightleftharpoons OH^{\ominus} + \underset{\underset{H}{|}\ \underset{CH_3}{|}}{C_6H_5-\overset{OH}{\underset{|}{C}}-CH-\overset{O}{\underset{|}{C}}-H}$$

$$\underset{\underset{H}{|}\ \underset{CH_3}{|}}{C_6H_5-\overset{OH}{\underset{|}{C}}-CH-\overset{O}{\underset{|}{C}}-H} \xrightarrow{\Delta} H_2O + C_6H_5-CH=\underset{\underset{CH_3}{|}}{C}-\overset{O}{\underset{|}{C}}-H$$

14. **A** $CH_3-\underset{\underset{Br}{|}}{CH}-CH_2-\underset{\underset{CH_3}{|}}{\overset{CH_3}{\overset{|}{CH}}}$ **B** $CH_3-\underset{\underset{OH}{|}}{CH}-CH_2-\underset{\underset{H}{|}}{\overset{CH_3}{\overset{|}{C}}}-CH_3$ **C** $CH_3-\overset{O}{\underset{|}{C}}-CH_2-\underset{\underset{H}{|}}{\overset{CH_3}{\overset{|}{C}}}-CH_3$

*15. A $C_6H_5-\underset{\underset{Cl}{|}}{CH}-CH_3$ **B** $C_6H_5-\underset{\underset{OH}{|}}{CH}-CH_3$ **C** $C_6H_5-\overset{O}{\underset{|}{C}}-CH_3$

*16. A $CH_3-CH_2-\underset{\underset{Br}{|}}{CH}-CH_2-CH_3$ **B** $CH_3-CH_2-\underset{\underset{OH}{|}}{CH}-CH_2-CH_3$

C $CH_3-CH_2-\overset{O}{\underset{|}{C}}-CH_2-CH_3$

*17. A $CH_3-CH_2-\overset{O}{\underset{|}{C}}-\underset{\underset{CH_3}{|}}{CH}-CH_3$ **B** $CH_3-CH_2-\underset{\underset{OH}{|}}{\overset{H\ \ CH_3}{\overset{|\ \ \ |}{C}-CH}}-CH_3$

C $CH_3-CH_2-CH=\underset{\underset{CH_3}{|}}{C}-CH_3$ **D** $CH_3-\overset{O}{\underset{|}{C}}-CH_3$ **E** $CH_3-CH_2-\overset{O}{\underset{|}{C}}-H$

*18. A $C_6H_5-CH=\underset{\underset{CH_3}{|}}{C}-CH_3$ **B** cyclohexyl-$\underset{\underset{CH_3}{|}}{\overset{H}{\overset{|}{CH_2-C}}}-CH_3$

C $C_6H_5-\overset{O}{\underset{|}{C}}-H$ **D** $CH_3-\overset{O}{\underset{|}{C}}-CH_3$ **E** $C_6H_5-\overset{O}{\underset{|}{C}}-OH$

Chapter 15

1. (b) $CH_3-CH_2-CH_2-\underset{\underset{Br}{|}}{CH}-\underset{\underset{Br}{|}}{CH}-\overset{O}{\underset{|}{C}}-OH$ **(c)** 4-NO_2-C_6H_4-$\overset{O}{\underset{|}{C}}-NH_2$

(i) CH$_3$—CH$_2$—CH$_2$—CH$_2$—O—C(=O)—CH(CH$_3$)—H

(l) 3,5-dinitrobenzoyl chloride (C$_6$H$_3$(NO$_2$)$_2$—C(=O)—Cl)

2. (a) 2,2-difluoropropanoic acid (e) isobutyryl chloride (f) N-phenylpropionamide

3. (a) CH$_3$—CH$_2$—CH(CH$_3$)—C(=O)—OH   (c) CH$_3$—CH$_2$—C(=O)—OH

(d) CH$_3$—CH$_2$—O—C(=O)—CH(CH$_3$)—CH$_2$—CH$_3$ (f) naphthalene-2-carboxylic acid   (h) benzoic acid

(j) CH$_3$—CH(Cl)—C(=O)—OH   (o) CH$_3$—CH(CN)—C(=O)—OH   (p) acetophenone (C$_6$H$_5$—C(=O)—CH$_3$)

4. (a) CaC$_2$ + H$_2$O ⟶ H—C≡C—H $\xrightarrow{H_2/Pt}$ CH$_2$=CH$_2$ $\xrightarrow{H_2O, H^\oplus}$ CH$_3$—CH$_2$OH $\xrightarrow{K_2Cr_2O_7, H_2SO_4}$ CH$_3$—COOH

CH$_3$—CH$_2$OH + CH$_3$COOH $\xrightarrow{H^\oplus}$ CH$_3$—CH$_2$—O—C(=O)—CH$_3$

(c) CaC$_2$ + H$_2$O ⟶ H—C≡C—H $\xrightarrow{H_2/Pt}$ CH$_2$=CH$_2$ $\xrightarrow{H_2O, H^\oplus}$ CH$_3$—CH$_2$OH $\xrightarrow{K_2Cr_2O_7, H_2SO_4}$

CH$_3$—COOH $\xrightarrow{P/Cl_2}$ Cl—CH$_2$COOH $\xrightarrow{CN^\ominus}$ CN—CH$_2$—COO$^\ominus$ $\xrightarrow{H_2O, H^\oplus}$ HOOC—CH$_2$—COOH

(e) CaC$_2$ + H$_2$O ⟶ H—C≡C—H $\xrightarrow{H_2/Pt}$ CH$_2$=CH$_2$ $\xrightarrow{HBr}$ CH$_3$—CH$_2$Br

H—C≡C—H + CH$_3$—CH$_2$Br $\xrightarrow{Na}$ CH$_3$—CH$_2$—C≡C—H $\xrightarrow{H_2/Pt}$

CH$_3$—CH$_2$—CH$_2$—CH$_2$OH $\xleftarrow[\text{2. H}_2\text{O}_2, \text{OH}^\ominus]{\text{1. B}_2\text{H}_6}$ CH$_3$—CH$_2$—CH=CH$_2$

$\downarrow$ K$_2$Cr$_2$O$_7$, H$_2$SO$_4$

CH$_3$—CH$_2$—CH$_2$—COOH $\xrightarrow{P/Cl_2}$ CH$_3$—CH$_2$—CH(Cl)—COOH $\xrightarrow[\text{then H}^\oplus]{\text{aq. NaOH}}$

CH$_3$—CH$_2$—CH(OH)—COOH

CHAPTER 16   445

(f) $CaC_2 + H_2O \longrightarrow H-C\equiv C-H \xrightarrow{H_2}{Pt} CH_2=CH_2 \xrightarrow{HBr} CH_3-CH_2Br \xrightarrow{CN^-}$

$CH_3CH_2CN \xrightarrow[H^+]{H_2O} CH_3-CH_2-COOH \xrightarrow{PCl_3} CH_3-CH_2-\overset{O}{\underset{\|}{C}}-Cl$

$C_6H_6 + CH_3-CH_2-\overset{O}{\underset{\|}{C}}-Cl \xrightarrow{AlCl_3} C_6H_5-\overset{O}{\underset{\|}{C}}-CH_2-CH_3 \xrightarrow[HCl]{Zn(Hg)} C_6H_5-CH_2-CH_2-CH_3$

(l) (i) $CO + H_2 \xrightarrow[cat.]{\Delta,\ press.} CH_3OH \xrightarrow{PCl_3} CH_3Cl$

(ii) benzene $\xrightarrow[AlCl_3]{CH_3Cl}$ toluene $\xrightarrow[Cl_2]{Fe}$ p-chlorotoluene $\xrightarrow[H_2SO_4]{K_2Cr_2O_7}$ p-chlorobenzoic acid

p-chlorobenzamide (CONH$_2$) $\xleftarrow{NH_3}$ p-chlorobenzoyl chloride (COCl) $\xleftarrow{PCl_3}$ p-chlorobenzoic acid (COOH)

5. (a) α–chlorobutyric acid

7. $CH_3-\overset{O}{\underset{\|}{C}}-NH_2 < CH_3-\overset{O}{\underset{\|}{C}}-OCH_2-CH_3 < CH_3-CH_2Br < CH_3-\overset{O}{\underset{\|}{C}}-F < CH_3-\overset{O}{\underset{\|}{C}}-Br$

Order is based on better leaving groups (anions of strong acids) and the influence of the carbonyl group which makes the acyl halides more reactive towards nucleophilic substitution than alkyl halides.

11. $C_6H_5-\overset{O}{\underset{\|}{C}}-O-CH_2-CH_3$

*12. A $CH_3-\overset{O}{\underset{\|}{C}}-CH_2-CH_2-\overset{O}{\underset{\|}{C}}-OH$ B $CH_3-\overset{O}{\underset{\|}{C}}-CH_2-CH_2-\overset{O}{\underset{\|}{C}}-Cl$

C $CH_3-\overset{O}{\underset{\|}{C}}-CH_2-CH_2-\overset{O}{\underset{\|}{C}}-H$

Chapter 16

1. (a) $CH_3-\underset{\underset{CH_3}{|}}{N}-CH_2-CH_3$ (c) $C_6H_5-\underset{\underset{CH_2-CH_3}{|}}{\overset{\overset{CH_3}{|}}{N}}$ (e) $CH_3-\underset{\underset{CH_2-CH_3}{|}}{\overset{\overset{CH_3}{|}}{\overset{\oplus}{N}}}-C_6H_5 + I^{\ominus}$

(g) 2-methylaniline (o-toluidine: benzene ring with CH₃ and NH₂ ortho)

3. (c) $CH_3-(CH_2)_6-NH_2$ (e) 4-methylaniline (p-toluidine)   (g) $CH_3-CH_2-NH_2$

(k) 4-methylbenzenediazonium chloride (p-tolyl-N₂⁺ Cl⁻)   (l) 4-methylbenzonitrile (p-tolyl-CN)   (r) $C_6H_5-SO_2-N(H)-C_6H_4-CH_3$ (m-tolyl)   *(s) 2-aminobenzoic acid (anthranilic acid)

(u) $CH_3-N(NO)-CH_2-CH_2-CH_3$

4. (a) $CH_3-CH_2-CH_2-Cl \xrightarrow{CN^\ominus} CH_3-CH_2-CH_2-CN \xrightarrow{LiAlH_4} CH_3-CH_2-CH_2-CH_2-NH_2$

(b) $CH_3-CH_2-CHBr-CH_3 \xrightarrow{aq.\ NaOH} CH_3-CH_2-CH(OH)-CH_3 \xrightarrow{K_2Cr_2O_7 / H_2SO_4} CH_3-CH_2-CO-CH_3 \xrightarrow{NH_3,\ H_2/Ni} CH_3-CH_2-CH(NH_2)-CH_3$

(c) $CH_3-CH_2-CH_2-CH_2-CH_2Br \xrightarrow{aq.\ NaOH} CH_3-CH_2-CH_2-CH_2-CH_2OH \xrightarrow{K_2Cr_2O_7,\ H_2SO_4} CH_3-CH_2-CH_2-CH_2-COOH \xrightarrow{PCl_3} CH_3-CH_2-CH_2-CH_2-COCl \xrightarrow{NH_3} CH_3-CH_2-CH_2-CH_2-CONH_2 \xrightarrow{NaOH / Cl_2} CH_3-CH_2-CH_2-CH_2-NH_2$

(d) $C_6H_5-CH_3 \xrightarrow{K_2Cr_2O_7 / H_2SO_4} C_6H_5-COOH \xrightarrow{PCl_3} C_6H_5-COCl \xrightarrow{NH_3} C_6H_5-CONH_2 \xrightarrow{NaOH / Cl_2} C_6H_5-NH_2$

5. (a) propionamide $\leq$ triphenylamine $<$ aniline $<$ n-propylamine $<$ cyclopentylamine $<$ di-n-propylamine **(b)** p-nitroaniline $<$ p-chloroaniline $<$ aniline $<$ p-toluidine $<$ p-methoxyaniline $<$ p-phenylenediamine

8. (a) Hinsberg Test or HNO_2 test **(c)** Hinsberg test

SOLUTIONS TO SELECTED PROBLEMS

[Reaction scheme: PhC(O)-N(⁻)-Br → Br⁻ + PhC(O)-N: —rearr.→ Ph-N=C=O]

[Reaction scheme: Ph-N=C=O + H₂O —OH⁻→ [Ph-NH-C(O)-OH] → Ph-NH₂ + CO₂↑ ; CO₃²⁻ ←OH⁻— CO₂]

*(b) No; Hofmann degradation can only be run on unsubstituted amides—need 2 H atoms on the N. ***(c)** Would get Ph-NH-C(O)-OCH₃ ester formed by the reaction of the isocyanate with CH₃OH. ***(d)** *p*-nitrobenzamide < benzamide < *p*-methoxybenzamide

11. *p*-BrCH₂-C₆H₄-CH₂-N(H)-CH₃

Chapter 17

1. (a) Cl–C*(H)(Br)–CH₃   **(b)** CH₃-CH₂-C(Cl)(H)-CH₃ and CH₃-CH₂-C(H)(Cl)-CH₃ (mirror images)

(c) Equal amounts of enantiomorphs, such as 50 % of each of the isomers in (b).

(d) H–C(OH)(COOH)–C(Cl)(H)–CH₃ and H–C(OH)(COOH)–C(H)(Cl)–CH₃   **(e)** H–C(OH)(COOH)–C(OH)(H)–COOH

(f)
(+)−HA
(−)−HA + (−)−B ⟶ (−)BH⊕ (+)A⊖ / (−)BH⊕ (−)A⊖ —H⊕→ (+) HA + (−)BH⊕ + (−)HA + (−)BH⊕

racemic modification of enantiomorphs | base | diasteroisomeric salts (separated by fractional recrystallization) | resolved enantiomorphs + base as a salt

(g) $HO^\ominus + C_6H_{13}-\overset{Br}{\underset{|}{C}H}-CH_3 \longrightarrow \left[HO\overset{\delta\ominus}{---}\underset{CH_3}{\overset{C_6H_{13}\quad H}{\underset{|}{\overset{|}{C}}}}\overset{\delta\ominus}{---}Br \right] \longrightarrow C_6H_{13}-\underset{OH}{\overset{H}{\underset{|}{\overset{|}{C}}}}-CH_3 + Br^\ominus$

(h) [Newman projection with H, Br on front and Cl, I on back]

2. (a) Has *cis* and *trans* isomers.

[cis-2-butene and trans-2-butene structures labeled *cis* and *trans*]

(b) No geometrical or optical isomers.
(c) Has optical isomers

[Two enantiomer structures with CH₃-CH₂-CH₂-C(Cl)-CH₂-CH₃ shown as mirror images]

(a pair of enantiomorphs)

(f) Has optical isomers

[Two biphenyl structures with F and CO₂H substituents shown as mirror images]

(a pair of enantiomers)

4. (a)

$$\underset{I}{\begin{array}{c} COOH \\ H_3C-C-H \\ Br-C-H \\ COOH \end{array}} \quad \underset{II}{\begin{array}{c} COOH \\ H-C-CH_3 \\ H-C-Br \\ COOH \end{array}} \quad \underset{III}{\begin{array}{c} COOH \\ H-C-CH_3 \\ Br-C-H \\ COOH \end{array}} \quad \underset{IV}{\begin{array}{c} COOH \\ H_3C-C-H \\ H-C-Br \\ COOH \end{array}} \quad \text{all are}$$

optically active. I and II are a pair of enantiomers. III and IV are a second pair of enantiomers. Equal amounts of I and II, or III and IV, can form a racemic mixture. I and III, II and III, II and IV, and I and IV are all pairs of diastereoisomers.

5. (a) *R*
6. (a) *S*; no bond to the asymmetric carbon atom was broken during the reaction, thus the configuration of the product doesn't change from that of the reactant. **(c)** *S*. Bond to asymmetric carbon atom is broken. Get complete inversion of configuration (back side attack by nucleophile) in S_N2 reactions. Product is (*S*)-2-methyl-2-amino-butanoic acid.
***8.** The carbonium ion formed in S_N1 reactions is partially shielded from the front side by the departing group which has not yet completely left this vicinity during the

450   **SOLUTIONS TO SELECTED PROBLEMS**

formation of the carbonium ion. Thus, the attacking nucleophile has a slightly better chance of reacting with the carbonium ion from the back side rather than from the front side of the molecule, where its approach is somewhat blocked. This results in the formation of unequal amounts of the two stereoisomers (*not* 50 % *d* and 50 % *l*) and partial racemization occurs.

*9. (a) $\text{(CH}_3\text{)(H)C=C(H)(CH}_3\text{)} \xrightarrow{\text{Br}_2/\text{CCl}_4} \text{CH}_3\text{-CH(+)-CHBr-CH}_3 \xrightarrow{\text{Br}^{\ominus}} \text{CH}_3\text{-CHBr-CHBr-CH}_3 + \text{CH}_3\text{-CHBr-CHBr-CH}_3$

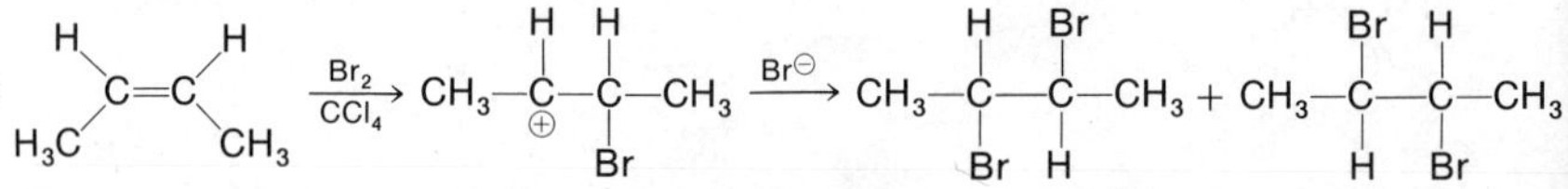

Get equal amounts of enantiomers (optically inactive) (racemic mixture)

(b) $\text{(H}_3\text{C)(H)C=C(H)(CH}_3\text{)} + \text{Br}_2 \xrightarrow{\text{CCl}_4} \text{H-CBr-CH(+)-CH}_3 \xrightarrow{\text{Br}^{\ominus}} \text{CH}_3\text{-CHBr-CHBr-CH}_3$

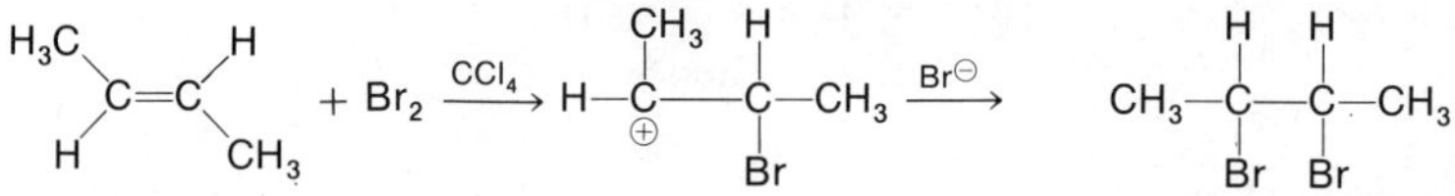

meso form (optically inactive)

These results are obtained since the reactions proceed thru the formation of carbonium ions and by *trans* addition of the bromine.

Chapter 18

1. (a) (D)-(+)-glucose — open chain Fischer projection: CHO, H–C–OH, HO–C–H, H–C–OH, H–C–OH, CH₂OH

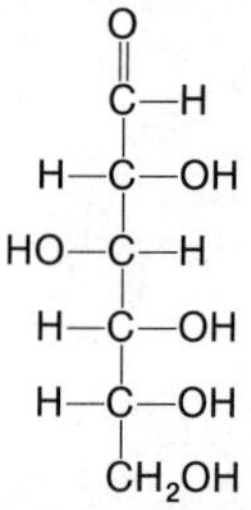

(b) maltose (cyclic structure shown)

(c) Starch or cellulose (see structures in text).

(d) CH₂OH, C=O, HO–C–H, CH₂OH

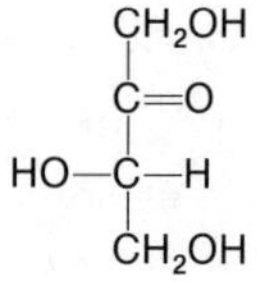

(e) 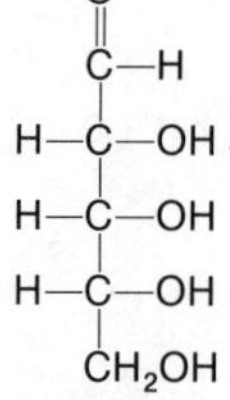 D–ribose: CHO, H–C–OH, H–C–OH, H–C–OH, CH₂OH

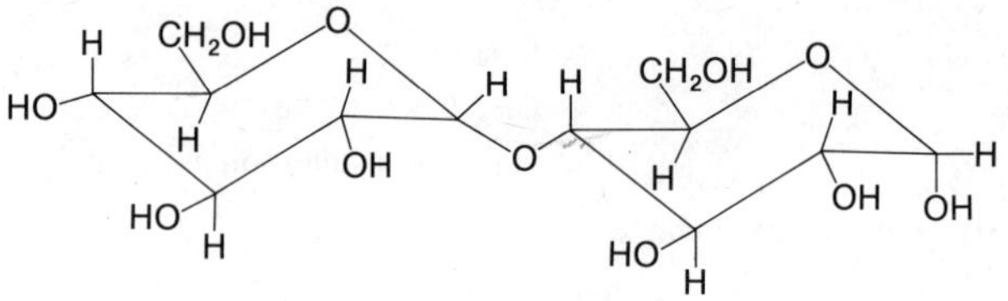

(f) CHO, H–C–OH, H–C–OH, HO–C–H, CH₂OH

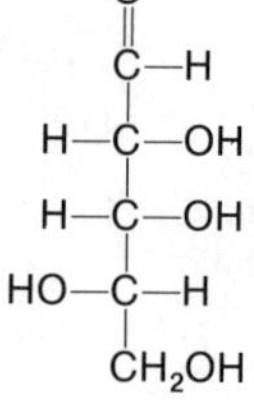

(g) CHO, H–C–OH, HO–C–H, H–C–OH, H–C–OH, CH₂OH

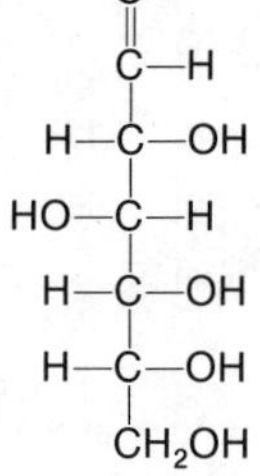

(h) sucrose (cyclic structure shown)

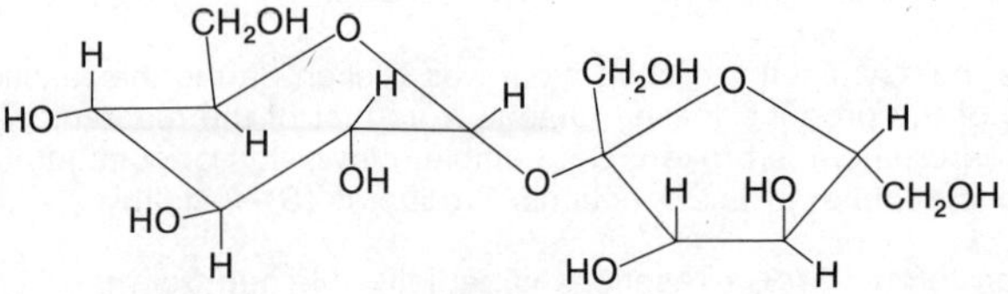

CHAPTER 18

(i)
```
    ┌───────────────┐
    H—C—O—CH₂—CH₃
    H—C—OH
    HO—C—H          O
    H—C—OH
    H—C────────────┘
    CH₂OH
```

(j)
```
       O
       ‖
       C—H
    H—C—OH
    H—C—OH    and
    H—C—OH
    CH₂OH
```
```
       O
       ‖
       C—H
    HO—C—H
    H—C—OH
    H—C—OH
    CH₂OH
```

(k)
```
    H   OH
     \ /
      C ──────┐
    H—C—OH
    HO—C—H    O
    HO—C—H
    H—C──────┘
    CH₂OH
    α-D-galactose
```
and
```
    HO   H
      \ /
       C ──────┐
    H—C—OH
    HO—C—H    O
    HO—C—H
    H—C──────┘
    CH₂OH
    β-D-galactose
```

(l)
```
    CH=N—NH—C₆H₅
    C=N—NH—C₆H₅
    HO—C—H
    H—C—OH
    H—C—OH
    CH₂OH
    D-glucosazone
```

2. (a)
```
        CH₂OH
          \
       ┌───O───┐
      HO       OH
       \  H   /
        \    /
     H   OH H
      \ /  \
       X    X
      / \  / \
     H   OH
```

3. (a)
```
    ┌───────────────┐
    H—C—OCH₂CH₃
    HO—C—H
    HO—C—H          O
    H—C—OH
    H—C────────────┘
    CH₂OH
```

(c)
```
                    O
                    ‖
              O     C—H
              ‖     |
         CH₃—C—O—C—H
              O    |
              ‖    |
         CH₃—C—O—C—H
                   |     O
                   |     ‖
              H—C—O—C—CH₃
                   |     O
                   |     ‖
              H—C—O—C—CH₃
                   |     O
                   |     ‖
              CH₂—O—C—CH₃
```

(f)
```
    H—C=N—NH—C₆H₅
    C=N—NH—C₆H₅
    HO—C—H
    H—C—OH
    H—C—OH
    CH₂OH
```

*5. The typical aldehyde reactions of glucose are due to a small amount of the open-chain aldehyde form of the sugar, which is used up rapidly as the reaction proceeds. Since most of the glucose molecules exist in the cyclic hemiacetal structure, the concentration of this open-chain structure is too low (less than 1 %) for certain reversible reactions, such as addition of NaHSO₃ to an aldehyde to occur.

8. α-D-glucose is a reducing sugar because the cyclic hemiacetal form is in equilibrium with the open-chain aldehyde form which has a free CHO group at carbon 1, available for reducing ability. Methyl-α-D-glucoside is an acetal which cannot open up to a free chain aldehyde structure. Thus, it has no reducing properties.

*9.

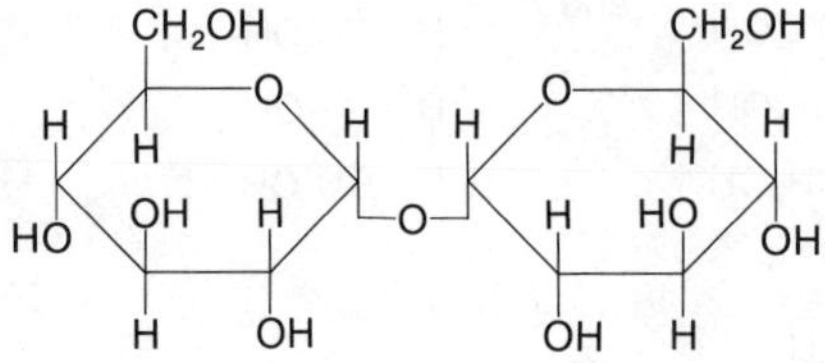

*10.

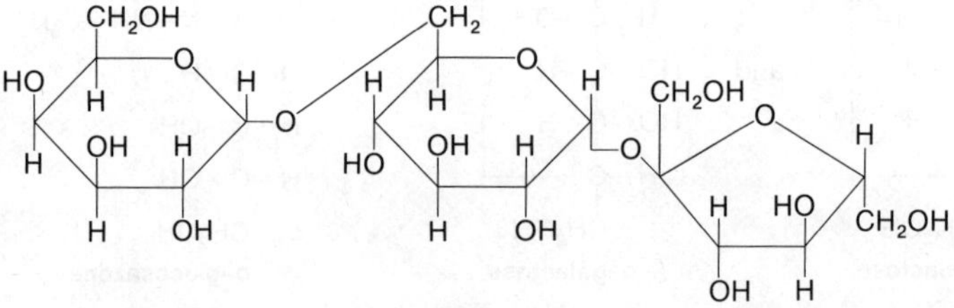

galactose   glucose   fructose

Would *not* reduce Fehling's solution, since no hemiacetal linkage or free CHO group present in molecule.

Chapter 19

1. (b) $CH_3-(CH_2)_{10}-CH_2-O-\overset{O}{\underset{\|}{C}}-(CH_2)_{16}-CH_3$ **(h)** $CH_2-O-\overset{O}{\underset{\|}{C}}-(CH_2)_{14}-CH_3$
                                          wax $CH-O-\overset{O}{\underset{\|}{C}}-(CH_2)_{16}-CH_3$

                                                                                                            $CH_2-O-\overset{O}{\underset{\|}{C}}-(CH_2)_{14}-CH_3$
                                                                                                                                      fat

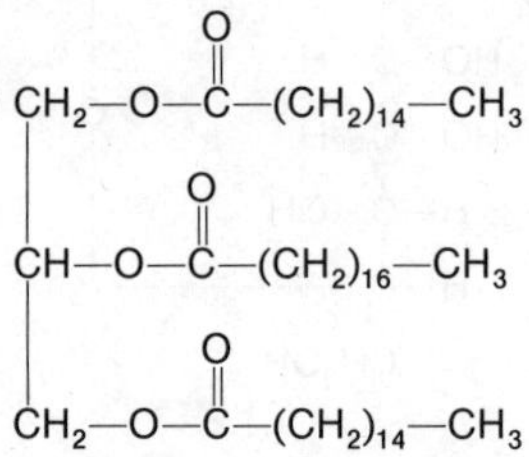

2. (a) glyceryl trilaurate (laurin)   **(c)** α_1-stearo-α,β-dimyristin
4. (a) 188.9
5. (a) 86.2
6. The greater the unsaturation (more double bonds) in the fat, the higher the numerical value of the iodine number.
*8. Six. The value is an average number since most carbon–carbon double bonds react quantitatively with iodine. The reaction is run several times on different samples of the fat, and an *average value* of the iodine number is obtained, from which the average number of double bonds in the fat can be determined.

CHAPTER 20

10. $CH_3-(CH_2)_{14}-\underset{\beta}{CH_2}-\underset{\alpha}{CH_2}-\overset{O}{\underset{\|}{C}}-OH + CoA-SH$

$\downarrow$

$H_2O + CH_3-(CH_2)_{14}-CH_2-CH_2-\overset{O}{\underset{\|}{C}}-S-CoA$
stearic acid thioester

$\xrightarrow{-H_2}$

$CH_3-(CH_2)_{14}-CH=CH-\overset{O}{\underset{\|}{C}}-S-CoA \xrightarrow{+H_2O} CH_3-(CH_2)_{14}-\underset{\underset{OH}{|}}{CH}-CH_2-\overset{O}{\underset{\|}{C}}-S-CoA$

$\downarrow -H_2$

$CH_3-\overset{O}{\underset{\|}{C}}-S-CoA \xleftarrow{CoA-SH} CH_3-(CH_2)_{14}\underset{\beta}{\overset{O}{\underset{\|}{C}}}-\underset{\alpha}{CH_2}-\overset{O}{\underset{\|}{C}}-S-CoA$
(acetyl CoA)   **β-keto ester**
$+$

$CH_3(CH_2)_{14}-\overset{O}{\underset{\|}{C}}-S-CoA \longrightarrow$ etc.
palmitic acid thioester

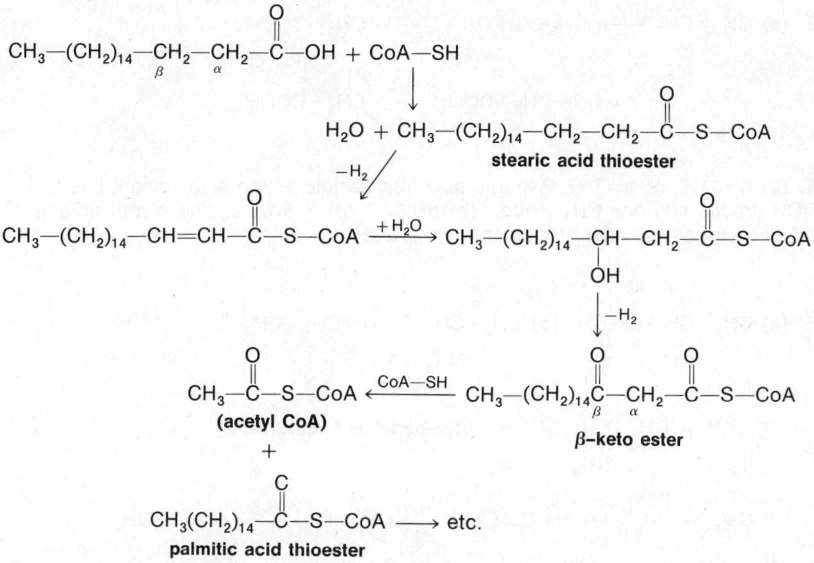

Chapter 20

1. (b) $\underset{OH}{\overset{}{CH_2}}-\underset{NH_2}{\overset{}{CH}}-\overset{O}{\underset{\|}{C}}-\overset{H}{\underset{}{N}}-\underset{COOH}{\overset{}{CH}}-CH_2-CH_2-S-CH_3$

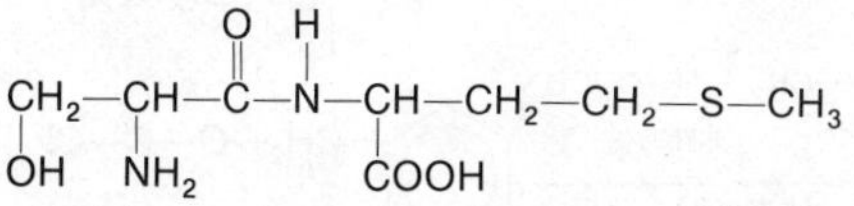

2. $CaC_2 + H_2O \longrightarrow H-C\equiv C-H \xrightarrow{H_2}{Pt} CH_2=CH_2 \xrightarrow{HBr} CH_3CH_2Br$

$\bigcirc + CH_3CH_2Br \xrightarrow{AlCl_3} \bigcirc-CH_2-CH_3 \xrightarrow{Cl_2}{h\nu} \bigcirc-\underset{Cl}{\overset{}{CH}}-CH_3$

$\downarrow$ alc. KOH

$\bigcirc-CH_2-CH_2-OH \xleftarrow[2.\ H_2O_2,\ OH^\ominus]{1.\ B_2H_6} \bigcirc-CH=CH_2$

$\xrightarrow[\Delta]{Cu}$

$\bigcirc-CH_2-\overset{O}{\underset{\|}{C}}-H \xrightarrow[NH_3]{HCN} \bigcirc-CH_2-\underset{CN}{\overset{NH_2}{\underset{|}{C}}}-H \xrightarrow{H_2O}{H^\oplus}$

$\bigcirc-CH_2-\underset{NH_2}{\overset{}{CH}}-COOH$

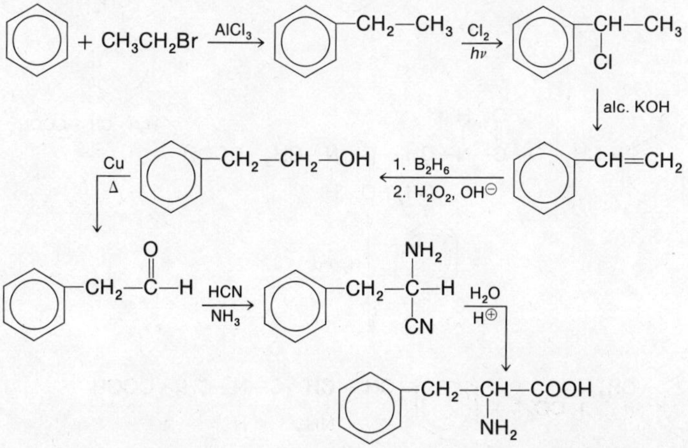

3. (a) 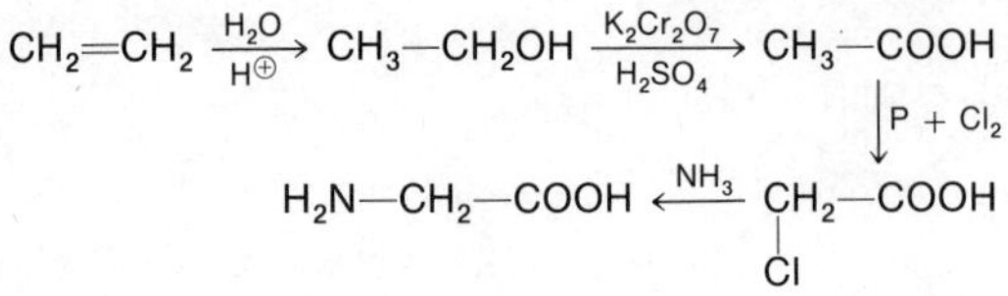 $CH_2=CH_2 \xrightarrow[H^{\oplus}]{H_2O} CH_3-CH_2OH \xrightarrow[H_2SO_4]{K_2Cr_2O_7} CH_3-COOH$

$\downarrow P + Cl_2$

$H_2N-CH_2-COOH \xleftarrow{NH_3} \underset{Cl}{CH_2}-COOH$

4. (a) pH < 7. pH = 2.77. Aspartic acid is an acidic amino acid since it has two COOH groups and one NH_2 group. **(b)** pH > 7. pH = 9.74. Lysine is a basic amino acid since it has two NH_2 groups and one COOH group.

5. (a) 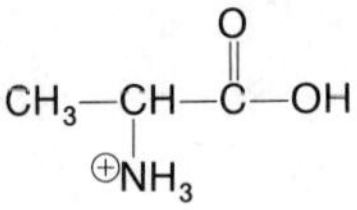 $CH_3-\underset{\overset{\oplus}{N}H_3}{CH}-\overset{O}{\overset{\|}{C}}-OH$ **(c)** $CH_3-\underset{NH_2}{CH}-\overset{O}{\overset{\|}{C}}-O-CH_2-CH_3$

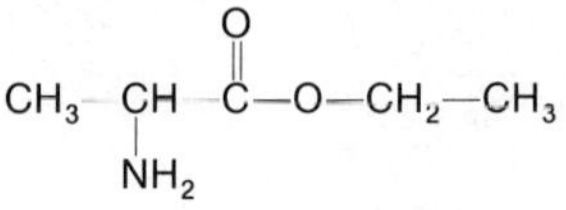

6. 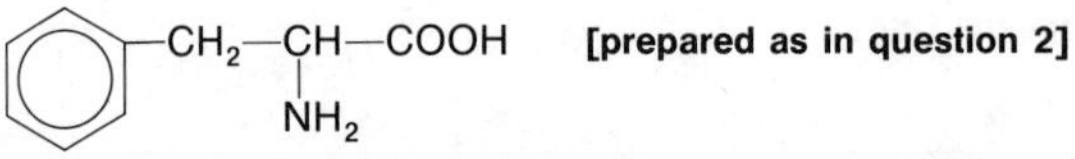 ⌬$-CH_2-\underset{NH_2}{CH}-COOH$ [prepared as in question 2]

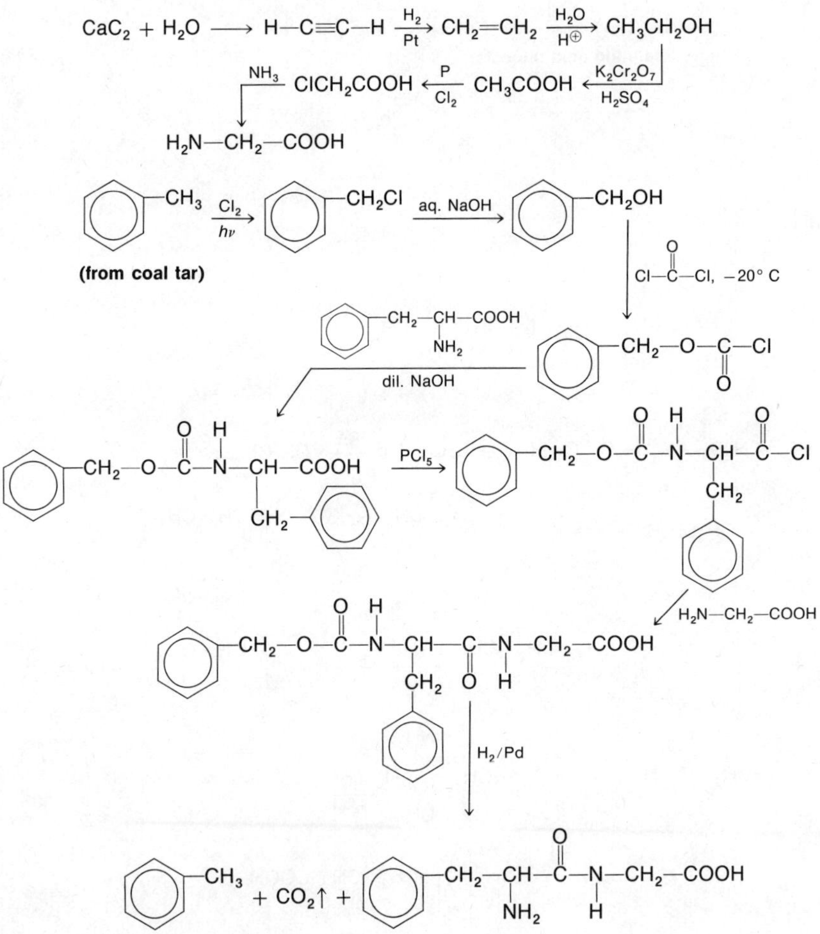

$CaC_2 + H_2O \longrightarrow H-C\equiv C-H \xrightarrow{H_2}{Pt} CH_2=CH_2 \xrightarrow[H^{\oplus}]{H_2O} CH_3CH_2OH$

$\xleftarrow{NH_3} ClCH_2COOH \xleftarrow[Cl_2]{P} CH_3COOH \xleftarrow[H_2SO_4]{K_2Cr_2O_7}$

H_2N-CH_2-COOH

⌬$-CH_3 \xrightarrow[h\nu]{Cl_2}$ ⌬$-CH_2Cl \xrightarrow{aq.\ NaOH}$ ⌬$-CH_2OH$

(from coal tar)

$\downarrow Cl-\overset{O}{\overset{\|}{C}}-Cl, -20°C$

⌬$-CH_2-\underset{NH_2}{CH}-COOH$ ⌬$-CH_2-O-\overset{O}{\overset{\|}{C}}-Cl$

dil. NaOH

⌬$-CH_2-O-\overset{O}{\overset{\|}{C}}-\overset{H}{N}-\underset{CH_2-⌬}{CH}-COOH \xrightarrow{PCl_5}$ ⌬$-CH_2-O-\overset{O}{\overset{\|}{C}}-\overset{H}{N}-\underset{CH_2-⌬}{CH}-\overset{O}{\overset{\|}{C}}-Cl$

$\swarrow H_2N-CH_2-COOH$

⌬$-CH_2-O-\overset{O}{\overset{\|}{C}}-\overset{H}{N}-\underset{CH_2-⌬}{CH}-\overset{O}{\overset{\|}{C}}-\overset{H}{N}-CH_2-COOH$

$\downarrow H_2/Pd$

⌬$-CH_3 + CO_2\uparrow +$ ⌬$-CH_2-\underset{NH_2}{CH}-\overset{O}{\overset{\|}{C}}-\overset{}{N}-CH_2-COOH$

*7. [Structure of phenylalanylvalylaspartylmethionylhistidine peptide]

phenylalanylvalylaspartylmethionylhistidine

9. oxaloacetic acid + glutamic acid ⇌ (enzyme) α-aminosuccinic acid + α-ketoglutaric acid

Chapter 21

1. (a) [Structure: guanosine — HOH₂C-ribose with guanine base, OH OH on ribose]

   (b) [Structure: uridine — HOH₂C-ribose with uracil base, OH OH on ribose]

   (d) [Structure: thymidine — HOH₂C-deoxyribose with thymine base, OH OH]

2. (a) [Structure: AMP — phosphate-O-CH₂-ribose with adenine base, HO OH on ribose]

3. [structure of trinucleotide with 5-methylcytosine and thymine bases, etc.]

5. xanthine, hypoxanthine [structures]

Chapter 22

3. (a) [ketone + NH₂OH → oxime]

(b) HCN → [cyanohydrin structure] or [alternative structure] **(d)** Zn(Hg)/HCl → [reduced structure]

*5. (a) $CH_3-\underset{|}{\overset{OH}{C}}-CH_2-\overset{*}{C}-OH$ with $-CH_2-$ and $-CH_2-OH$ branches (labels 14 on two carbons)

(b) $CH_2=\underset{|}{\overset{CH_3}{C}}-CH_2-CH_2-O-\underset{|}{\overset{O}{\overset{\|}{P}}}-O-\underset{|}{\overset{O}{\overset{\|}{P}}}-OH + \overset{*}{C}O_2\uparrow$ (label 14)

(c) $CH_3-\underset{14}{C}(CH_3)=CH-CH_2-CH_2-\underset{14}{C}(CH_3)=CH-CH_2OH$ (d) [structure with CH_2OH and three 14 labels]

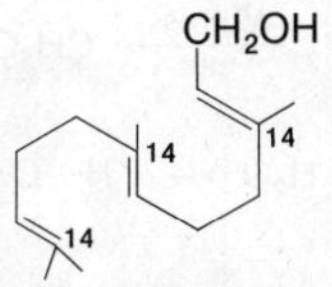

*6. (a) More reactive than benzene towards electrophilic substitution. The presence of a hetero atom such as N, O, S in the ring with an electron pair to donate into the π cloud of the ring activates the ring, makes it aromatic in character, and very reactive towards electrophilic reagents. **(b)** The 2 position is most susceptible to electrophilic attack as can be seen from resonance structures.

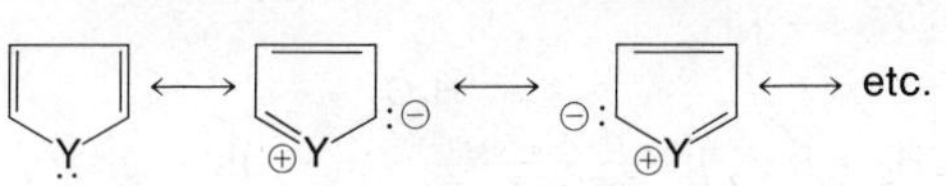

[resonance structures of heterocycle Y] ⟷ [⊕Y with :⊖] ⟷ [⊖: with ⊕Y] ⟷ etc.

*8. (a) [benzene] $\xrightarrow{H_2SO_4, \Delta}$ [C$_6$H$_5$SO$_3$H] $\xrightarrow{NaOH, \Delta}$ [C$_6$H$_5$ONa$^\oplus$] $\xrightarrow{H^\oplus}$

[2-nitro-phenol with CHO ortho, OH] $\xleftarrow{CHCl_3 / NaOH}$ [o-nitrophenol, O$_2$N, OH] $\xleftarrow{dil.\ HNO_3,\ 20°C}$ [phenol, OH]

$\downarrow Ac_2O\ |\ NaAc,\ \Delta$

[coumarin with NO$_2$ substituent]

$CaC_2 + H_2O \longrightarrow H-C\equiv C-H \xrightarrow{H_2/Pt} CH_2=CH_2 \xrightarrow{H_2O/H^\oplus} CH_3CH_2OH$

$CH_3-\underset{O}{\overset{\|}{C}}-ONa \xleftarrow{NaOH} CH_3COOH \xleftarrow{K_2Cr_2O_7 / H_2SO_4}$
(NaAc)

$\downarrow PCl_3$

$CH_3-\underset{O}{\overset{\|}{C}}-O-\underset{O}{\overset{\|}{C}}-CH_3 \xleftarrow{NaAc} CH_3-\underset{O}{\overset{\|}{C}}-Cl$
(Ac$_2$O)

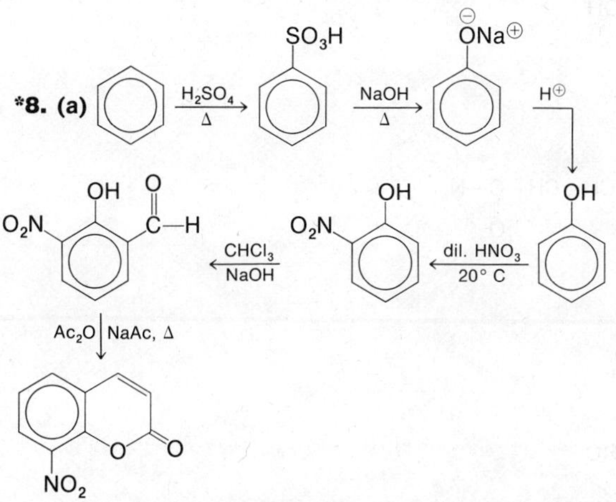

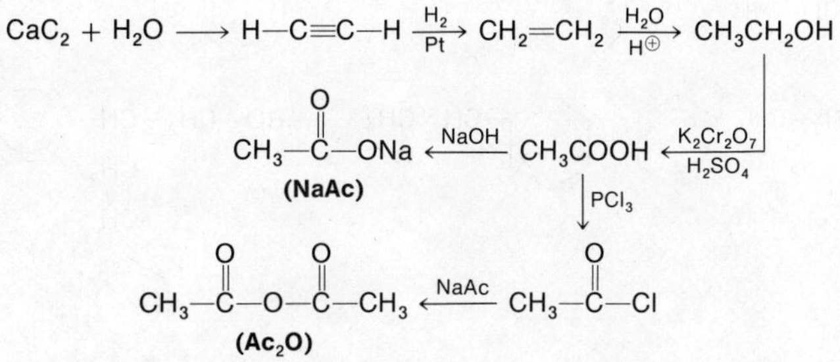

(b) [benzene] $\xrightarrow{CH_3Cl / AlCl_3}$ [toluene] $\xrightarrow{HNO_3 / H_2SO_4}$ [p-nitrotoluene, NO$_2$] $\xrightarrow{Sn / HCl}$ [p-toluidine, NH$_2$]

$\downarrow HNO_3, H_2SO_4$

[nitrobenzene, NO$_2$]

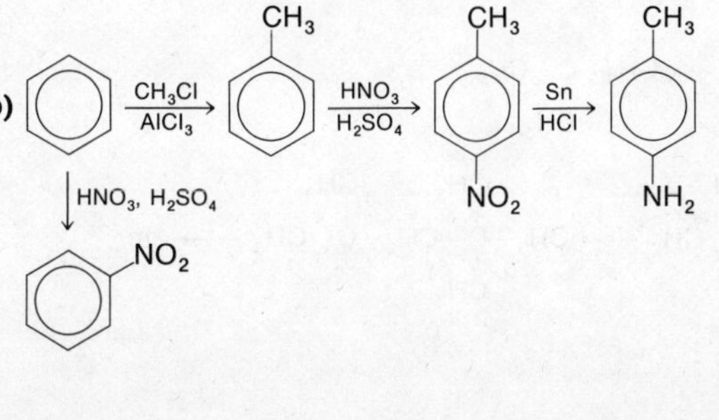

SOLUTIONS TO SELECTED PROBLEMS

$$CO + H_2 \xrightarrow[\text{cat.}]{\Delta,\ \text{press.}} CH_3OH \xrightarrow{PCl_3} CH_3Cl$$

$$CaC_2 + H_2O \longrightarrow H-C\equiv C-H \xrightarrow[CH_3Cl]{Na} CH_3-C\equiv C-H \xrightarrow{H_2/Pt}$$

$$HO-CH_2-CH=CH_2 \xleftarrow{\text{aq. NaOH}} Cl-CH_2-CH=CH_2 \xleftarrow[600°C]{Cl_2} CH_3-CH=CH_2$$

↓ HOCl

$$HO-CH_2-\underset{\underset{Cl}{|}}{\overset{\overset{OH}{|}}{CH}}-CH_2 \xrightarrow{\text{aq. NaOH}} HO-CH_2-\underset{\underset{OH}{|}}{CH}-CH_2-OH$$

[Structural diagram: p-toluidine (CH₃-C₆H₄-NH₂) + CH₂-CH-CH₂ (OH, OH, OH) + nitrobenzene, with H₂SO₄, FeSO₄, Δ → 6-methylquinoline]

*9. [Piperine structure: methylenedioxyphenyl-CH=CH-CH=CH-C(=O)-N-piperidine]

Chapter 23

1. (a) $R-O-O-R \xrightarrow{\Delta\ \text{or}\ h\nu} 2RO\cdot$

$RO\cdot + $ [C₆H₅-CH=CH₂] $\longrightarrow R-O-CH_2-\dot{C}H-$[C₆H₅]

$RO-CH_2-\dot{C}H-$[C₆H₅] $+ $ [C₆H₅-CH=CH₂] $\longrightarrow RO-CH_2-CH(-C_6H_5)-CH_2-\dot{C}H-C_6H_5$ etc.

(d) $CH_3-\underset{\underset{CH_3}{|}}{C}=CH_2 + H^\oplus \longrightarrow CH_3-\underset{\underset{CH_3}{|}}{\overset{\oplus}{C}}-CH_3$

$CH_3-\underset{\underset{CH_3}{|}}{\overset{\oplus}{C}}-CH_3 + CH_3-\underset{\underset{}{|}}{\overset{\overset{CH_3}{|}}{C}}=CH_2 \longrightarrow CH_3-\underset{\underset{CH_3}{|}}{\overset{\overset{CH_3}{|}}{C}}-CH_2-\overset{\oplus}{\underset{\underset{CH_3}{|}}{C}}-CH_3 \longrightarrow$ etc.

3.

$$3 \underset{\text{phthalic anhydride}}{\text{C}_6\text{H}_4(\text{CO})_2\text{O}} + 2\,\text{CH}_2\text{—CH—CH}_2 \xrightarrow{\Delta,\ -\text{H}_2\text{O}}$$
$$\quad\quad\quad\quad\quad\quad\quad \text{OH}\ \ \text{OH}\ \ \text{OH}$$

etc. —O—CH₂—CH(—O—C(=O)—C₆H₄—C(=O)—O—CH₂)—CH₂—O— etc. (polyester with pendant ester branches to phthalate groups)

*4. There is most likely some probability that the polymers be formed with regularly alternating groups in the mechanism because of the geometry of intermediate formed during the reaction. Regularly alternating groups would also tend to reduce the steric interactions between the groups in the polymer if they all were on the same side of the extended polymer chain.

Chapter 24

2. τ_{max} would be at a longer wavelength than τ_{max} for $CH_2=CH_2$ or $CH_2=CH—CH=CH_2$. The actual value for the τ_{max} of 1,3,5-hexatriene is 258 mμ.

3. No. They are both ketones and have very similar molecular structures. There would be some differences in the fingerprint region of their IR spectra, which would be difficult for a novice spectroscopist to interpret. NMR spectroscopy would best differentiate between the two compounds. 2-Pentanone would have four NMR absorption signals, a singlet, triplet, sextet, and another triplet, with the area under the respective peaks being in the ratio 3:2:2:3. The NMR spectrum of 3-pentanone would have two NMR absorption signals, a triplet and a quartet, with relative areas under the peaks in the ratio of 6:4.

5. CH_3F protons. The electronegative F atom draws electrons away from the H atoms, making the H atoms in CH_3F more deshielded than the H atoms in CH_4.

*6. i-Propyl bromide—a septet (7) peaks at $\delta = 4$-4.5 and a doublet at $\delta = 1.7$. Ethyl bromide—a quartet at $\delta = 3.2$-3.7 and a triplet at $\delta = 1.5$-1.9.

8. $CH_3—CH_2—CH(OH)—CH_3$ (2-butanol).

*9. The $m/e = 106$ is the molecular ion peak, which is the molecular weight of ethylbenzene. The $m/e = 91$ peak is due to the benzyl, $C_6H_5—CH_2^{\oplus}$, carbonium ion (very stable species due to resonance). $m/e = 78$ is due to benzene, C_6H_6; and $m/e = 77$ is due to the phenyl ion, $C_6H_5^{\oplus}$.

*11. $CH_3—\overset{O}{\underset{\|}{C}}—CH_3$, acetone.

*12. $C_6H_5—CH=CH—\overset{O}{\underset{\|}{C}}—H$, cinnamaldehyde.

*13. $CH_3—CH_2—CH_2—NH_2$, n-propylamine.

Index

Abietic acid, 374
Absolute alcohol, 164
Acetaldehyde, 115, 132, 198, 209, 217, 344
   acetal formation, 211
   aldol condensation, 218–20
   cyanohydrin, 209
   halogenation, 216
   preparation, 132, 200, 202
Acetals, 210–11
   sugar series, 311–14
Acetamide, 244, 265
Acetanilide, 263, 270
Acetate ion,
   biogenesis of terpenes, 378–79
Acetic acid, 33, 43–44, 47, 52–53, 229–30, 242
   dimer, 232
   preparation, 234
   reactions, 202, 237, 344
Acetic anhydride, 221, 239, 242, 270, 309
Acetoacetaldehyde, 216
Acetone, 33, 115, 117, 132, 198, 216
   aldol condensation, 219
   oxime, 212
   preparation, 132, 166, 202, 241
Acetophenone, 198, 246
Acetylacetone, 216
Acetyl chloride, 236, 239, 242, 244
Acetyl group, 230
Acetylcholine, 269–70
Acetylene, 15–17, 131–35, 202
   hydration, 132, 133
Acetylenes, *see* Hydrocarbons, unsaturated
Acetylides, 133–34
Acetylsalicylic acid, 250
Acid anhydrides, 231, 239–40, 242–46
Acid derivatives, 231, 236–46
   summary of reactions, 244
Acid strength, 38–44
   binary acids, 39–40
   carboxylic acids, 231
   ternary acids, 40–44
Acrylic acid, 248
   addition of HBr, 248
Acrylonitrile, 132, 395
Acyl group, 229, 231
Acyl halides, 231, 236–37, 239, 242, 244–46

Acylation
   alcohols, 243–44
   amines, 270, 347
Addition,
   1,4–, 122–23, 248
Addition reactions, 65, 98–114, 131–33
   aldehydes and ketones, 207–14
   alkynes, 131–33
   cis, 109–12
   free-radical addition of HBr to alkenes, 107–08
   free-radical mechanism, 99–100
   ionic, 103–07
   ionic mechanism, 98–99
   nucleophilic, 207–14
   nucleophilic mechanism, 207–08
   trans, 98–99
Adenine, 362–66, 368
Adipic acid, 230, 398
Adrenal cortex hormones, 377–78
Alanine, 341
   preparation, 249, 344
   reaction with nitrous acid, 347
   transamination, 358
Albumin, 355
Alcohols, 161–65, 167–74
   acidity, 168–69
   classification, 162
   dehydration, 118–19
   nomenclature, 162
   oxidation, 172–73, 199–200, 232
   physical properties, 167
   polyhydric, 164–65
   preparation, 107, 110, 163–65, 191–93, 214
   primary, 162, 192–93, 199, 206, 214, 232
   reactions, 168–73, 210–11, 347
   secondary, 162, 200, 206, 214
   sulfur analogs, 173–74
   tertiary, 162, 214
Alcoholysis, 243–44
Aldehydes, 197–222
   addition reactions, 207–14
   halogenation, 216
   nomenclature, 198
   oxidation, 204–06, 232
   preparation, 172, 199–202
   reactions, 204–22
   reduction, 206–07, 264
   tests, 205–06

INDEX

Aldol, 219
Aldol condensation, 218–20
  mechanism, 219
  mixed, 220
Aldopentoses, 307
Aldose, 305
Aliphatic, definition, 138
Alkali fusion of sulfonates, 166
Alkaloids, 298, 386–87
Alkanes, 21, 74–85
  chemical properties, 78–81
  physical properties, 78
  preparation, 81–82
Alkenes, 21–22, 93–121
  chemical properties, 97–117
  nomenclature, 93–94
  physical properties, 96–97
  preparation, 117–20
Alkenes, see Hydrocarbons, unsaturated
Alkoxides, 168–69, 180–81
Alkyl fluorides, 190
Alkyl groups, 29, 75–78
Alkyl halides, 189–92
  displacement reactions, 179–81, 190–91
  hydrolysis, 163
  preparation, 104–06, 169–71, 190
  reduction, 81
Alkyl hydrogen sulfates, 107, 172
Alkyl iodides, 190
Alkyl isocyanate, 266
Alkyl lithium compounds, 82
Alkyl nitrates, 171
Alkylation
  ammonia, amines, 260–61
  aromatic hydrocarbons, 146–49, 155, 166
Alkynes, 21, 130–36
  acidic hydrogen, 133–34
  addition reactions, 131–33
  nomenclature, 130–31
  preparation, 134–35, 191
Alkynes, see Hydrocarbons, unsaturated
Allene, 289
Allyl alcohol, 165
Allyl chloride, 114
Allylic halogenation, 114–15
Allylic rearrangement, 380
Aluminum chloride, 146–49
Amides, 231, 242, 244, 262, 265–66
  preparation, 233, 240, 270
Amination of α-halogenated acids, 249, 343–44
Amines, 22, 48, 256–75, 278
  acylation, 270
  base strength, 259–60
  classification, 257
  dissociation constants, 260
  heterocyclic, 275, 362–63, 385–91
  nomenclature, 257–58
  physical properties, 259

  preparation, 191, 260–68
  primary, 257–58, 262–65, 271
  protection, 273–74
  reactions, 268–75
  salts, 258–59, 268
  secondary, 257–58, 263–64, 272
  tertiary, 257–58, 263, 272
  tests, 271–72
Amino acids, 340–54, 358
  acidic, 341
  amphoterism, 345–46
  analysis, 348–52
  analyzer, 349
  basic, 341
  classification, 341–43
  deamination, 358
  essential, 341
  nomenclature, 341–43
  optical activity, 340
  preparation, 209, 249, 343–45
  properties, 345–46
  reactions, 346–47
  structure, 340–43
Ammonia, 37–38, 48, 186, 211–12, 359
  alkylation, 260–61
  reaction with aryl halides, 261
Ammonium bisulfide, 262
Ammonium cyanate, 21
Ammonolysis, 243–44
AMP (adenine-5′-monophosphate), 365
Amylopectin, 323
Amylose, 323
Angstrom unit, 402
Angular methyl groups, 375
Aniline, 53–54, 258, 261
  alkylation, 261
  bromination, 273
  diazotization, 272, 275
  nitration, 274
  preparation, 262, 267–68
  resonance, 54
  in Skroup synthesis, 389
Anilinium chloride, 259
Anisole, 178, 181, 183
Anomers, 312
Anthracene, 140
Anthranilic acid, 278
Antibiotics, 371, 390–91
Antiknock fuels, 113
Antimalarials, 389
Apollo-15 mission, 3
Apollo-Saturn V, 60
D-Arabinose, 328
Arachidic acid, 331
Arachidonic acid, 331
Arenes, 138–56
Arginine, 343
Aromatic hydrocarbons, see Hydrocarbons, aromatic
Aryl groups, 141

Aryl halides, 144–45, 155, 166, 267–68
Arythallium compounds, 153–55
   arylthallium di-trifluoroacetate, 154–55
Ascorbic acid, 383
Aspartic acid, 343
Aspirin, 250
Asymmetric carbon atom, 284, 288, 290
Asymmetry, 286–88
Atom, structure, 4
   electronic configuration, 4–7
   electron orbital configuration, 6–7
   unpaired electron, 7
ATP (adenosine-5′-triphosphate), 326–27, 337, 365
Atropine, 388
Aufbau principle, 4
Auxochrome, 279
Axial bonds, 88
Azo dyes, 279

Baeyer strain theory, 85–87, 184
Baeyer test for double bonds, 116
Bakelite, 399
Basic strength, 38, 259–60
Bathochrome, 279
Beeswax, 331, 333
Benedict's test, 205, 316
Benzal chloride, 146, 200–01
Benzaldehyde, 198
   phenylhydrazone, 213
   preparation, 200, 245
   reactions, 200–22
Benzamide, 242
Benzene, 32, 49–51, 138–39
   acylation, 187
   alkylation, 146–49, 166
   halogenation, 144–45, 155
   nitration, 143
   preparation, 277
   resonance, 51
   structure, 139
   sulfonation, 144
   thallation, 154
Benzenediazonium chloride, 272, 275–78
Benzenesulfonic acid, 144
Benzenesulfonyl chloride, 144, 271
Benzoic acid, 229–30, 237, 242
   p-amino, 275
   preparation, 149, 233
Benzonitrile, 276
Benzophenone, 198
Benzotrichloride, 146, 233
Benzoyl bromide, 237, 243
Benzoyl chloride, 246, 347
Benzoyl group, 231
Benzyl alcohol, 222, 352
Benzyl chloride, 146, 345
Benzyl chloroformate, 352
Benzyne, 267–68, 277–78
Bergmann synthesis, 352–54

Beri-beri, 390
Berry, R. S., 277
Beryllium chloride, 11
Bijvoet, J. M., 293
Biochemical resolution, 298
Biot, Jean-Baptiste, 286
Biphenylene, 278
Biphenyls, 289
Bisulfite addition compounds, 213–14
Bloch, K., 382
Boat conformation, 87–88
Bond, 7
   axial, 88
   covalent, 7–8
   coordinate covalent, 9, 38
   double, 8, 13–15
   equatorial, 88
   formation, 10
   ionic, 7–9
   multiple, 8
   pi, 13–17, 404
   polar, 9
   sigma, 13–17, 404
   triple, 8, 15–17
Bond cleavage,
   heterolytic, 64, 100
   homolytic, 64, 103
Boron trichloride, 11, 37–38
Boron-trichloride-ammonia addition compound, 37
Bromic acid, 42
Bromination, *see* Halogenation
Bromoacetaldehyde, 216
p-Bromoaniline, 273
Bromobenzene, 139, 276
2-Bromobutane, 290
   Fischer projection formulas, 291
Bromonium ion, 100–03
β-Bromopropionic acid, 248
m-Bromotoluene, 277
Bronsted–Lowry Theory, 36–37, 143
Brown, Herbert C., 109, 206
Brucine, 298
1,3-Butadiene, 94, 111–12, 122, 396, 405
Butanal-3-ol, 33
Butane, 24, 29, 75–76
   conformations, 85
Butanoic acid, *see* Butyric acid
Butanols, *see* Butyl alcohols
Butanone,
   IR spectrum, 409
Butenes, 94, 120
Butter fat, 331
Butyl alcohol
   normal, 170, 409
   secondary, 162
   tertiary, 68, 169–70, 412
i-Butylamine, 263
n-Butyl chloride, 170
t-Butyl chloride, 170

INDEX

i-Butyl group, 76–77
n-Butyl group, 76–77
sec.-Butyl group, 76–77
t-Butyl group, 30, 55, 67, 70, 76–77
n-Butyl mercaptan, 174
2-Butyne, 131
n-Butyraldehyde, 198
n-Butyric acid, 48, 229, 230

Caffeine, 390
Cahn, R. S., 294
Calcium carbide, 135
Calvin, Melvin, 325
Camphor, 373
Cane sugar, *see* Sucrose
Cannizzaro reaction, 221–22
Carbamic acid, 266
Carbanions, 64, 215, 217–21, 246, 344, 396
Carbenes, 111
Carbobenzoxy group, 352
Carbohydrates, 304–27
Carbon atom, 21
  electronic structure, 21
  primary, 77
  quaternary, 77
  secondary, 77
  tertiary, 77
  tetrahedral, 12, 287–88
Carbon–carbon bonds
  double, 49, 94
  single, 49
  triple, 131
Carbon dioxide, 8, 234, 250, 359
Carbonium ions, 64, 68–70, 105–06, 118, 146, 148, 163, 245–46, 248, 396, 417–18
Carbon monoxide, 164
Carbon tetrachloride, 79
Carbonyl group, 197, 202–04, 406
Carboxylic acids, 22, 42, 53, 228–44, 246–51
  amino, *see* Amino acids
  derivatives, 231, 236–46
  electrophilic aromatic substitutions, 246
  halogenated, 247–48
  α-hydroxy, 209, 248–49
  α-keto, 358
  nomenclature, 228–30
  phenolic, 249–50
  physical properties, 232
  preparation, 204, 232–34, 249
  reactions, 234–40, 246–51
  salts, 231, 235–36, 239
Carboxyl group, 22, 228
Carboxylate ion, 51, 231
Carboxypeptidase, 351
β-Carotene, 123, 405–06

Caryophyllene, 374
Castor oil, 331–32
Catalysts, 61
  Ziegler, 114, 397
Catechol, 163
Catnip, 373
Cellobiose, 321
Cellosolve, 185
Cellulose, 324
Cerotic acid, 331
Cetyl palmitate, 334
Chair conformation, 87–89
Chemical kinetics, 60–61
Chevreul, Michel Eugène, 330
Chiral center, 288
Chirality, 288
Chloral, 210
  hydrate, 210
Chloric acid, 42
Chlorination, *see* Halogenation
Chloroacetic acid, 247, 249, 344
Chlorobenzene, 51
  preparation, 145
  reactions, 154
2-Chlorobutanoic acid, 48
3-Chlorobutanoic acid, 48
4-Chlorobutanoic acid, 48
3-Chloro-2-butanol,
  Fischer projection formulas, 295
  stereochemistry, 295
Chlorocyclohexane, 31
3-Chlorocyclohexene, 94
Chlorofluoroiodomethane, 294
Chloroform, 79
Chlorohydrins, 106, 165
2-Chloro-2-iodopropane, 132
2-Chloro-5-methylheptane, 30
1-Chloro-1-phenylethane,
  preparation, 189
Chlorophyll, 325, 385
1-Chloropropane, *see n*-propyl chloride
2-Chloropropane, *see i*-propyl chloride
3-Chloropropene, 33
Chloroquine, 389
Chlorosulfite ester, 171
Chlorotoluenes, 145
Cholesterol, 376–77, 382
Choline, 269
Chromatography, 349
Chromones, 384
Chromophore, 279
Cinnamaldehyde, 198, 220
Cinnamic acid, 221
Cis-2-butene, 95–96
  from dehydration of *n*-butyl alcohol, 119
Cis-1,2-dichlorocyclobutane, 96
Cis-1,2-dichloroethene, 97
Cis isomers, 95–96, 331

Cis-trans isomerism, 94–97, 119, 131–32, 250, 375
Citric acid cycle, 326
Claisen condensation, 246, 379
Clemmensen reduction, 207
Clostridium acetobutylicum, 202
Cocaine, 298, 388
Codeine, 389
Coenzymes, 326, 336
   acetyl CoA, 336–37, 366, 379
Collagen, 355
Color, 278–79, 405–06
Configuration, 67, 292
   absolute and relative, 292–95, 299–301
   amino acids, 340–41
   D, 292–93, 305–07
   inversion, 67, 300–01
   L, 292–93, 305–06, 340–41
   R, 294–95, 300–01
   retention, 301
   S, 294–95, 300–01
   sugars, 305–08
Conformations,
   boat, 87–88
   carbohydrates, 314
   chair, 87–89
   cyclohexane, 87–89
   eclipsed, 83–84, 87
   skew, 84–85
   staggered, 83, 87
   trans (anti), 84–85
Conjugate acid, 36–37, 42
Conjugate addition, 121–22
Conjugate base, 36–37, 42
Conjugated system, 14, 49, 121–22, 404–06
Corey, E. J., 82
Cortisone, 378
Cotton, 304
Coumarin, 383–84
Coupling,
   alkyl halides with dialkyllithium compounds, 82
   alkyl halides with sodium, 82
   diazonium salts, 278
Cracking, 80–81
Crafts, James M., 146
Cresols, 163
Crisco, 333
Crotonaldehyde, 206, 219
Crotyl alcohol, 206
Cumene, see Isopropylbenzene
Cumene hydroperoxide, 166
Cyanides, 191, 233, 263, 306
Cyanohydrins, 208–09
Cyclic compound, 31
Cycloalkanes, 85–89
Cyclobutane, 86
Cyclohexane, 86–89, 139, 375–76
Cyclohexanol, 170

Cyclohexanone, 213
Cyclohexene, 409
Cyclonite, 212
Cyclopentadiene, 59
Cyclopentadienyl anion, 141
Cyclopentane, 86
Cyclopentanone, 241
Cyclopropane, 85–86
Cyclopropyl chloride, 86
Cysteine, 342
Cystine, 342
Cytidine-5′-monophosphate, 369
Cytosine, 362–63, 366, 368

Dacron, 398–99
Dalton, John, 3
   atomic theory, 4
Decane, 29
Dehydration,
   alcohols, 118–19
   2-methyl-2-butanol, 119
Dehydrogenation of alcohols, 200
Dehydrohalogenation, 120, 134–135
Democritus, 3
Denaturation of proteins, 312
Deoxyadenosine, 364–65
Deoxyribonucleic acids (DNA), 366–69
   base-pairing, 367
   bases, 363
   double helix structure, 361, 367–68
   functions in cell, 368–69
   primary structure, 364–67
   in protein synthesis, 368
   replications, 368–69
   secondary structure, 367–68
   section of chain, 366
   tertiary structure, 368–69
2-Deoxy-D-ribose, 307, 362, 364–66
Detergents, 236, 335
Dextrorotatory substances, 285, 291–92
Dextrose, see Glucose
Diastereoisomers, 295–97, 306, 311
Diazomethane, 111
   formation of methylene from, 111
Diazonium ion, 275
Diazonium salts, 271–72, 275–78
   coupling, 278
Diazotization, 272, 275
Diborane, 109–11
1,3-Dibromobenzene, 32
1,2-Dibromoethane, 101
1,3-Dibromopropane, 86
Dicarboxylic acids,
   preparation, 249
Dichloroacetic acid, 47
p-Dichlorobenzene, 155
Dichlorocarbene, 111, 201
Dichlorodifluoromethane, 190
1,2-Dichloroethane, 97
1,1-Dichloroethene, 93

INDEX

1,2-Dichloropropane, 103
2,2-Dichloropropane, 132
Diels-Alder reaction, 111–12
Dienes, 111, 122
Diethyl ether, 178–79, 182
Diethylene glycol, 185
Diisopropyl ketone, 209
Dimethylamine, 260
N,N-dimethylaniline, 261, 272
   p-nitroso compound, 272
1,3-Dimethylcyclopentane, 31
Dimethyl ether, 24–26
Dimethyl sulfate, 181, 319, 321
Dimethyl sulfoxide, 184
Dimethyl terephthalate, 398–99
2,4-Dinitrofluorobenzene, 195, 349–50
Dipeptide, 348, 352
Dipolar ions, 274
Dipole moment, 9, 96–97
Direct dyes, 279
Disaccharides, 305, 318–22
Disulfides, 174
Diterpenes, 374
DNA molecule, electron micrograph, 46
Dodecane, 29
Dow Process, 165
Drying oils, 334
Dyes, 277–80
Dynamite, 171

E_1, 117
E_2, 117
Eclipsed conformations, 83–84, 87
Edman procedure, 350–51
$E = h\nu$, 402–03
Eicosane, 29
Electromagnetic spectrum, 402–03
Electron cloud, 4
Electronegativity, 9
Electronic configuration,
  excited state, 404
  ground state, 404
Electronic effects, 46–54
  acidity and basicity, 52–54
Electrons,
  delocalized, 14
  localized, 14
Electrophile, 65, 142, 144–46
Electrophilic addition,
  bromine to ethene, 100–03
Electrophoresis, 346
Elimination-addition mechanism, 267
Elimination reaction, 65, 117–20
Enantiomers, 288, 291–92
Energy levels, 4–5
Energy profile diagram, 63, 66, 69
Enol, 132–33, 215–16, 233, 248, 363–64
Enolate anion, 215–16
Enzymes, 298
Epimers, 318

Epoxides, 184–86
Equatorial bonds, 88
Ergosterol, 377
D-Erythrose, 306
Esterification, 237
  mechanism, 238–39
Esters, 172, 231, 237–39, 242–44
  inorganic, 171–72
  glycerol, 330, 332–33
  hydrolysis, 242
  poly-, 393, 397–99
Estrogens, 377
Estrone, 377
Ethane, 29–30, 75, 81
  conformations, 83–84
  preparation, 81
Ethanoic acid, see Acetic acid
Ethanol, see Ethyl alcohol
Ethanolamine, 186, 257
Ether, see Diethyl ether
Ethers, 178–86
  nomenclature, 178
  preparation, 179–81, 191
  properties, 181–82
  reactions, 182–83
  sulfur analogs, 183
Ethyl acetate, 237, 243, 246
Ethyl acetoacetate, 246
Ethyl alcohol, 24–26, 52, 162, 164
  azeotrope with water, 164
  reactions, 118, 170, 179, 200, 202, 218, 237, 347
Ethylamine, 262–63
Ethyl bromide, 81, 104, 170
Ethyl chloride, 97, 146–47, 411
Ethyl ether, see Diethyl ether
Ethyl group, 30, 55, 67, 70, 75–76
Ethylbenzene, 146–47, 149, 155, 189, 405, 413
Ethylene, 12–13, 94
  preparation, 118, 179
  reactions, 103–04, 109, 114, 184–85
Ethylene bromohydrin, 102
Ethylene chlorohydrin, 185
Ethylene dibromide, 101–02
Ethylene glycol, 116, 165, 185, 398–99
Ethylene oxide, 184–86, 192–93
Ethyne, see Acetylene
Explosives
  cyclonite, 212
  dynamite, 171
  nitroglycerine, 171
  picric acid, 166
  TNT, 155

Faraday, Michael, 138
Farnesol, 374
Farnesyl pyrophosphate, 381
Fats, 330–33
  analysis, 335–36

Fats (Continued)
  hydrogenation, 333
  hydrolysis, 332–33
  iodine number, 336
  metabolism, 336–38
  saponification number, 335
Fatty acids, 331
  biosynthesis, 337
Fehling's test, 205, 316
Fibroin, 355
Fischer, Emil, 305, 311, 313, 317–18
Fischer-Kiliani synthesis, 305–06
Fischer projection formulas, 289–93, 295–96
Flavones, 384
Fluorobenzene, 276
Formaldehyde, 198, 214
  polymers, 399
  preparation, 201–02
  reaction with ammonia, 212
  uses, 202, 206
Formic acid, 228, 230
Formyl group, 231
Free radicals, 64–65, 70–72, 78–81, 160, 395
Freons, 189–90
Friedel, Charles, 146
Friedel-Crafts alkylation, 146–49, 155, 166
  alkenes, 149, 155
  ketone synthesis, 245–46
  limitations, 147–48
  mechanism, 146, 148
  polyalkylation, 147
  rearrangement, 148
D-(−)-Fructose, 307, 315–16, 318, 322
Fumaric acid, 250–51
Functional group, 22
Furan, 314
Furanoses, 314–16
  α-D-2-deoxyribo, 315
  β-D-2-deoxyribo, 362
  fructo, 315–16, 322
  β-D-ribo, 362

Gabriel phthalimide synthesis, 264–65
D-(+)-Galactose, 308, 320
Gasoline, 80–81, 113
Gem-dihalides,
  hydrolysis, 200
  preparation, 132
Genes, 361, 369
Geraniol, 380–81
  pyrophosphate, 381
Globulins, 354
Glucaric acid, 309–10
Gluconic acid, 308–10
D-(+)-Glucose, 308–15, 320–22
  α and β forms, 312–15
  evidence for cyclic structure, 310–14
  mutarotation, 312

osazone, 318
pentaacetate, 309–10
stereochemistry, 309–10
structure, 308–15
Glucosides, 313–14
Glutamic acid, 343, 358
Glutaric acid, 230
Glyceraldehyde, 292–93, 305–06
Glycerides, 330, 332
Glycerol
  esters, 330, 332–33
  from fats, 333
  nitration, 171
  preparation, 165
  in Skraup synthesis, 389
Glyceryl trinitrate, 171
Glycine, 341
  preparation, 344
  reaction with benzoyl chloride, 347
  reaction with ethyl alcohol, 347
Glycogen, 324
Glycols, 116
Glycosides, 317, 362–64
Glycylalanine, 352
  synthesis, 352–54
Glyptal, 400
Gomberg, Moses, 160
Grain alcohol, see Ethyl alcohol
Grignard, Victor, 191–92
Grignard reagent, 191–93
  acid synthesis, 233–34
  alcohol synthesis, 192–93, 214
  hydrolysis, 81, 192
  saturated hydrocarbon, synthesis, 81, 192
Guanine, 362–63, 368
Guncotton, 287

Halide exchange reaction, 190–91
Haloform reaction, 216–17
Halogen compounds, 189–93
Halogenation
  acids, 247–48
  aldehydes, 216–18
  alkanes, 71, 78–80
  alkenes, 103–04, 114–15
  alkynes, 131
  aromatic hydrocarbons, 144–46
  ketones, 218
  saturated hydrocarbons, 71, 78–80
  unsaturated hydrocarbons, 97, 100–04, 121–22, 131
Halohydrin, 106
Halonium ion, 144
Haworth projection formulas, 315–16, 320–24
Heisenberg uncertainty principle, 4
Hell-Volhard-Zelinsky reaction, 247
Hemiacetal, 210–11, 311, 314
Hemiketal, 210–11

INDEX

Hemin, 354–55
Hemoglobin, 355
Heptane, 29, 190
Heroin, 389
Heterocyclic compounds, 138, 275, 362–63, 382–91
Hexamethylenediamine, 398
Hexamethylenetetramine, 212
Hexane, 29
Hexose, 305
Hinsberg test, 271
Hippuric acid, 347
Histidine, 343
von Hofmann, A. W., 256, 265
Hofmann rearrangement, 265–66
   mechanism, 266
Hogarth engraving of Beer Street, 161
Homologous series, 75
Hormones
   adrenal cortical, 377–78
   sex, 377–78
House, Herbert, 82
Hückel $4n+2$ rule, 140–41
Humulene, 374
Hybridization, 10
   sp, 11, 15
   sp^2, 11–13, 49–50, 202–04, 306, 347
   sp^3, 12
Hydration of acetylene, 132–33
Hydration af alkenes, 106–07
Hydrazine, 213
Hydrazones, 213
1,2-Hydride shift, 148
Hydride transfer, 222
Hydroboration, 109–11
Hydrocarbons, 21
   aromatic, 138–56
   saturated, 21, 74–89
      cyclic, 85–89
   unsaturated, 21, 93–136
      acetylenic, 21, 130–36
      olefins, 21, 93–121
Hydrogen bonding, 46, 167–68, 259, 355–57, 361, 367–68
Hydrogen chloride molecule, 9
Hydrogen fluoride, 39–40
Hydrogen iodide, 40, 182–83
Hydrogen molecule, 8
Hydrogenation
   carbonyl compounds, 206
   fats and oils, 333
   unsaturated hydrocarbons, 109, 131–32
Hydrohalogenation,
   alkenes, 104–06
Hydrolysis
   alkyl halides, 163
   carboxylic acid derivatives, 242–44
   cyanides, 233
   gem-dihalides, 200
   gem-trihalides, 233

Grignard reagents, 81, 192
Hydroxyl group, *see* Alcohols and Phenols
β-Hydroxyaldehydes,
   preparation, 219
3-Hydroxybutanal, 218–19
β-Hydroxyketones,
   preparation, 219
Hydroxylamine, 212
Hydroxy acids, 209
Hydroxyproline, 343
Hypohalous acids, addition to olefins, 106
Hypophosphorous acid, 276–77

I.U.P.A.C. system of nomenclature, 29–31
Imide, 264
Imine, 264
Indigo, 279
Indole, 258, 386–87
Inductive effect, 41–44, 46–48, 208
Infrared spectroscopy (IR), 406–08
   absorption bands, 406–08
   C-H bending frequency, 406
   C-H stretching frequency, 406
   C-O stretching frequency, 406
   N-H stretching frequency, 406
   O-H stretching frequency, 406
   Fingerprint region, 407
   spectra, 408–09
      n-butyl alcohol, 409
      butanone, 409
      cyclohexene, 409
      toluene, 408
   wave number, 406
Ingold, Sir Christopher, 294
Inorganic esters, 171–72
Insulin, 350
Iodic acid, 42
Iodine number, 336
Iodobenzene,
   preparation, 154–55, 276
Iodoform, 217–18
   test, 217–18
2-Iodo-4-ethyl-5,6-dimethyloctane, 31
Ionic reactions, 64, 68–70
Isobutane, 24
Isobutylene, 94
Isoelectric point, 345–46
Isoleucine, 342
Isomerism, 24–26
   branched chain, 24
   conformational, 83
   functional group, 24–26
   geometric, 94–96
   optical, 284–301
   positional, 24
   stereo, 284–301
Isooctane, 113
Isopentenyl pyrophosphate, 380
Isoprene, 123, 372, 380
   unit, 123, 372–75, 406

Isopropyl
  alcohol, 107, 110, 218
  bromide, 82, 104, 107
  hydrogen sulfate, 107
Isopropylbenzene, 148, 166
Isopropylidene moiety, 416
Isoquinoline,
  reduced ring system, 389
Isotopic tracer, 238–39, 267–68, 378–82

K_a, 38–39, 44, 47–48, 52
K_b, 38, 48, 53, 55, 260
Kekulé, Friedrich August, 49, 142
  formulas for benzene, 49–50, 139
Keratin, 355–56
Ketals, 210–11
Ketene, 111
Keto form, 133, 215–16, 248, 363–65
α-Ketoglutaric acid, 358
Ketohexose, 307
Ketones, 197–222
  halogenation, 218
  nomenclature, 198
  preparation, 173, 199–202
  reactions, 204–22
  reduction, 206–07, 264
Ketose, 305
Khorana, Gobind, 369
Kinetic control, 122–23
Knoop beta-oxidation theory, 336–37
Kolbe reaction, 249–50
Kornblum, Nathan, 277
Krebs cycle, 326

Lactic acid,
  oxidation, 249
  preparation, 209, 249, 293, 347
  stereochemistry, 292–293
Lactone, 373, 383–84
Lactose, 319–20
Lanosterol, 374–75, 381–82
Lauric acid, 230, 331
Le Bel, J. A., 287–88
Lecithin, 269, 331
Leucippus, 3
Leucine, 342
Levorotatory substances, 285, 291–92
Lewis acid, 37–38, 55, 65, 144, 146, 148
Lewis base, 37–38, 48, 55, 65, 148
Lewis electron-dot formulas, 7
Lewis, Gilbert Newton, 6
Lewis salt (acid-base complex), 38, 55, 148
Lewis theory, 37–38
Limonene, 373
Linoleic acid, 331
Linolenic acid, 331
Linoleum, 334
Linseed oil, 331, 334
Lipids, 329–338
  metabolism, 336–38

Lipmann, Fritz, 336
Lithium aluminum hydride, 206, 262–63
Lithium dialkyl copper compounds, 82
Longifolene, 374
Lucas test, 170
Lucite, 395, 397
Lynen, F., 382
Lysergic acid, 387
  diethylamide (LSD), 387
Lysine, 343

Malachite green, 280
Maleic acid, 250–51
Maleic anhydride, 112, 251
Malonic acid, 230, 249
Malonic ester, 344
  preparation of phenylalanine, 345
  substituted, 191, 344–45
Maltose, 320
Mandelic acid, 196
D-(+)-Mannose, 308, 318
Markovnikov's Rule, 104–06, 248
  applied to alkenes, 104–07
  applied to alkynes, 131–33
Markovnikov, Vladimir, 105
Mass Spectrometry, 416–18
  base peak, 497
  t-butyl carbonium ion, 418
  CH_3 fragment, 417–18
  fragmentation processes, 417–18
  mass spectrum, 417–18
    methyl alcohol, 417–18
    4-methylheptane, 417
    neopentane, 418
    n-octane, 417
  m/e ratio, 416
  molecular ion, 417
  parent ion, 417
McKillop, Alexander, 153
Mechanisms, 61–62
  addition to carbonyl compounds, 207–08, 210–12, 214
  addition to conjugated systems, 122–23
  addition to unsaturated hydrocarbons, 98–103, 105–13
  electrophilic aromatic substitution, 142–48
  esterification, 238–39
  halogenation of saturated hydrocarbons, 70–71, 79
  nucleophilic substitution, 65–70, 163, 180–83, 234–35, 238–39
  polymerization, 113, 395–96
Menthol, 373
Mercaptans, 173–74, 191
Mercurochrome, 280
Merrifield, R. B., 354
  solid-phase technique, 354
Mesitylene, 141
Meso form, 296

INDEX

Meta, 141, 149
Metabolism
  fats, 336–38
  carbohydrates, 325–26
  proteins, 358–59
Meta-directing groups, 150–51
Methane, 11–12, 29–30, 74
  halogenation, 71, 79–80
  reactions, 71, 79–80
Methanol, see Methyl alcohol
Methionine, 342
Methyl alcohol, 42, 66, 185, 243
  catalytic dehydrogenation, 202
  industrial preparation, 164
  mass spectrum, 417–18
  reaction with sodium, 103
Methylamine, 48, 53–54, 257, 260, 265
N-Methylaniline, 261
Methyl benzoate, 243
  thallation, 154
Methyl t-butyl ether,
  preparation, 181
Methyl cellosolve, 185
Methyl chloride, 71, 79, 411
Methylcyclopropane, 111
1-Methylcyclohexene, 31
Methyl glucosides, 314
Methyl group, 30, 55, 67, 70, 75, 375
Methyl radical, 71, 79
Methyl salicylate, 250
Methylene chloride, 79
Methylene group, 75, 207
Methylethyl ether,
  preparation, 180
4-Methylheptane, 417
2-Methylhexane, 82
Methyl ketone,
  test, 217–18
Methyl methacrylate, 395, 397
2-Methylpropane, 33, 76
Mevalonic acid, 379–80
"Mickey Finn," 210
Micron, 403
  milli, 403
Mohr-Sasche principle, 89
Molecular models, 22
  FMO models, 22
Monochloroacetic acid, 43–44, 47
Monochromatic light, 285–86
Monomer, 393–95
Monosaccharides, 305, 307–18
  determination of ring size, 319
Monoterpenes, 372–373
Mordants, 279
Morphine, 298, 389
Mutarotation, 312
Myoglobin, 356
Myosin, 355
Myrcene, 373
Myricyl palmitate, 334

Myristic acid, 331, 337

Naphthalene, 140, 240
Naphthols, 163
β-Naphthylamine, 259
Natta, Giulio, 397–98
Natural gas, 80
Natural products, 371–91
NBS, 114–15
Neopentane, 418
Neopentyl group, 30
Neptalactone, 373
Newman projection formulas, 83–85
Newton, Sir Isaac, 3
Nicol prism, 285
Nicotine, 387–88
Nicotinic acid, 387–88
Ninhydrin, 348–49
Nitration
  aniline, 273–74
  benzene, 143
  mechanism, 143–44
  phenol, 173
  toluene, 155
Nitric acid, 37, 41–42, 143, 171
Nitriles, see Cyanides
m-Nitroaniline, 262
p-Nitroaniline, 274
Nitrobenzene
  preparation, 143
  reduction, 262
  Skraup synthesis, 389
m-Nitrobenzoic acid,
  preparation, 151
o-Nitrobenzoic acid,
  preparation, 151
Nitrobromobenzenes,
  preparation, 150
Nitrochlorobenzenes, 165–66
Nitro compounds,
  reduction, 262
Nitrogen heterocycles, 275, 362–63, 385–91
Nitroglycerine, see Glyceryl trinitrate
Nitronium ion, 143–44
p-Nitrophenol, 163, 165, 173
N-Nitrosoamine, 272
Nitrous acid, 271–72, 275, 347
Nomenclature, 28–34
  acetylenes, 130–31
  acids, 228–30
  alcohols, 162
  aldehydes, 198
  alkanes, 29, 75
  alkenes, 93–94
  alkyl groups, 75–76, 78
  amines, 257–58
  aromatic hydrocarbons, 141
  ethers, 178
  glycerides, 332

Nomenclature *(Continued)*
  I.U.P.A.C., 29–34
  ketones, 198
  olefins, 93–94
  phenols, 163
Nonane, 29
Normal alkalene, 75, 77
Nuclear Magnetic Resonance Spectroscopy (NMR), 408, 410–16
  applications in solving complex structural problems, 416
  chemical shifts, 411–13
  delta (δ) scale, 413
  deshielding of protons, 411
  equivalent protons, 411
  multiplicity of peaks, 415
  non-equivalent protons, 411
  number of signals, 411–413
    ethyl chloride, 411
    methyl chloride, 411
    i-propyl chloride, 411
    n-propyl chloride, 411
  peak area and number of protons, 412–15
    t-butyl alcohol, 412
    ethylbenzene, 413
    integration, 413
  positions of signals, 411–13
    t-butyl alcohol, 412
  shielding of protons, 411
  spectra, 401, 412, 414–16
    t-butyl alcohol, 412
    1,2-0-isopropylidene-3-0-benzoyl-5-deoxy-β-L-arabinose, 416
  splitting of signals, spin-spin coupling, 415
  tau (τ) scale, 413
  tetramethylsilane, 411, 413
Nucleic acids, 361–70
Nucleophile, 65, 163, 190–91, 207, 211, 234–35
Nucleophilic substitution, 65–70, 163, 166, 179–83, 190–91
Nucleoproteins, 361, 368
Nucleoside, 363–65
Nucleotide, 365–69
Nylon, 3, 393, 398

Octane, 29, 418
Oil
  bay, 373
  castor, 331–32
  citronella, 374
  cloves, 374
  coconut, 331
  cottonseed, 331
  eucalyptus, 373
  geranium, 380
  hops, 374
  lemon, 373
  linseed, 331, 334
  olive, 332
  peanut, 331
  peppermint, 373
  pine, 374
  rose, 193
  soybean, 334
  sperm, 329
  tung, 334
  wintergreen, 250
Oil paints, 334
Oilcloth, 334
Oils, 330–33
  analysis, 335–36
  drying, 334
  hardening, 333
  rancidity, 334
Olefins, 21
Olefins, *see* Hydrocarbons, unsaturated
Oleic acid, 331
Olein, 333
Oleomargarine, 333
Oligosaccharides, 305
Olive oil, 332
Optical activity, 285, 288–89
Optical isomerism, 284–301
Optical resolution, 297
Orbitals, 4–7, 10–15
  atomic, 10–12
  molecular, 12–17
  pi, 5, 7
  polynuclear, 14–16
  s, 4–5
  shape, 4–5
Organic, 20
  chemistry, 21
Orlon, 132, 395
Ortho, 141, 149
Ortho-para-directing groups, 150–51
Osazones, 317–18
Oxalic acid, 20, 230
Oxidation, 199
  alcohols, 172–73, 199–200, 232
  aldehydes, 204–06, 232
  alkanes, 80
  alkenes, 115–17
  aromatic side-chain, 149
  saturated hydrocarbons, 80
  unsaturated hydrocarbons, 115–17
Oximes, 212, 263
Oxonium ion, 118
Oxy-acetylene blow torch, 130, 135
Oxygen heterocycles, 383–84
Oxytocin, 351
Ozone, 115
Ozonolysis, 115–16

Palmitic acid, 230, 331, 337
Palmitin, 332
Paper chromatography, 349

INDEX

Para, 141, 149
Para red, 279
Paraffins, see Hydrocarbons, saturated
Pasteur, Louis, 286-87, 294, 297
Pauli exclusion principle, 4
Pellagra, 388
Penicillin, 390-91
Pentane, 29, 75, 78
Pentanes, boiling points of, 78
Pentose, 305
1-Pentyne, 134
Peptide bond, 240, 347-48, 353
   geometry of, 347-48
Peracetic acid, 184
Perchloric acid, 41-42
Perfluoroheptane, 190
Perfumes, 197
Perkin reaction, 221, 383-84
Phenanthrene, 140
Phenetole, 183
Phenobarbital, 389-90
Phenols, 52, 161-63, 165-66, 169, 173-74, 183, 242
   polymerization, 399
   preparation, 165-66, 276
   reactions, 169, 173, 180-81, 183, 201, 384
Phenoxides, 169, 180, 201
Phenyl group, 141
Phenylalanine, 341-42, 345
2-Phenylethanol, preparation, 193
Phenylhydrazine, 213
   2, 4-dinitro-, 213
Phenylhydrazones, 213
Phenylisothiocyanate, 350
Phenylketonuria, 341
Phenylthiohydantoin, 350-51
Phosgene, 352
Phosphates, 172
Phospholipids, 329
Photochemical reactions, 61
Photon, 402
Photosynthesis, 325
Phthalic acid, 230, 232, 240
Phthalic anhydride, 240, 264
Phthalimide, 264-65
Pi (π) bond, 13-17, 404
Picric acid, 166
Pinacol rearrangement, 177
α-Pinene, 373
Piperidine, 388
Piperine, 392
Piperitone, 373
Planck, Max, 402
Plastics, see Polymers
Polar covalent compound, 9
Polarimeter, 285-86
Polarization, 14, 100
Polarized light, 284-85

Polyalkenes, 121-24
Polyamide, 398
Polyesters, 393, 397-99
Polyethylene, 113-14, 397
Polyhydric alcohols, 164-65
Polymer, definition, 112, 393
   addition, 394-98
   atactic, 398
   average chain length, 394
   average molecular weight, 394
   Bakelite, 399
   characteristics, 393-94
   co-, 395-97
   condensation, 398-99
   cross-linked, 399
   Dacron, 398-99
   ethylene, 113-14, 397
   isotactic, 398
   Lucite, 395, 397
   natural, 305, 322-24, 340, 352-59, 361-70
   natural rubber, 123
   Nylon, 3, 393, 398
   Orlon, 132, 395
   phenol-formaldehyde, 399
   physical properties, 394
   polyamides, 398
   polyesters, 393, 397-99
   stereochemical control of addition, 397-98
   styrene, 155, 395, 397
   syndiotactic, 398
   synthetic, 393-99
   synthetic rubber, 396
   Teflon, 113
   Vinyl acetate, 395
   viscosity, 394
Polymerization, 112
   addition, 394-98
   alkenes, 112-14, 394-98
   anionic, 396
   cationic, 396
   co-, 395-97
   condensation, 398-99
   ethylene, 113-14, 397
   free radical, 113, 395
   ionic, 113, 396
   isobutylene, 112-13
   isoprene, 123
   methyl methacrylate, 395, 397
   styrene, 155, 395, 397
   tetrafluoroethylene, 113
Polypeptides, 347-59
   naming, 354
   sequence of amino acids, 349, 351-52
   structure determination, 348-52
   synthesis, 352-54
Polypropylene, 114, 397-98
Polysaccharides, 305, 322-24
Polystyrene, 155, 395, 397

Polyvinyl acetate, 395
Porphyrin, 354–55, 385
Prelog, V., 294
Priority,
   absolute configuration, 294–95
Progesterone, 377
Proline, 343
Proof, 164
Propane, 29, 75
Propanone, see acetone
Propene, see propylene
Propionic acid, 43–44, 230, 233, 241
Propionyl group, 231
i-Propyl alcohol, 162
   preparation, 107
i-Propyl chloride, 24, 411
i-Propyl group, 30, 55, 67, 70, 76
n-Propyl group, 30, 76
n-Propyl
   alcohol, 110, 162
   bromide, 104, 107
n-Propylbenzene, 140
n-Propyl chloride, 24, 33, 148, 411
Propylene, 33, 94
   polymerization, 114, 397–98
   preparation, 120
   reactions, 103–08, 110–11, 114, 397–98
Propylene chlorohydrin, 106
Propyne, 131–34
Prosthetic group, 354
Proteins, 347–59
   chemical behavior, 357
   classification, 354–55
   conjugated, 354
   denaturation, 357
   fibrous, 355
   globular, 355
   metabolism, 358–59
   structure, 355–57
Purine, 362–63, 390
Pyran, 314
Pyranoses, 314
   fructo, 316
   galacto, 320
   gluco, 314, 320–24
Pyridine, 140, 258, 387–88
Pyrimidine, 258, 362–363, 390
Pyrrole, 258, 385
Pyruvic acid, 249, 358

Quanta, 402
Quaternary ammonium hydroxide, 269
Quaternary ammonium salt, 259, 261, 268–69
Quercetin, 384
Quinine, 298, 389
Quinoline, 258, 388–89
Quinuclidine, 59

R,
   definition, 78
Racemic mixtures, 291
   resolution, 291, 297–98
Raffinose, 328
Rate of reaction,
   activation energy, 62–63
   effective collisions, 62
   factors influencing, 61
   rate-determining step, 61
   theory, 62–64
Reactivation mechanism, 61–62
   activated complex, 62
   reaction intermediate, 62
   transition state, 62–63
Rearrangements
   Hofmann, 265–66
Reducing sugars, 316, 320
Reduction, 199
Reductive amination, 264
Reimer-Tiemann reaction, 201, 383–84
Replacement, see Substitution
Reserpine, 387
Resolution of racemic mixtures, 291, 297–98
Resonance, 49–51
Resonance effects, 46–47, 151
Resonance form, 50
Resonance hybrid, 50–52, 139
Resonance structures,
   acetate ion, 53
   acetic acid, 53
   allylic carbonium ion, 248
   amides, 240, 348
   aniline, 54
   anilinium ion, 54
   benzene, 51
   carbonyl group, 204
   carboxylate ion, 51, 53
   chlorobenzene, 51
   diazonium ion, 275
   electrophilic aromatic substitution, 152–53
   phenol, 52
   phenoxide ion, 52
Riboflavin, 390
Ribonucleic acids (RNA), 369–70
   base-pairing, 369
   bases, 363
   messenger, 369–70
   protein synthesis, 369–70
   ribosomal, 369–70
   structure, 369
   transfer, 369–70
D-Ribose, 307, 362
β-D-ribose-3,5-diphosphate, 362
Ribosomes, 369–70
Ricinoleic acid, 331
Roberts, John D., 267
Rosenmund reduction, 245

INDEX

Rubber, 123

Salicylaldehyde,
   preparation, 201, 384
   synthesis of coumarin, 384
Salicylic acid, 230, 250
Sandmeyer reaction, 276
Sanger, F., 349–50
Sanger method, 349–350
Saponification, 242, 332–33
   number, 335
Saran, 393
   wrap, 93
Saturated hydrocarbons, see Hydrocarbons, saturated
Saytzeff rule, 119–20
SBR, 396
Scheele, Karl Wilhelm, 20
Semicarbazide, 213
Semicarbazone, 213
Serine, 341–42
Serotonin, 387
Sesquiterpenes, 374
Sex hormones, 377–78
Sigma (σ) bond, 13–17, 404
Silvering of mirrors, 202, 205–06
Skraup synthesis, 388–89
S_N1 reaction, 68–70
   alcohols, 169–70
   alkyl halides, 190–91
   carbonium ions, 69–70
   cleavage of ethers, 182–83
   electronic effects, 69–70
   factors effecting, 70
   mechanism, 68, 70
   preparation of alcohols, 163
S_N2 reaction, 66–68
   alcohols, 169–70
   alkyl halides, 190–91, 260–61
   cleavage of ethers, 182
   mechanism, 66
   preparation of alcohols, 163
   steric effects, 67–68
   transition state, 68, 300
   Walden inversion, 300–01
   Williamson ether synthesis, 180
Soap, 236, 333–35
Sodium acetate, 235, 237, 239
Sodium acetylide, 133–34
Sodium alkoxides, 168–69, 180–81
Sodium alkylaryl sulfonate, 335
Sodium benzoate, 222
Sodium bisulfite addition reactions, 213–14
Sodium borohydride, 206
Sodium ethoxide,
   preparation, 169
Sodium lauryl sulfate, 172, 335
Sodium phenoxides, 165
Sodium stearate, 236, 334

Solvolysis, 243
Sorbitol, 310
Spatial Formulas, 22
   C_4H_{10}, 23
   representation, 22–23
Specific rotation, 286
Spectrophotometer, 404
Spectrum, 404
Spermaceti, 331, 333
Spirane, 289
Squalene, 124, 374–75, 381
Staggered conformations, 83, 87
Starch, 322–23
Stearic acid, 230, 331
Stearin, 333
Stereochemistry,
   addition polymers, 397–98
   formation of 1-bromo-2-methylbutane, 300
   formation of 2-chlorobutane, 299
   formation of 2-octanol, 300–01
   glucose, 309–10
Stereoselectivity, 298
Stereospecific reaction, 298
Steric effects, 54–56, 67–68, 260
Steric hindrance, 54–56, 163, 208–09
Steroids, 329–30, 375–82
Stiles, M., 277
Strecker amino acid synthesis, 209, 344
Structure of proteins, 355–57
   α-helix arrangement, 355–356
   hydrogen bonding, 355–57
   primary, 355
   quaternary, 356
   secondary, 355
   sheet or β-structure, 356–57
   tertiary, 356
Strychnine, 298, 387
Styrene, 140, 155, 184, 395–96
Styrene oxide, 184
Substituted malonic ester,
   preparation, 191, 344–45
Substitution reactions, 65
   alcohols, 169–70
   alkanes, 78–80
   alkenes, 114–15
   alkyl halides, 190–91, 260–61
   aromatic alkyl side chain, 144–45
   carboxylic acids and derivatives, 234–35, 237–44
   mechanism of electrophilic aromatic, 142–43
   nucleophilic, 66–70, 163, 166, 179–83, 190–91, 234–35, 237–44, 300–01
   orientation in electrophilic aromatic, 149–153
Succinic acid, 230
Sucrose, 321–322
Sugars,
   fermentation, 164

Sugars (*Continued*)
  non-reducing, 316, 321–22
  reducing, 316, 320
Sulfa drugs, 274
Sulfanilamide, 274–75
Sulfanilic acid, 274
Sulfhydryl group, 107
Sulfides, 183
Sulfonamides, 270–71
Sulfonation, 144
Sulfone, 183
Sulfoxide, 183
Synthesis, 120–21

TNT, 155
Tartaric acid, 296
Tautomerism, 133, 215, 363–64
Taylor, Edward C., 153
Teflon, 113
Template theory, 298
C-Terminal acid, 348
  determination by selective enzymatic hydrolysis, 351–52
N-Terminal acid, 348
  determination by Edman procedure, 350–51
  determination by Sanger method, 349–50
Terminal residue analysis, 349–52
Terpenes, 372–75
  biogenesis, 378–81
Testosterone, 378
1,1,2,2,-Tetrachloroethane, 131
Tetrahedral carbon, 12, 287–88
2,2,3,3-Tetramethylbutane, 30
Tetramethylsilane, 411, 413
Thallation, 153–55
Thermodynamic control, 122–23
Thermodynamics, 60
Thiamine, 390
Thiazole ring, 390–91
Thioalcohols, 173–74, 191
Thioesters, 337
Thioethers, 183
Thiols, 173–74
Thionyl chloride, 171, 236
Threonine, 342
D-Threose, 306
L-Threose, 307
Thymidine, 365
Thymine, 362–66, 368
α-Tocopherol, 383
Tollens' test, 205–06, 308
Toluene, 140
  IR spectrum, 408
  nitration, 155
  NMR, 410
  ring chlorination, 145
  side chain chlorination, 145-46, 200, 233
  side chain oxidation, 149
m-Toluic acid, 230
o-Toluic acid,
  preparation, 234
o-Toluidine, 258
Trans, *see* Cis-trans isomerism
Transamination, 358
Trans-2-butene, 95–96
  from dehydration of n-butyl alcohol, 119
Trans-1,2-dichlorocyclobutane, 96
Trans-1,2-dichloroethene, 97
Trans isomers, 95–96
Transesterification, 243
Trans ring fusion, 375–77
Trehalose, 328
Trialkylborane, 110
Tribromoacetaldehyde, 216
2,4,6-Tribromoaniline, 273
2,4,6-Tribromophenol, 173
Trichloroacetic acid, 47
Tridecane, 29
3,3,3-Trifluoropropene,
  reaction with HBr, 106
Trimethylamine, 260
Trimethylboron, reaction with amines, 55–56
Trimethylmethane, *see* 2-methylpropane
2,4,6-Trinitroaniline, 261
2,4,6-Trinitrotoluene, 155
Triphenylmethane, 140
  dyes, 279–80
Triphenylmethyl free radical, 160
Triterpenes, 374–75
Tropylium ion, 141
Tryptophan, 342, 386
Turpentine, 373–74
Tyrosine, 342

Ultraviolet-Visible Spectroscopy (UV-VIS), 404–06
  1,3-butadiene, 405
  β-carotene, 406
  ethylene, 405
  τmax, 405
  molar extinction coefficient, 405
Undecane, 29
α,β-Unsaturated aldehydes,
  preparation, 219
Unsaturated hydrocarbons, *see* Hydrocarbons, unsaturated
α,β-Unsaturated ketones,
  preparation, 219
Uracil, 362–63, 366
Urea, 21, 359, 399
Urotropine, 212

Valine, 342
Vanillin, 198
van't Hoff, Jacobus Hendricus, 62, 284, 287–88

rule, 295–96
Van Slyke detection of amino nitrogen, 347
Vat dye, 279
Vicinal dihalide, 103–04, 134–35
Vinyl acetate, 395
Vinyl chloride, 114
Vitamin A, 124, 374
Vitamin B_1, 390
Vitamin B_2, 390
Vitamin B_{12}, 385
Vitamin C, 383
Vitamin D, 321
Vitamin E, 383

Walden inversion, 300
Walden, Paul, 300
Watson and Crick hypothesis, 361, 367–68

Wave numbers, 403
Waxes, 333
Weizmann, Chaim, 202
Williamson, Alexander, 179–80
Williamson ether
  synthesis, 179–81, 183, 191
Wöhler, F., 21
Wolff-Kishner reaction, 207
Wood alcohol, see Methyl alcohol
Woodward, Robert Burns, 372, 385
Wurtz, Charles Adolphe, 82
Wurtz reaction, 82

Xylenes, 147, 232, 240

Ziegler, Karl, 113, 397–98
  catalysts, 114, 397
  -Natta technique, 397–98
Zwitterion, 274, 345